Mia's MATH SERIES

AP PRECALCULUS

Mia's AP Precalculus

발　행　2023년 9월 15일 초판 1쇄
저　자　소미혜
발행인　최영민
발행처　헤르몬하우스
주　소　경기도 파주시 신촌로 16
전　화　031-8071-0088
팩　스　031-942-8688
전자우편　hermonh@naver.com
출판등록　2015년 3월 27일
등록번호　제406-2015-31호

ISBN　979-11-92520-61-2 (53410)

Why? Mia's AP Precalculus

AP Precalculus에 필요한 모든 개념은 물론 실전 능력까지 한 번에 잡는다

지난 10년간 결과로 증명된 학교 GPA관리 및 5월 AP 시험대비용 최적의 교재

'이해하기 쉬운 개념 + 다양한 example 문제 + AP style problem

3중 그물망구조로 AP Precalculus에 필요한 모든 토픽 및 개념과 실전 능력을 한 번에 잡는다!

이해하기 쉽고 친근한 이미지를 활용하여 어려운 수식을 빠르게 이해!

Ti 84, Ti nspire CAS(CX도 가능) 계산기의 필수 사용법 정리

1. '이해하기 쉬운 개념 + 다양한 example 문제 + AP style problem'

   그 동안의 AP Calculus 교재에서는 볼 수 없었던 삼중 그물망구조로 AP Calculus에 필요한 모든 토픽 및 개념과 실전 능력을 한 번에 잡는다!

2. 스스로 빈칸을 채워가며 개념을 꼼꼼히 공부할 수 있도록 설계한 교재

3. 그저 그런 교재가 아니다!

   지난 10년간 현장에서 수많은 학생들에게 결과로 증명된 학교 GPA 및 5월 AP 시험 대비용 최적의 교재

Preface

Precalculus의 뜻을 살펴보면 '이전의(before)~'의 뜻을 가진 'pre'와 미분(differentiation)과 적분(Integration)을 배우며 수학의 꽃이라 불리는 calculus(미적분학)가 합쳐져서 '미적분학을 배우기 전에 꼭 필요한 내용을 과목'이라고 해석을 할 수 있습니다. 실제로 precalculus에서의 내용을 마스터 해야만 그 다음단계의 수학 calculus를 순조롭게 들을 수 있습니다. 또한, AMC, AIME 등과 같은 미국고등수학경시대회들은 precalculus를 시험범위에 포함시키고 있습니다. 이 때문에 precalculus는 미국고등심화수학을 배우기전에 꼭 통과해야 하는 관문이고, 저자는 precalculus가 미국고등학교 수학의 '핵심'이라 생각합니다.

그 동안 Precalculus 내용만 출제되는 시험이 없었습니다. 그리하여 SAT와 AP시험을 주관하는 College Board에서는 Precalculus에 대한 시험인 'AP Precalculus' 시험을 새롭게 출시하여 2024년 5월부터 시험을 응시할 수 있게 되었습니다. 저자는 이 시험이 많은 학생들로 하여금 Precalculus를 더 체계적으로 그리고 더 깊이 공부할 수 있는 기회가 될 것이라 생각합니다. 이를 위해 학생들이 쉽고 재미있게 공부할 수 있고, 기본기를 튼튼하게 다질 수 있는 다양한 문제들이 담긴 저만의 AP precalculus 교재를 제작하기 시작했습니다.

교재 제작을 위해 collgeboard의 자료뿐만 아니라 수 많은 precalculus textbook들의 내용들과 문제들을 연구하였습니다. 그리고 학원 및 개인지도를 통해 알게 된 국내외 많은 국제학교/외국인학교 학생들의 Precalculus GPA관리를 지도하면서 각 학교 선생님들의 가르치는 방식과 내용들을 벤치마킹(Benchmarking) 하였습니다. 또한 학생들이 어려워하는 부분들을 분석하고, 학생들의 의견들을 참고하며 교재를 지속적으로 개선하며 교재수준을 향상시켰습니다.

이 책을 통해 학생들이 AP precalculus라는 과목이 어렵지만 충분히 'Doable'하다는 것을 깨닫고, 문제를 해결했을 때의 즐거움을 느끼며, 깊이 있는 문제들을 다루면서 사고력과 응용력이 한층 더 깊어지길 바랍니다.

마지막으로 이 책을 함께 만들기 위해 애써준 사랑하는 남편TY와 잘생긴 아들 주원이, 그 동안 함께해 준 고마운 학생들, 그리고 헤르몬 하우스 관계자 여러분과 마스터프렙의 권주근 대표님께 감사의 마음으로 드립니다. 그리고 무엇보다도 소중한 기회를 주신 하나님께 감사와 찬양을 올려드립니다.

Mia Mihye So

Mia's AP Precalculus의 특징

1. 교재 내용, 문제에 대한 쉽고 명쾌한 Mia쌤의 설명, 해설강의는 유학 인터넷 강의 전문 사이트인 마스터프렙 (www.masterprep.net)에 마련되어 있습니다.
2. '이해하기 쉬운 개념 + 다양한 example 문제 + AP style Problem' 삼중 그물망 구조로 개념과 실전연습을 한번에 잡아줍니다. 어려운 개념들을 쉽게 배우고 다양한 example 문제로 연습을 한 뒤, 배운 개념에 대한 AP style Problem(기출유형문제)으로 실전에 적용하는 연습까지 완벽한 개념정리를 완성시킬 수 있습니다.
3. **스스로 빈칸을 채워가며 개념을 꼼꼼하게 공부**할 수 있게 설계하였습니다. 빈칸의 답은 페이지 하단에 배치하여 학생들이 필요 시 바로 참고할 수 있습니다.
4. college board에서 주최하는 AP Precalculus 시험에 나오는 모든 토픽을 커버하였고, 배워놓으면 유용하면서 난이도가 높고 깊이 있는 내용 및 문제들도 담아 *(star)표시로 표기하였습니다.
5. 이해하기 쉽고 친근한 이미지를 활용하여 어려운 수식을 빠르게 이해할 수 있도록 작성하였습니다. 꼭 암기해야 할 개념, 공식은 회색 shade박스 안에 정리하였습니다.
6. 계산기 Ti 84, Ti nspire CAS (CX도 가능)를 처음 사용하는 학생들을 위한 필수 계산기 사용법을 정리하였습니다. 계산기가 필요한 실전문제들도 풀어보면서 계산기 사용법도 정복할 수 있습니다.

◈ 기호 정리

* (star): 심화 문제

$\mathbb{R}$: Real numbers	$\cap$: and
$\mathbb{Z}$: Integers	∞ : infinity
$\mathbb{N}$: Natural numbers	$\therefore$: Therefore
$\cup$: or	$\because$: Since

AP Precalculus 5월 시험구성 및 내용

1) AP Precalculus Exam Date

매해 college board 사이트(https://apcentral.collegeboard.org)에 공지

2) AP Precalculus Exam Format

Section 1:	40 Multiple Choice	62.5% of Exam Score	
	Part A : 28 Questions	80min	No calculator
	Part B : 12 Questions	40min	Calculator
Section 2:	4 Free Response	37.5% of Exam Score	
	Part A : 2 Questions	30min	Calculator
	Part B : 2 Questions	30min	No calculator

3) AP Precalculus Course Overview

Units		Exam Weighting
Unit 1 :	Polynomial and Rational Functions	30-40%
Unit 2 :	Exponential and Logarithmic Functions	27-40%
Unit 3 :	Trigonometric and Polar Functions	30-35%
Unit 4 :	Parametric Curve, Vectors, and Matrices	Not assessed on the AP Exam

Unit 4 describes additional topics that teachers may include based on state or local requirements

저자소개

Mia (소미혜) 선생님은 지난 10년 이상을 유학 수학 현장에서 다양한 학생들과 호흡하면서 최적화된 미국 수학 및 국제학교 수학에 대한 솔루션을 제공해온 수학 전문가이다.

압구정 미국수학 전문강사라는 타이틀이 위의 노력들을 통해서 자연스럽게 얻게 된 선생님의 별칭이다.

미국에서 인증된 수학전문강사 (Texas 8-12 미국수학교사자격증 content exam + PPR exam 통과)로 관련된 전문자격증을 소지하고 있으며, 특히, 해외 엄마들 사이에 입 소문난 실력파 강사이다.

Precalculus, AP calculus AB BC, AP Statistics, SAT 1 2 math, IB Math 등에서 12년 이상의 경력을 가지고 있다. 또한 한국수능수학 강의 경력도 4년 이상을 가지고 있어서 한국

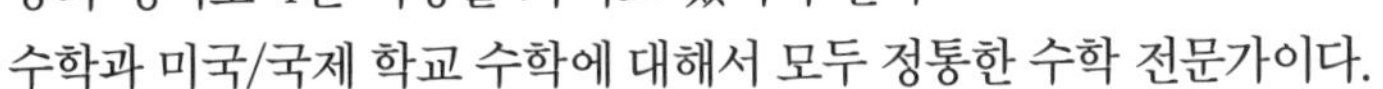
수학과 미국/국제 학교 수학에 대해서 모두 정통한 수학 전문가이다.

- ▷ 8-12 Texas Mathematics Teacher Certificate (content exam + PPR exam 통과)
- ▷ (현) No.1 유학 인터넷 강의 사이트 마스터프렙(www.masterprep.net) 수학강사
- ▷ (전) IBAdvance IB, sat 수학대표강사
- ▷ (전) 해커스유학 미국수학강사
- ▷ (전) PSU Edu AP, SAT 수학강사
- ▷ 미국텍사스고등학교, 국내국제고등학교 수학교사 경력 6년
- ▷ 수능수학강의 경력 4년
- ▷ 용인외대부고 , 경기외고 , KIS, 제주KIS, SIS, 청라달튼, 브랭섬홀, 일본, 싱가포르, 베트남 국제학교 등 학생들의 온라인/오프라인 개인지도
- ▷ College Board certification for AP Calculus AB, BC
- ▷ College Board certification for AP Statistics

Contents

1. Functions

2. Polynomial and Rational Functions

3. Exponential and Logarithmic Functions

Contents

4. Trigonometry Definition and Graphs

5. Trigonometry Identities

6. Trig Equations and Inequalities

7. Polar Curve

(※Ch8–Ch11 is NOT assessed on the AP Exam)

8. Parametric Equation

9. Conic Section

10. Vector

11. Matrices

Part 1

Functions

1.1 Functions

1. Relation

x → f → y or f(x)

A ① ________________ is a set of pairs of input and output values.

It ② ____________ an **input** to an **output**.

You can write a relation in many ways..such as

Mapping | Table | Order Pairs | Graph

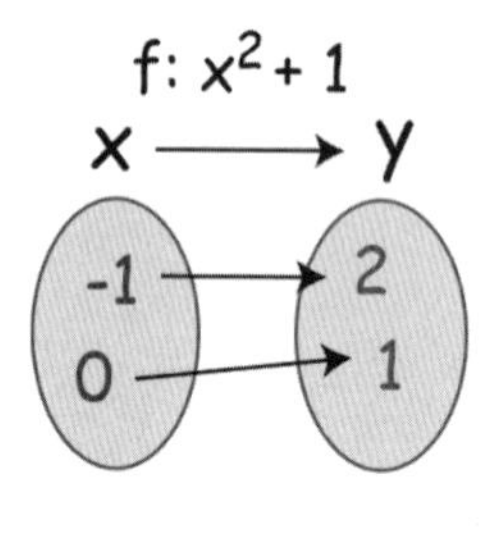

x	y
-1	2
0	1
1	2

$(x, y) = \{(-1,2), (0,1)\}$

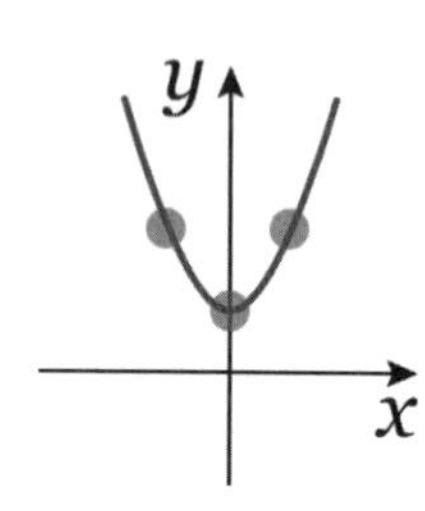

2. Function

※ **Function**

x → f → y or f(x)

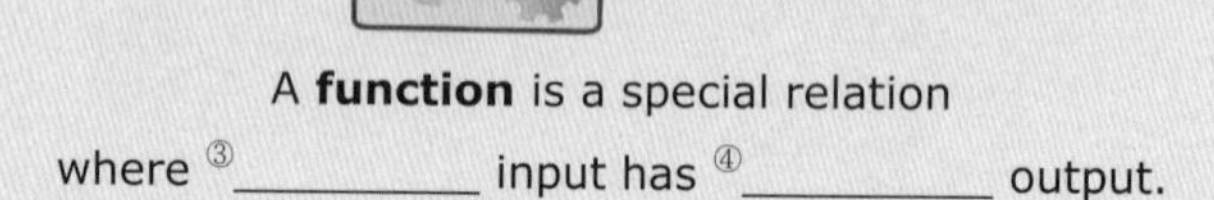

A **function** is a special relation

where ③ __________ input has ④ __________ output.

x —f→ y: a, b, c → 1, 2

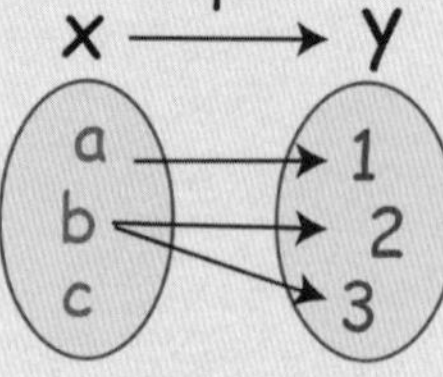

⑤ ______________ ⑥ ______________

Blank : ① relation ② relates ③ each ④ one ⑤ function ⑥ Not a function

Input (=① ____________ =② ________________ variable) is the set of X.

Output (=③ ____________ =④ ________________ variable) is the set of Y.

Easy way to check if a relation is a function?

① different ⑤____ coordinate

② ⑥__________________ test

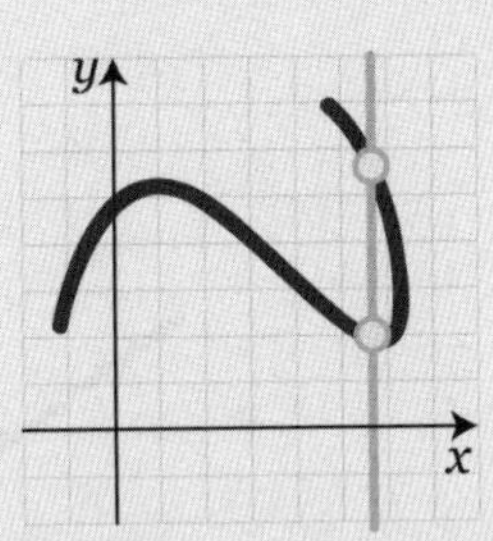

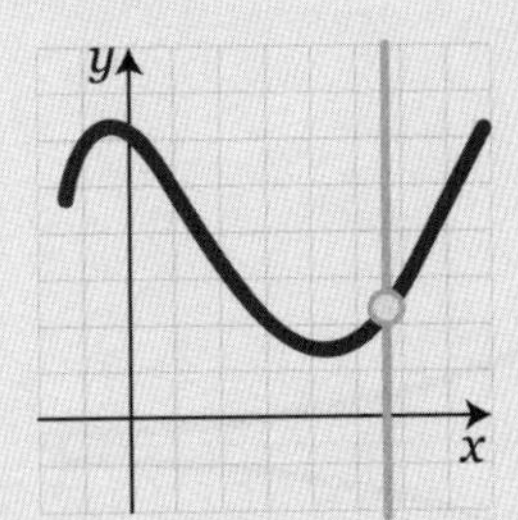

Not a function !   Function !

If a vertical line crosses the curve twice or more, then it is NOT a function.

※ **Evaluating** (E-'*value*'-ate) **Functions**

$f(x) = x^2 + 1$

"x" is Just a Place-Holder

$f(@) = @^2 + 1$

$f(2x) =$ ⑦ [   ]

☺ $(2x)^2 \neq 2x^2$

EXAMPLE 1. Determine whether each relation is a function or not.

①

②

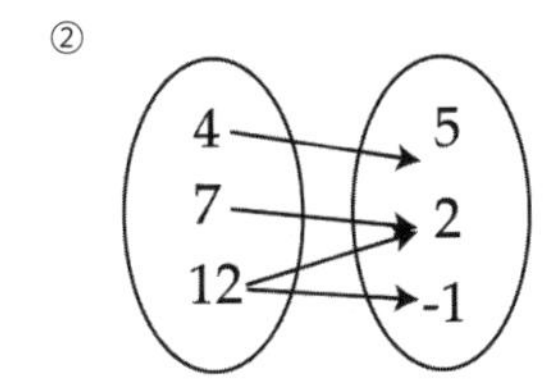

Blank : ① domain ② independent ③ range ④ dependent ⑤ x ⑥ vertical line ⑦ $(2x)^2+1$

③ $\{(2,-3),(-1,4),(3,5),(8,-11)\}$

④ $\{(-1,3),(0,1),(2,-3),(-1,0)\}$

⑤ $\{(0,2),(-1,3),(12,-30),(0,4),(2,3)\}$

⑥ $\{(1,2),(2,1),(4,4),(0,4),(3,4)\}$

⑦

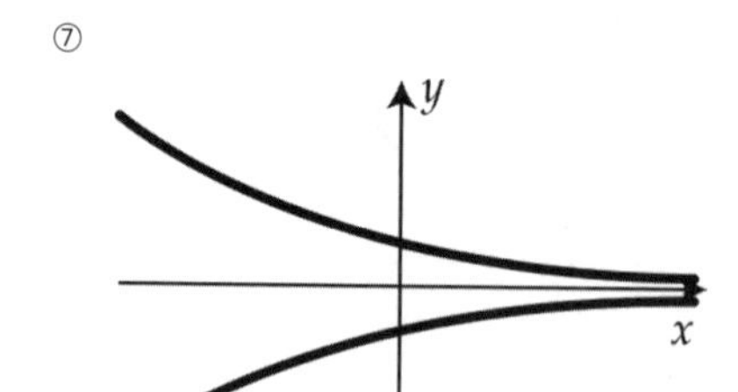

⑧

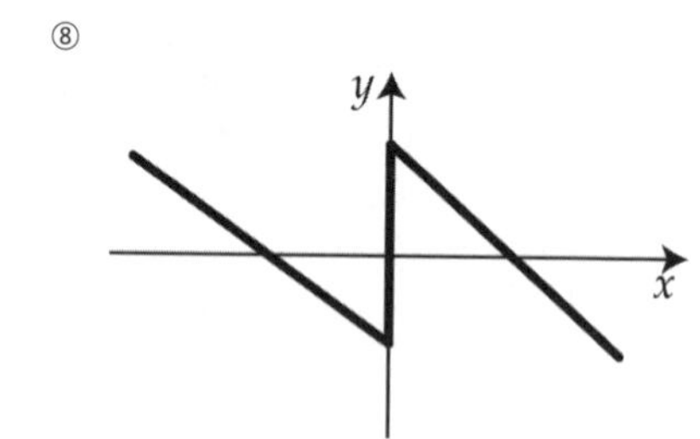

⑨

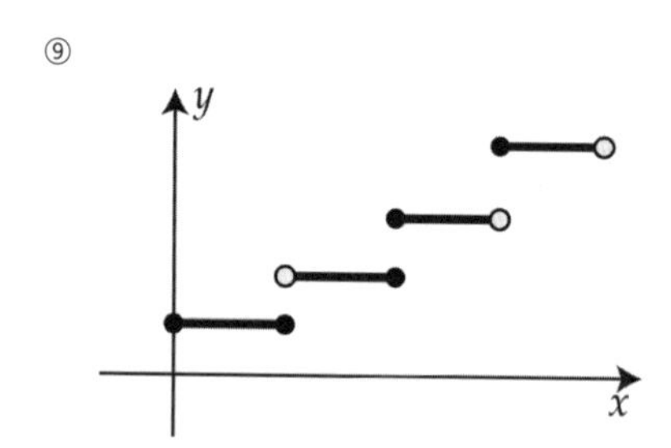

⑩

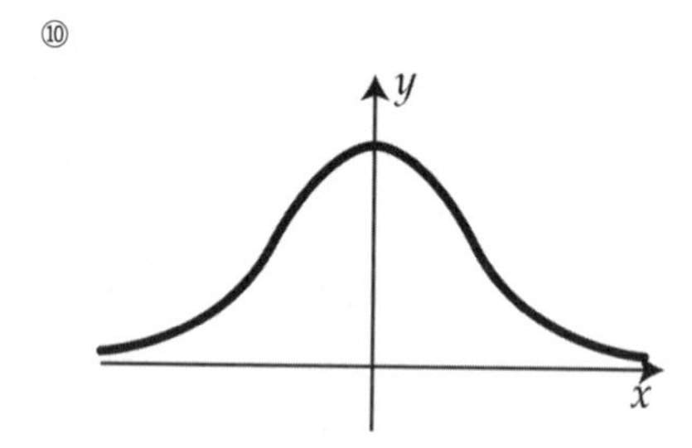

⑪

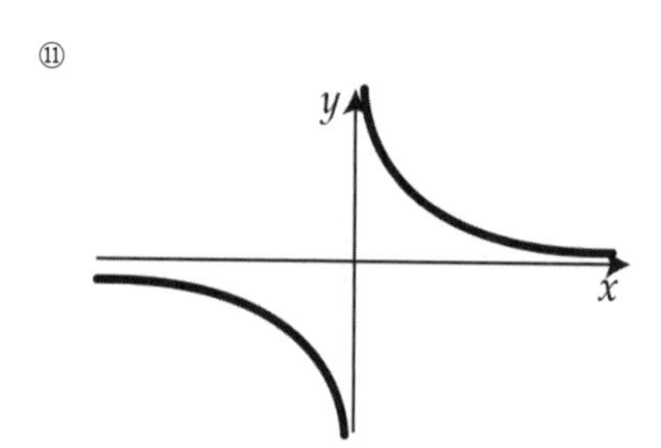

⑫

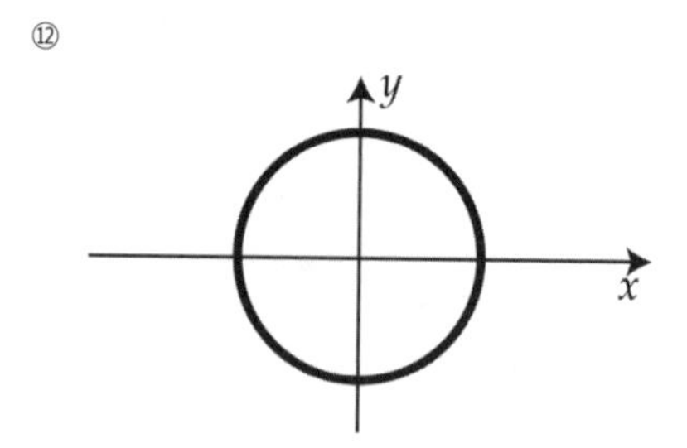

EXAMPLE 2. Evaluate $f(2)$, $f(2+h)$, and $\frac{f(2+h)-f(2)}{h}$ for given function.

① $f(x)=x^2+1$

② $f(x)=2x-1$

③ $f(x)=\frac{x}{x+1}$

④ $f(x)=\frac{1}{x-1}$

⑤ $f(x)=\sqrt{x+1}$

3. Domain and Range

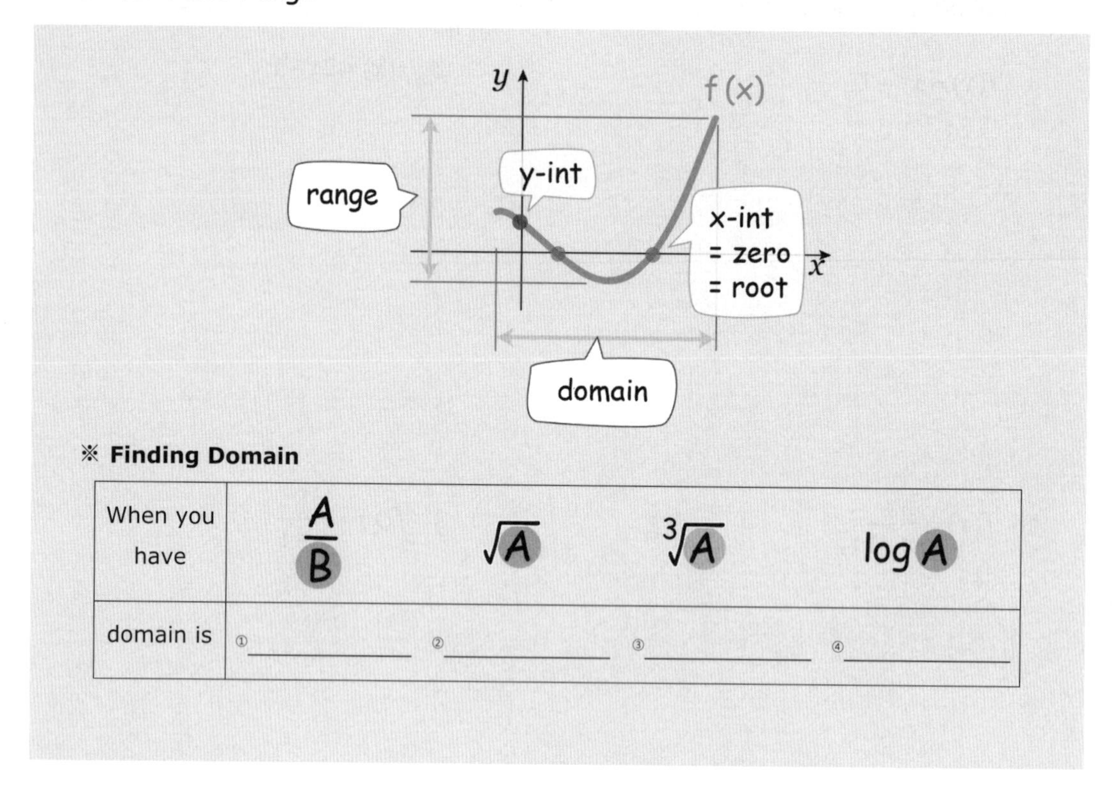

※ **Finding Domain**

When you have	$\frac{A}{B}$	$\sqrt{A}$	$\sqrt[3]{A}$	$\log A$
domain is	① ________	② ________	③ ________	④ ________

EXAMPLE 3. Find the domain of each of the following functions:

① $f(x) = x^2 + 5x$

② $h(x) = \sqrt{4 - 3x}$

③ $g(x) = \dfrac{3x}{x^2 - 4}$

④ $f(x) = \dfrac{x+4}{x^2 - 5x - 6}$

Blank : ① $B \neq 0$ ② $A \geq 0$ ③ all real numbers (R) ④ $A > 0$

⑤ $f(x)=\dfrac{x}{x^2+4}$

⑥ $f(x)=\dfrac{x-1}{x^2+2x+3}$

⑦ $F(x)=\dfrac{\sqrt{3x+12}}{x-5}$

⑧ $g(x)=\dfrac{\sqrt{x-2}}{x-7}$

⑨ $f(x)=\dfrac{(2x-1)^2}{\sqrt{x-5}}$

⑩ $f(x)=\dfrac{x^2}{\sqrt{x-1}}$

⑪ $f(x)=\sqrt{x^2-3x-4}$

⑫ $f(x)=\sqrt[3]{x^2-4}$

⑬ $f(x)=\dfrac{2}{\sqrt{4-x^2}}$

⑭ $f(x)=\dfrac{2}{\sqrt[3]{x^2-5}}$

⑮ $f(x)=\dfrac{1}{x-2}+\sqrt{x+3}$

⑯ $f(x)=\log_3(x-3)$

※ Interval Notation VS Inequality Notation

For 'Interval Notation',

we use [ ] a square bracket when we include the end values

we use ( ) a round bracket when we don't include the end values

(for ∞ or - ∞, always use round bracket ( ))

Number line	Inequality Notation	Interval Notation
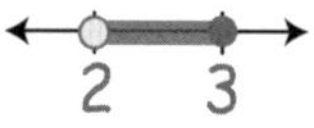	$2 < x \le 3$	(Not include 2) (include 3) $(2, 3]$
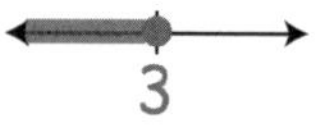	$x \le 3$	$(-\infty, 3]$
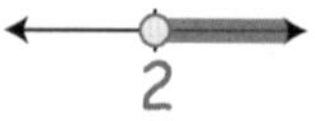	$x > 2$	$(2, \infty)$

EXAMPLE 4. For each of the following graphs, find the domain and range:

①

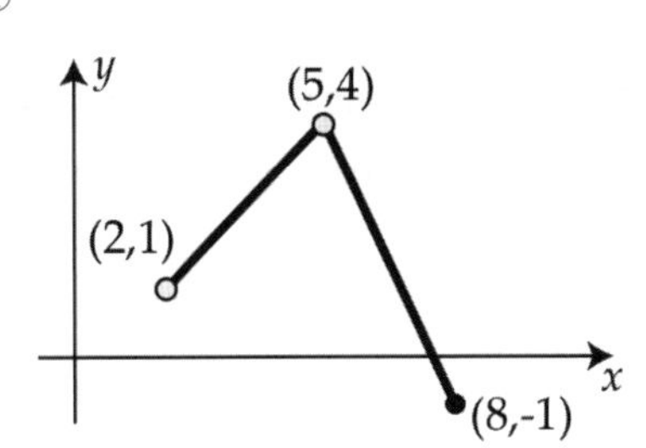

②

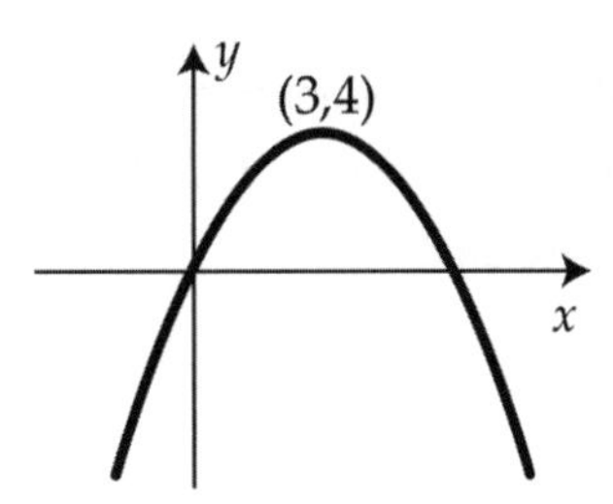

③

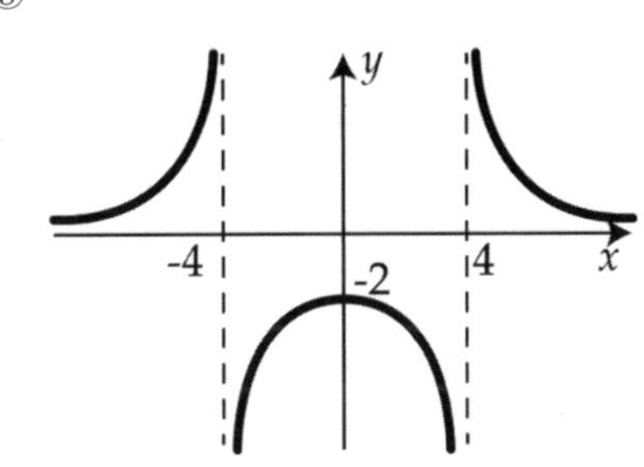

④

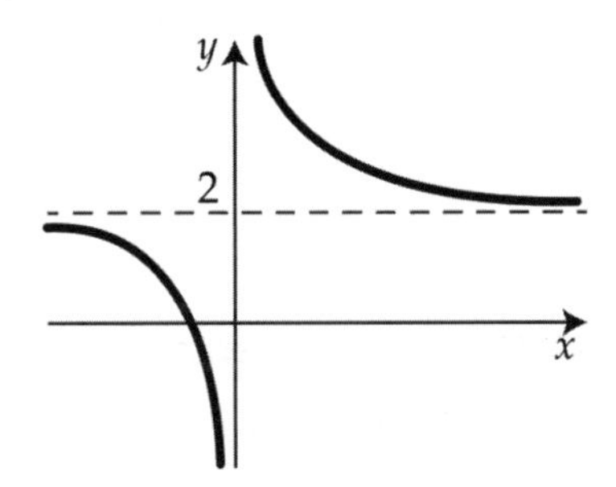

4. Implicit VS Explicit Function

Explicit: The dependent variable(y) and independent variable(x) are separated on opposite sides.	**Implicit**: The dependent variable(y) and independent variable(x) are not separated on opposite sides.
"y = function of x".	"function of y and x = something else".
$y = \pm\sqrt{4-x^2}$ 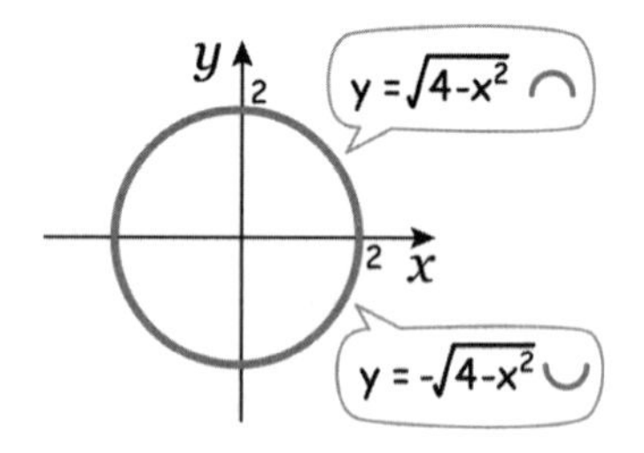	$x^2 + y^2 = 4$ 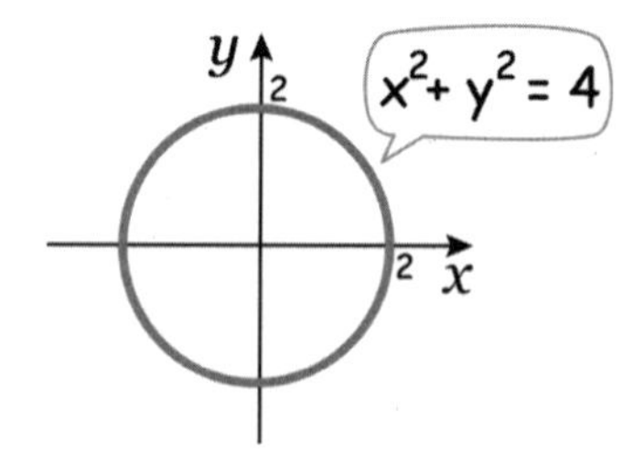

EXAMPLE 5. Convert the given implicit function to explicit function. Determine whether the equation defines y as a function of x.

① $x^2 + 2y = 4$

② $x^2y - y = 1$

③ $x^2 + (y+2)^2 = 4$

④ $\frac{x^2}{4} - \frac{y^2}{9} = 1$

⑤ $x-y^3=0$

⑥ $x-y^4=0$

EXAMPLE 6. *Find $f(x)$ when;

① $f(x+1)=4x^2+x$

② $f(2x)=x^2-1$

③ $f(2x-1)=\log 4x$

AP Style Problem

1. Which of the following is a function?

   I. $x^2 - y = 4$

   II. $x - y^3 = 4$

   III. $\sqrt{y} - x = 4$

   IV. $x^2 + y^2 = 4$

A) I, II

B) I, III

C) I, II, III

D) I, II, IV

2. Which of the following is NOT in the domain of the function $f(x) = \dfrac{\log(1-x)}{\sqrt{2x+8}}$?

A) 0.5

B) 0

C) -2

D) 4

3. If $f(x)=2+\sqrt{4-x}$, which of the following is NOT a value of $f(x)$?

A) 1

B) 2

C) 4

D) 5

4.* If $f(x^2-2x)=2x^2-4x+5$, then what is $f(8)$ =?

A) 2

B) 8

C) 16

D) 21

1.2 Rate of Change

☺ Reminder:

Slope shows the ① ____________ of a line.

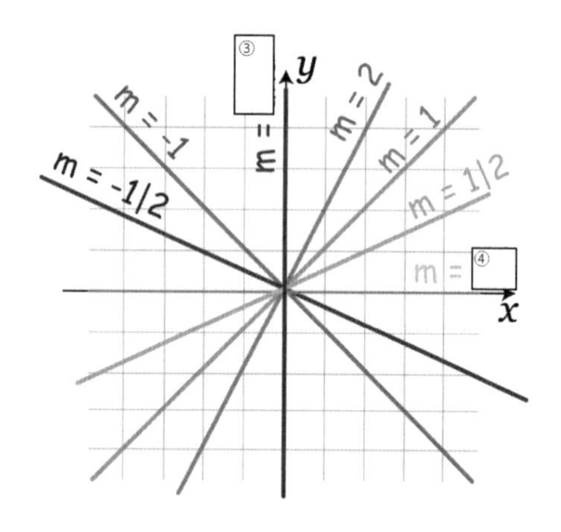

slope = ② ______ of change

$$= \frac{\text{rise}}{\text{run}} = \frac{\Delta y}{\Delta x} = \frac{y_2 - y_1}{x_2 - x_1}$$

1. Average rate of change

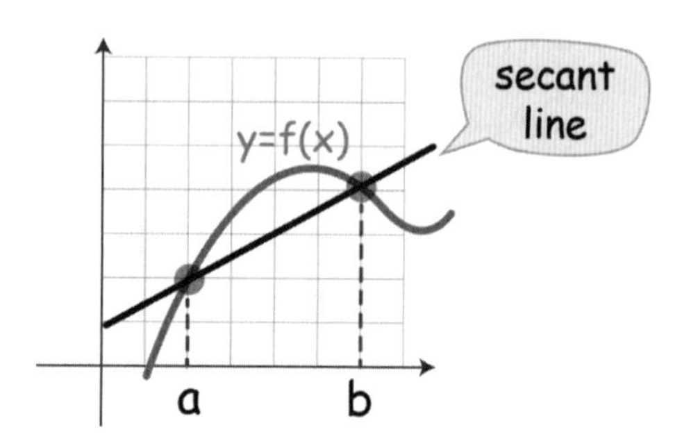

The line that crosses two points on a curve is called a ⑤ ____________ line.

The **slope of the** ⑥ ____________ **line** will be ;

⑦ ______________________________

We call it the '⑧ ____________ **rate of change'**.

※ **Average Rate of Change**

The **slope of the secant line** that goes through two points **a** and **b**

=Average rate of change =

$$\frac{f(b) - f(a)}{b - a}$$

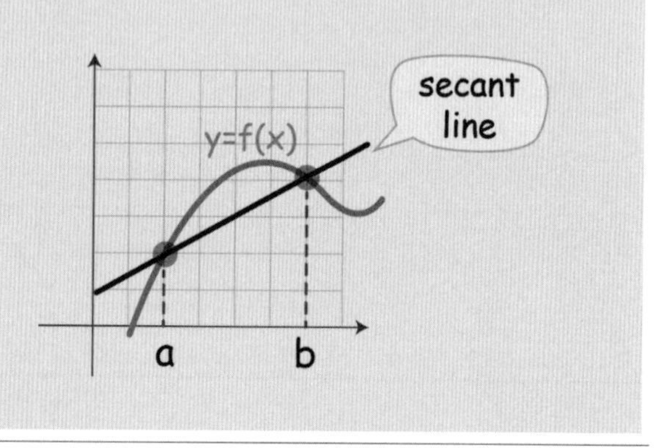

Blank : ① steepness ② rate ③ undefined ④ 0 ⑤ secant ⑥ secant ⑦ $\frac{f(b) - f(a)}{b - a}$ ⑧ Average

EXAMPLE 1. Find the average rate of change over the given interval.

① $f(x)=x^2+3x$, [2, 6]

② $f(x)=2x^2-4x+3$, [-2, 5]

③ $f(x)=\dfrac{x+3}{x}$, [1, 7]

④ $f(x)=\dfrac{x+5}{x-4}$, [1, 7]

⑤ $f(x)=\sqrt{x+4}$, [5, 8]

⑥ $f(x)=\sqrt{x-1}$, [2, 6]

EXAMPLE 2. For the given table and graph, find the following;

x	-5	-1	0	4	9	12
$f(x)$	9	8	-1	-8	-5	7

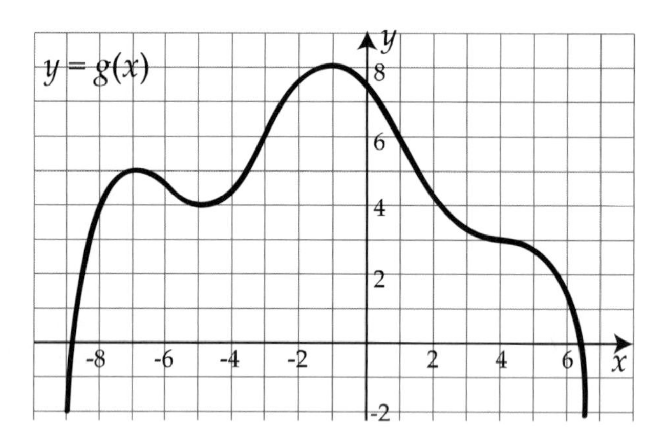

① average rate of change of f over $-5 \le x \le 0$:

② average rate of change of f over $4 \le x \le 12$:

③ average rate of change of g over $-3 \le x \le 4$:

④ average rate of change of g over $-1 \le x \le 1$:

⑤ average rate of change of g over $-8 \le x \le -5$:

⑥ average rate of change of g over $-7 \le x \le -3$:

2. Average rate of change of Polynomials

	Linear Function	Quadratic Function
Table looks like;	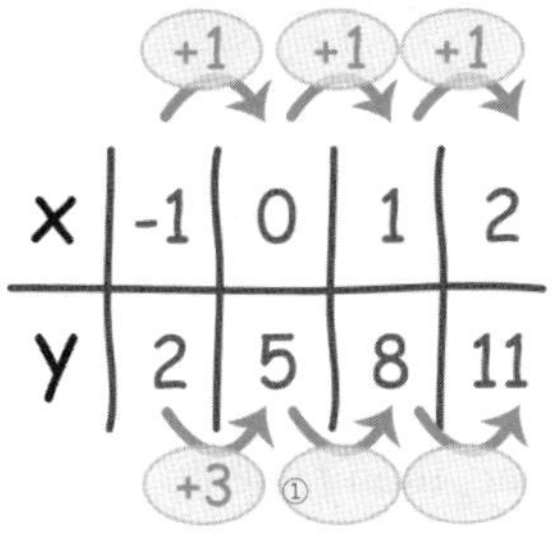 +1 +1 +1 x | -1 | 0 | 1 | 2 y | 2 | 5 | 8 | 11 +3 ①	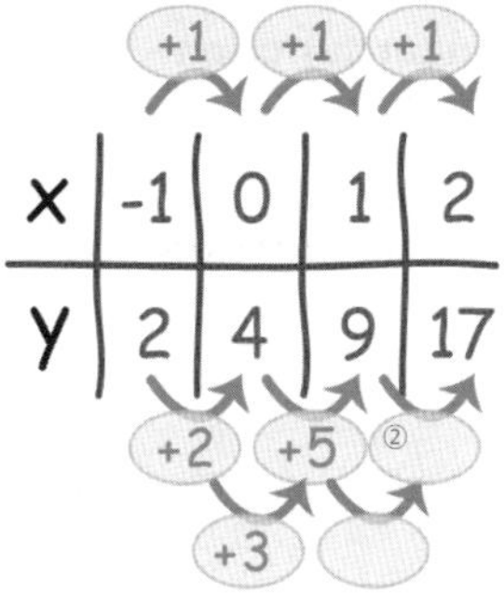 +1 +1 +1 x | -1 | 0 | 1 | 2 y | 2 | 4 | 9 | 17 +2 +5 ② +3
Graph looks like;	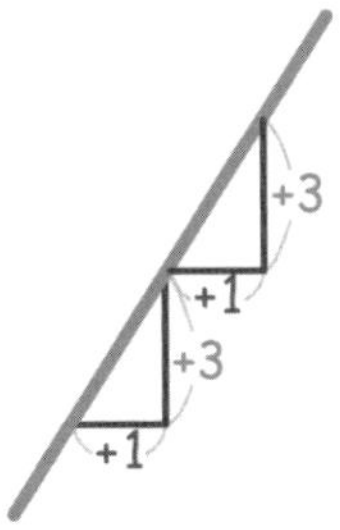 +3 +1 +3 +1	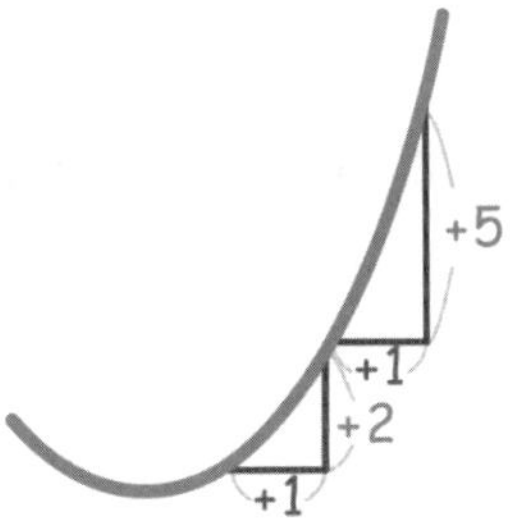 +5 +1 +2 +1
Difference	First difference of y is ③________ over the equal length of x.	First difference of y is ④________ over the equal length of x.
	Second difference of y is ⑤________ over the equal length of x.	Second difference of y is ⑥________ over the equal length of x.
Average rate of change	Average rate of change is a ⑦________. ① if it is positive :⑧________ ② if it is negative :⑨________	Average rate of change is ⑩________ ① if it is increasing :⑪________ ② if it is decreasing:⑫________
	The rate of change of the average rates of change is ⑬________	The rate of change of the average rates of change is ⑭________

Blank : ① +3 +3 ② +8 +3 ③constant (=same number) ④ linear ⑤ 0 ⑥ constant ⑦ constant ⑧ increasing ⑨ decreasing ⑩ linear ⑪ open up(=concave up) ⑫ open down(=concave down) ⑬ 0 ⑭ constant

※ **Average rate of change of Polynomials**

Constant function —— : Average rate of change is 0

Linear function ／ : Average rate of change is constant

= Rate of change of the average rate of change is 0

Quadratic function ∪ : Average rate of change is linear

= Rate of change of the average rate of change is constant

If the n th difference of y is a same number, then f will be nth degree polynomial.

EXAMPLE 3. Determine whether f is linear or quadratic. Describe the shape (increasing or decreasing or open up or open down).

①

x	0	2	4	6
$f(x)$	1	4	7	10

②

x	-8	-2	4	10
$f(x)$	10	7	4	1

③

x	-7	-5	-3	-1
$f(x)$	0	4	6	6

④

x	4	7	10	13
$f(x)$	1	3	6	10

⑤

x	-2	0	1	3
$f(x)$	5	-3	-4	0

⑥

x	-6	-3	-2	0
$f(x)$	-12	-15	-16	-18

⑦

x	-4	-1	1	4
$f(x)$	-9	-12	-16	-25

AP Style Problem

1. On which interval is the average rate of change the least?

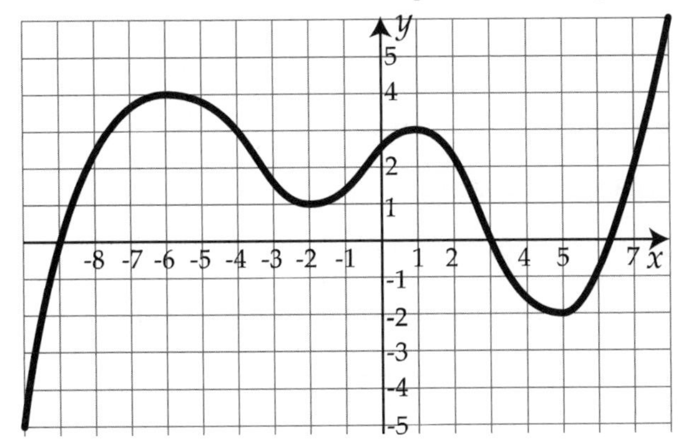

A) $-9 \le x \le -2$

B) $-2 \le x \le 1$

C) $1 \le x \le 5$

D) $5 \le x \le 7$

2. The average rate of change of f is given over the intervals. In which interval is the function f decreases by 9?

x	$0 \le x \le 3$	$3 \le x \le 7$	$10 \le x \le 13$	$15 \le x \le 20$
average rate of change	3	-3	-3	3

A) $0 \le x \le 2$

B) $3 \le x \le 7$

C) $10 \le x \le 13$

D) $15 \le x \le 20$

3. The values of a function f are given as shown. What can we conclude about the function f?

x	0	2	5	9
$f(x)$	1	9	27	59

A) f is quadratic since average rate of change is constant.

B) f is linear since average rate of change is constant.

C) f is linear since rate of change of the average rates of change is 0.

D) f is quadratic since rate of change of the average rates of change is constant.

4. The values of a function g are given as shown. What can we conclude about the function g?

x	0	2	3	6
$g(x)$	5	-1	-4	-13

A) g is linear since average rate of change is linear

B) g is quadratic since average rate of change is constant.

C) g is linear since rate of change of the average rates of change is 0.

D) g is quadratic since rate of change of the average rates of change is 2.

5. The average rate of change of quadratic function f is given over the intervals.

x	$0 \le x \le 2$	$2 \le x \le 4$
average rate of change	2	-1

What is the average rate of change of f in the interval $8 \le x \le 10$?

A) -7

B) -10

C) -17

D) -21

6. Which of the following could be true about f?

x	-2	-1	0	1	2
$f(x)$	-360	-150	-40	0	0

A) f is linear which means f has degree of 1.

B) f is quadratic which means f has degree of 2.

C) f is cubic which means f has degree of 3.

D) f is quartic which means f has degree of 4.

1.3 Analyzing Functions

1. Even and Odd Functions

※ **Even and Odd Functions**

Even function

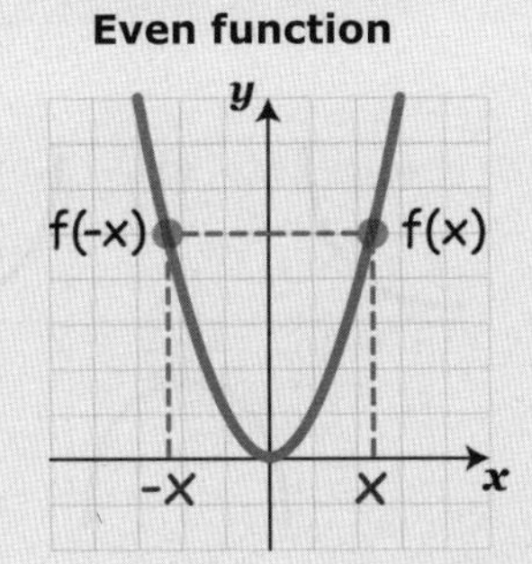

$$f(-x) = f(x)$$

symmetric over ① ____________

ex) $y = x^2$

$y = x^4 - 3x^2 + 3$

$y = \cos x$

$y = |x|$

Odd function

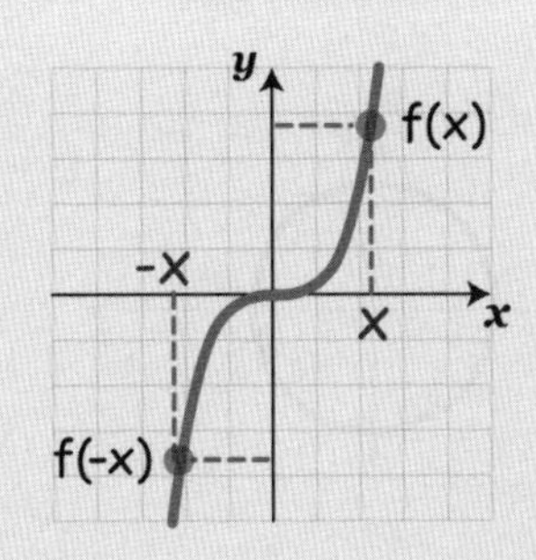

out

$$f(-x) = -f(x)$$

symmetric over ② ____________

ex) $y = x^3$

$y = x^7 - 8x^5 + x$

$y = \sin x$

$y = \tan x$

EXAMPLE 1. Determine whether each of the following functions is even, odd, or neither.

①

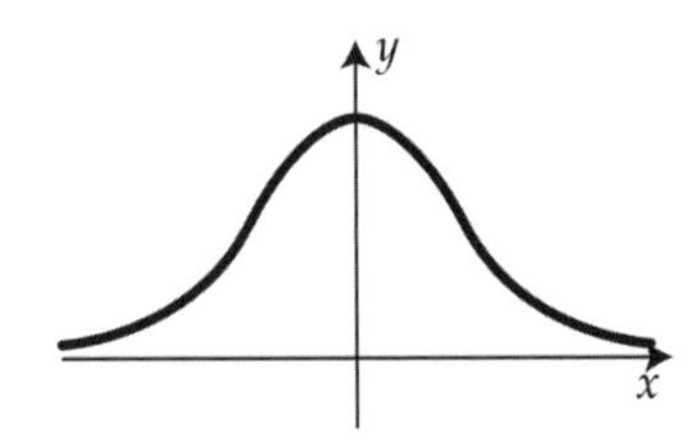

②

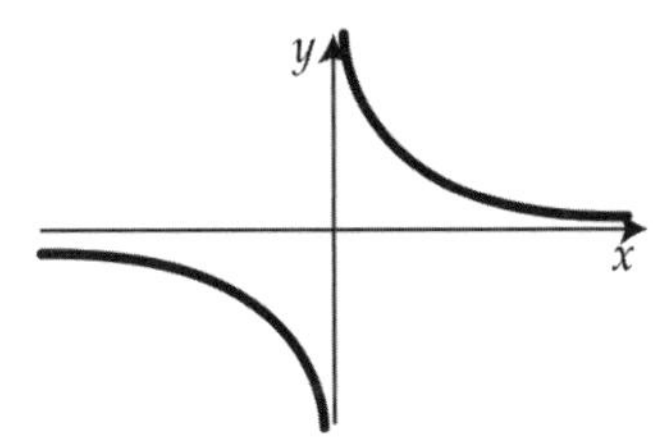

Blank : ① y axis ② origin

③

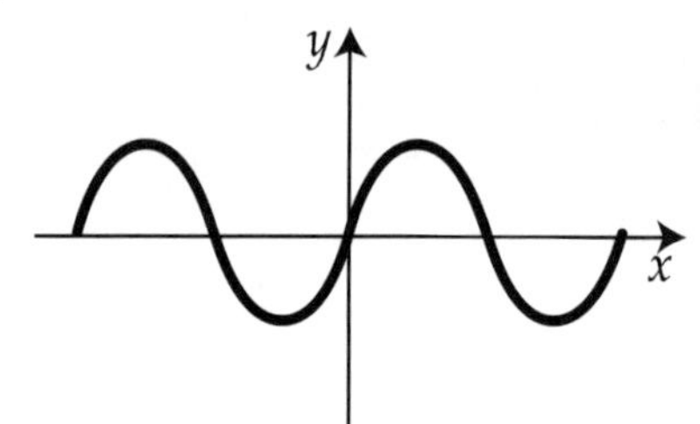

④

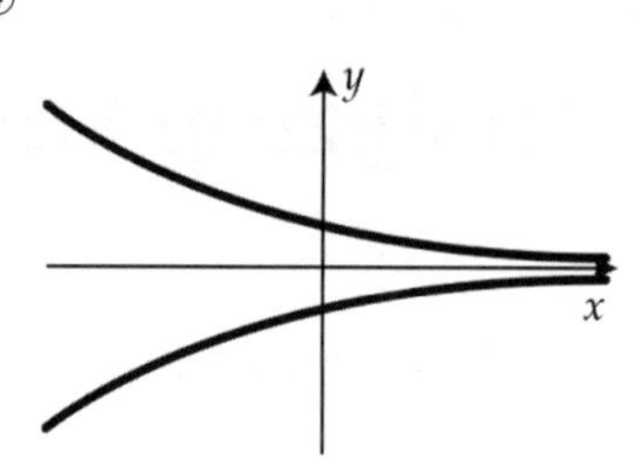

⑤

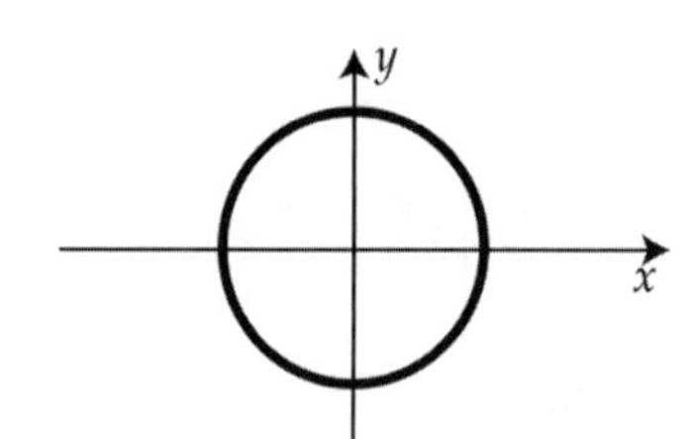

⑥

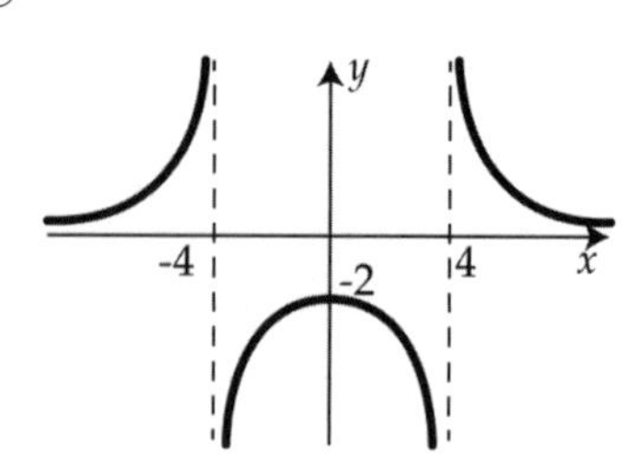

⑦

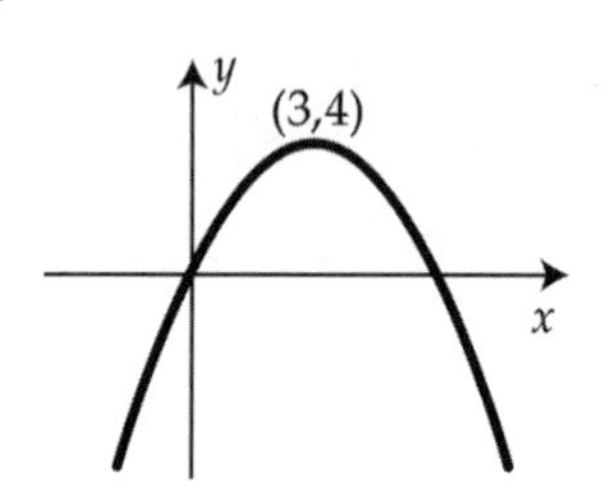

EXAMPLE 2. Determine whether each of the following functions is even, odd, or neither.

☺ Tip: Find ①____________ ,

If you get the ②____________ function back (=$f(x)$) the function is **even**.

If *all of the signs are changed* (= -$f(x)$) the function is **odd**.

① $f(x) = -x^4 + 2x^2 - 5$

② $f(x) = x^5 + 3x^3 - x$

Blank : ① f(-x) ② original

③ $f(x) = -3x^7 + x$

④ $f(x) = x^{100} + 2x^{10} - 8$

⑤ $f(x) = -6x^4 + 2x - 8$

⑥ $f(x) = 3x^{10} + x + 7$

⑦ $g(x) = |x| + 2$

⑧ $g(x) = \sqrt{x-2}$

⑨ $h(x) = \sqrt[3]{x}$

⑩ $f(x) = \sqrt[3]{x^2 + 2}$

⑪ $f(x) = -\dfrac{x^3}{x^2 + 2}$

⑫ $g(x) = \dfrac{x}{|x|}$

⑬ $g(x) = x|x^2 - 2|$

⑭ $h(x) = |x - 2|$

2. Increasing and Decreasing

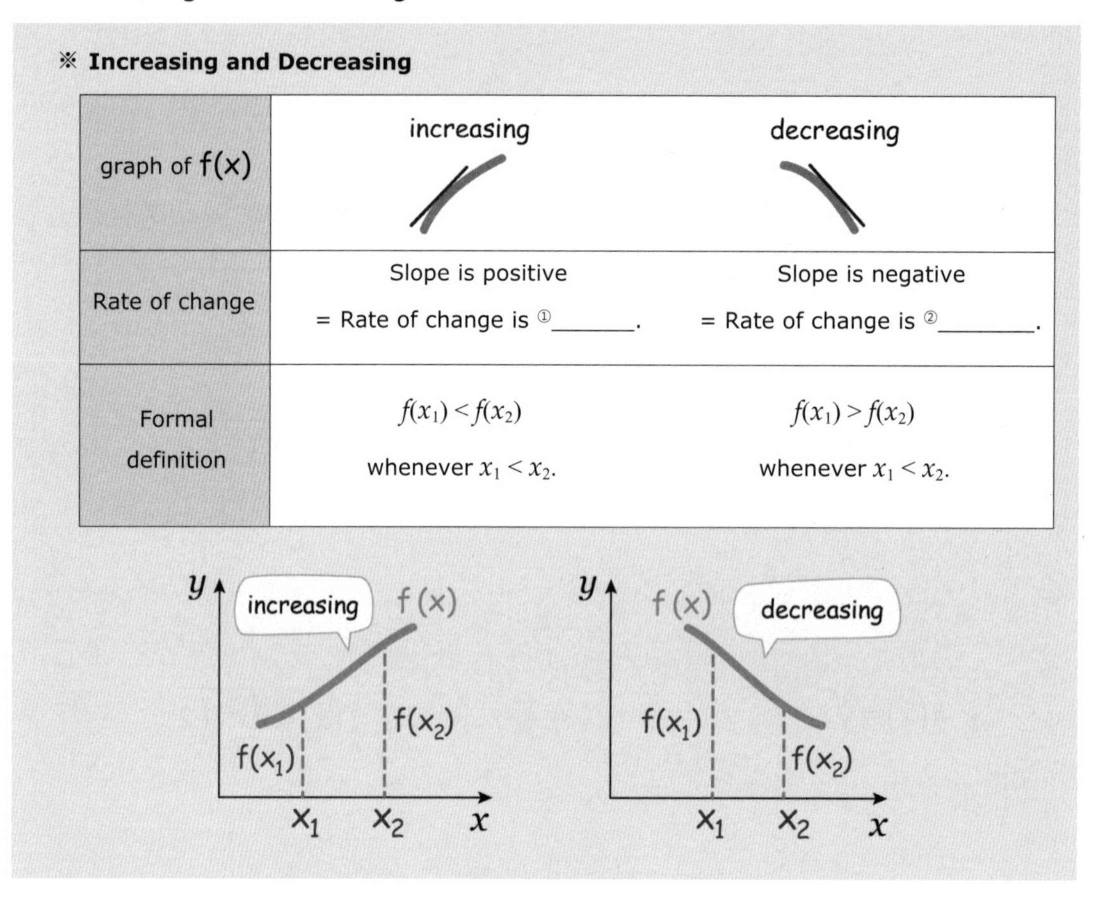

※ **Increasing and Decreasing**

graph of f(x)	increasing	decreasing
Rate of change	Slope is positive = Rate of change is ①______.	Slope is negative = Rate of change is ②_______.
Formal definition	$f(x_1) < f(x_2)$ whenever $x_1 < x_2$.	$f(x_1) > f(x_2)$ whenever $x_1 < x_2$.

EXAMPLE 3. Fill in the blank with positive or negative.

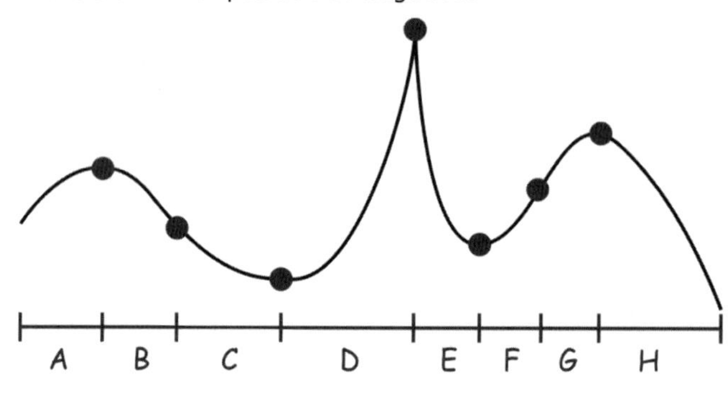

intervals	A	B	C	D	E	F	G	H
Rate of change								

Blank : ① positive ② negative

3. Maximum and minimum

1) Relative (=Local) maximum and minimum

A ①_______________ maximum is a point that is higher than the points beside it on both sides, and a relative minimum is a point that is lower than the points beside it on both sides.

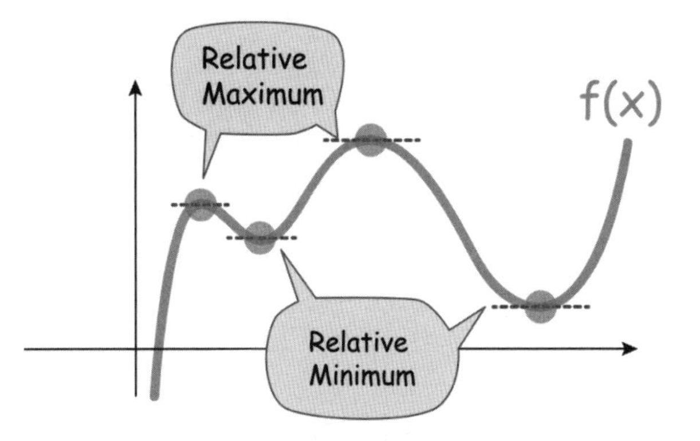

※ **Relative** (=**Local**) **maximum and minimum**

graph of f(x)	Relative Maximum	Relative Minimum
Rate of change	Rate of change changes from pos to neg	Rate of change changes from neg to pos
Formal definition	f has relative maximum at c if $f(c) \geq f(x)$ for x in a certain interval containing c.	f has relative minimum at c if $f(c) \leq f(x)$ for x in a certain interval containing c.

Blank : ① relative

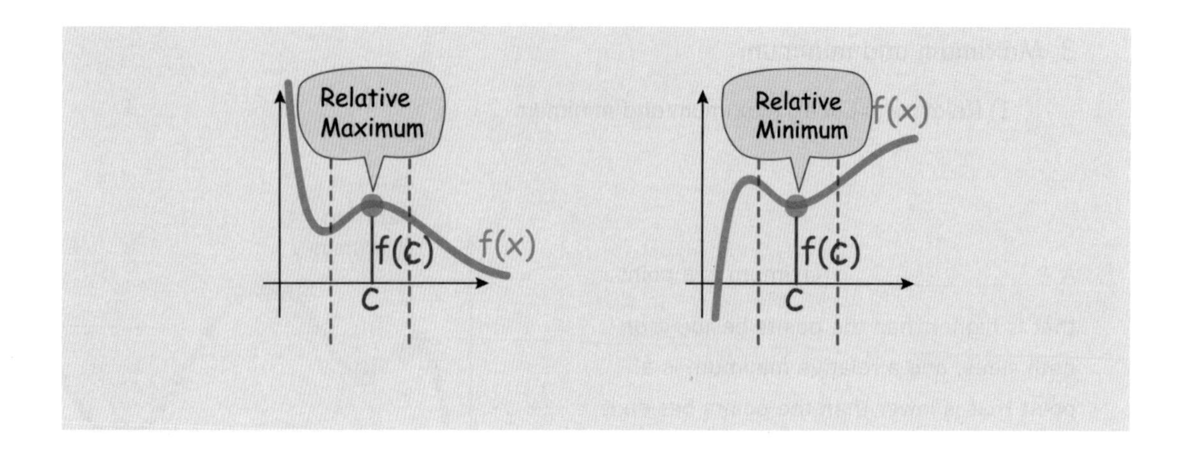

2) Absolute(=Global) maximum and minimum

The maximum or minimum *over the* ① ____________ *function* or *given* ②____________ is called absolute (=global) maximum or minimum.

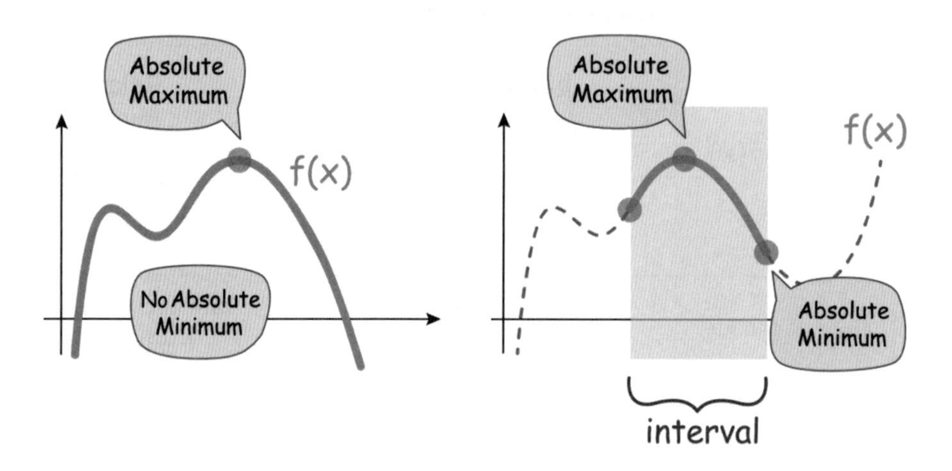

● *note*

Relative = Local, Absolute = Global

The plural of Maximum is ③_________

The plural of Minimum is ④________

Maxima and Minima are collectively called ⑤__________

Blank : ① entire ② interval ③ Maxima ④ Minima ⑤ Extrema

EXAMPLE 4. Using the graph of $f(x)$ and $g(x)$ given, find the following if they exist.

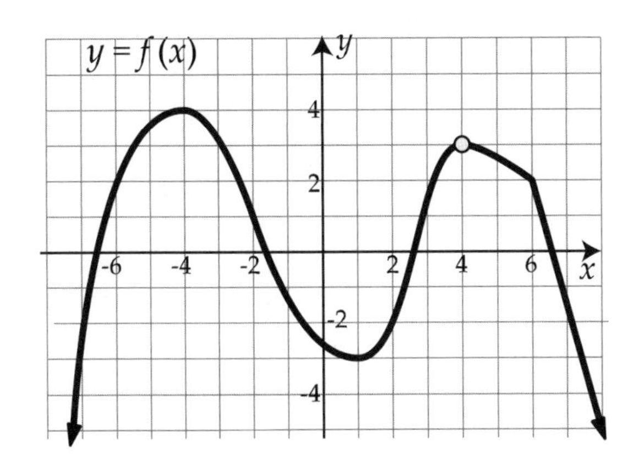

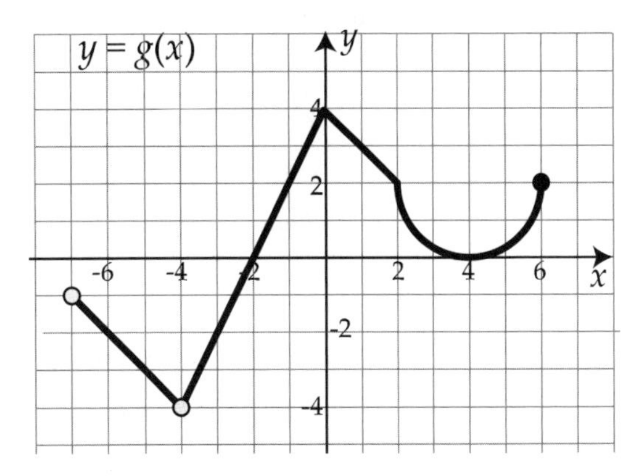

① Find the relative extrema of f.

② Find the relative extrema of g.

③ Find the absolute extrema of f.

④ Find the absolute extrema of g.

⑤ Find the absolute extrema of f in the interval $-2 \le x \le 2$

⑥ Find the absolute extrema of g in the interval $-2 \le x \le 2$

⑦ Find the interval where $f(x_1) > f(x_2)$ whenever $x_1 < x_2$.

⑧ Find the interval where $g(x_1) < g(x_2)$ whenever $x_1 < x_2$.

⑨ Find the interval where the rate of change of f is positive.

⑩ Find the interval where the rate of change of g is negative.

4. Concavity

Concave up:① ___________

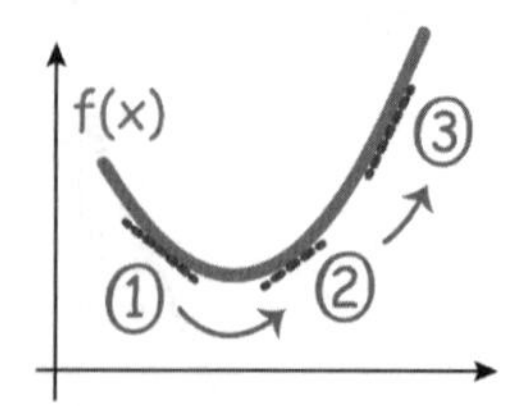

When the function **concaves upward,** the rate of change continually (③decreases/increases).

Concave down:② ___________

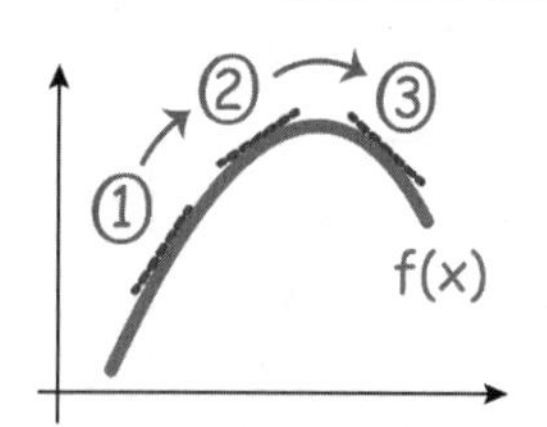

When the function **concaves downward,** the rate of change continually (④decreases/increases).

※ **Concavity**

graph of $f(x)$	Concave Upward	Concave Downward
Rate of change	slope is increasing = Rate of change is ⑤________. = f has increasing rate	slope is decreasing = Rate of change is ⑥________. = f has decreasing rate
Rate of change of rate of change	Rate of the rate of change is a ⑦___________	Rate of the rate of change is a ⑧___________

Blank : ① ∪ shape ② ∩ shape ③ increases ④ decreases ⑤ increasing ⑥ decreasing ⑦ positive ⑧ negative

EXAMPLE 5. Fill in the blank with increasing or decreasing.

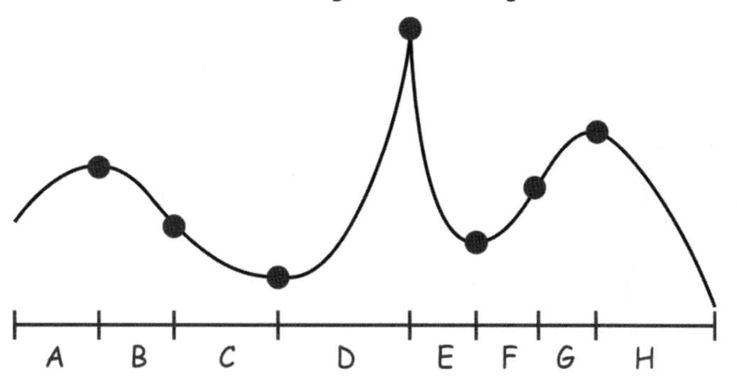

intervals	A	B	C	D	E	F	G	H
Rate of change								

5. Inflection point

※ **inflection Point**

An **inflection Point** is where a curve *changes* from concave upward to concave downward or vice versa. (where concavity changes)

Inflection Point

EXAMPLE 6. Find the number of points of inflection.

①

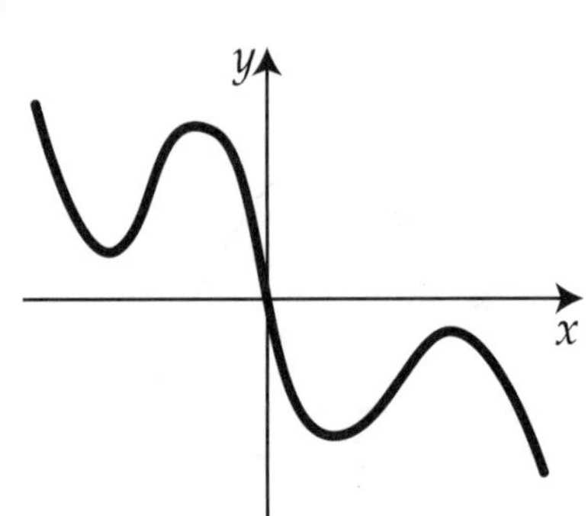

②

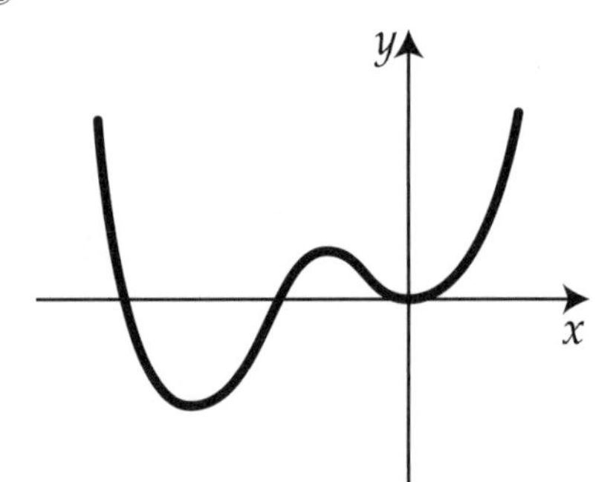

③

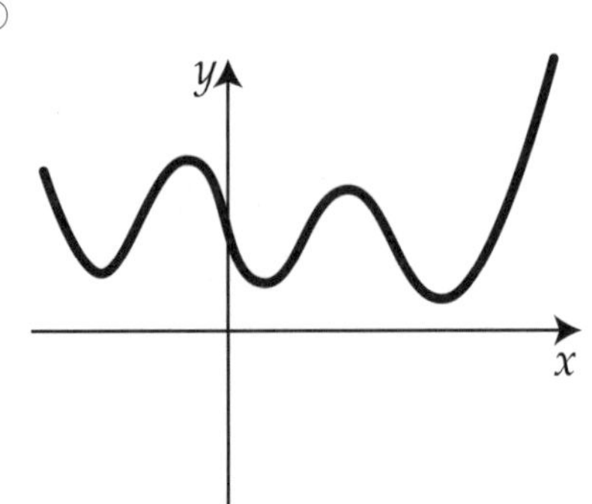

④

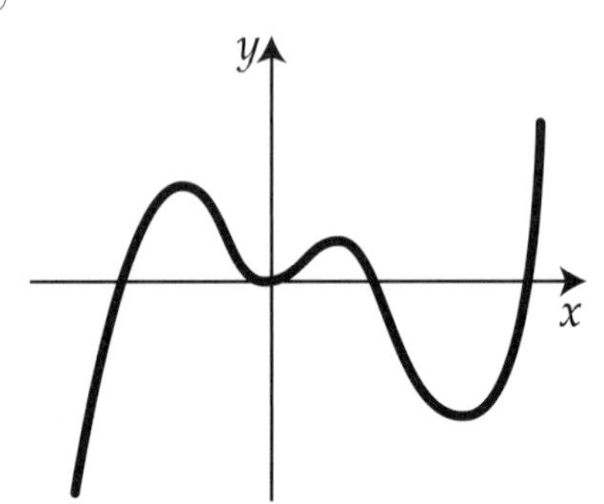

☺ Tip:

rate is increasing ◡		rate is decreasing ◠	
increasing	decreasing	increasing	decreasing

EXAMPLE 7. Graph of f is given over the interval [0, 16]. Find the interval for the following graph of f;

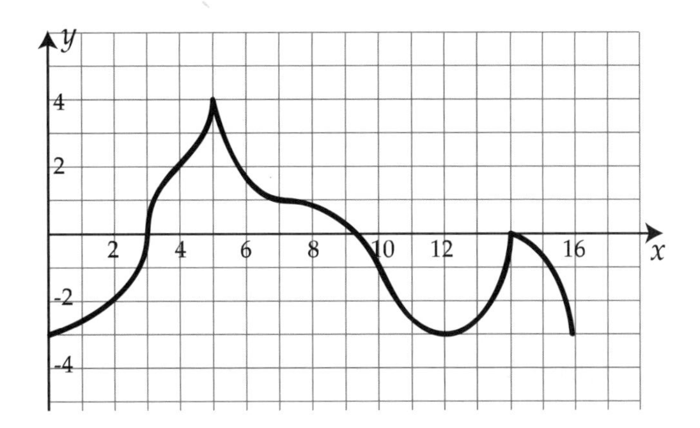

① Where f is increasing at increasing rate.

② Where f is decreasing at decreasing rate

③ Where f is decreasing at increasing rate.

④ Where f is increasing at decreasing rate

⑤ Where the rate is positive and decreasing

⑥ Where the rate is negative and decreasing

AP Style Problem

1. If $f(x)$ is an even function and $g(x)$ is an odd function, $f(3) = 1$ and $g(2) = -3$, then

   $f(-3) + g(-2) =$

   A) -2

   B) 2

   C) 4

   D) -4

2. The average rate of change of f is given other the intervals.

x	$0 \le x \le 3$	$3 \le x \le 6$	$6 \le x \le 9$	$9 \le x \le 12$
average rate of change	positive, decreasing	positive, increasing	negative, increasing	negative, decreasing

1) Which of the following is true about the interval $0 \le x \le 3$

   A) f is increasing and concave up

   B) f is increasing and concave down

   C) f is positive and concave up

   D) f is positive and concave down

2) What is the interval on which the rate of rate of change is negative?

   I. $0 \le x \le 3$

   II. $3 \le x \le 6$

   III. $6 \le x \le 9$

   IV. $9 \le x \le 12$

   A) I, II   B) I, III

   C) II, III   D) I, IV

3) If $f(x)$ is a continuous function, find the value of x of the point of inflection of f in the interval $0 \le x \le 12$.

A) $x = 6$ B) $x = 3, 9$

C) $x = 3, 6$ D) $x = 3, 6, 9$

3. The temperature outside a house during a 14-hour period is given by $T(x)$. Of the following, on which interval is the rate of the temperature positive and decreasing?

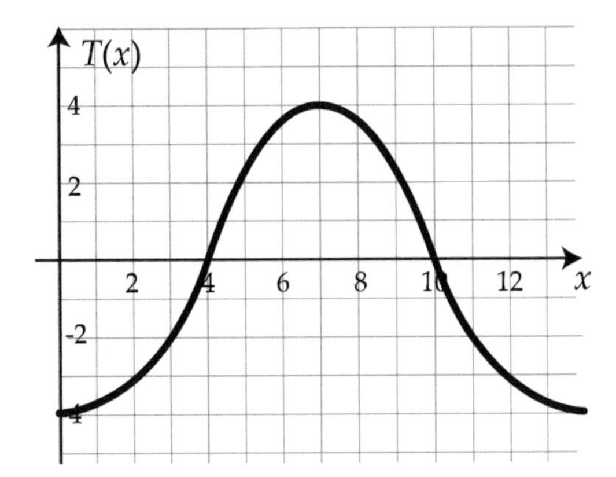

A) $0 \le x \le 4$

B) $4 \le x \le 7$

C) $7 \le x \le 10$

D) $10 \le x \le 13$

4. The rate of fuel consumption, in gallons per minute, recorded during an airplane flight is given by $F(x)$. Of the following, on which interval is the rate of fuel consumption positive and decreasing?

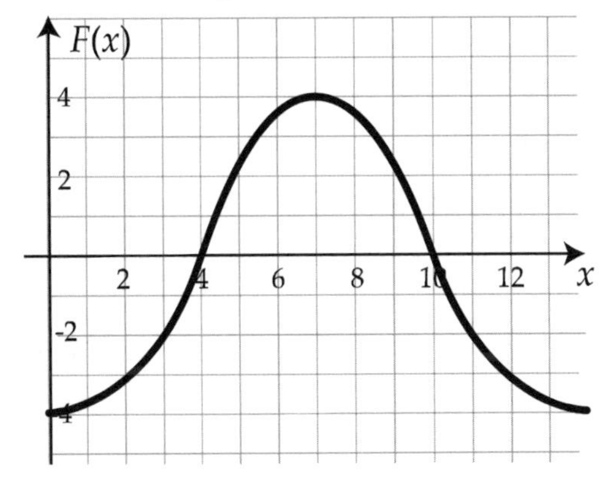

A) $0 \le x \le 4$

B) $4 \le x \le 7$

C) $7 \le x \le 10$

D) $10 \le x \le 13$

5. The graph of rate of change of f is given above. Of the following, on which interval is f concave down?

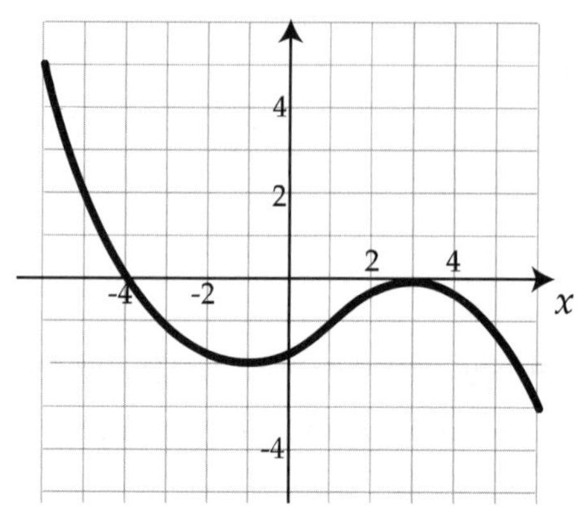

A) $-6 \le x \le 1$

B) $-1 \le x \le 3$

C) $1 \le x \le 6$

D) $-6 \le x \le -1$ or $3 \le x \le 6$

6.* $f(x)$ is an even function and $g(x)$ is an odd function. Which of the following $h(x)$ is an odd function?

I. $h(x) = f(x) + g(x)$

II. $h(x) = [g(x)]^2$

III. $h(x) = g(g(x))$

A) I

B) III

C) II, III

D) I, III

1.4 Piecewise Functions

1. Thirteen Basic Functions

Constant function

$y = c$

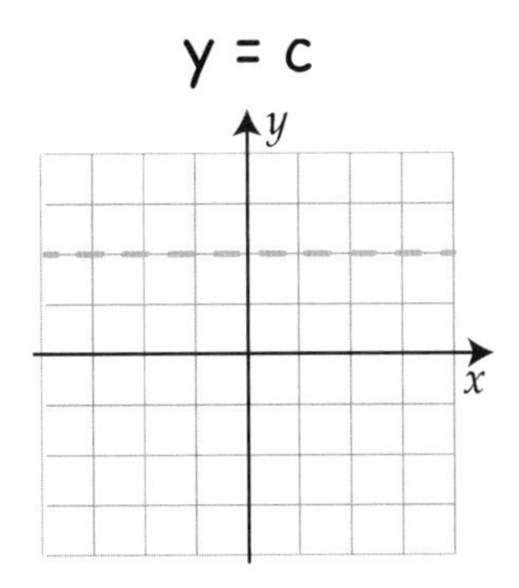

Linear function

$y = x$

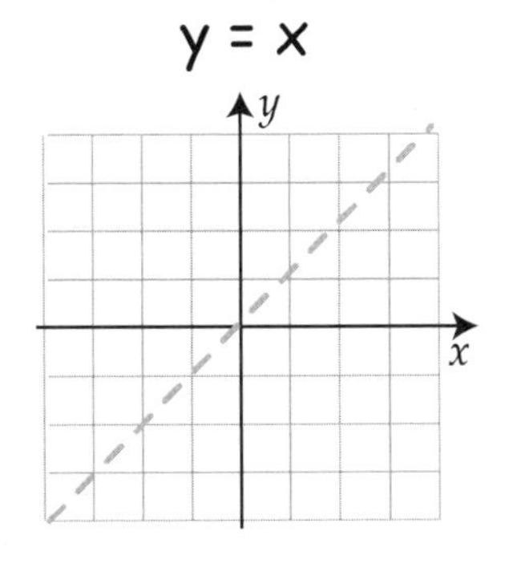

Quadratic function

$y = x^2$

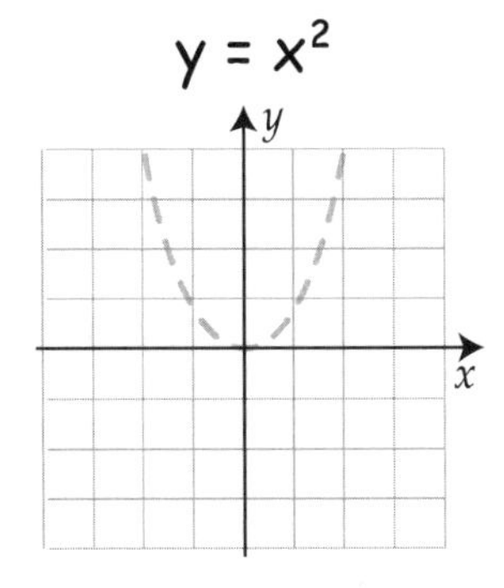

Cubic function

$y = x^3$

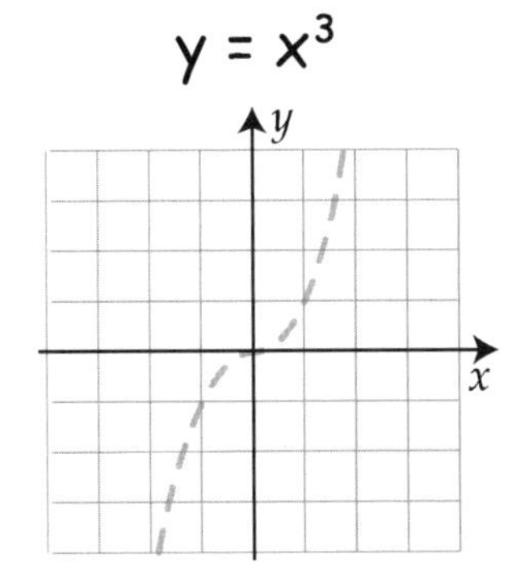

Square root function

$y = \sqrt{x}$

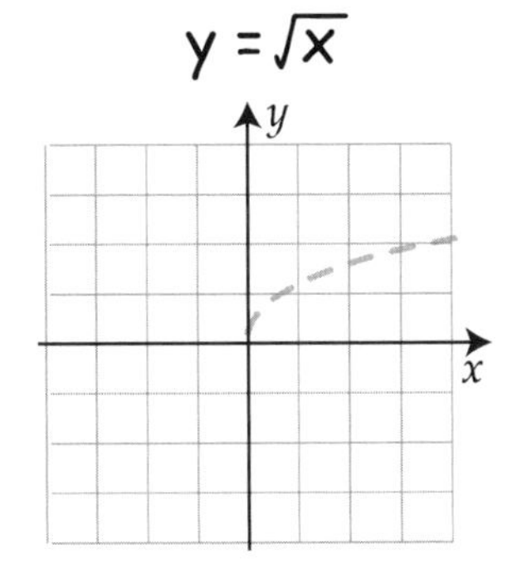

Cube root function

$y = \sqrt[3]{x}$

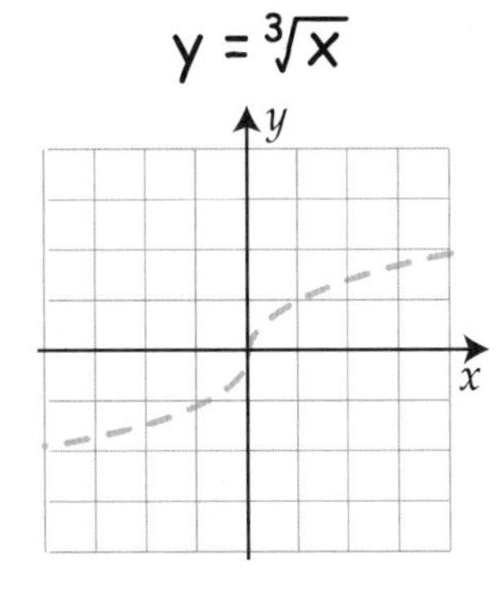

Reciprocal function

$y = \frac{1}{x}$

Absolute value function

$y = |x|$

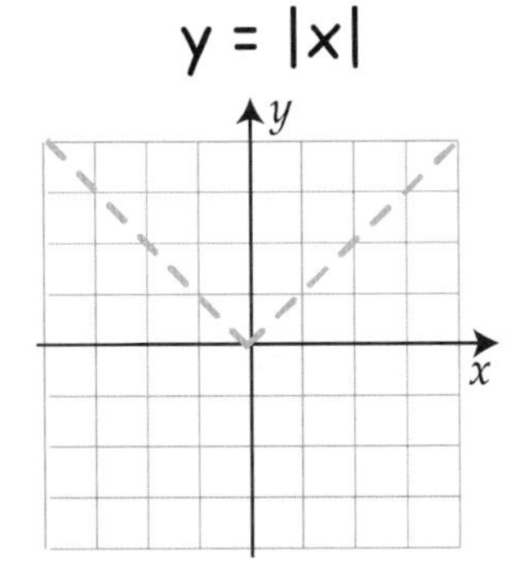

Exponential and Logarithmic function

$y = 2^x \quad y = \log_2 x$

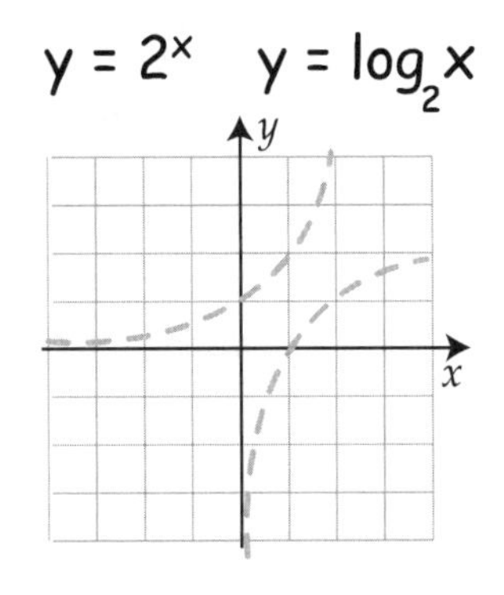

Sine function

y = sin x

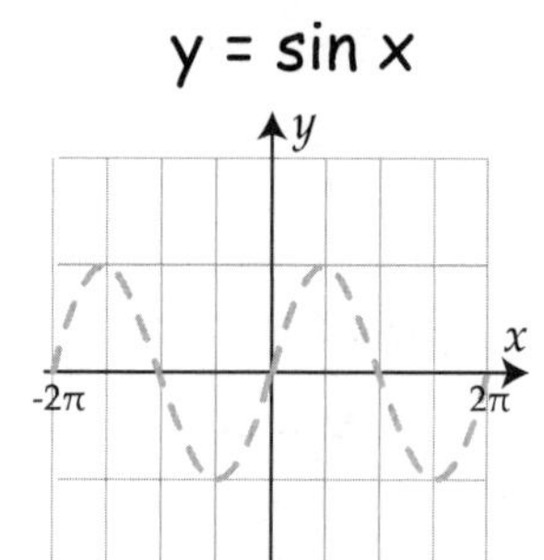

Cosine function

y = cos x

Tangent function

y = tan x

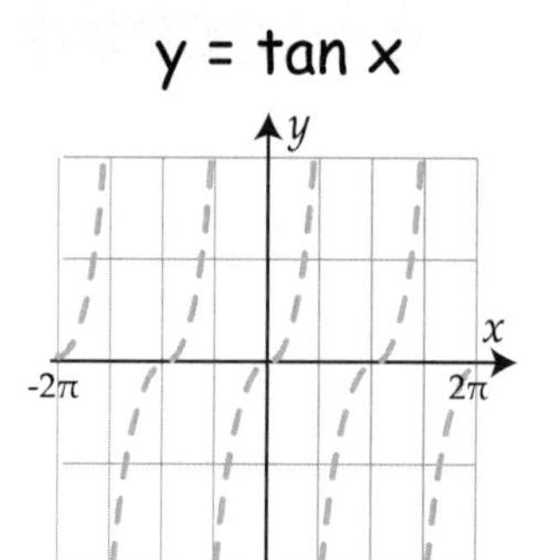

Which one satisfy $f(-x) = f(x)$?

Which one satisfy $f(-x) = -f(x)$?

2. Continuity

Continuous Function: a single unbroken graph.

(that you could draw without lifting your pen from the paper)

Discontinuity: a graph which is not continuous

When we have;

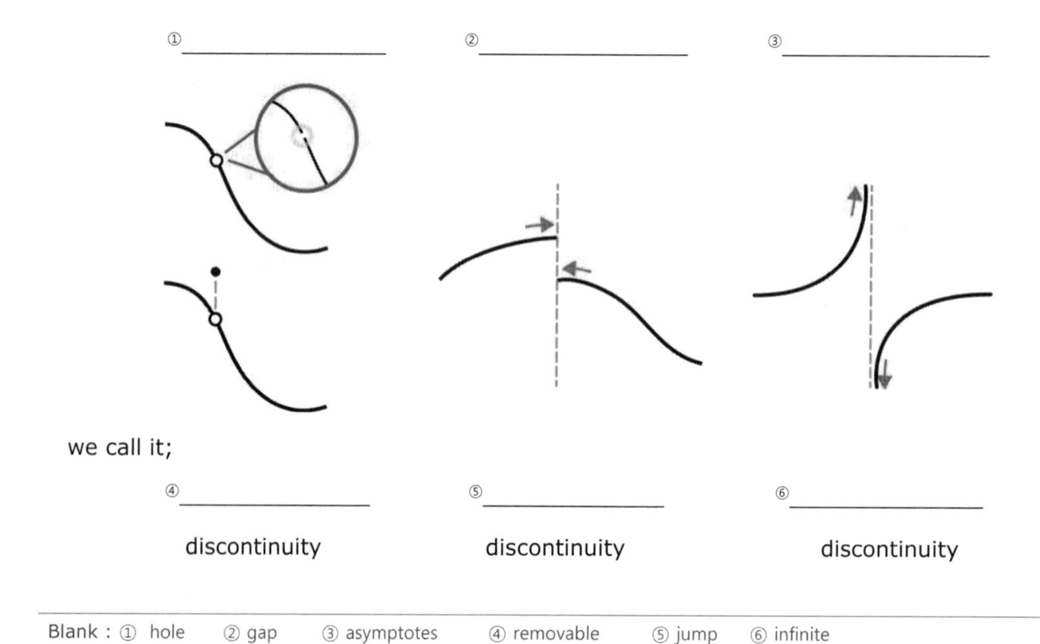

① ______________ ② ______________ ③ ______________

we call it;

④ ______________ discontinuity

⑤ ______________ discontinuity

⑥ ______________ discontinuity

Blank : ① hole ② gap ③ asymptotes ④ removable ⑤ jump ⑥ infinite

3. Piecewise-defined Functions

: a function made up in pieces.

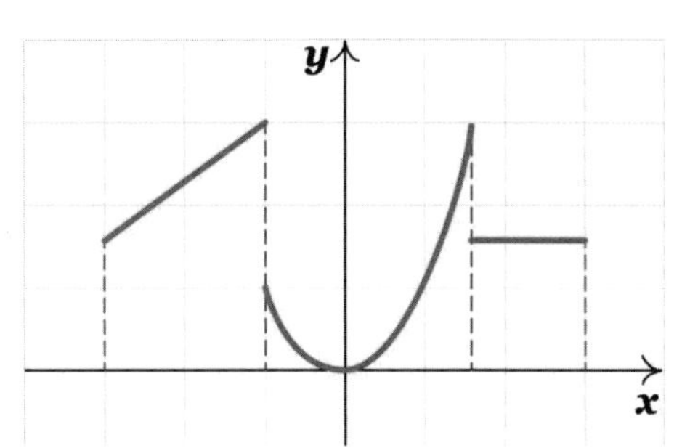

EXAMPLE 1. Graph the function. State the domain and range. Is it continuous? If not, what kind of discontinuity it is?

① $f(x)=\begin{cases} 4, & \text{if } x \le 1 \\ x, & \text{if } x > 1 \end{cases}$

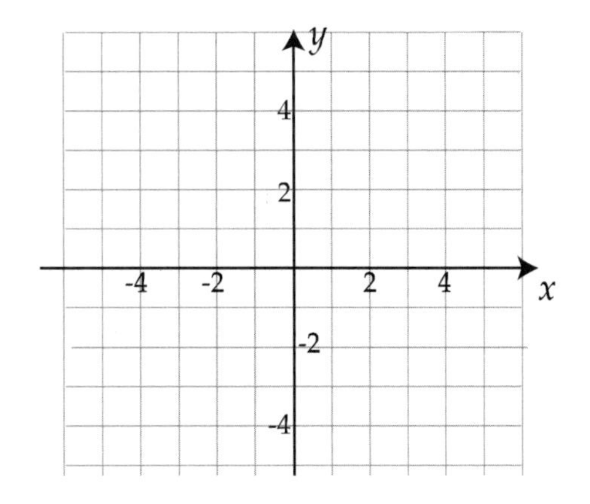

② $h(x)=\begin{cases} \sqrt{x}, & \text{if } x \ge 4 \\ x-2, & \text{if } x < 4 \end{cases}$

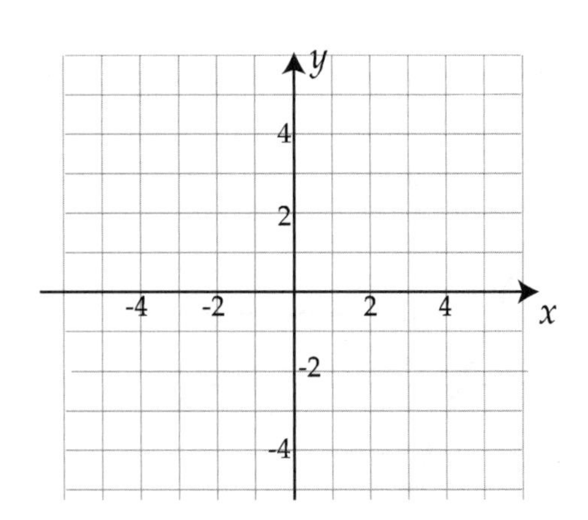

③ $g(x)=\begin{cases} 3, & \text{if } x=0 \\ \sqrt[3]{x}, & \text{if } x\neq 0 \end{cases}$

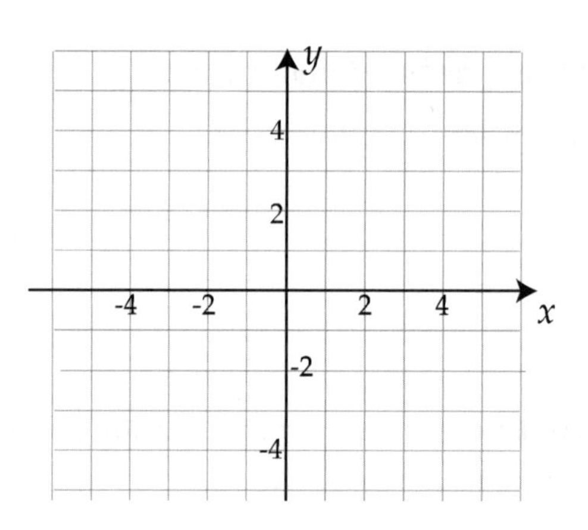

④ $f(x)=\begin{cases} x, & \text{if } x<0 \\ 2, & \text{if } x=0 \\ x^2, & \text{if } x>0 \end{cases}$

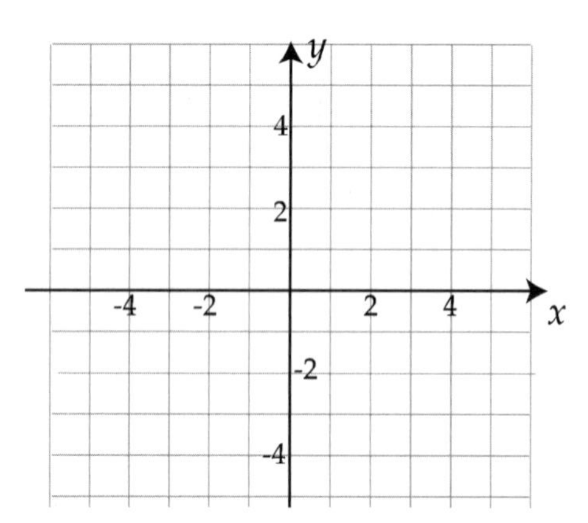

EXAMPLE 2. Give a rule for the piecewise-defined function.

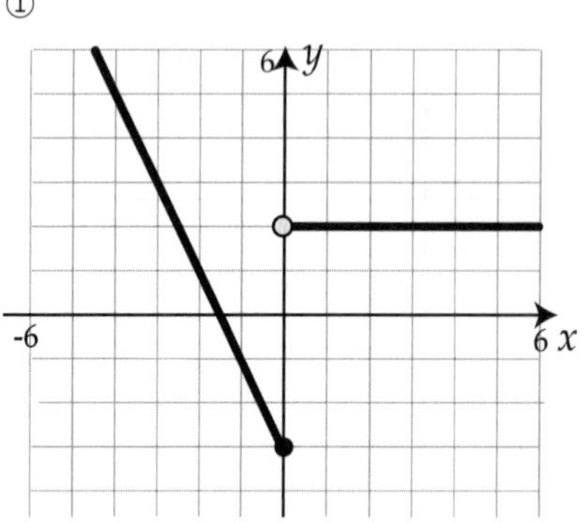

②

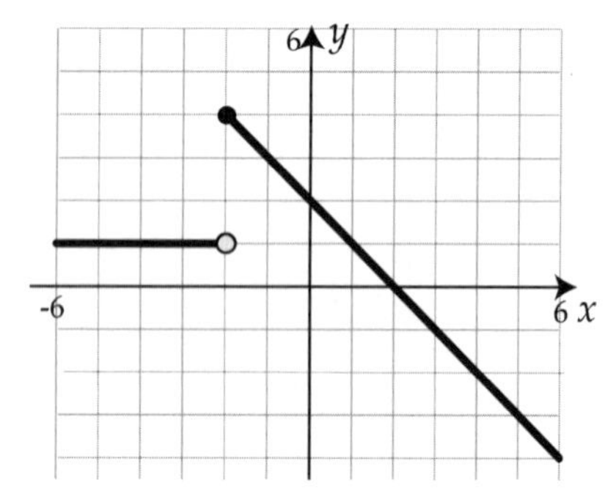

③

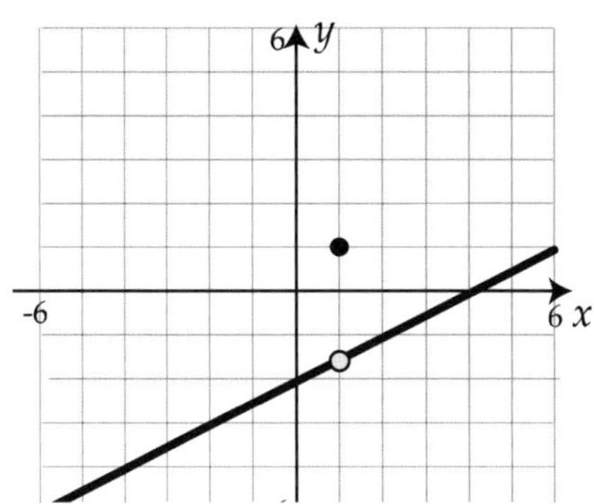

④

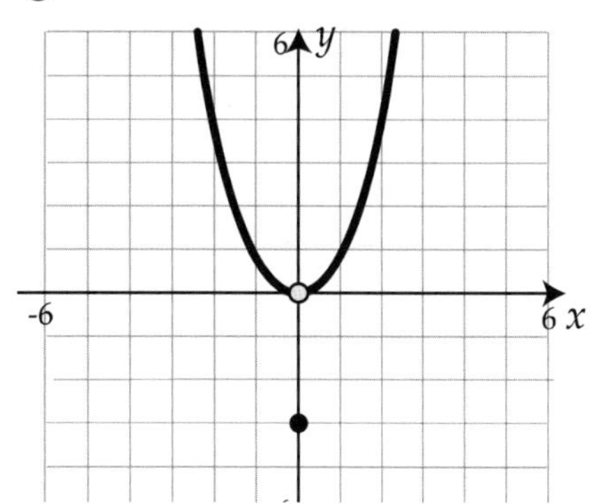

⑤

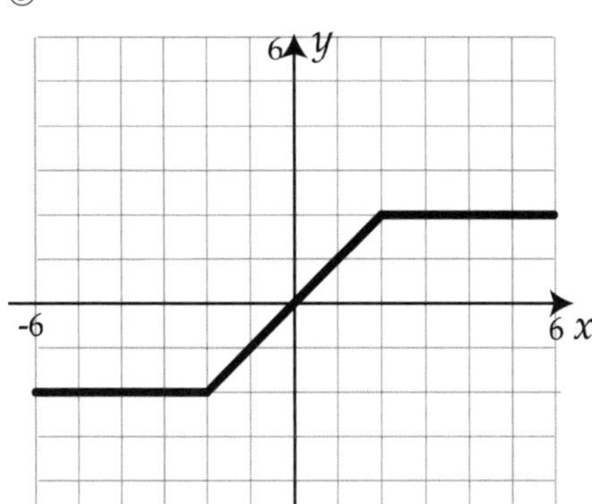

EXAMPLE 3. Evaluate the function.

① Find $f(0)$, $f(1)$, and $f(3)$ if $f(x)=\begin{cases} 2, & \text{if } x \le 1 \\ x, & \text{if } x > 1 \end{cases}$

② Find $g(27)$, $g(-2)$, and $g(0)$ if $g(x)=\begin{cases} \frac{1}{x}, & \text{if } x<0 \\ \sqrt[3]{x}, & \text{if } x\geq 0 \end{cases}$

③ Find $f(-4)$, $f(1)$, and $f(2)$ if $f(x)=\begin{cases} -2x+1, & \text{if } x<1 \\ 2, & \text{if } x=1 \\ x^2, & \text{if } x>1 \end{cases}$

EXAMPLE 4. For what value of b will f(x) be continuous?

$$f(x)=\begin{cases} -x+4, & x>3 \\ 2-\sqrt{x-b}, & x\leq 3 \end{cases}$$

AP Style Problem

1. If a piecewise defined function f(x) is

$$f(x)=\begin{cases} x+1 & ,x\le 2 \\ f(x-2) & ,x>2 \end{cases}$$

, then what is $f(1)+f(8)=?$

A) 3

B) 5

C) 7

D) 9

1.5 Transforming Function

1. Transformation of the function

※ **Translation**

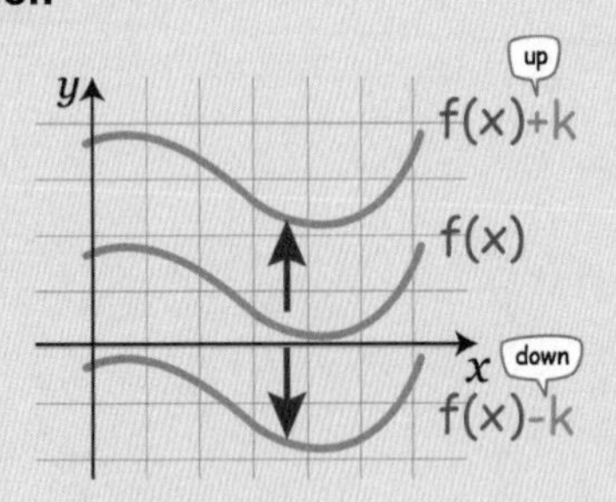

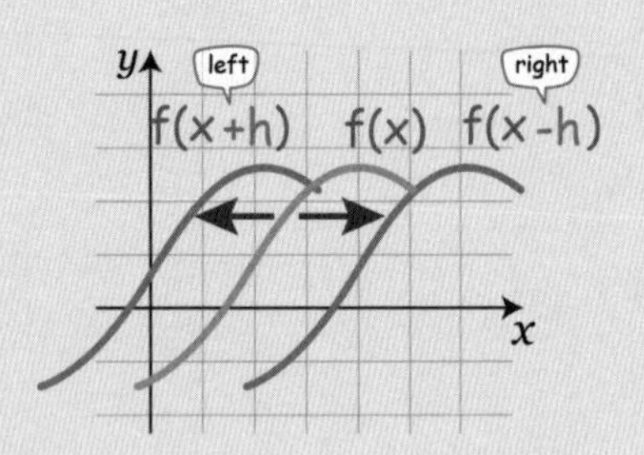

y = f(x) + k

: Vertically Shifted

①________ k units

(k>0)

y = f(x - h)

: Horizontal Shift

to the ②________ h units

(h>0)

※ **Reflection**

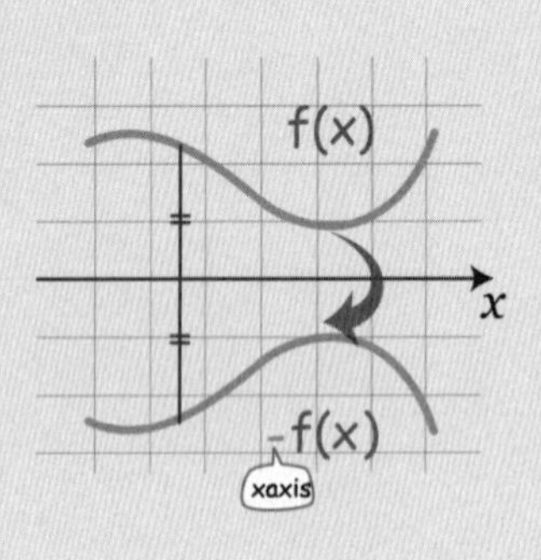

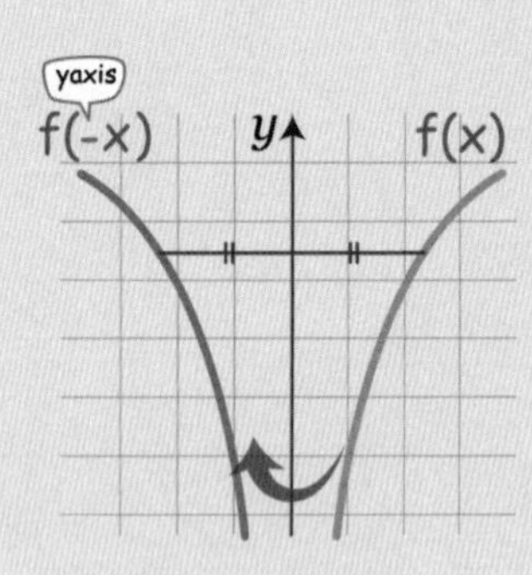

y = -f(x)

: reflected over ③_____-axis

y = f(-x)

: reflected over ④_____-axis

Blank : ① up ② right ③ x ④ y ⑤ 2 ⑥ shrink ⑦ 1/2 ⑧1/2 (=0.5) ⑨ stretch ⑩ 2

EXAMPLE 1. $f(x)$ is given. Graph the transformed function

① $y=f(x+2)-3$

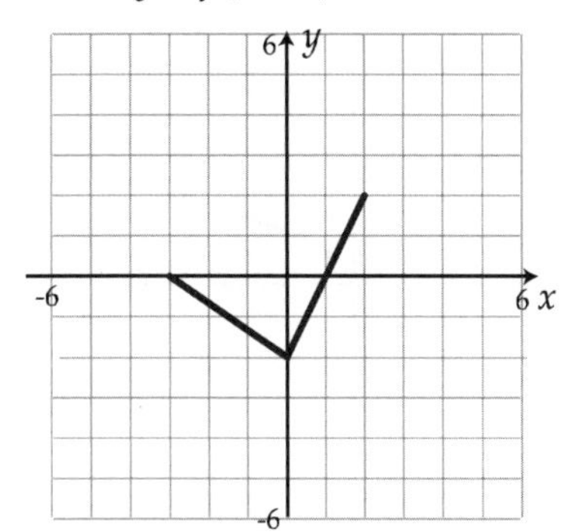

② $y=\sqrt{x-1}+2$

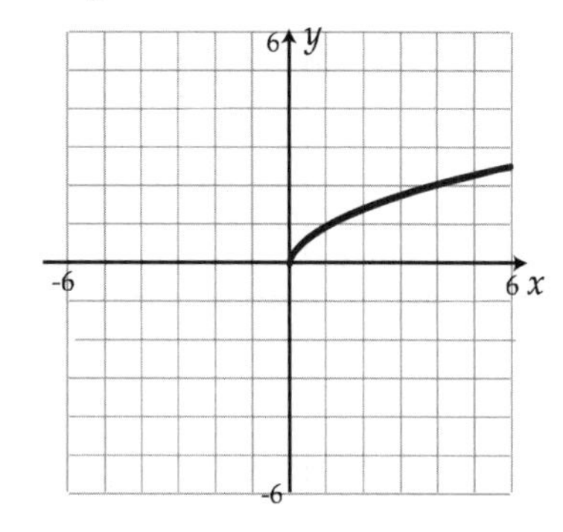

③ $y=-f(x)-1$

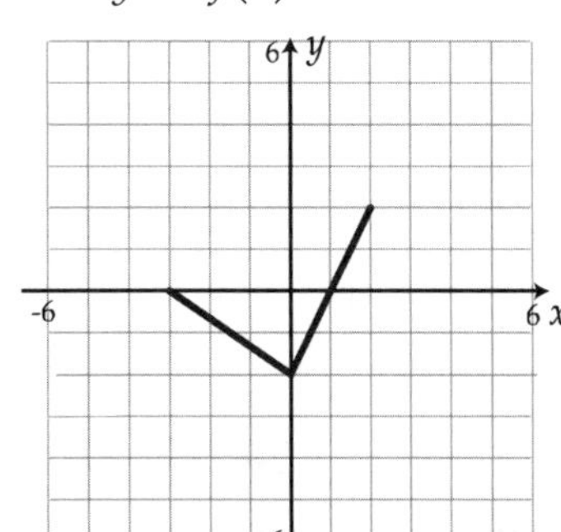

④ $y=\sqrt{-x}+2$

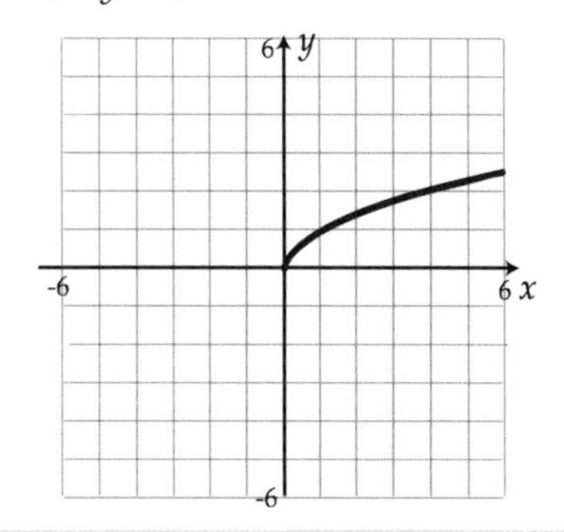

※ **Dilation**

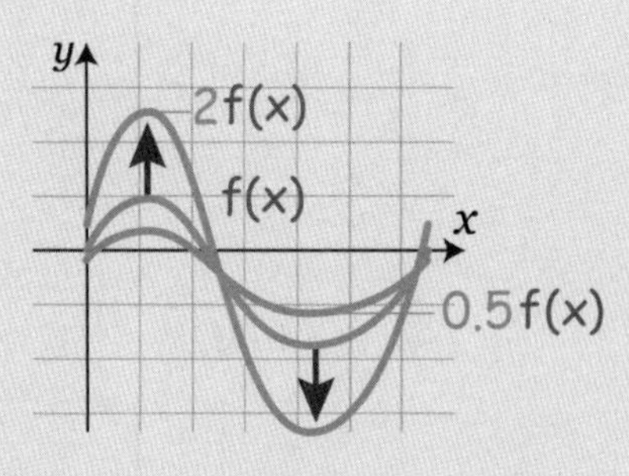

y
f(2x) f(x) f(0.5x)

$y = 2f(x)$

: vertically stretch

by the factor ⑤_____

$y = f(2x)$

: horizontally ⑥_________

by the factor ⑦_____

$y = 0.5f(x)$

: vertically shrink

by the factor ⑧_____

$y = f(0.5x)$

: horizontally ⑨_________

by the factor ⑩_____

EXAMPLE 2. Graph the transformed function

① $y = \dfrac{f(x)}{2}$

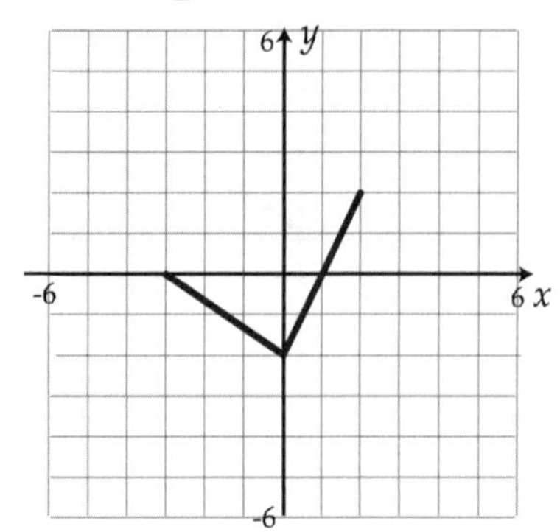

② $y = 2\sqrt{x}$

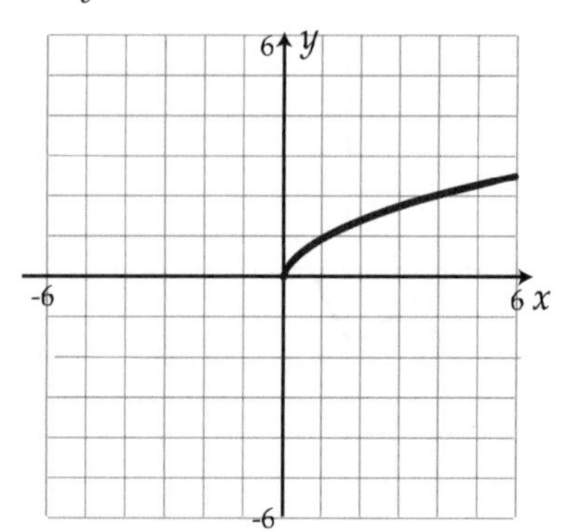

③ $y = f(2x)$

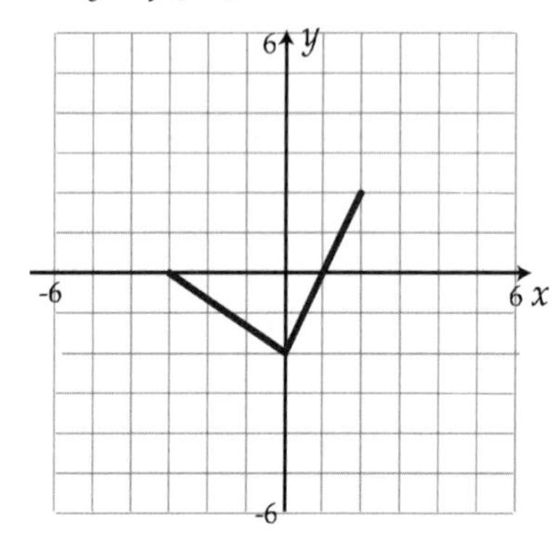

④ $y = \sqrt{\dfrac{x}{3}}$

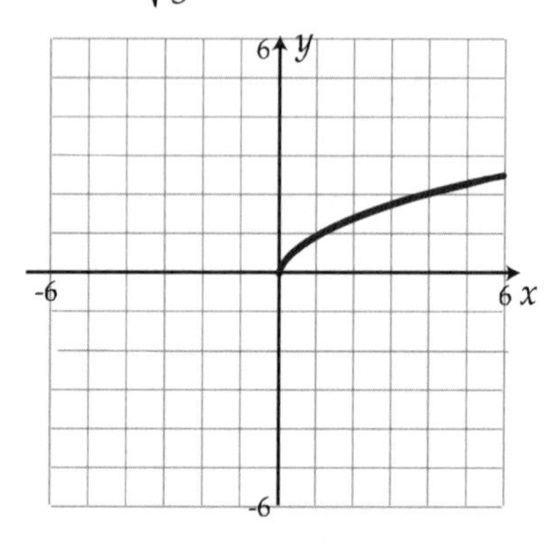

⑤ $y = f\left(\dfrac{x}{2}\right) - 1$

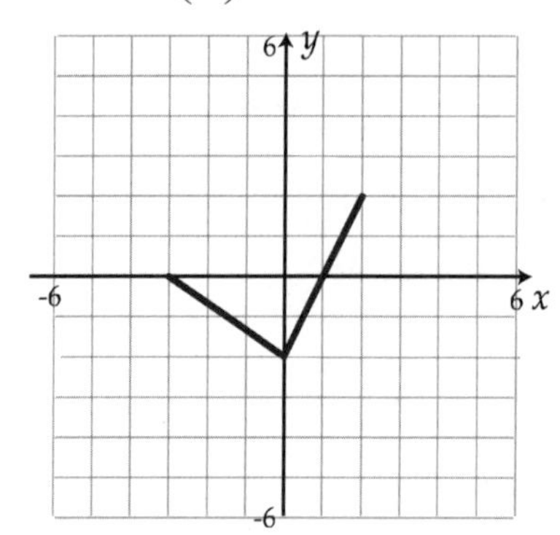

⑥ $y = 1 + \sqrt{2x}$

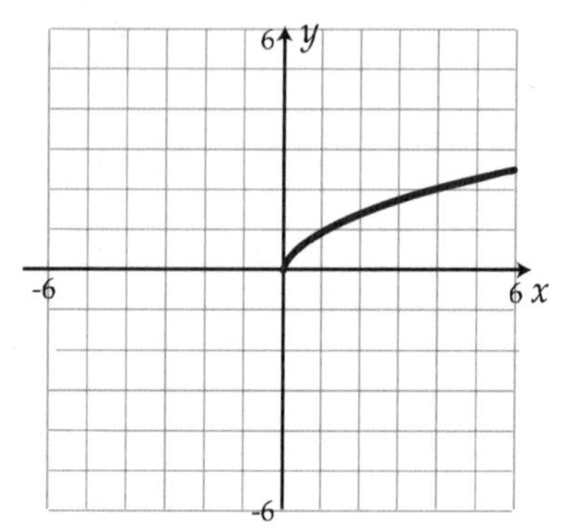

2. Transformation of the function

$y = f(x)$	original function (parent function)
$y = f(x) + k$	Vertically Shifted up k units
$y = f(x) - k$	Vertically Shifted down k units ($k>0$)
$y = f(x + h)$	Horizontal Shift to the left h units
$y = f(x - h)$	Horizontal Shift to the right h units ($h>0$)
$y = -f(x)$	reflected over x-axis
$y = f(-x)$	reflected over y-axis
$y = af(x)$, $a > 1$	vertically stretch by the factor a
$0<a<1$	vertically shrink by the factor a
$y = f(bx)$, $b > 1$	horizontally shrink by the factor $1/b$
$0<b<1$	horizontally stretch by the factor $1/b$

★ Few things to Remember

1) Order of Transformation is determined by order of operations.
Reflect or stretch or compress → translation (shifting)

2) When you have $a\,f(bx+c) + d$? ① ____________________

☺ All in one (If a, b > 1, c, d >0)

refl. by x axis / ver. stretch by a / refl. by y axis / hor. stretch by 1/b / Left c units / Up d units

$$y = f(x) \longrightarrow y = -a\,f(-b(x+c)) + d$$

Blank : ① change to $a\,f\left(b\left(x+\frac{c}{b}\right)\right)+d$ (factor b out from bx+c)

EXAMPLE 3. Graph the transformed function

① $y=2f(x)-2$

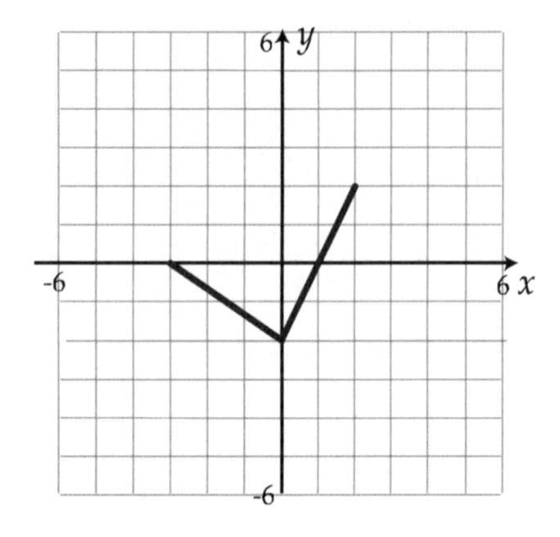

② $y=f(x-1)+3$

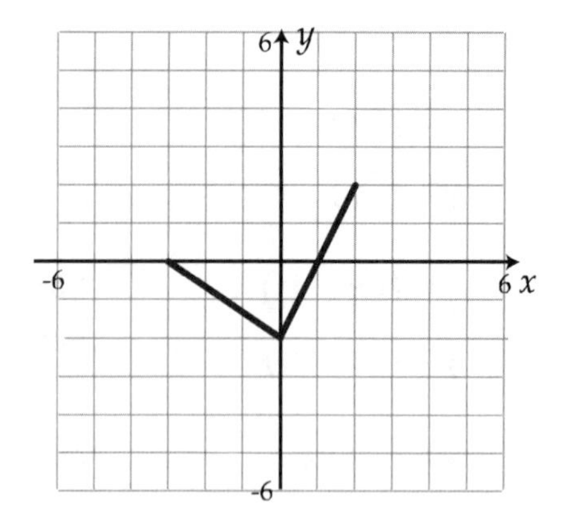

③ $y=f(-x-1)$

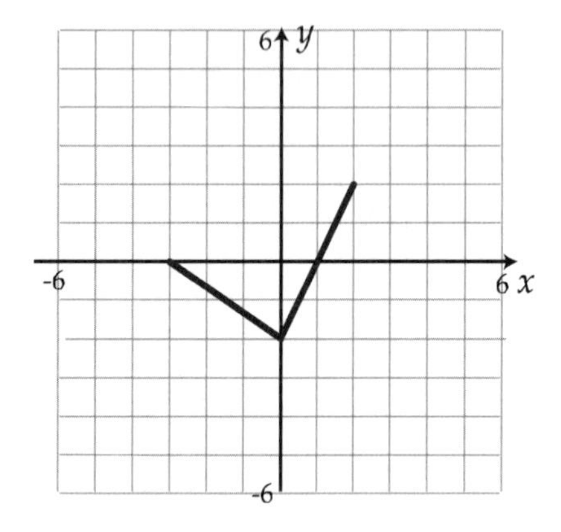

④ $y=f\left(-x+2\right)$

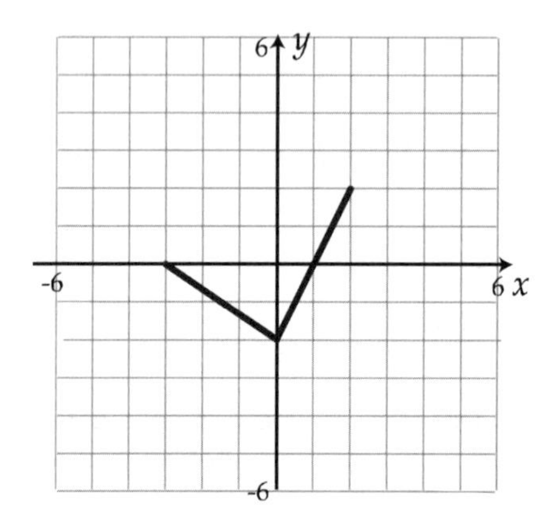

⑤ $y=-f\left(2x-4\right)$

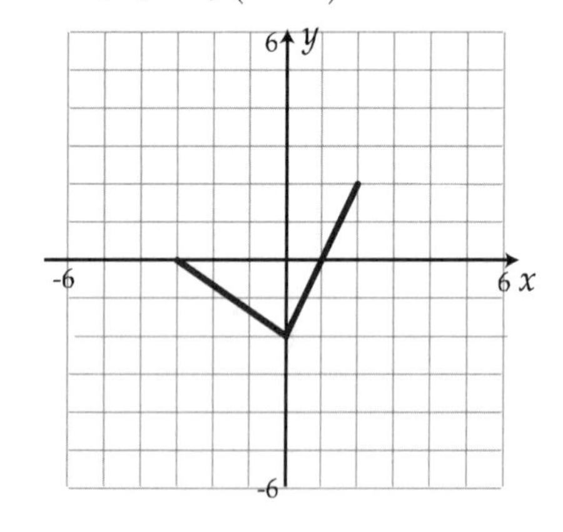

⑥ $y=f\left(2x+2\right)$

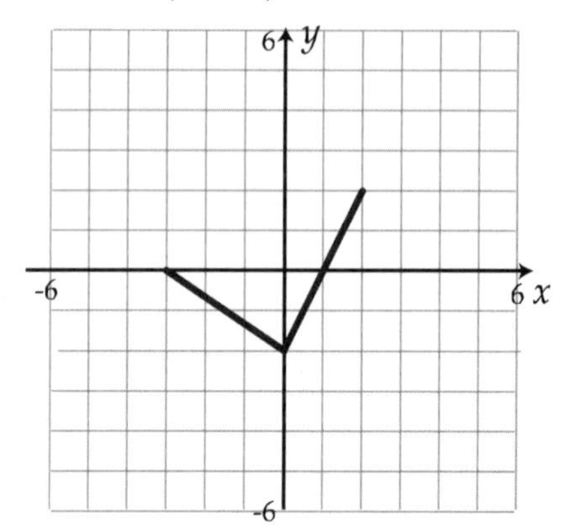

⑦ $y = f(0.5x+1)$

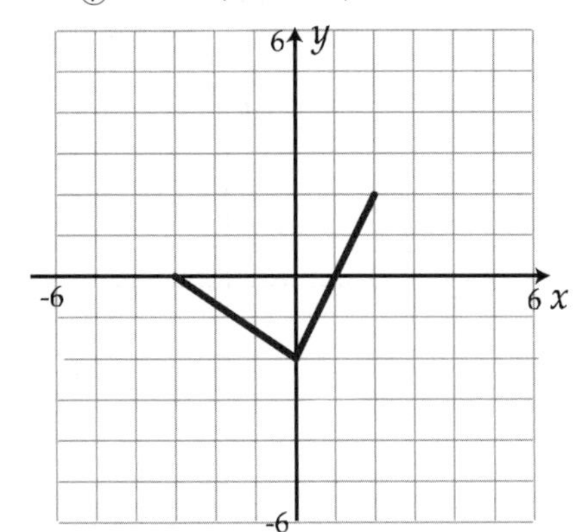

⑧ $y = f(0.5x)+1$

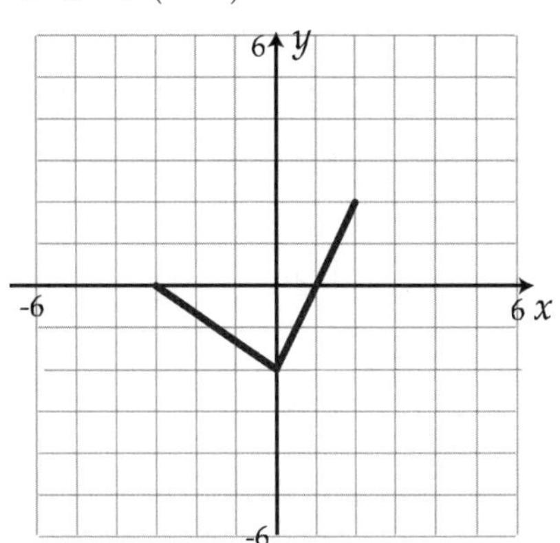

EXAMPLE 4. Graph the Function

① $f(x) = -x^2 + 4$

② $f(x) = -\sqrt{-x} + 2$

③ $f(x) = -2\sqrt[3]{x} - 2$

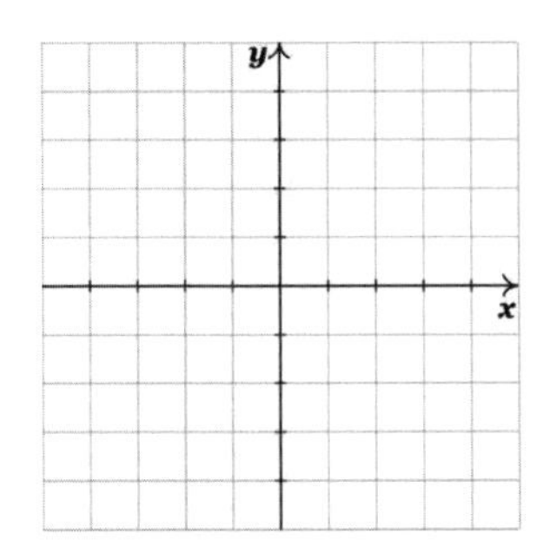

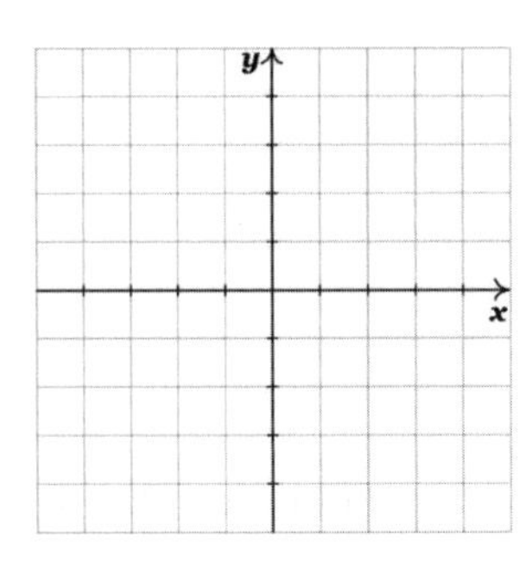

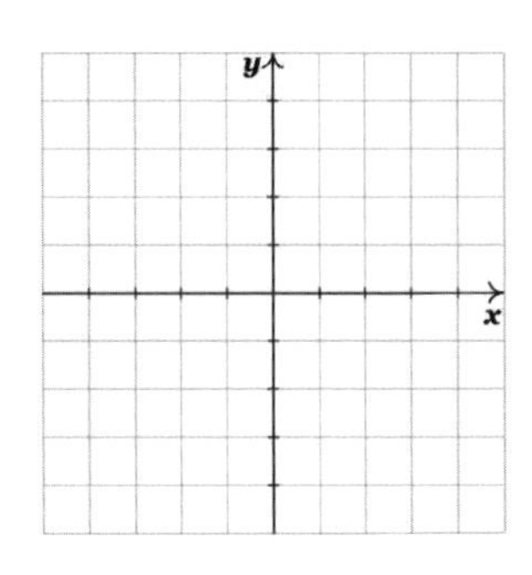

④ $f(x) = -(x-2)^3 - 1$

⑤ $f(x) = -2\sqrt{3-x} - 1$

⑥ $f(x) = |2-x| + 3$

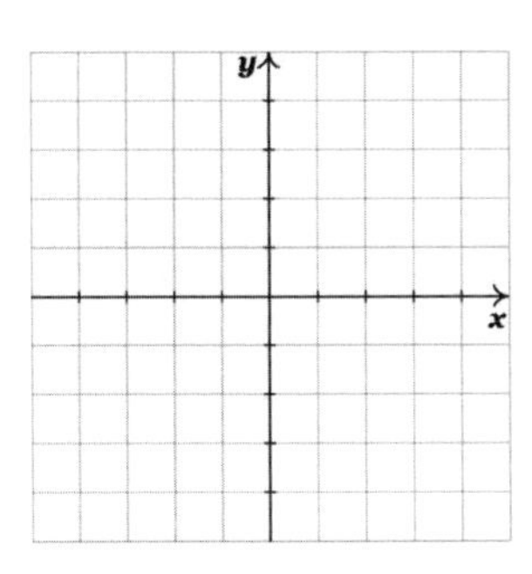

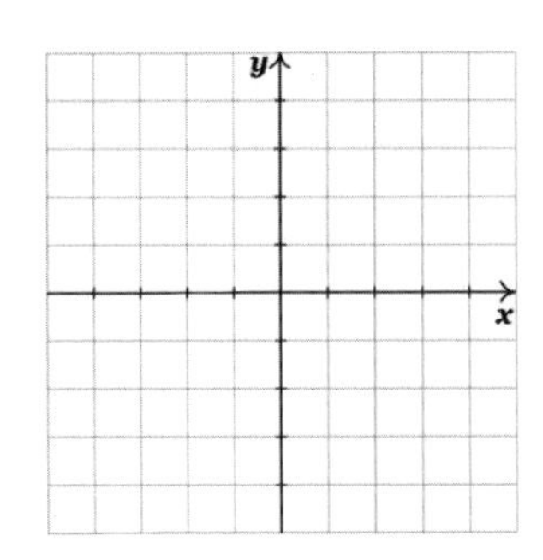

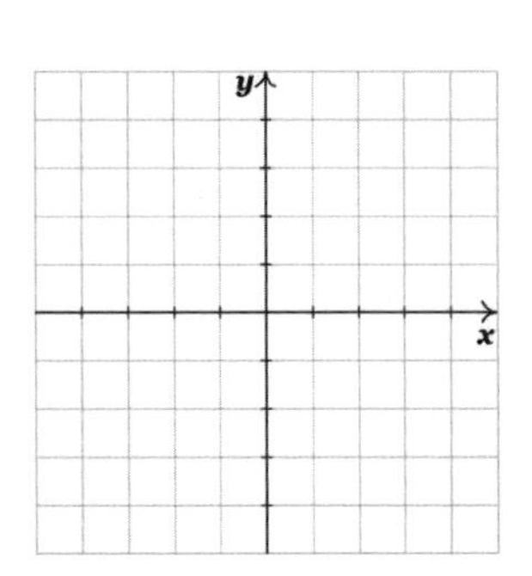

⑦ $f(x)=|2x-4|$

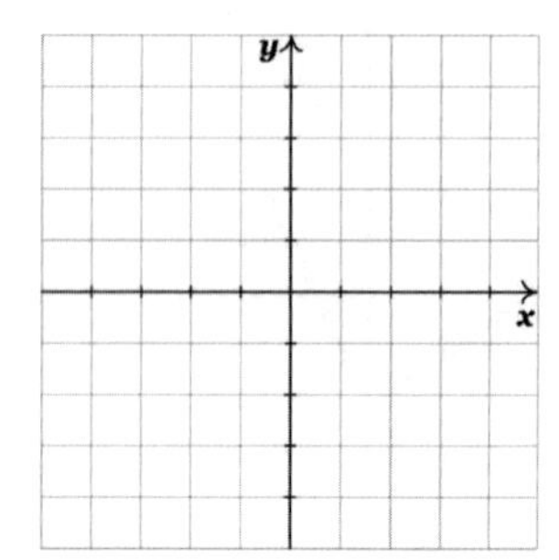

⑧ $f(x)=\sqrt{-3x-18}$

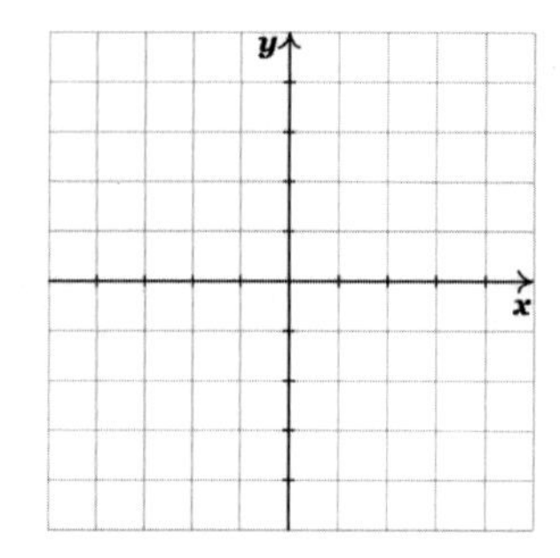

⑨ $f(x)=-\dfrac{1}{x-2}+3$

⑩ $f(x)=1-\dfrac{1}{x+2}$

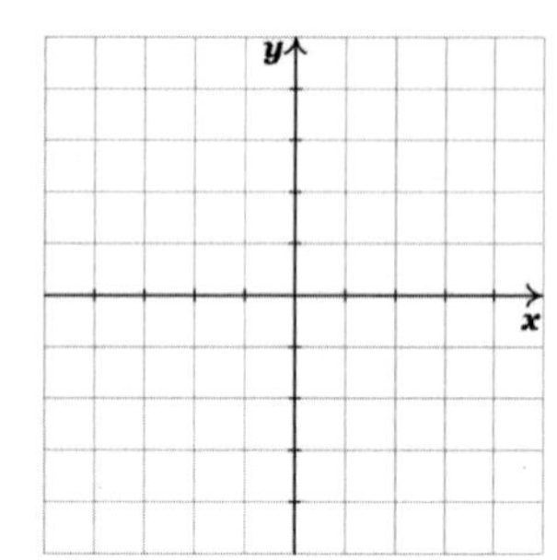

⑪ $f(x)=2e^{-x-1}+3$

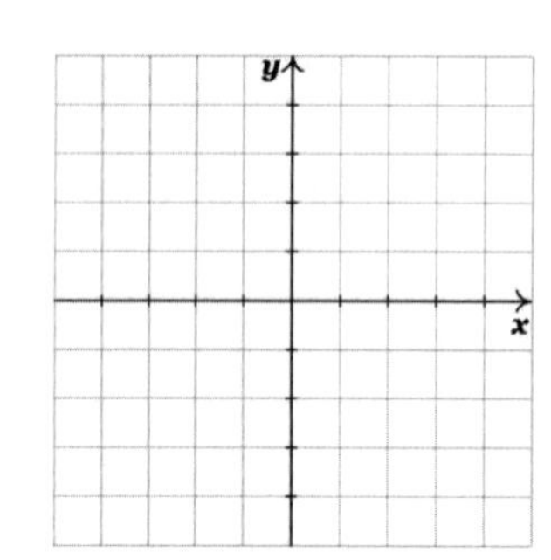

⑫ $f(x)=e^{2-x}$

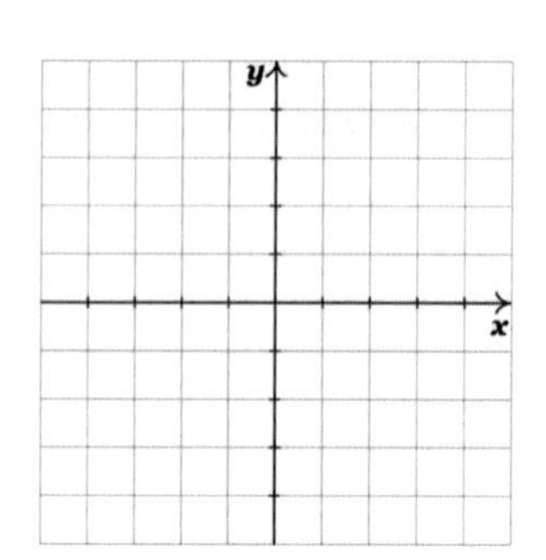

⑬ $f(x)=\log(0.5x+1)+2$

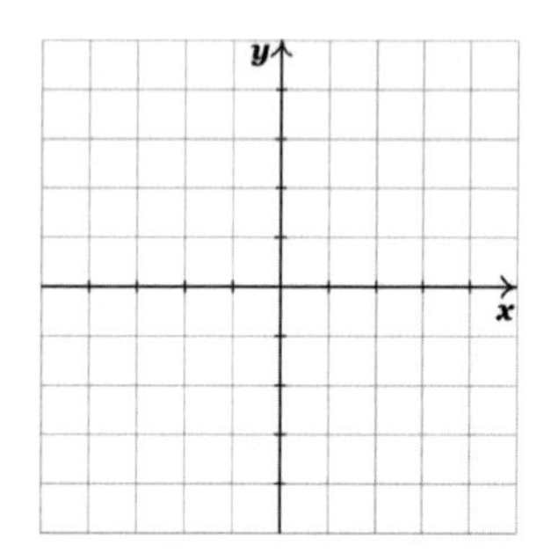

⑭ $f(x)=\log(2x+1)$

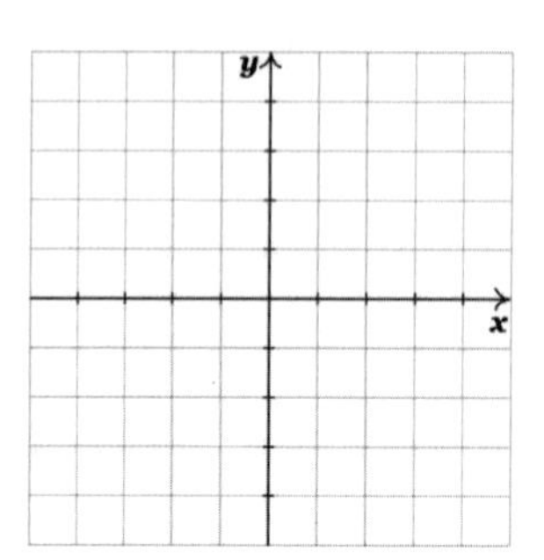

$e \approx 2.71...$ and e^x has a similar shape with 2^x .

(We will learn this more is Chapter 3)

EXAMPLE 5. The point $A(4, -2)$ is on $f(x)$. Write the coordinates of the image of A on

① $2f(x)$

② $\frac{1}{2}f(x)$

③ $\frac{1}{2}f(x-2)$

④ $2f(x+1)$

⑤ $f(2x-2)$

⑥ $f(2x)-3$

⑦ $f(-x-1)$

⑧ $f(-x+2)$

EXAMPLE 6. If $f(x)=\sqrt{x}+2$, find the function $g(x)$ which shows the graph of $f(x)$ after;

☺ Tip :

vertical change : (up / down / reflect x/ vertical dilation)	change of outside	change of ENTIRE f(x)	same direction
horizontal change : (right / left / reflect y/ horizontal dilation)	change of inside	change ONLY x	opposite direction

① shifted horizontally left 3 followed by a shifted vertically down 4 .

② shifted vertically up 5 followed by a shifted horizontally right 2.

③ shifted vertically up 4 followed by a vertical dilation of scale factor $\frac{1}{2}$.

④ vertical dilation of scale factor 2 followed by a shifted horizontally left 1.

⑤ shifted horizontally right 1 and vertically down 2 followed by a horizontal dilation of scale factor $\frac{1}{2}$

⑥ horizontal dilation of scale factor 2 followed by shifted vertically down 3 and horizontally left 2

⑦ shifted horizontally left 2 followed by a reflection through the y axis followed by a vertical dilation of scale factor 3.

⑧ vertical dilation of scale factor 2 followed by reflection through the x axis and followed by a shifted horizontally left 2.

⑨ reflection through the x axis and horizontal dilation of scale factor $\frac{1}{3}$ followed by a shifted horizontally right 1 and vertically down 1 .

3. *Absolute Value transformations

The definition of absolute value is :

$$|x| = \begin{cases} \boxed{①} , & x \geq 0 \\ \boxed{②} , & x < 0 \end{cases}$$

"flip back to positive"

EXAMPLE 7. Express the given function as a piecewise function.

① $f(x) = |x-2|$

② $f(x) = |2-x|$

③ $f(x) = \dfrac{|x-1|}{x-1}$

④ $f(x) = \dfrac{|x|}{x}$

⑤ $f(x) = |x^2 - 3x - 4|$

⑥ $f(x) = |x^2 - x|$

Blank : ① x ② −x

⑦ $f(x)=|3-x|+x+2$

※ **Absolute Value transformations**

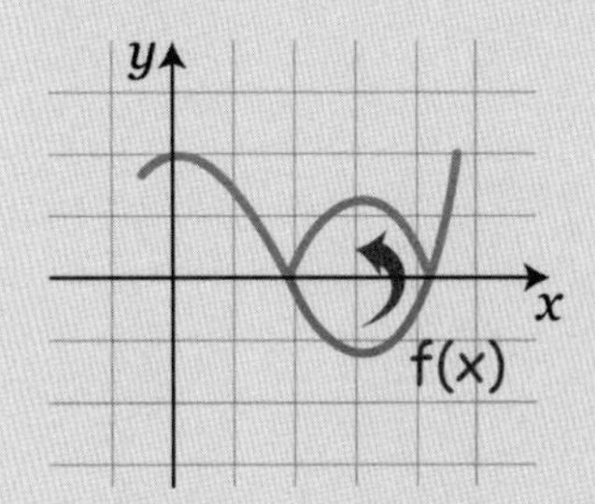

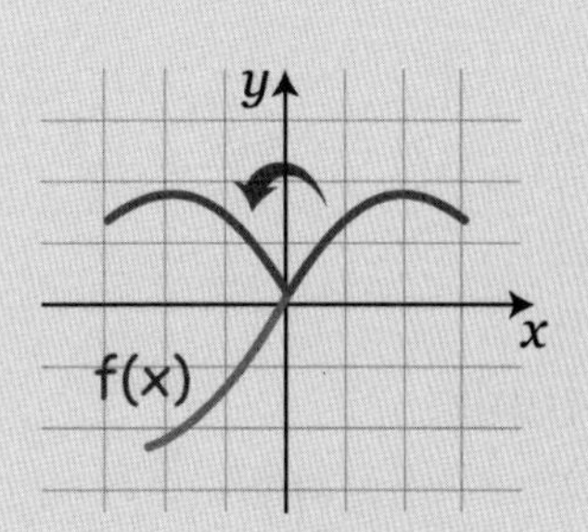

$y = |f(x)|$

: Fold up along the x axis

$y = f(|x|)$

: copy the right side and create a mirror image on the left (like 'decalcomanie')

EXAMPLE 8. Graph the transformed function

① $y=|f(x)|$

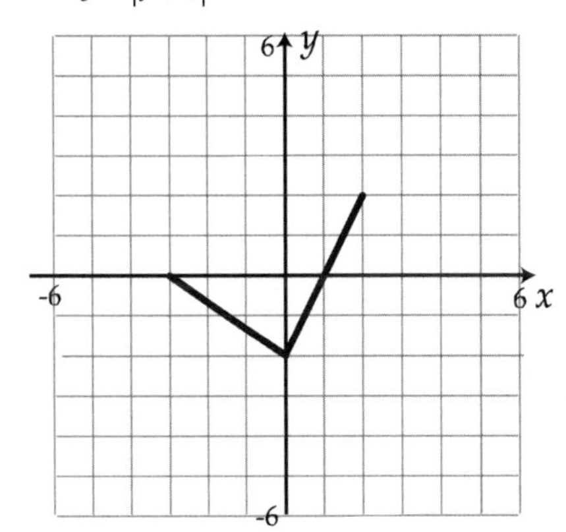

② $y = f(|x|)$

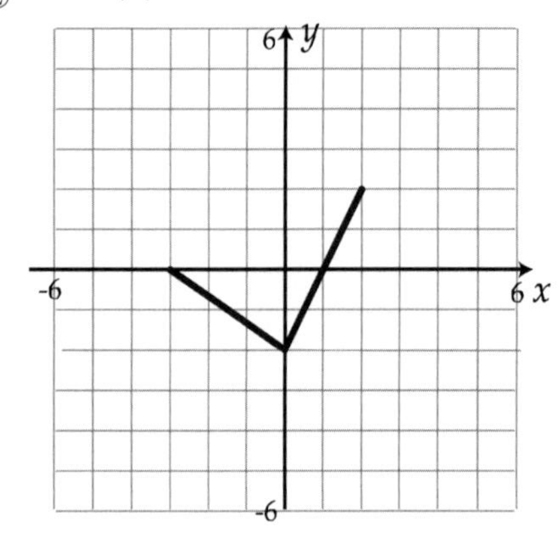

③ $y = |f(|x|)|$

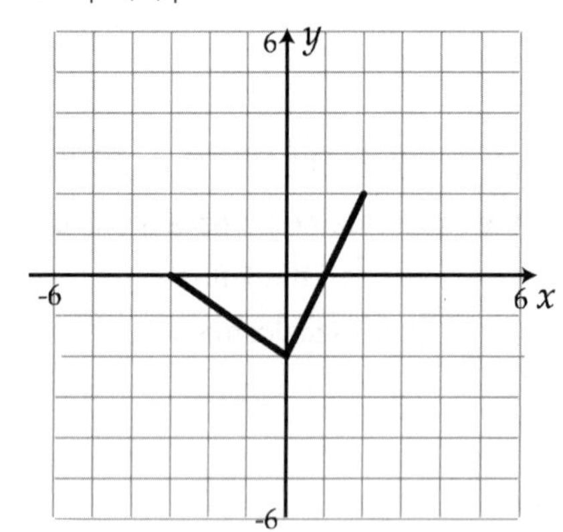

AP Style Problem

x	-1	1	4	7
$f(x)$	1	3	2	-1

1. The graph of $g(x)$ is the result of a vertical dilation of f by the factor of 2, then horizontal dilation by the factor of 4, followed by translation of 5 units to the left and 2 units up. What is $g(-1)$?

A) 1

B) 4

C) 6

D) 8

2. The graph of $f(x)$ has domain $[0, 4]$ and the range $[0, 2]$. What is the domain and range of $g(x) = -f\left(\frac{x}{2}\right) - 3$?

A) domain $[0, 2]$ and the range $[1, -3]$.

B) domain $[0, 2]$ and the range $[-5, -3]$.

C) domain $[0, 8]$ and the range $[-5, -3]$.

D) domain $[0, 8]$ and the range $[1, -3]$.

3. The function $f(x)$ is given by $f(x) = -2x^2 + 3$. The graph of g is a horizontal translation by 4 followed by vertical translation by -1 followed by vertical dilation by the factor 2. What is $g(x)$?

A) $g(x) = -4(x-4)^2 + 4$

B) $g(x) = -4(x+4)^2 + 4$

C) $g(x) = -4(x+4)^2 + 2$

D) $g(x) = -4(x-4)^2 + 2$

☺ Tip: horizontal translation by 4 = horizontal translation of 4 units to the ①________

horizontal translation by -4 = horizontal translation of 4 units to the ② ________

4. The functions f and g are defined as $g(x) = f(-x+1)$. Which of the following steps of transformations maps the graph of f to the graph of g?

A) Reflection over y axis followed by horizontal translation by 1

B) Reflection over y axis followed by horizontal translation by -1

C) Reflection over x axis followed by horizontal translation by 1

D) Reflection over x axis followed by horizontal translation by -1

5. When $f(x) = 3x^2 + 18x + 26$, $f(x-a)$ is an even function. Find the value of a.

A) 1

B) -1

C) 3

D) -3

Blank : ① right ② left

6. The functions f and g are given by $f(x)=\sqrt{x}-1$ and $g(x)=2\sqrt{x}-1$. Which of the following steps of transformations maps the graph of f to the graph of g?

A) Vertical dilation by the factor of 2

B) Vertical dilation by the factor of 0.5

C) Horizontal dilation by the factor of 4

D) Horizontal dilation by the factor of 0.25

1.6 Composing Functions

1. Composite Functions

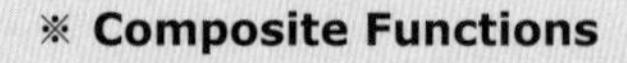

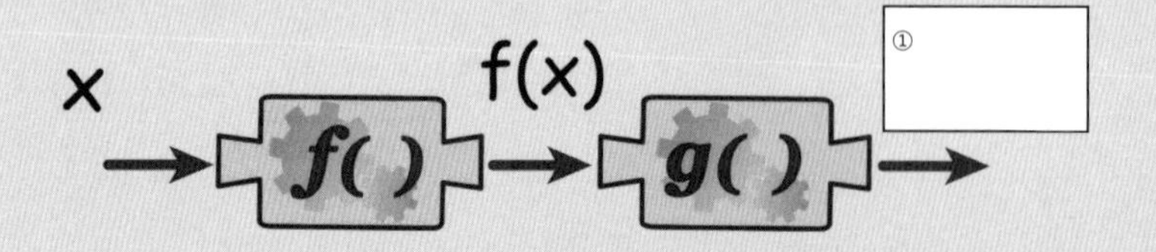

The result of f is sent through another function g.

outer function / inner function

Composition of g and f : $(g \circ f)(x) = g(f(x))$

composed with

We read it 'g composed with f'.

EXAMPLE 1. $f(x) = 2x + 3$ and $g(x) = \sqrt{x}$, then what is

① $(f \circ g)(x)$

② $(g \circ f)(x)$

③ $(f \circ f)(x)$

④ $(g \circ g)(x)$

Blank : ① g(f(x))

⑤ $(g \circ f)(3)$

⑥ $(f \circ g)(1)$

⑦ $(f \circ f \circ g)(x)$

EXAMPLE 2. Use the table to find;

x	1	2	4	10	12
$f(x)$	-4	6	11	3	13

x	-5	-4	3	10	11
$g(x)$	4	10	6	2	12

① $(f \circ g)(-5)$

② $(g \circ f)(1)$

③ $(f \circ g \circ f)(4)$

④ $(f \circ g \circ g)(-4)$

EXAMPLE 3. Use the table to find;

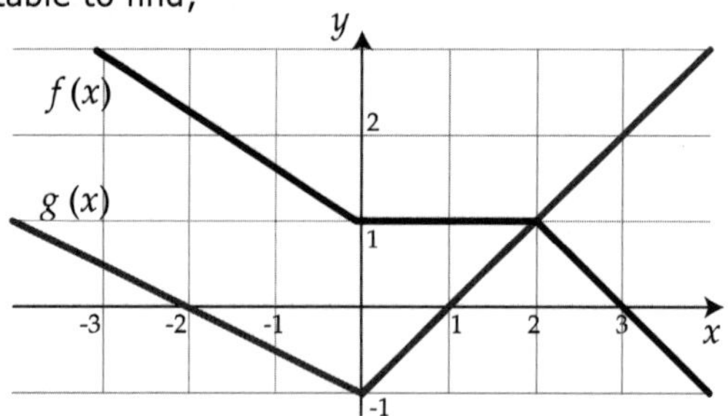

① $(f \circ g)(-2)$

② $(g \circ f)(1)$

③ $(g \circ f \circ f)(2)$

④ $(f \circ g \circ f)(2)$

EXAMPLE 4. Find $(f \circ g)(x)$ and $(g \circ f)(x)$. Find the domain of each composite function.

☺ Tip ☞ You must find the intersection of

Domain of ①_______________ function and domain ②_____________ function!

ex) Domain of $(f \circ g)(x)$ = domain of ③__________ ∩ domain of ④________

① $f(x)=\dfrac{x}{x+3}$, $g(x)=\dfrac{2}{x}$

② $f(x)=\dfrac{2}{x}$, $g(x)=\dfrac{4}{x+4}$

Blank : ① result (=composite) ② inner ③$(f \circ g)(x)$ ④ $g(x)$

③ $f(x)=\sqrt{x}$, $g(x)=\sqrt{1-x}$

④ $f(x)=\sqrt{x}$, $g(x)=\sqrt{x-2}$

⑤ $f(x)=x^2$, $g(x)=\sqrt{x}$

⑥ $f(x)=x^2$, $g(x)=\sqrt{x-3}$

EXAMPLE 5. For a given function h(x), find its decomposition into simpler functions if $h(x)=f(g(x))$. (There can be more than one answer)

① for $h(x)=\sqrt{x^2+1}$

② for $h(x)=\dfrac{1}{(x^3+x+1)^2}$

③ for $h(x)=(x+1)+2\sqrt{x+1}+3$

④ for $h(x)=(x-3)^2+2(x-3)-5$

AP Style Problem

1. Let the function g is given by $g(x)=\dfrac{x}{x+2}$. What is the domain of the function h where

$h(x) = g(g(x))$?

A) All real number.

B) All real number except -2

C) All real number except $-\dfrac{4}{3}$

D) All real number except -2 and $-\dfrac{4}{3}$

2. If $(f\circ g)(x)=16x+1$ and $f(x)=4x+5$, find $g(x)$.

A) $g(x)=4x+1$

B) $g(x)=4x-19$

C) $g(x)=4x-1$

D) $g(x)=4x+19$

3. If $(f\circ g)(x)=16x+1$ and $g(x)=4x+5$, find $f(x)$.

A) $f(x)=4x-1$

B) $f(x)=4x-19$

C) $f(x)=4x+1$

D) $f(x)=4x+19$

4. Graph of $f(x)$ is given below. Find the number of distinct x that satisfies $(f \circ f)(x) = 4$.

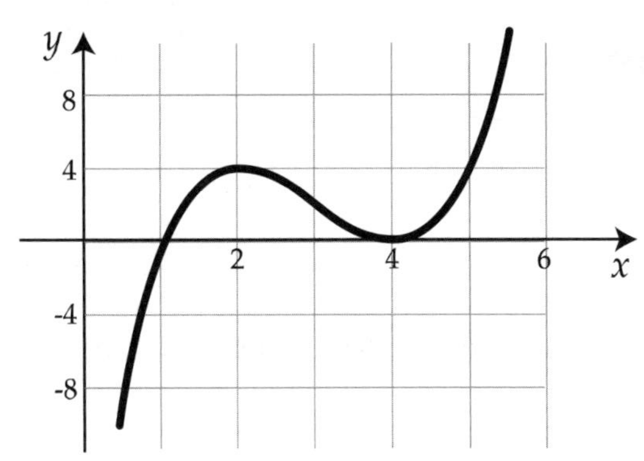

A) 1

B) 2

C) 4

D) 8

5.* Given that we can denote

$$\underbrace{f \circ f \circ f \circ \cdots \circ f}_{n\ times} \text{ as } f^n,$$

then if $f(x) = 1 + x$, then what is $f^{10}(x)$?

A) $x + 10$

B) $10x + 1$

C) $10x + 10$

D) $x + 1$

1.7 Inverse Function

1. One to One Function

※ **One-to-one Function**

: every member of "X" has **its own unique** matching member in "Y".

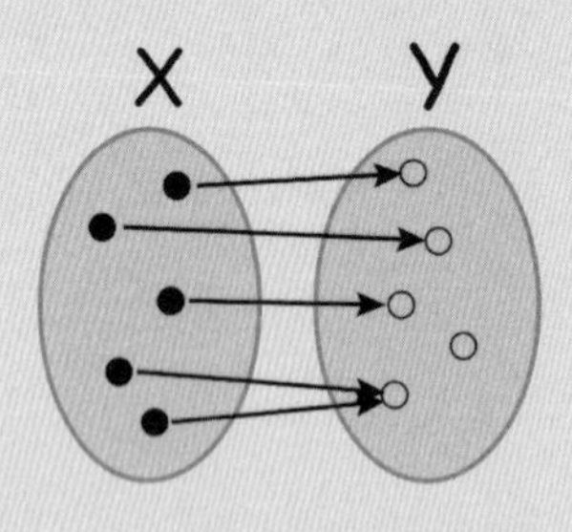

X Y

① ____________________ ② ____________________

To be a 'One to one function';

① ③__________ has to be different! ($f(x_1) \neq f(x_2)$ whenever $x_1 \neq x_2$)

② Satisfy the ④____________________ test and ⑤____________________ test

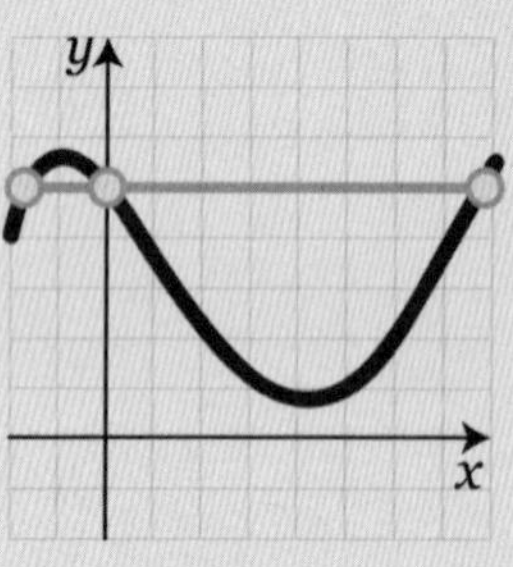

Not One-to-one function !

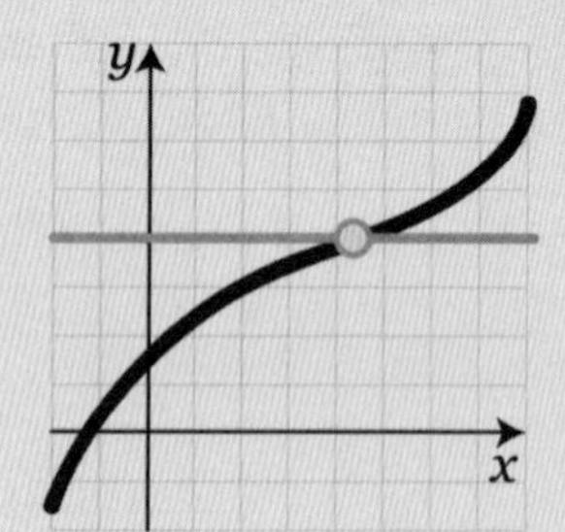

One-to-one Function !

If a horizontal line crosses the curve twice or above, then it is NOT one to one function.

Blank : ① function But NOT One-to-one ② One-to-one function ③ x, y ④ vertical line ⑤ horizontal line

EXAMPLE 1. Determine whether it is a one-to-one function, just a function, or not a function.

① $\{(3,-1),(2,2),(-1,-11),(5,-12)\}$

② $\{(3,-1),(-1,2),(3,-11),(5,-12)\}$

③ $\{(-2,0),(3,-.5),(1,9),(2,0)\}$

④ $\{(0,1),(1,2),(3,4),(5,1)\}$

⑤ $\{(2,2),(2,1),(2,.5),(2,.5)\}$

⑥ $\{(1,2),(2,3),(3,4),(4,5)\}$

⑦

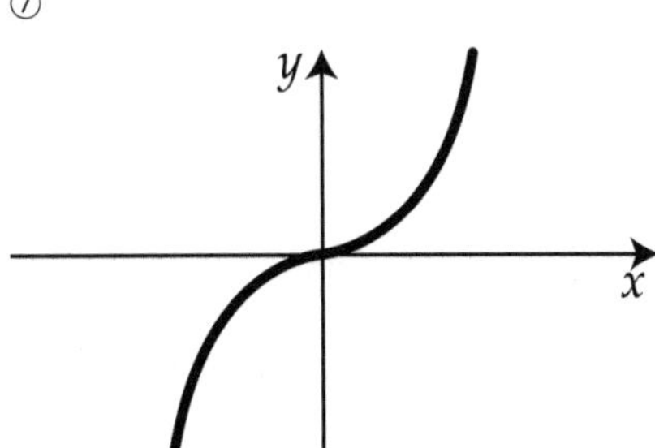

⑧

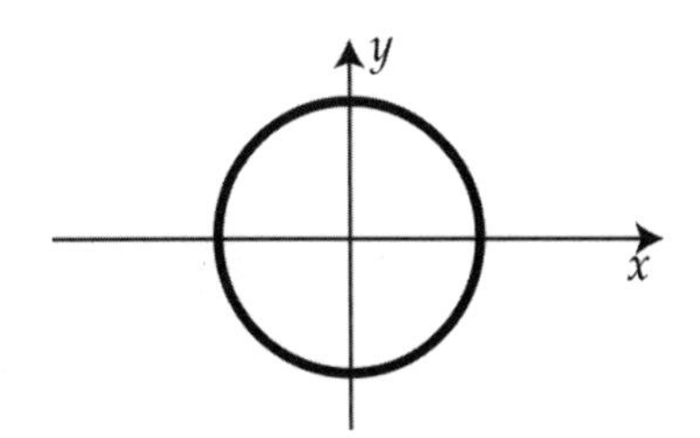

⑨

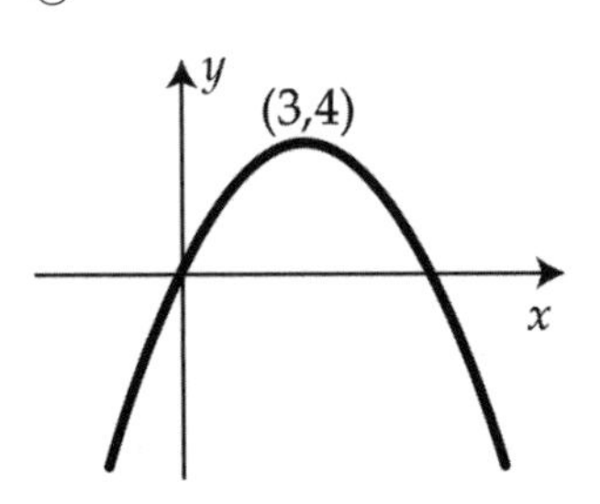

⑩

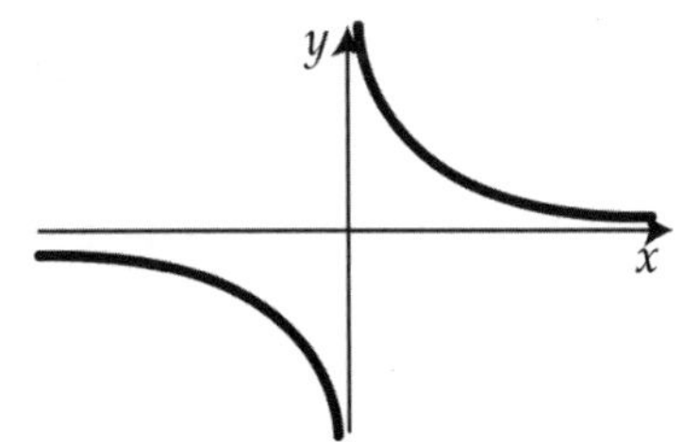

⑪

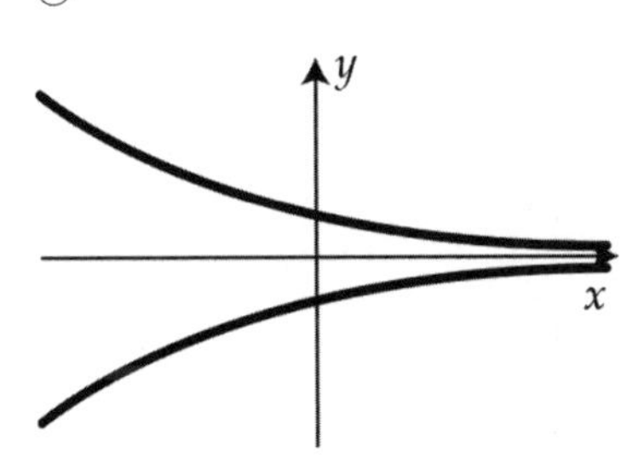

⑫

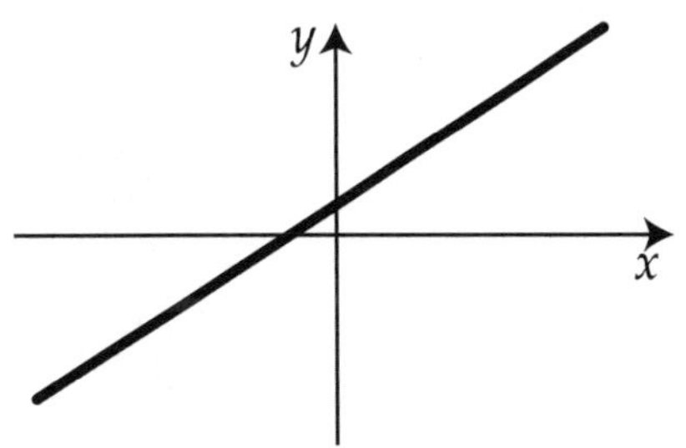

2. Inverse Function

※ **Inverse function**

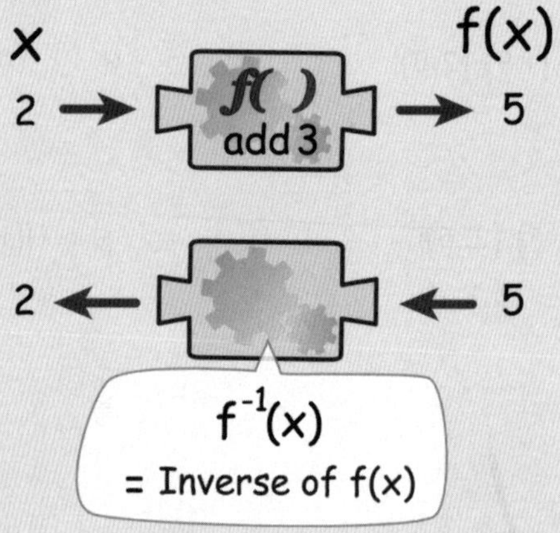

x → f(x)

2 → [f( ) add 3] → 5

2 ← [ ] ← 5

$f^{-1}(x)$
= Inverse of f(x)

When we **switch *input(x)*** and ***output(y)*** of a function $f(x)$,

we will have ① ________ which is the "② ______________ Function of $f(x)$"

- If $f(a)=b$, then ③ __________. ex) $f(2)=5 \Leftrightarrow 2=f^{-1}(5)$

- $f^{-1}(x) \neq \dfrac{1}{f(x)}$ (-1 is not a power, it is a 'inverse' notation)

$f^{-1}(x)$?? We switch x and y.

EXAMPLE 2. Use the table to find;

x	1	2	4	10	12
$f(x)$	-4	6	11	3	13

x	-5	-4	3	10	11
$g(x)$	4	10	6	2	12

Blank : ① $f^{-1}(x)$ ② inverse ③ $a = f^{-1}(b)$

① $f(2)$

② $g(-4)$

③ $f^{-1}(3)$

④ $g^{-1}(12)$

⑤ $(f \circ g^{-1})(2)$

⑥ $(g^{-1} \circ f^{-1})(6)$

⑦ $(f \circ f^{-1})(11)$

⑧ $(g^{-1} \circ g)(4)$

EXAMPLE 3. Use the table to find;

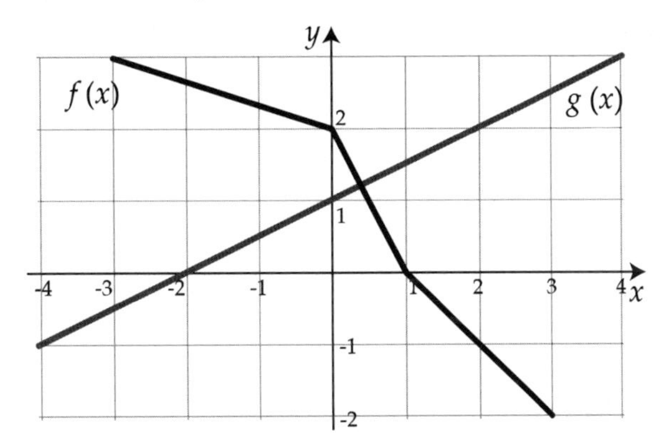

① $f(2)$

② $g(0)$

③ $f^{-1}(2)$

④ $g^{-1}(0)$

⑤ $(f^{-1} \circ g)(-2)$

⑥ $(f^{-1} \circ g)(0)$

⑦ $(g^{-1} \circ f)(2)$

⑧ $(g \circ f^{-1})(-1)$

⑨ $(f \circ f^{-1})(-1)$

⑩ $(g \circ g^{-1})(-1)$

3. Finding the Inverse Algebraically

※ How to find the inverse function

① Write $y = f(x)$.

② **Switch x and y.**

③ Solve for y.

EXAMPLE 4. Find the inverse of a function.

① $f(x) = 6x + 7$

② $f(x) = \frac{4}{x}$

③ $f(x) = \frac{1}{2}(2x-3)^3 + 2$

④ $f(x) = 2\sqrt[3]{x+1}$

⑤ $f(x) = x^2 + 1,\ x \geq 0$

⑥ $f(x) = x^2 + 1,\ x < 0$

⑦ $f(x) = (x-2)^2 + 3,\ x \leq 2$

⑧ $f(x) = 2(3-x)^2,\ x \leq 3$

⑨ $f(x)=\dfrac{2x+6}{-7x-3}$

⑩ $f(x)=\dfrac{-3x+4}{x-2}$

⑪ $f(x)=\dfrac{2}{3+x}$

⑫ $f(x)=\dfrac{-3x}{x+2}$

※ Inverse of Rational Functions shortcut

$$y^{-1}=\left(\frac{ax+b}{cx+d}\right)^{-1}=\frac{dx-b}{-cx+a}$$

Switch a,d and negate(put -) on b,c.

EXAMPLE 5. Find the inverse function and determine whether the inverse represents a function.

① $f=\{(6, 0), (-1, 1), (-3, 2), (-5, 3)\}$

② $g=\{(-7, 5), (-5, 7), (6, 5), (-6, 9)\}$

Only **ONE-TO-ONE function** has an inverse function!
If f is NOT a one-to-one?
Restrict the domain(cut the graph) and make a one to one!

EXAMPLE 6. Does the function have an inverse function? If not, then restrict its domain and find the inverse of the function with the restricted domain.

① $y=(x-2)^3$

② $y=\sqrt{x-2}$

③ $y=(x+2)^2-1$

④ $y=(x-1)^2+2$

⑤ $*y=|x+1|$

⑥ $*y=x^2+2x+2$

4. Facts about Inverse Function

1. If $f(a) = b$, then ①________________.

2. Only **ONE-TO-ONE function** has an inverse function!

If f is NOT a one-to-one?

Restrict the domain(cut the graph) and make a one to one!

3. The domain of f is the ②__________ of f^{-1}

and the range of f is the ③____________ of f^{-1}.

4. If the point (a, b) lies on graph of f ,

then the point ④__________ must lie on the graph of f^{-1}.

5. $(f \circ f^{-1})(x) =$ ⑤____ and $(f^{-1} \circ f)(x) =$⑥______.

6. f^{-1} is a reflection of the graph of f in the line ⑦________.

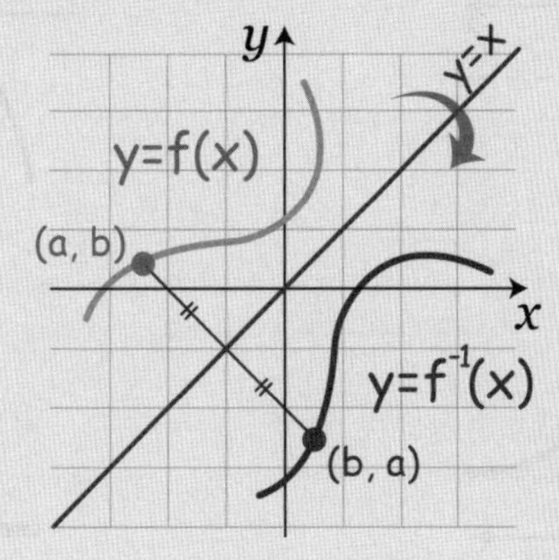

Blank : ① $a = f^{-1}(b)$ ② range ③ domain ④ (b, a) ⑤ x ⑥ x ⑦ y = x

EXAMPLE 7. Draw the graph of the inverse function $f^{-1}(x)$.

①

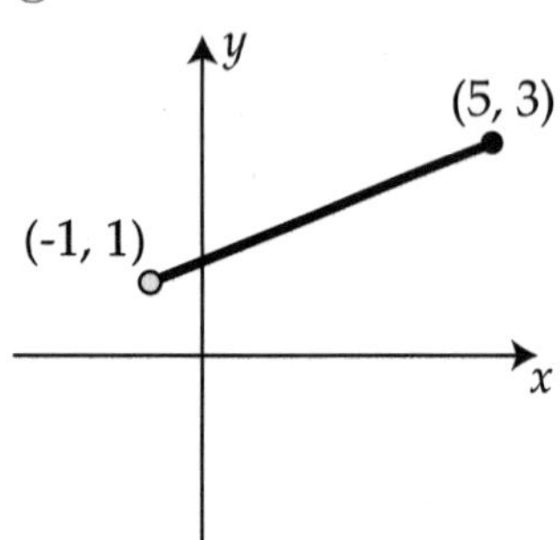

②

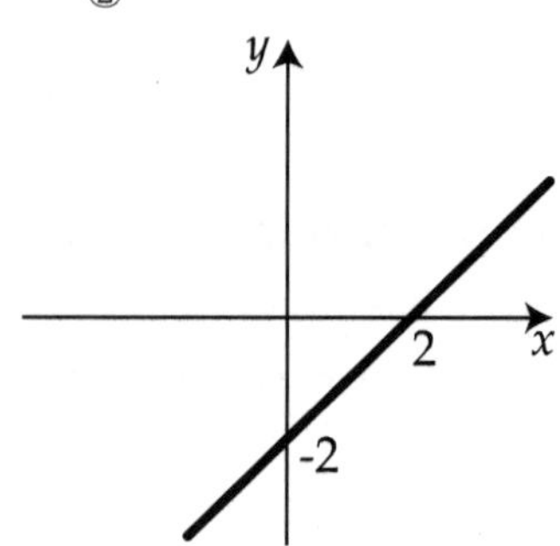

③

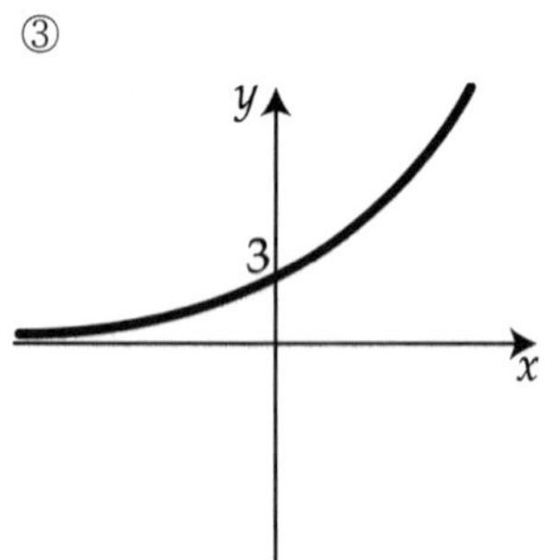

④

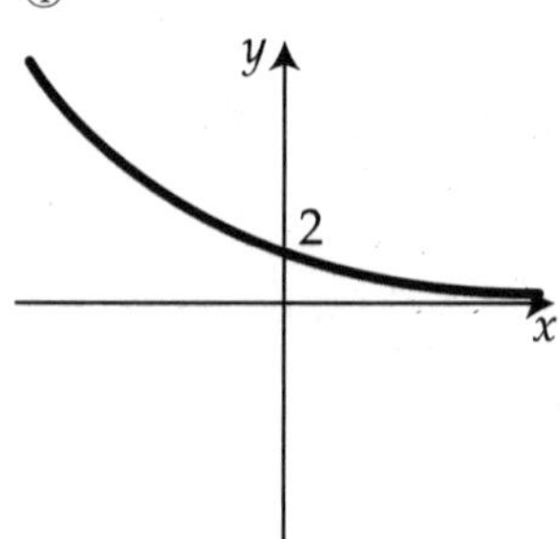

⑤

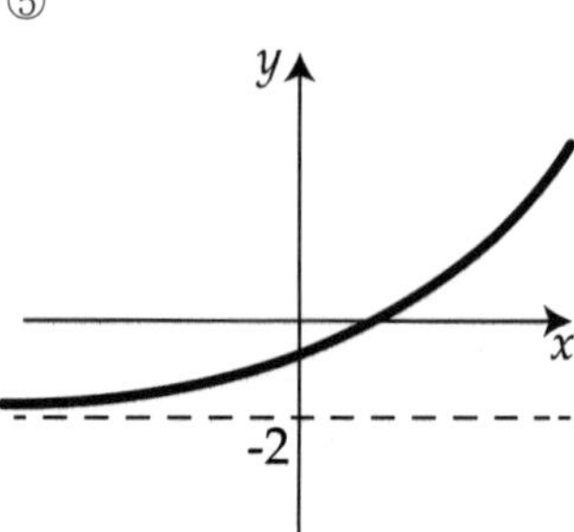

⑥

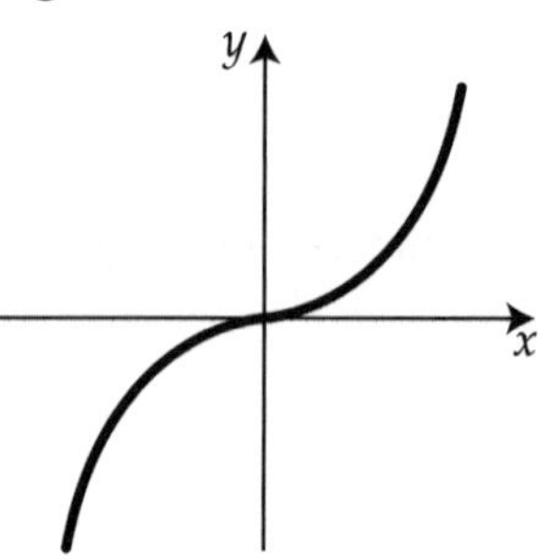

⑦

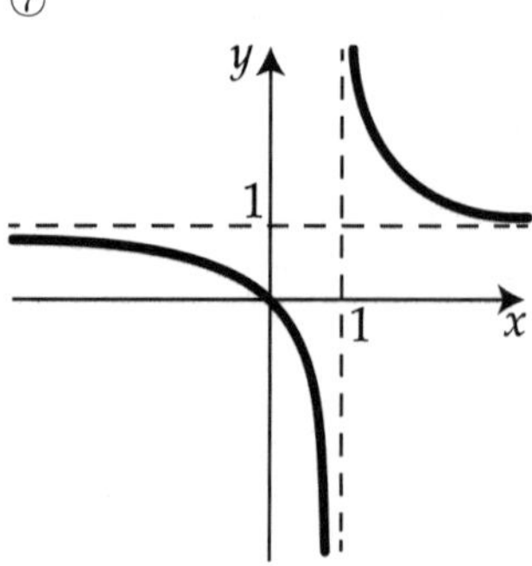

⑧

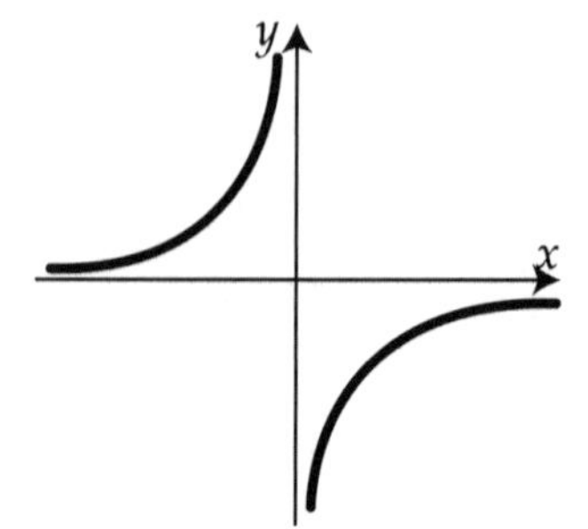

EXAMPLE 8. Does the function have an inverse function? If not, then restrict its domain and graph the inverse of the function with the restricted domain.

①

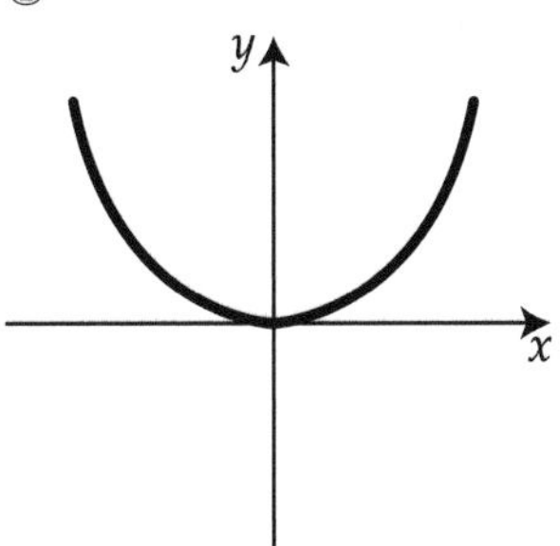

②

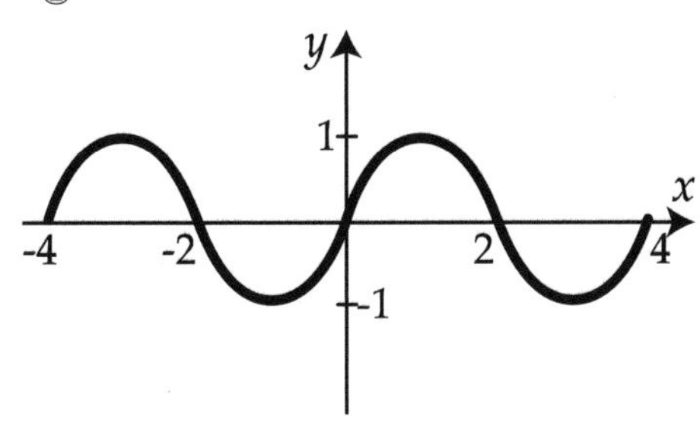

EXAMPLE 9. Determine i) the domain of the function, ii) the range of the function, iii) the domain of the inverse, and iv) the range of the inverse.

☺ Tip ☞ $f^{-1}(x)$?? We switch x and y.   = switch ①____________ and ②______________.

① $f(x)=\dfrac{3x+1}{x+2}$

Blank : ① domain ② range

② $f(x)=x^2, x<0$

③ $f(x)=\sqrt{x+2}$

④ $f(x)=x^3+2$

5. Self-Inverse Function

A function that satisfy

$f(x)=f^{-1}(x)$ or $f(f(x))=x$

is called **self-inverse function**.

To be a self-inverse function, the function should be symmetrical about the line y=x.

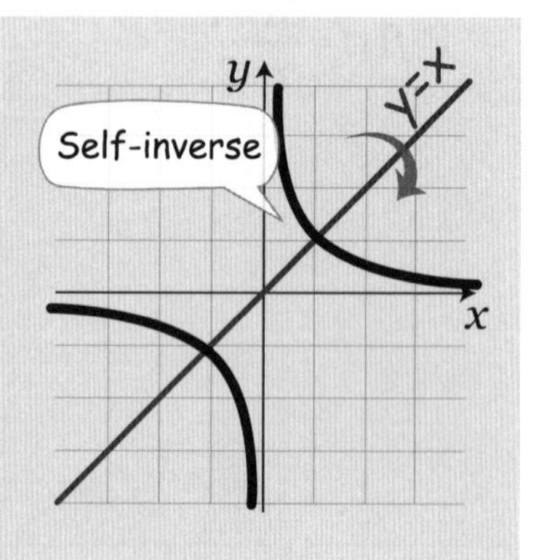

EXAMPLE 10. Find the function which satisfies $f(x) = f^{-1}(x)$.

a) $y = x$

b) $y = -x$

c) $y = 2x$

d) $y = -x + 2$

e) $y = -2x$

f) $y = \frac{1}{x}$

g) $y = -\frac{1}{x}$

h) $y = \frac{2}{x-1} + 1$

i) $y = \frac{1}{x-1}$

AP Style Problem

x	-1	0	1	2
$f(x)$	4	-1	2	0
$g(x)$	3	2	1	-1

1. The values for the function f and g are given in the table. What is $(f \circ f^{-1} \circ g^{-1})(2)$?

A) 2

B) 0

C) -1

D) 1

2. The graph of $f(x)$ has domain $[0, 3]$ and the range $[-2, 2]$. What is the domain and range of $g(x) = f^{-1}(x) + 2$?

A) domain $[-2, 2]$ and the range $[-2, 1]$.

B) domain $[-2, 2]$ and the range $[2, 5]$.

C) domain $[-2, 4]$ and the range $[0, 5]$.

D) domain $[-2, 1]$ and the range $[0, 1]$.

3. What could be the function that satisfies the following?

The difference of the input is linear every time the output increases by 1.

A) $f(x) = x^2 + 2$

B) $f(x) = \sqrt{x} + 2$

C) $f(x) = 2x + 1$

D) $f(x) = \frac{1}{x}$

4. Which of the following satisfy $f(f(x)) = x$?

I. $f(x) = -x$

II. $f(x) = \frac{1}{x}$

III. $f(x) = \frac{1}{x-1} + 1$

A) I, II

B) I, III

C) II, III

D) I, II, III

5. * If $f\left(\dfrac{x+1}{x-1}\right)=2x$, then what is $f^{-1}(6)=$?

A) 1

B) 2

C) 3

D) 6

6. * A piecewise function $f(x)$ is given. What could be the value of a so that f(x) has an inverse?

$$f(x)=\begin{cases} ax, & x\geq 0 \\ -x^2, & x<0 \end{cases}$$

A) -1

B) -0.5

C) 0

D) 1

1.8 Limit

1. Limit

① What is the value of y when x**=2**?

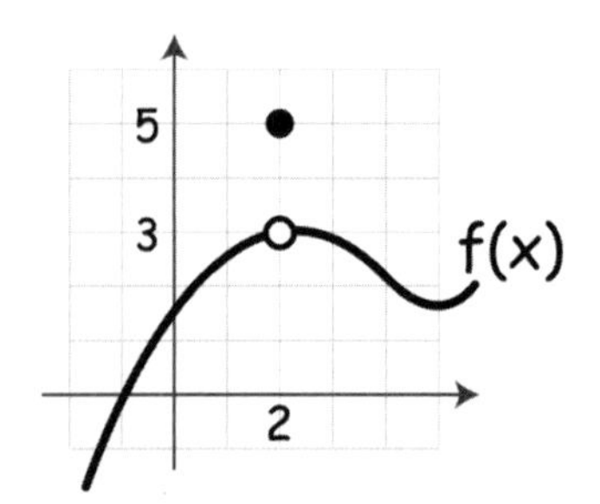

② What is the value of y when x ***approaches* to 2**?

When x = 2, the y value is 5. We can write this with function notation; f(2) = 5.

When x approaches to 2, the y value will be about 3. (y approaches to 3)

We can write this using 'limit' notation: $\lim_{x \to 2} f(x) = 3$

※ **Notation of a Limit**

$$\lim_{x \to 2} f(x) = 3$$

when x approaches to 2

f(x) approaches to 3

When you see "limit", think "approaching"

2. One-sided Limit

When x approaches to 2, we can approach in two different directions; from the negative (-) direction (from the left), and from the positive (+) direction (from the right).

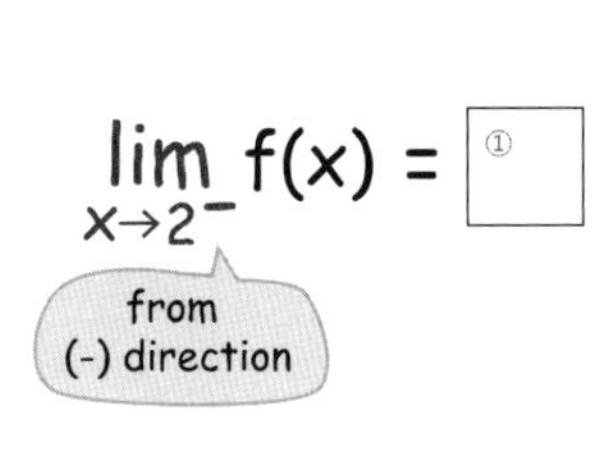

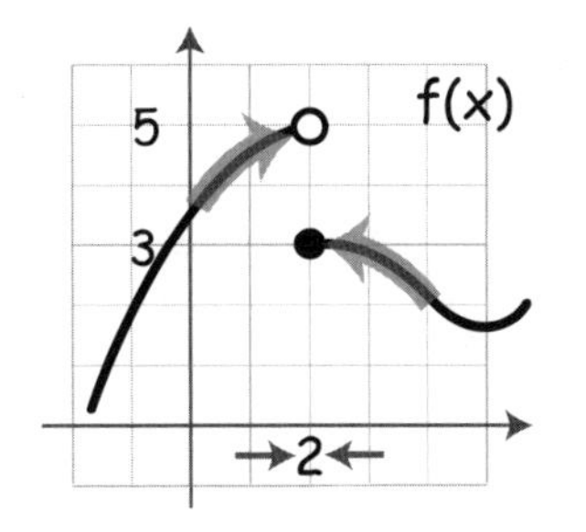

$\lim_{x \to 2^+} f(x) =$ ②

from (+) direction

When x approaches to 2
from the **negative direction**,
f approaches 5.
= the **left-hand limit**(-) is 5

When x approaches to 2
from the **positive direction**,
f approaches 3.
= the **right-hand limit**(+) is 3

Then

$\lim_{x \to 2} f(x)$ ____________③

When you have $\lim_{x \to 2} f(x)$ (with no direction sign) then you should check *both directions*. If the left-hand limit and the right-hand limit are different, then we can say the limit 'Does not exist.' So it will be $\lim_{x \to 2} f(x) = DNE$.

※ Existence of a Limit

If *f* is a function and c and *L* are real numbers,

then $\lim_{x \to c} f(x) = L$ if and only if both the left- and right- limits are equal to *L*.

$$\lim_{x \to c^+} f(x) = \lim_{x \to c^-} f(x) = \lim_{x \to c} f(x)$$

from (+) from (-) from both

The limit ④__________.

Blank : ① 5 ② 3 ③ Does not exist (DNE) ④ Exists

EXAMPLE 1. Evaluate the function. And find the one-sided limit, two sided limit at each point if they exist.

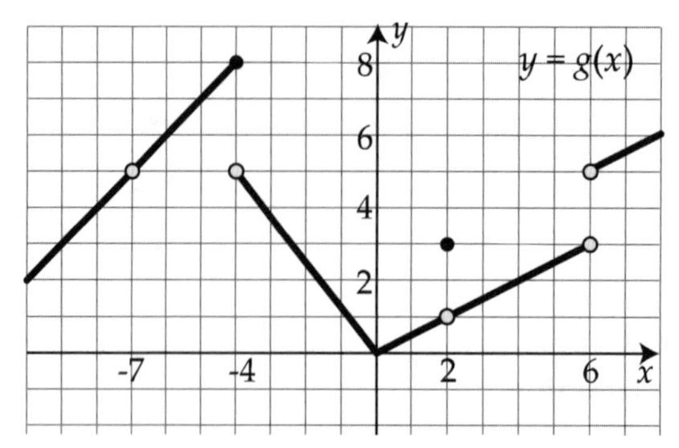

① i) $g(2)$

ii) $\lim_{x \to 2^+} g(x)$

iii) $\lim_{x \to 2^-} g(x)$

iv) $\lim_{x \to 2} g(x)$

② i) $g(-4)$

ii) $\lim_{x \to -4^+} g(x)$

iii) $\lim_{x \to -4^-} g(x)$

iv) $\lim_{x \to -4} g(x)$

③ i) $g(6)$

ii) $\lim_{x \to 6^+} g(x)$

iii) $\lim_{x \to 6^-} g(x)$

iv) $\lim_{x \to 6} g(x)$

④ i) $g(-7)$

ii) $\lim_{x \to -7^+} g(x)$

iii) $\lim_{x \to -7^-} g(x)$

iv) $\lim_{x \to -7} g(x)$

3. Limit with infinity

※ **Limit with infinity**

$x \to \infty$ means x becomes a large positive number (On the graph, x goes to the right endlessly...)

$x \to -\infty$ means x becomes a large negative number (On the graph, x goes to the left endlessly...)

$\lim\limits_{x \to -\infty} f(x) =$ ①

As x gets larger in (-) direction

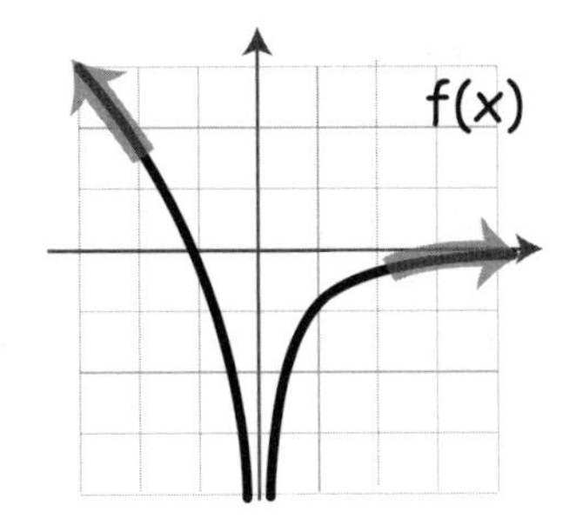

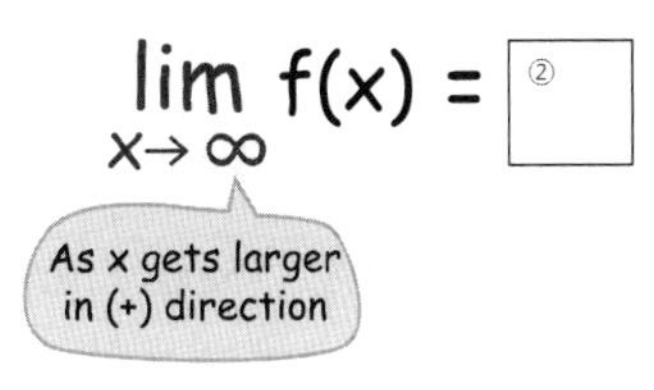

EXAMPLE 2. Find the limit.

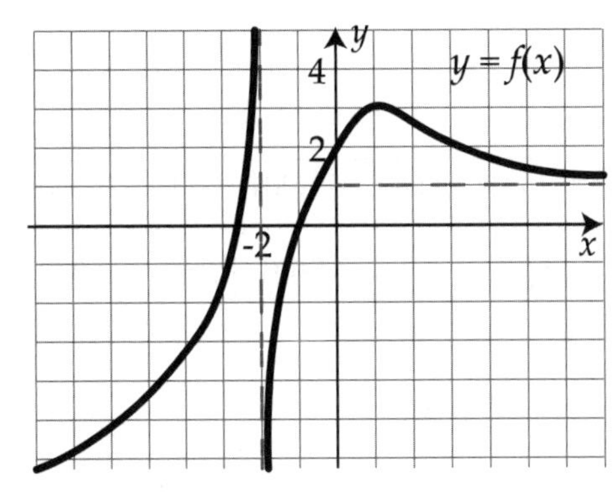

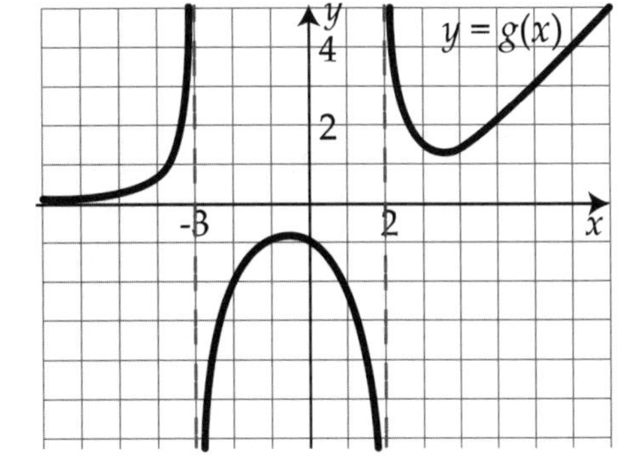

① $\lim\limits_{x \to \infty} f(x)$

② $\lim\limits_{x \to \infty} g(x)$

③ $\lim\limits_{x \to -\infty} f(x)$

④ $\lim\limits_{x \to -\infty} g(x)$

⑤ $\lim\limits_{x \to -2^{+}} f(x)$

⑥ $\lim\limits_{x \to -3^{+}} g(x)$

⑦ $\lim\limits_{x \to -2^{-}} f(x)$

⑧ $\lim\limits_{x \to -3^{-}} g(x)$

⑨ $\lim\limits_{x \to 2^{+}} g(x)$

⑩ $\lim\limits_{x \to 2^{-}} g(x)$

Blank : ① ∞ ② 0

EXAMPLE 3. Find the limit from the given table.

①

x	-0.1	-0.001	0	0.999	0.9
$f(x)$	-5.8	-5.001	und	-4.999	-4.1

$\lim_{x\to 0} f(x)$

②

x	4.99	4.999	5	5.001	5.01
$h(x)$	2.2	2.001	8	1.999	1.75

$\lim_{x\to 5} h(x)$

③

x	-3.1	-3.001	-3	-2.999	-2.9
$g(x)$	7.99	7.999	-5	8.001	8.01

$\lim_{x\to -3} g(x)$

④

x	-11.1	-11.001	-11	-10.999	-10.9
$k(x)$	-4.1	-4.001	und	-3.999	-3.1

$\lim_{x\to -11} k(x)$

⑤

x	0.9	0.999	1	1.001	1.1
$f(x)$	90	9999	und	-9999	-90

$\lim_{x\to 1^+} f(x)$ and $\lim_{x\to 1^-} f(x)$

⑥

x	8.99	8.999	9	9.001	9.01
$h(x)$	-77	-9999	und	9999	77

$\lim_{x\to 9^+} h(x)$ and $\lim_{x\to 9^-} h(x)$

4. Evaluating Limits when x→constant

① Direct substitution(Just put the values in)

If f is a function, and c, a are any real number, then

$$\lim_{x \to c} f(x) = f(c) \qquad \lim_{x \to c} a = a$$

② Cancelation

When you do the substitution and get $\frac{0}{0}$, you can factor out the numerator (or denominator) and cross out.

$$\lim_{x \to 1} \frac{x^2 - 1}{x - 1} = \lim_{x \to 1} \frac{\boxed{①}}{(x-1)} = \lim_{x \to 1}(x+1) = 2$$

③ Rationalization(use conjugate)

$$\lim_{x \to 4} \frac{2-\sqrt{x}}{4-x} = \lim_{x \to 4} \frac{(2-\sqrt{x})\boxed{②}}{(4-x)\boxed{③}} = \lim_{x \to 4} \frac{4-x}{(4-x)(2+\sqrt{x})} = \lim_{x \to 4} \frac{1}{(2+\sqrt{x})} = \frac{1}{4}$$

EXAMPLE 4. Evaluate.

☺ Tip: Always try substitution 1st!

① $\lim_{x \to 2}\left(4x^2 + 3\right)$

② $\lim_{x \to 3} \frac{\sqrt{x+1}}{x-4}$

Blank : ① (x+1)(x-1) ② 2+√x ③ 2+√x

③ $\lim_{t\to2}\frac{t^2-3t+2}{t^2-4}$

④ $\lim_{x\to2}\frac{-x^2-x+6}{x-2}$

⑤ $\lim_{x\to0}\frac{\sqrt{x+3}-\sqrt{3}}{x}$

⑥ $\lim_{x\to9}\frac{3-\sqrt{x}}{x-9}$

5. Evaluating Limits when $x\to\infty$

※ Finding limit when x→∞

: ① ________________ the degree of the numerator and denominator

Deg of Top < Deg of Bottom $\lim_{x\to\infty}\frac{2x+3}{3x^2+1} = 0$: always 0

Deg of Top = Deg of Bottom $\lim_{x\to\infty}\frac{2x^2+3}{3x^2+1} = \frac{2}{3}$: ratio of the leading coefficient

Deg of Top > Deg of Bottom $\lim_{x\to\infty}\frac{2x^3+3}{3x+1} = \infty$: always ∞ or -∞

Blank : ① compare

EXAMPLE 5. Find the limit.

① $\lim_{x\to\infty}\frac{x-2}{2x^2-4x+3}$

② $\lim_{x\to\infty}\frac{2x^3-4x^2+100}{2x^2+3x^3}$

③ $\lim_{x\to\infty}\frac{4x^4+2x^3+8x-2}{x^2+3}$

④ $\lim_{x\to\infty}\frac{x^4+4x+3}{x^3-7x^2+10}$

⑤ $\lim_{x\to\infty}\frac{3x^2-2x+1}{x^2+2}$

⑥ $\lim_{x\to\infty}\frac{x}{x^2-1}$

AP Style Problem

1. The graph of f(x) is given. What would be true about f(x)?

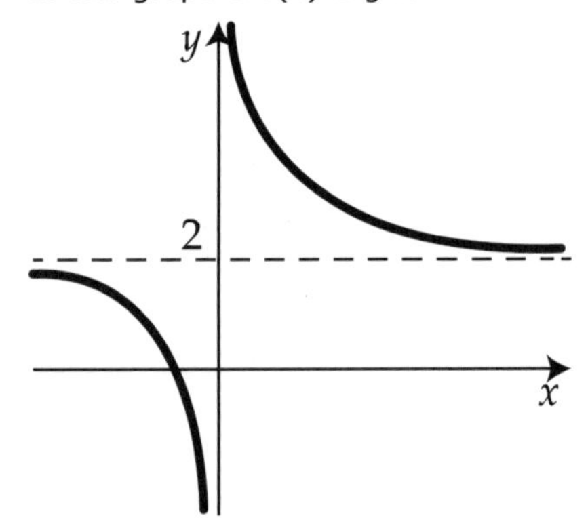

A) $\lim_{x \to 0^-} f(x) = \infty$ and $\lim_{x \to -\infty} f(x) = 0$

B) $\lim_{x \to 0^-} f(x) = \infty$ and $\lim_{x \to \infty} f(x) = 2$

C) $\lim_{x \to 0^+} f(x) = \infty$ and $\lim_{x \to -\infty} f(x) = 0$

D) $\lim_{x \to 0^+} f(x) = \infty$ and $\lim_{x \to \infty} f(x) = 2$

2. If $a \neq 0$, then $\lim_{x \to a} \frac{x^4 - a^4}{x^2 - a^2}$ is

A) $6a^2$

B) a^2

C) 0

D) $2a^2$

Part 2

Polynomial and Rational Functions

2.1 Polynomial Functions

1. Polynomial

poly- means ① ____________

-nomial means ② __________

※ A **polynomial** function of degree n :

$$P(x) = a_n x^n + a_{n-1} x^{n-1} + \cdots + a_1 x + a_0$$

(Leading Coefficient: a_n; Degree: n; Leading Term: $a_n x^n$; Constant Term: a_0)

where n is a nonnegative integer and $a_n \neq 0$.

$a_n, a_{n-1}, \ldots, a_1, a_0$ is called the **coefficients**.

a_0 is called the **constant coefficient** or **constant term**.

$a_n x^n$ is the **leading term** and a_n is the **leading coefficient**.

ex)

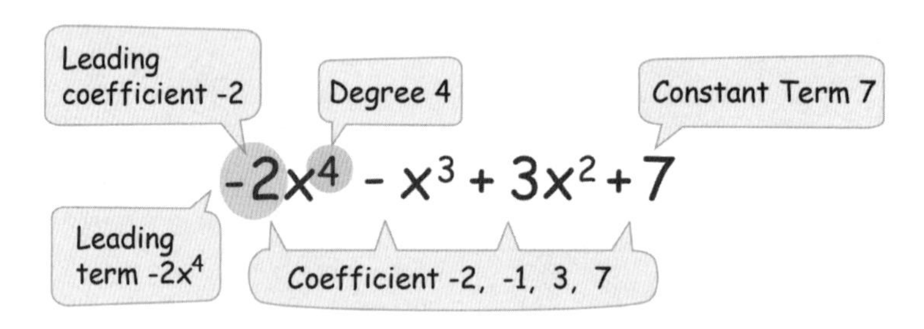

Blank : ① many ② terms

Degree of Polynomial: highest power that appears

Polynomial or not?

① no division by a variable.

② Coefficient is a ①__________.

③ exponents can only be nonnegative integers (②____________)

EXAMPLE 1. Determine which of the following is polynomial function. For those that are, state the degree.

① $f(x) = 2 - 3x^4$

② $f(x) = \sqrt{x}$

③ $g(x) = \dfrac{x}{x^2 - 1}$

④ $f(x) = -7$

⑤ $k(x) = -2x^3(x-1)^2$

⑥ $f(x) = \dfrac{x}{3}$

⑦ $h(x) = \sqrt[3]{x^2}$

⑧ $k(x) = \pi$

Blank : ① real number ② 0,1,2,3,...

2. Shape of Polynomial Functions

Each graph, based on their degree, has a different shape and characteristics.

when $a > 0$	when $a < 0$
Linear function $y = ax + b$ 	
Quadratic Function $y = ax^2 + bx + c$ 	
Cubic Function $y = ax^3 + bx^2 + cx + d$ 	
Quartic Function $y = ax^4 + bx^3 + cx^2 + dx + e$	

We can find ;

End Behavior when degree is odd : $a > 0$ ①____________ $a < 0$ ②____________

End Behavior when degree is even : $a > 0$ ③____________ $a < 0$ ④____________

Number of inflection point of cubic Function : ⑤____________

Number of inflection point of quartic Function : ⑥____________

※ Degree and inflection point

If $f(x)$ is a polynomial function of degree n,

then the graph of $f(x)$ has at most ⑦__________ inflection points

Blank : ① ↓ ↑ (down to up) ② ↑ ↓ (up to down) ③ ↑ ↑ (both up) ④ ↓ ↓ (both down) ⑤ 1 ⑥ 0 or 2 ⑦ n-2

3. Graphing Polynomial

1) End behavior

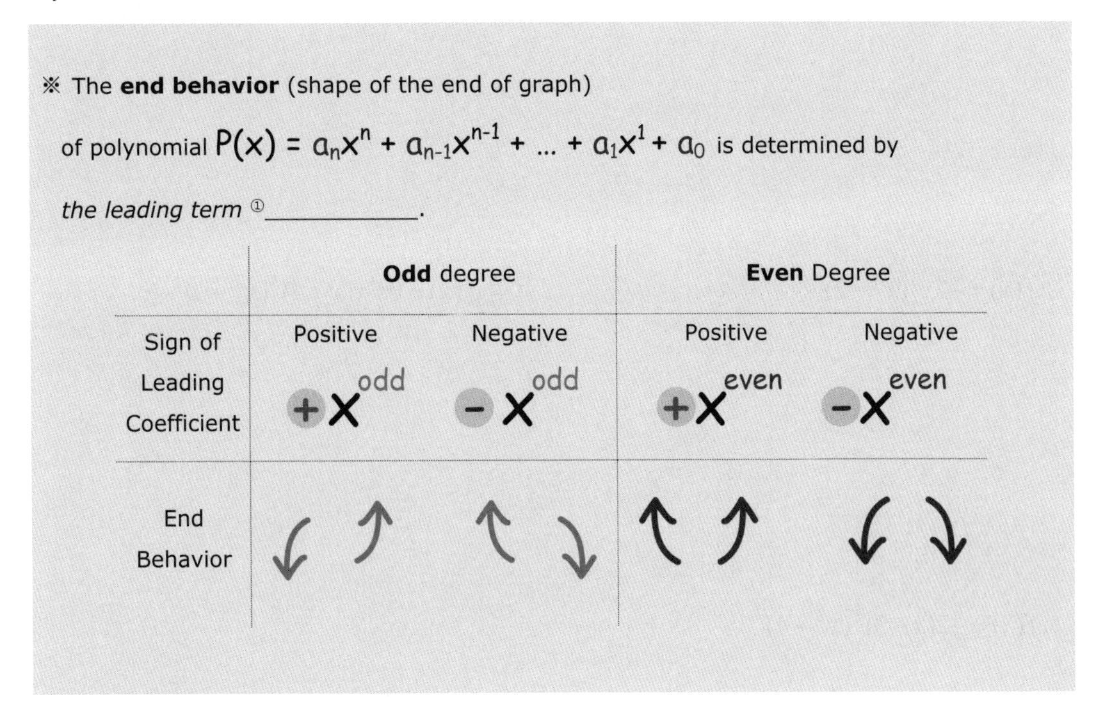

※ The **end behavior** (shape of the end of graph)

of polynomial $P(x) = a_nx^n + a_{n-1}x^{n-1} + \dots + a_1x^1 + a_0$ is determined by

the leading term ①__________.

	Odd degree		**Even** Degree	
Sign of Leading Coefficient	Positive	Negative	Positive	Negative
	$+x^{odd}$	$-x^{odd}$	$+x^{even}$	$-x^{even}$
End Behavior	↙ ↗	↖ ↘	↖ ↗	↙ ↘

EXAMPLE 2. Determine the end behavior. Find $\lim_{x \to \infty} f(x)$ and $\lim_{x \to -\infty} f(x)$.

☺ Tip:

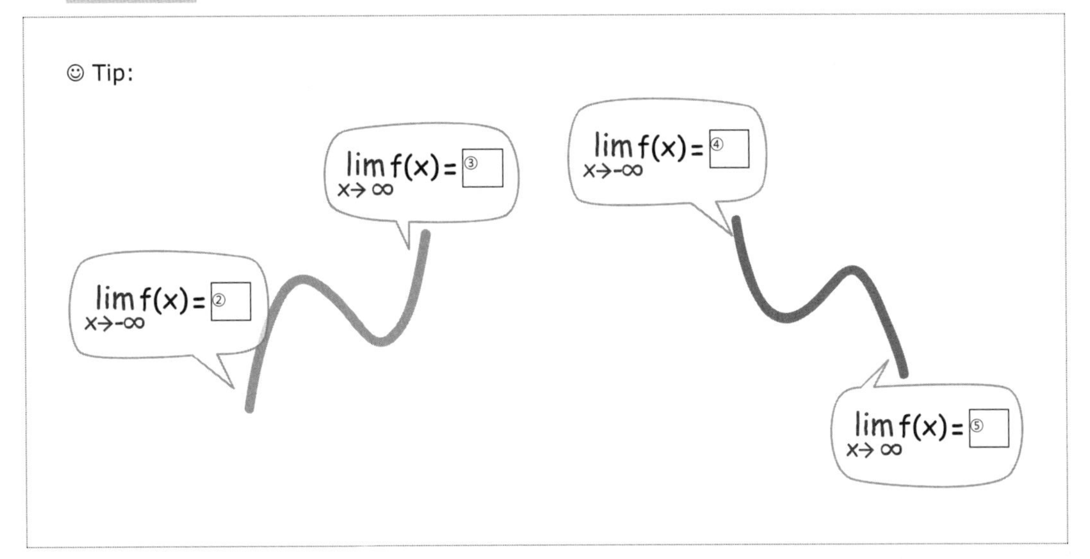

Blank : ① a_nx^n ② $-\infty$ ③ ∞ ④ ∞ ⑤ $-\infty$

① $f(x)=4(x+4)(x+3)^3$

② $f(x)=-(x-3)^2(x-1)^3$

③ $f(x)=-2x^3(x^2-2)^2$

④ $f(x)=0.5x(x+4)^3(x^2-2)^2$

⑤ $f(x)=-2(x-3)^2(x^2+4)^3$

⑥ $f(x)=3x^2(x^3-2)^3(x^2+1)$

2) Zeros of Polynomial

① When we have a ①__________ **form** of a polynomial,

then we can find the zeros(=roots = x intercept) using *zero product property*.

$$X \cdot Y = 0 \text{ then } X = 0 \text{ or } Y = 0$$

② Multiplicity: how often a root appears.

ex) $(x-1)^2$: The root(zero) x = 1 appears twice, so the multiplicity is ②____.

Odd multiplicity	**Even** multiplicity
$(x-r)^{odd}$	$(x-r)^{even}$
or	or
r r	r r
③__________the x axis at r.	④__________the x axis at r.

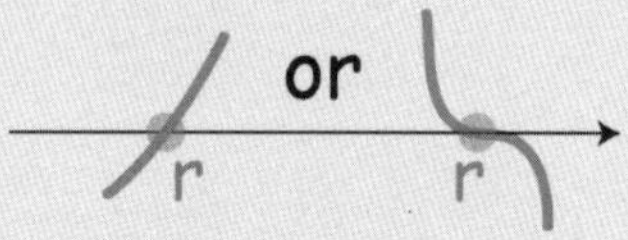

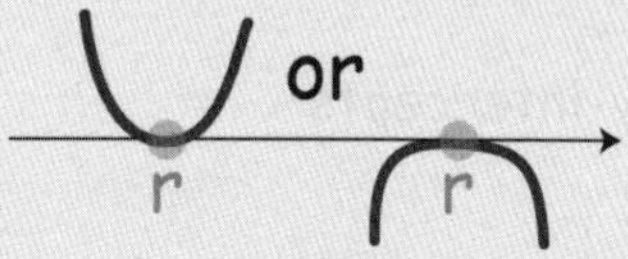

ex)

multiplicity

$$P(x) = -2(x-1)^2(x+2)^3$$

P(x) ⑤(crosses/touches) at x = 1, and ⑥(crosses/touches) at x = -2.

Blank : ① factored ② 2 ③ cross ④ touch ⑤ touches ⑥ crosses

③ Shape of the graph near zero

: 3 or more multiplicities <u>flattens out</u> near zero.

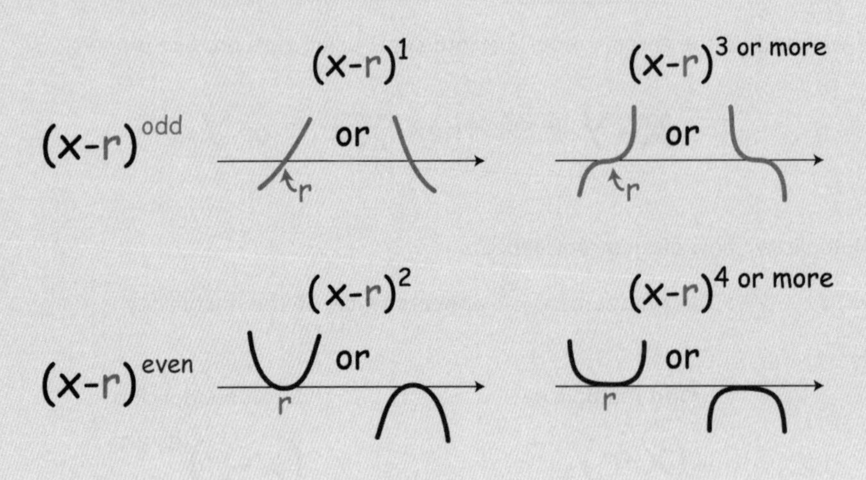

2) Finding y-intercepts

For every polynomial function, <u>constant</u> term is the y-intercept.
(or plug in x = 0)

EXAMPLE 3. For the polynomial,

a) determine the end behavior,

b) determine whether the graph crosses or touches the x-axis at each x -intercept.

c) find the y intercept.

d) Sketch the graph briefly.

① $f(x)=(x-2)(x+3)(2x-1)$

② $f(x)=(x+1)^2(x-2)$

③ $f(x)=-2x^2(x-2)$

④ $f(x)=-2x^3(x-2)^2$

⑤ $f(x)=-2(x+2)^2(x-1)^4$

⑥ $f(x)=(x-1)^2(x-5)^3$

⑦ $f(x)=-2x^2(x-1)^3(x+6)^4$

⑧ $f(x)=x(x-\sqrt{3})^3(x-2)^4$

⑨ $f(x)=x^3(x+2)^3(x-3)^4$

⑩ $f(x)=-2x^2(x+1)^3(x-3)^2$

⑪ $f(x)=x(x^2+1)(x^2-5)$

⑫ $f(x)=x^3(x^2+2x+3)(x+2)$

⑬ $f(x)=(x-1)^2(x^2+2x+5)^3$

⑭ $f(x)=-x(x-1)^2(x^2+3x+3)$

EXAMPLE 4. Factor the polynomial and use the factored form to find the zeros. Then sketch the graph.

$$a^2 - b^2 = (a \ ① \ b)(a \ \bigcirc \ b)$$

$$a^3 - b^3 = (a \ ② \ b)(a^2 \ \bigcirc \ ab \ \bigcirc \ b^2)$$

$$a^3 + b^3 = (a \ ③ \ b)(a^2 \ \bigcirc \ ab \ \bigcirc \ b^2)$$

① $f(x) = x^3 - x^2 - 2x$

② $f(x) = x^4 + 2x^3 - 8x^2$

③ $f(x) = x^3 + x^2 - x - 1$

④ $f(x) = x^3 - 2x^2 - 4x + 8$

⑤ $f(x) = -2x^4 - x^3 + 3x^2$

⑥ $f(x) = x^5 + 8x^2$

Blank : ① + − ② − + + ③ + − +

⑦ $f(x)=x^4+x^3-x-1$

EXAMPLE 5. Find lowest order polynomial equation for each of these graphs.

①

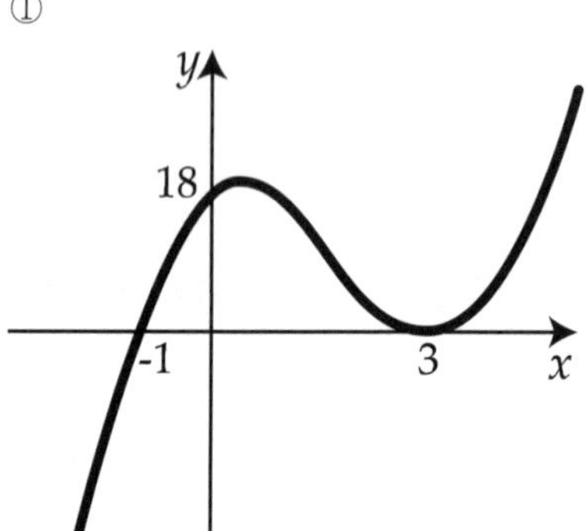

②

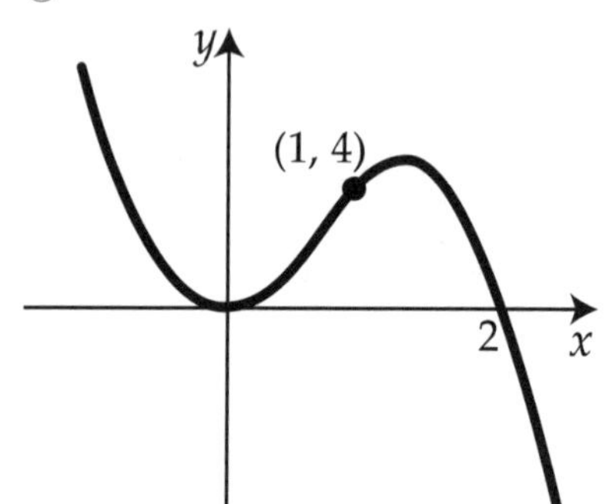

③

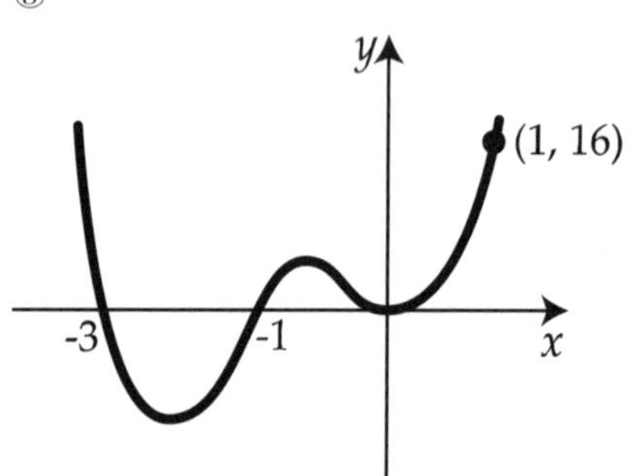

④

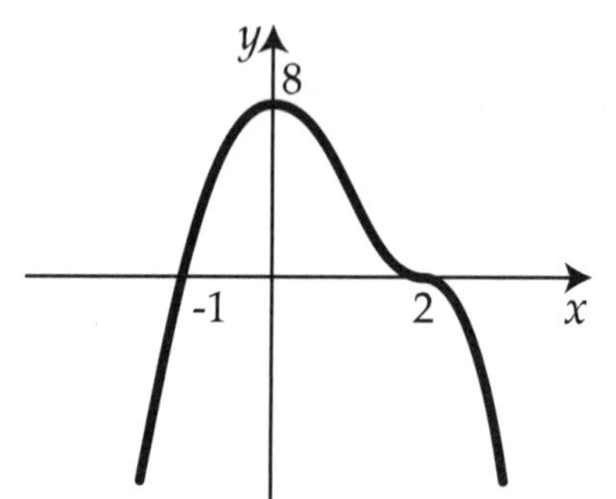

⑤

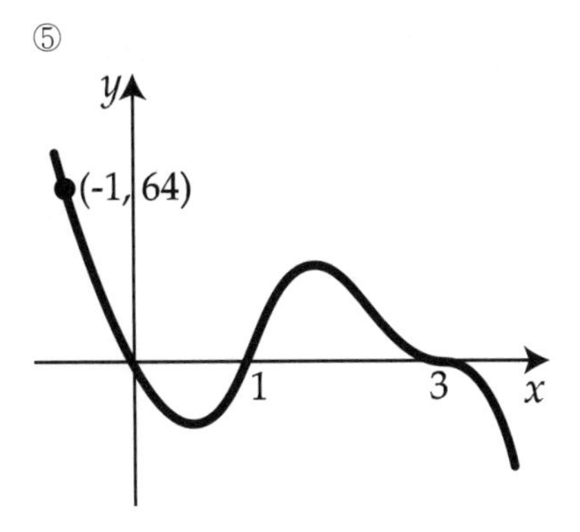

AP Style Problem

1. Which of the following polynomial $p(x)$ has exactly three distinct real zeros?

A) $p(x)=(x^2-4)(x^2+4)$

B) $p(x)=(x^4-1)(x-3)$

C) $p(x)=(x^2+4x+4)(x+2)$

D) $p(x)=(x^2+4x+7)(x+2)$

2. The polynomial $P(x)=ax^4+bx^2+c$ where $a>0, b\neq 0$ and $c\neq 0$. Which of the following is true about the graph of P?

I. $\lim_{x\to\infty} P(x)=\infty$

II. $\lim_{x\to-\infty} P(x)=-\infty$

III. $P(x)$ has absolute maximum.

IV. $P(x)$ has absolute minimum.

A) I, II

B) I, III

C) II, III

D) I, IV

3. When a is a real number, m is odd and n is even, which of the following could be the graph of $y=a(x-2)^m(x+1)^n$?

A)

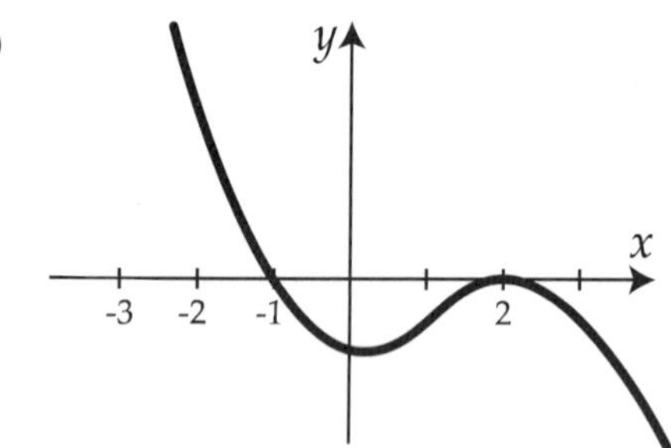

B)

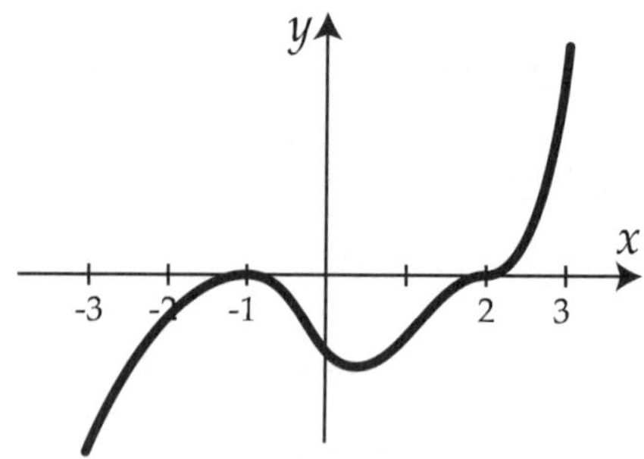

C)

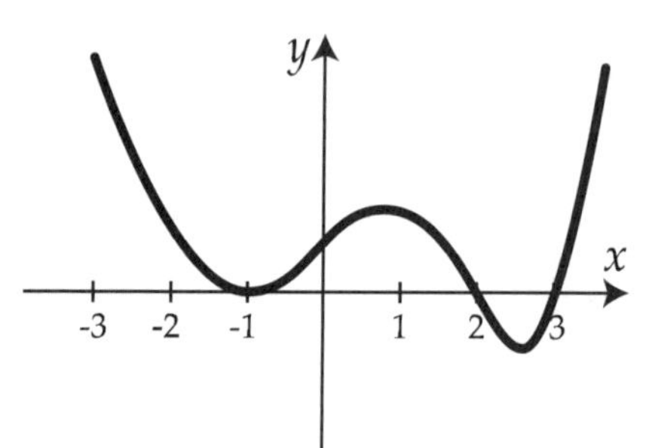

D)

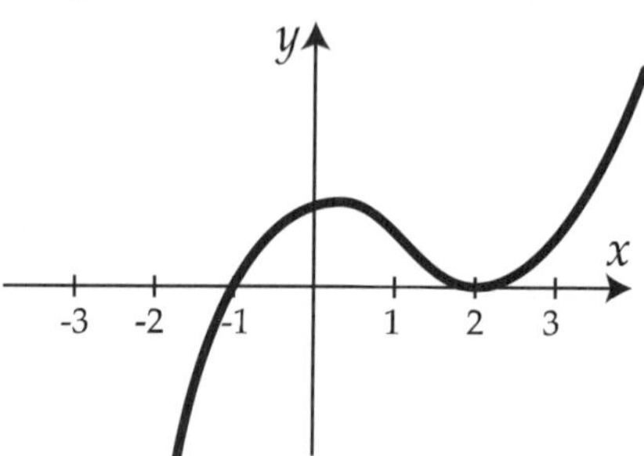

2.2 Dividing Polynomials

1. Dividing Polynomial

When you divide a polynomial P(x) by another polynomial D(x), we get a quotient polynomial ①________ and a remainder polynomial ②_________.

$$\begin{array}{r} Q(x) \\ D(x)\,\overline{)\,P(x)} \\ \vdots \\ \hline R(x) \end{array}$$

We can rewrite this as;

dividend, divisor, Quotient, Remainder

$$P(x) = D(x)\cdot Q(x) + R(x)$$

or

dividend, divisor

$$\frac{P(x)}{D(x)} = Q(x) + \frac{R(x)}{D(x)}$$

Notice that the **degree of the remainder** is always *less than* the **degree of the divisor**.

2. Long Division of Polynomial

This is a method similar to division for *Numbers*.

Divide $2x^3 - 7x^2 + 5$ by $x - 3$.

> ※ **Be sure that**
>
> Rewrite the polynomial P(x) from highest to lowest exponent.
>
> When you have a missing term, include the missing terms with a coefficient of ③_________.

Blank : ① Q(x) ② R(x) ③0

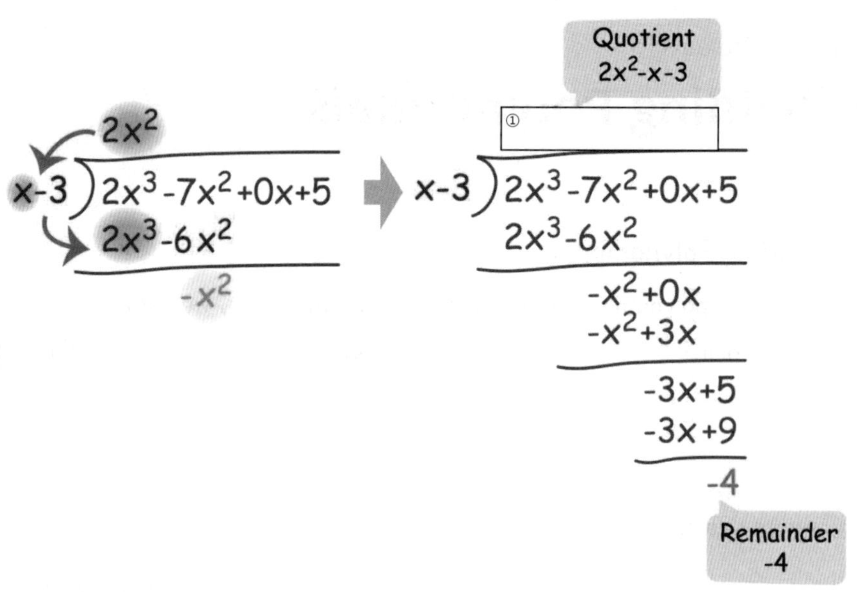

You can write the result as ;

② ______________________________

EXAMPLE 1. Use long division to perform the division.

① $\dfrac{x^3-9x^2+27x-27}{x^2-3}$

② $\dfrac{x^3+6x+2}{x^2-2x+2}$

Blank : ① $2x^2-x-3$ ② $2x^2-x-3+\dfrac{-4}{x-3}$

③ $\dfrac{4x^3-3}{x+5}$

④ $\dfrac{3x^4-5x^3-10x+2}{x^2+x+2}$

3. Synthetic Division

Synthetic division is a quick method of dividing polynomials;

it can be used *only* when we divide a polynomial by ①__________________________.

Divide $2x^3-7x^2+5$ by $x-3$ using synthetic division.

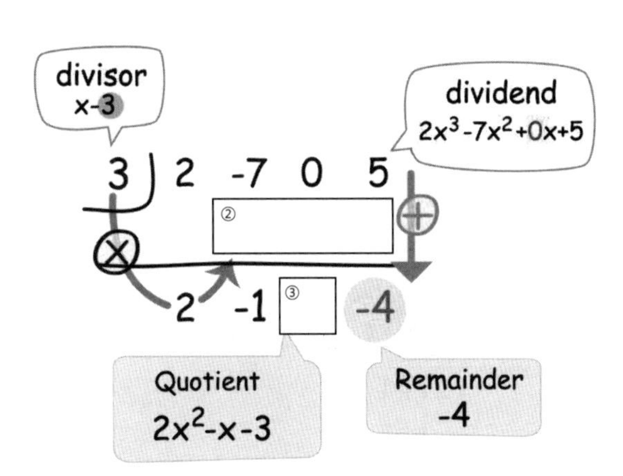

Blank : ① x – c (linear) ② 6 -3 -9 ③ -3

EXAMPLE 2. Use synthetic division to perform the division.

① $\dfrac{x^3-9x^2+27x-27}{x-3}$

② $\dfrac{3x^3+5x^2-4x+1}{x+4}$

③ $\dfrac{x^3-8x+2}{x+2}$

④ $\dfrac{5x^3+x^2-3}{x+3}$

⑤ $\dfrac{x^4-5x^2-10}{x-4}$

⑥ $\dfrac{2x^4-10x^3+8}{x-2}$

⑦ $\dfrac{2x^3+4x^2-2x+3}{2x-1}$

⑧ $\dfrac{6x^3-10x^2+5x+2}{2x-3}$

4. The Remainder Theorem

※The Remainder Theorem

When you divide a polynomial $P(x)$ by $(x-c)$, the remainder R will be ①______.

☺ Proof:

The polynomial $P(x)$ can be expressed as; ② ______________________

And when we plug in $x=c$; ③______________________

EXAMPLE 3. Find the remainder using remainder theorem.

Just calculate ④________!

① $\dfrac{2x^3-3x^2+9x+1}{x-2}$

② $\dfrac{x^3-x^2+x+5}{x+1}$

③ $\dfrac{x^{10}+3x^7-7x+6}{x-1}$

④ $\dfrac{x^{117}+3x^{10}-3x+1}{x+1}$

Blank : ① P(c) ② P(x) = (x − c)·Q + R ③ P(c) = (c − c)·Q + R ⇒ R = P(c) ④ P(c)

EXAMPLE 4. For each polynomial, evaluate $P(c)$.

① $P(x)=5x^4+30x^3-40x^2-36x+14$.

Find $P(-7)$.

② $P(x)=2x^4-21x^3-30x^2+8x-100$.

Find $P(12)$.

③ $P(x)=3x^3+4x^2-2x-1$.

Find $P\left(\frac{2}{3}\right)$.

④ $P(x)=x^3-x+1$.

Find $P\left(\frac{1}{4}\right)$.

EXAMPLE 5. If the polynomial $f(x)=2x^3-kx^2-5x+2$ is divided by $x-2$, the remainder is 2. What is the value of k?

5. Factor Theorem

When you divide a polynomial $P(x)$ by $(x-c)$, what happen when remainder is 0? ①

※ **Factor Theorem**

When $P(c) = 0$,

then $(x- c)$ is the **factor** of the polynomial P.

then polynomial P is **divisible by** $(x- c)$.

EXAMPLE 6. Use the Factor Theorem to determine whether the function $f(x) = x^4 + x^3 - 19x^2 + 11x + 30$ has a factor;

① $x-1$

② $x+1$

③ $x-2$

④ $x+2$

⑤ $x-3$

Blank : ① $P(x) = (x - c) \cdot Q \Rightarrow (x - c)$ is factor of $P(x)$.

EXAMPLE 7. If a polynomial P(x) = $x^2 + kx - 8$ has a factor of $x+3$, then what is the value of constant k?

EXAMPLE 8. If the polynomial $f(x) = 2x^3 - kx^2 - 5x + 2$ is divisible by $x-2$, what is the value of k?

AP Style Problem

1. The function P has degree of 4. If $P(-2)=0$, which of the following is true?

I. $x + 2$ is the factor of $P(x)$.

II. $\dfrac{P(x)}{x+2}$ is a quadratic polynomial.

III. $x = -2$ is one of the real zeros of $P(x)$.

IV. $P(x)$ is divisible by $x + 2$.

A) I, IV

B) II, III

C) I, II, III

D) I, III, IV

2. *The polynomial $x^2-(k+1)x-4$ has a factor $(x-k+1)$. Find k.

A) -1

B) 1

C) 2

D) -2

2.3 Real Zeros of Polynomial

1. The Rational Zeros Theorem

The **possible rational zeros** of the polynomial

$P(x) = a_n x^n + a_{n-1} x^{n-1} + \ldots + a_1 x^1 + a_0$ is

$$\text{Possible Rational Zeroes of } P(x) = \frac{\text{Factors of constant term } (a_0)}{\text{Factors of Leading Coefficient } (a_n)}$$

ex) Find the possible rational zeros of $f(x) = 2x^3 + x^2 - 13x + 6$.

$$\textit{possible Rational Zeros of } f(x) = \frac{\textit{factor of } ①\_\_}{\textit{factor of } ②\_\_} = ③\_\_\_\_\_\_\_\_$$

EXAMPLE 1. List the possible rational zeros of the polynomial function.

① $f(x) = 6x^4 + 4x^3 - 3x^2 + 2$

② $f(x) = 2x^4 - 5x^2 + 5x - 8$

③ $f(x) = 5x^4 + 6x^3 - 2x^2 - 8$

④ $f(x) = 4x^3 - 2x^2 + x + 7$

Blank : ① 6 ② 2 ③ $\pm 1, \pm 2, \pm 3, \pm 6, \pm \frac{1}{2}, \pm \frac{3}{2}$

2. Descartes' Rule of Signs

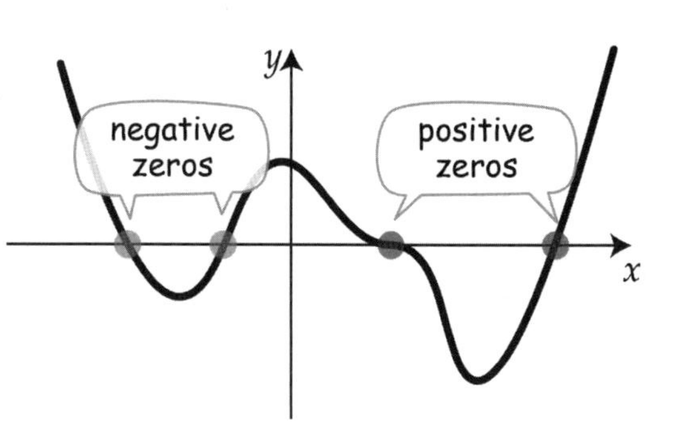

Descartes rule of signs determines the *'possible'* number of positive and negative zeros

First, rewrite the polynomial P(x) from highest to lowest exponent (ignore any missing terms) Then, count how many times there is a change of sign.

※ **Descartes' Rule of Sign**

Number of positive roots equals the number of sign changes of P(x) , or minus even integers (-2, -4, ..).

Number of negative roots equals the number of sign changes of P(-x) , or minus even integers.

$$P(x) = +2x^5 - 3x^4 - 3x^3 + 4x^2 - x - 1$$

3 Sign changes : 3 or 1 positive roots

$$P(-x) = -2x^5 - 3x^4 + 3x^3 + 4x^2 + x - 1$$

2 Sign changes : 2 or 0 negative roots

EXAMPLE 2. Use Descartes' Rule of Signs to determine how many positive and how many negative real zeros the polynomial can have.

① $P(x) = -x^3 - x^2 - x + 2$

② $P(x) = x^4 - 3x^3 + 2x^2 - x - 3$

③ $P(x) = x^5 + 2x^3 - x^2 + 5x - 1$

④ $P(x) = 3x^7 - 8x^5 + 4x^4 - 4x^3 - 4x^2 + 2$

3. Finding All Rational Zeros

EXAMPLE 3. Factor the polynomial $f(x) = 2x^3 + x^2 - 13x + 6$, and find all real zeros.

i) Find the *first zero*

(Using possible rational zeros, find the first number that makes f = 0)

$f(1) =$

$f(-1) =$

$f(2) =$

ii) Divide (using synthetic division)

iii) Repeat and find all zeros

EXAMPLE 4. Factor the polynomial and find all real zeros.

① $P(x)=x^3+3x^2-4x-12$

② $P(x)=2x^3+5x^2+x-2$

③ $P(x)=2x^3+4x^2+5x+3$

④ $P(x)=2x^3-7x^2+9x-6$

⑤ $P(x)=x^4-11x^2-18x-8$

⑥ $P(x)=x^4+8x^3+14x^2-8x-15$

2.4 Fundamental Theorem of Algebra

1. Complex Number

In algebra 2, we've learned about complex number.

A **complex Number** is a combination of a ①__________ number and an ②______________ number.

real part | imaginary part | $a + bi$ | $\sqrt{-1}$ | complex number

ex) $3+2i$, -2 (imaginary part is 0) , $-3i$ (real part is 0) ...

Conjugate of $a+bi$ is③ _______________.

2. Fundamental Theorem of Algebra

We have two different roots (=zeros)

① real roots and ② imaginary roots (roots with imaginary number i)

All together we call it **complex roots.**

complex roots { real roots (no i) = x intercept — $2, 0.5, \sqrt{3}$ ...
imaginary roots (with i) — $2i + 3$, $-4i$...

※ **Fundamental Theorem of Algebra**

① Every polynomial equation has at least one **complex roots.**

② Any polynomial of **degree n** has **n complex roots.**

Blank : ① real ② imaginary ③ a – bi

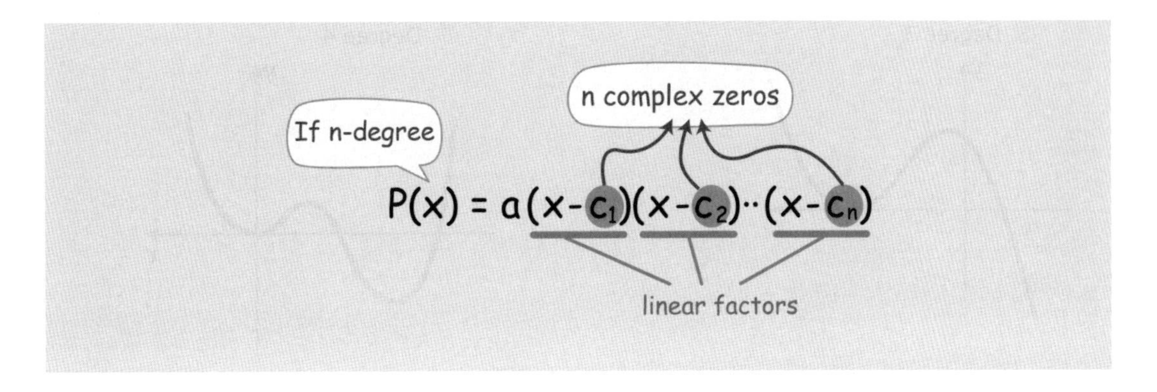

EXAMPLE 1. Determine the number of distinct real roots and imaginary roots for each function.

> You can see your real roots in the graph (= ①_______________)
> but you cannot see your imaginary roots in your graph.

① Degree 2

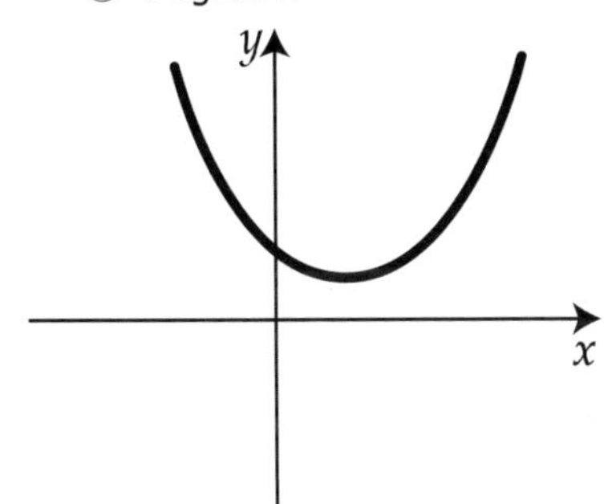

② Degree 2

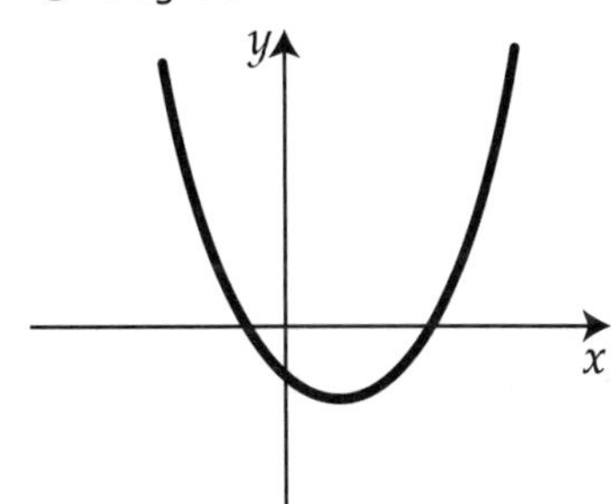

③ Degree 3

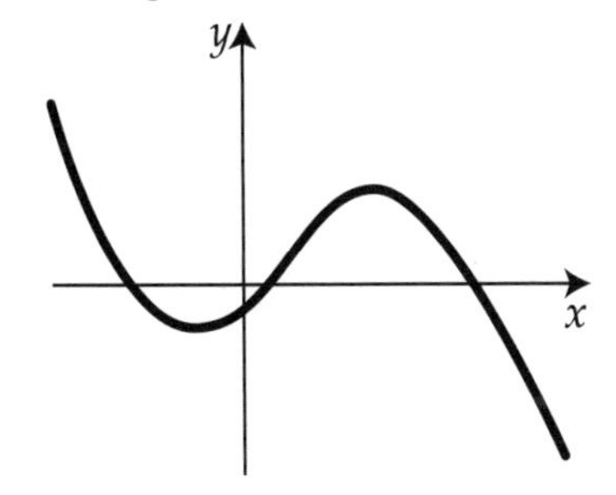

④ Degree 3

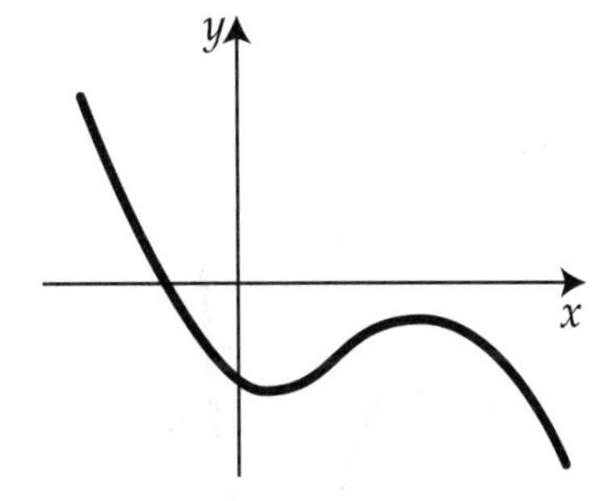

Blank : ① x intercept

⑤ Degree 3

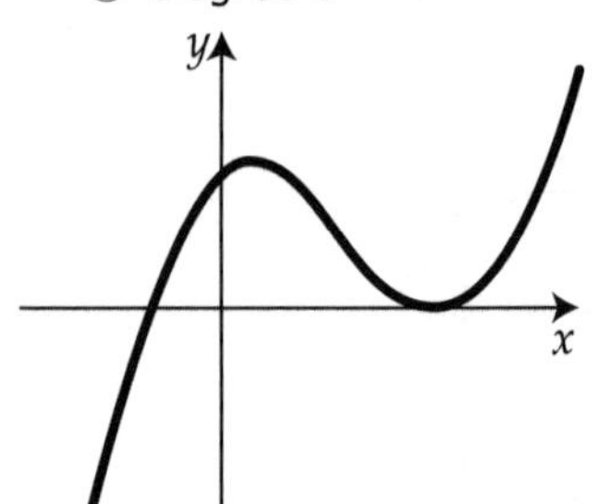

⑥ Degree 4

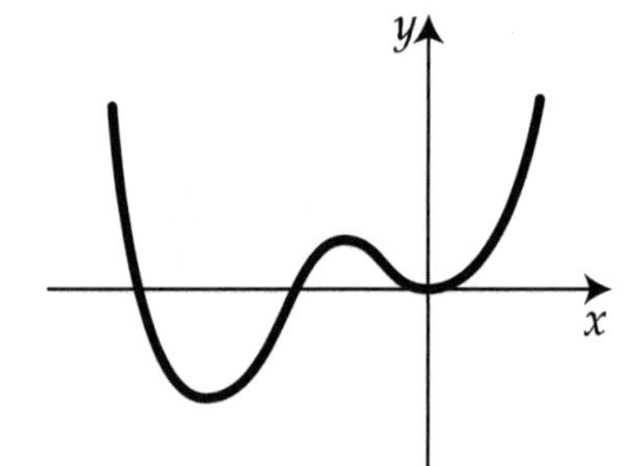

⑦ Degree 4

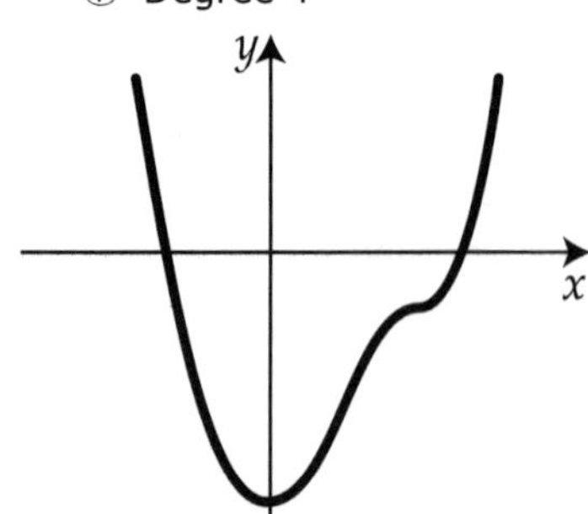

⑧ Degree 6

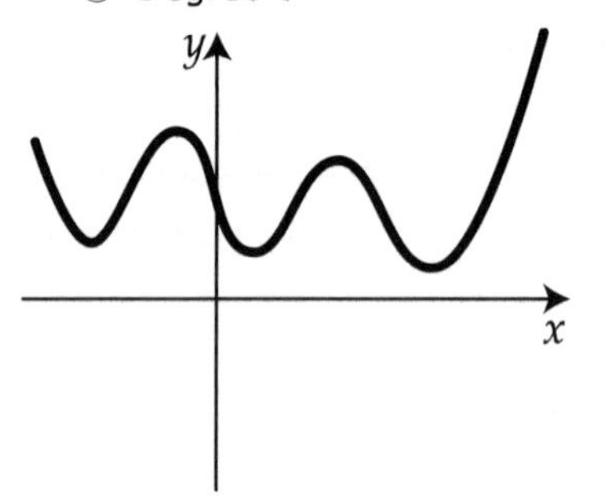

⑨ Degree 5

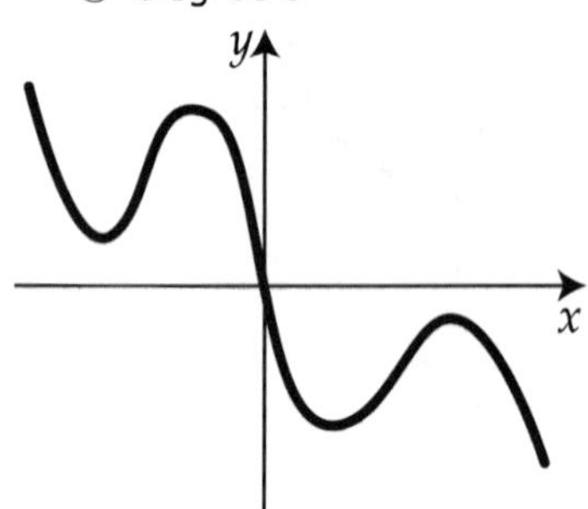

⑩ Degree 5

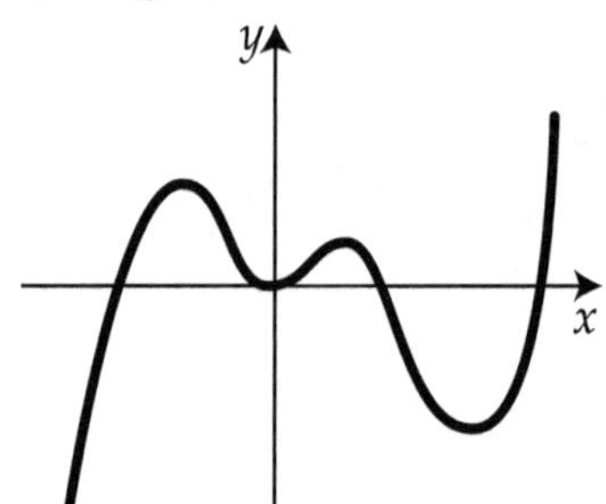

If we factor the polynomial **completely**, then we will find the *imaginary zeros*.

EXAMPLE 2. Factor f(x) completely and find all complex zeros.

① $f(x)=x^3+x$

② $f(x)=x^4+4x^2$

③ $f(x)=x^3+5x^2+10x+8$

④ $f(x)=x^3-3x^2+5x-3$

⑤ $f(x)=x^3-3x^2+2x-6$

⑥ $f(x)=x^4+9x^2+8$

⑦ $f(x)=x^4+4x^2-32$

3. Conjugate Zeros Theorem

※ **Conjugate Zeros Theorem**

Complex Roots always come in pairs!

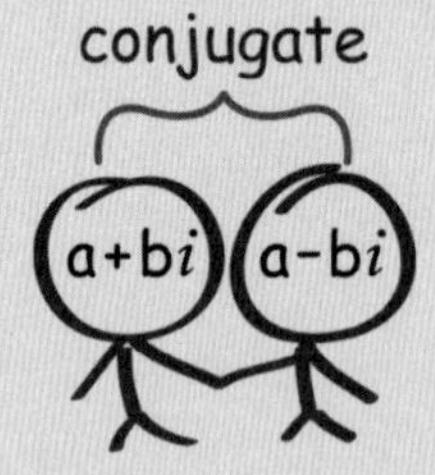

When $a+bi$ is the root, $a-bi$ is also the root.

EXAMPLE 3. Write a polynomial function of least degree with integer coefficients that has the given zeros.

① $1+2i$

② $3+i$

③ $-2, 2+i$

④ $-1, 3i$

⑤ $-2, i, -2i$

EXAMPLE 4. The polynomial $f(x) = x^3 - 7x^2 + 25x - 39$ has one zero as $2 - 3i$, find the other zeros.

AP Style Problem

1. According to the Fundamental Theorem of Algebra, which of the following polynomial has exactly 6 roots?

A) $x^5 - 8x^4 + 4x^3 + x^2 - 3x + 1$

B) $(x^2 + 4x + 3)(x^3 - x + 4)$

C) $(x^3 + 5x^2 - 7x + 4)^2$

D) $6x^4 - 8x + 2x - 10$

2. The polynomial has zeros $-i,\ 2+i,\ 1$ and -3. What is the least possible degree of the polynomial?

A) 4

B) 6

C) 7

D) 8

3. Find a polynomial with least degree that has a zeros 4 and $5i$.

A) $x^3-4x^2+25x-100$

B) $x^3-4x^2-25x+100$

C) $x^2-9x+20$

D) x^2+x-20

2.5 Rational Functions

1. Rational Function

Rational Function:

① __________ of two polynomials

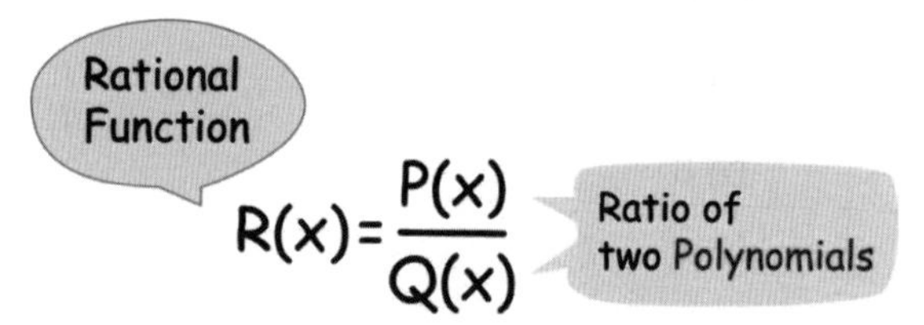

2. Reciprocal Function $y=\frac{1}{x}$

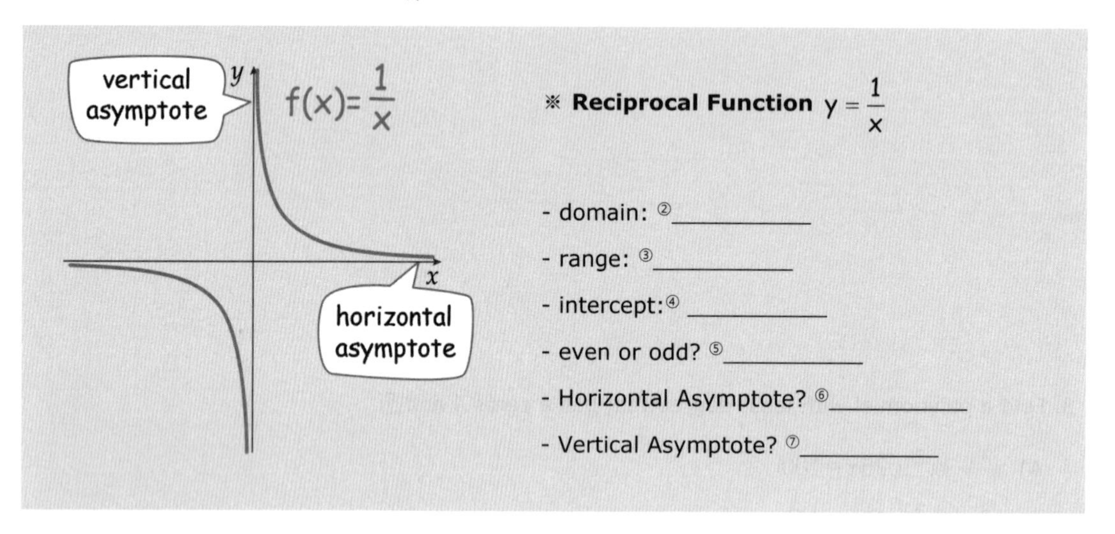

※ **Reciprocal Function** $y=\frac{1}{x}$

- domain: ② __________
- range: ③ __________
- intercept: ④ __________
- even or odd? ⑤ __________
- Horizontal Asymptote? ⑥ __________
- Vertical Asymptote? ⑦ __________

EXAMPLE 1. Graph the rational function using transformation.

① $R(x)=\frac{3}{x-1}+2$

② $R(x)=\frac{4}{x+2}-1$

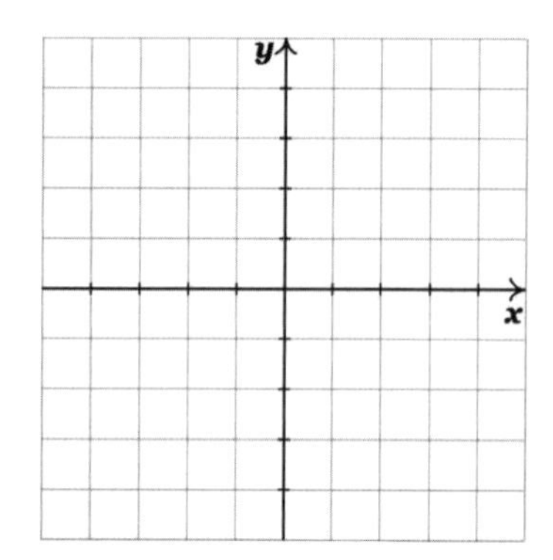

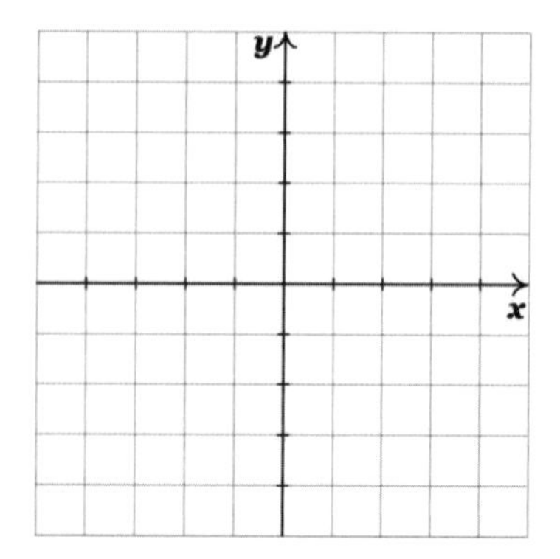

③ $R(x) = -\dfrac{1}{x+1} + 1$

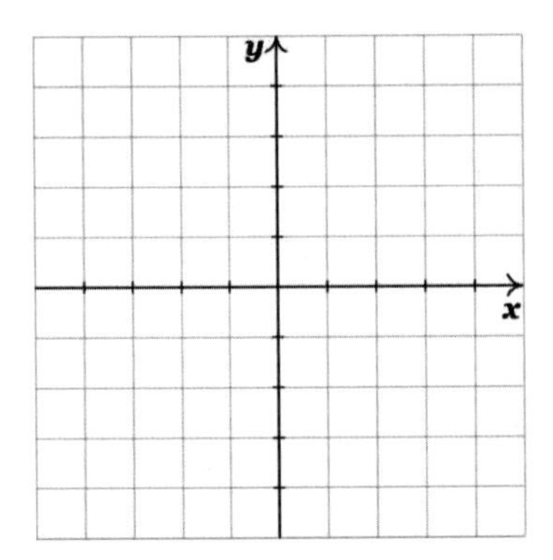

④ $R(x) = -\dfrac{2}{x} - 1$

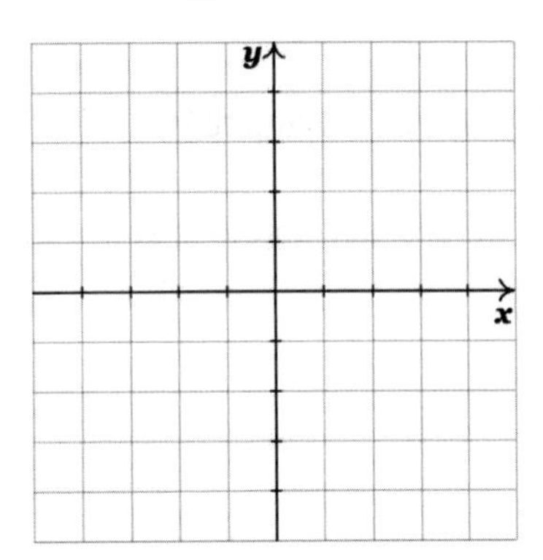

⑤ $R(x) = \dfrac{3x+2}{x-1}$

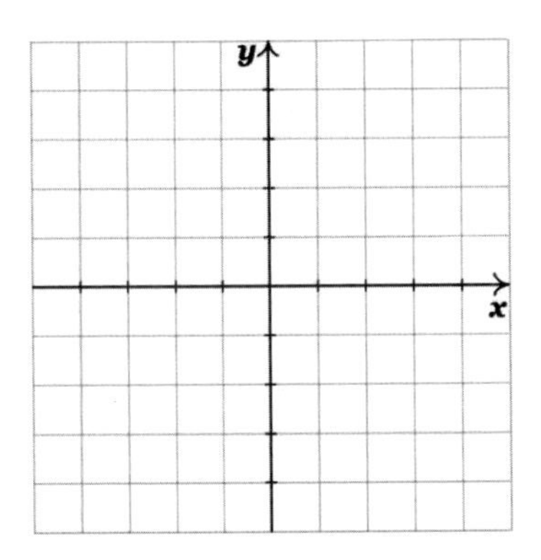

⑥ $R(x) = \dfrac{2x-1}{x+1}$

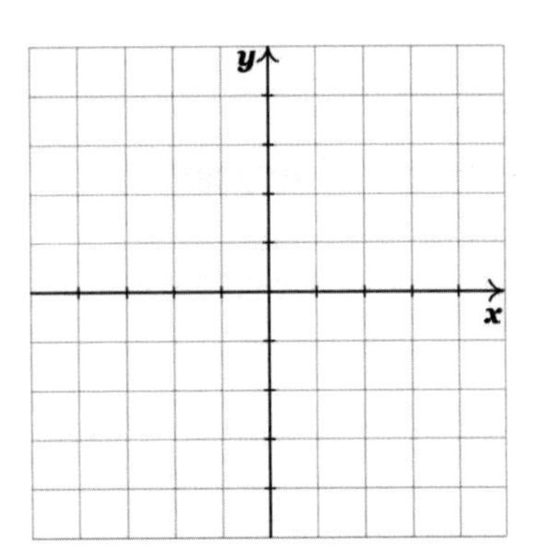

Blank : ① ratio ② All real numbers $x \neq 0$ ③ All real numbers $y \neq 0$ ④ none ⑤ odd ⑥ $y = 0$ ⑦ $x = 0$

3. Definition of Asymptotes

①________________: A line that a curve approaches as the graph goes towards infinity

1) 'Vertical' Asymptote

When vertical asymptote is at x = a;

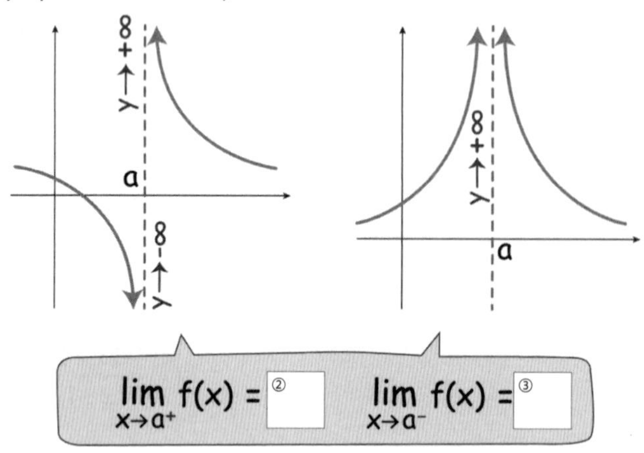

2) 'Horizontal' Asymptote :

When horizontal asymptote is at y = b;

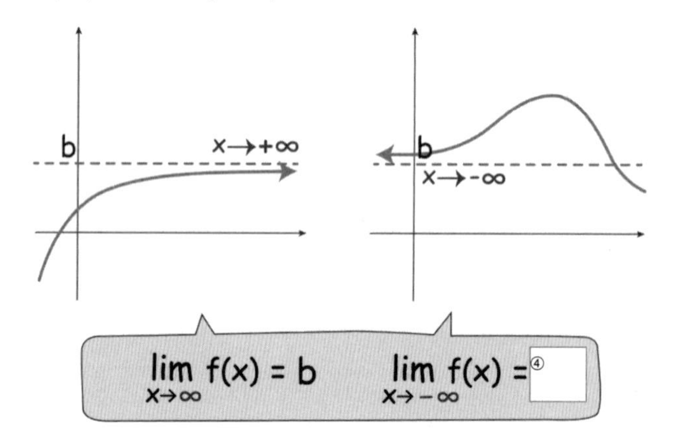

Blank : ① asymptotes ② ±∞ ③ ±∞ ④ b

4. Graphing Rational Functions

1) Finding the Domain and Asymptotes

※ **Rational Functions**

Rational Function

$$R(x)=\frac{P(x)}{Q(x)}$$

Ratio of two Polynomials

① Finding domain :① ________________ (②Reduce / Do not reduce)

② Vertical Asymptotes :③ ________________ (④Reduce / Do not reduce)

③ Hole: When you **reduce (x-h)** from the top and the bottom,

then there will be a hole at ⑤ __________.

④ Horizontal or Oblique (=slant) Asymptotes

: ⑥ ________________ the degrees of the top and the bottom

Deg of Top < Bottom $\frac{2x+3}{3x^2+1}$ ⑦ [ ]

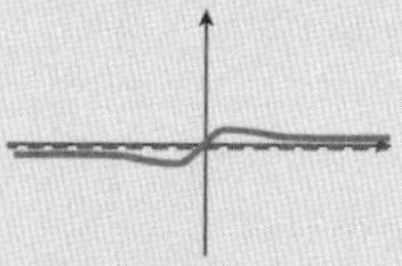

Deg of Top = Bottom $\frac{2x^2+3}{3x^2+1}$ ⑧ [ ] ratio of the leading coefficient

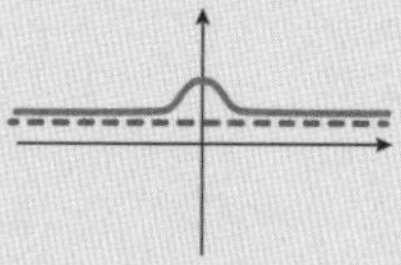

Deg of Top > Bottom (1 greater) $\frac{3x^2+3}{3x+1}$ $y=x-\frac{1}{3}$ oblique asymptote

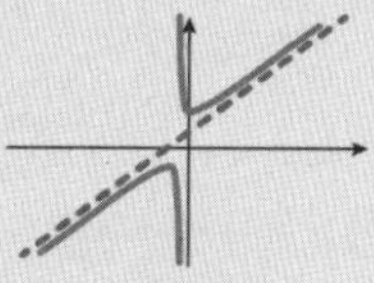

Deg of Top > Bottom (2,3..greater) $\frac{2x^3+3}{3x+1}$ ⑨ [ ]

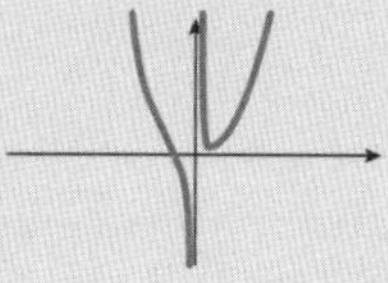

Blank : ① $Q(x)\neq0$ ② Do not reduce ③ $Q(x)=0$ ④ Reduce ⑤ $x = h$ ⑥ compare ⑦ $y = 0$ ⑧ $y=\frac{2}{3}$ ⑨ none

You can find oblique asymptote by

using ①________ ___________.

oblique asymptote $y = x - \frac{1}{3}$

$$
\begin{array}{r}
x - \frac{1}{3} \\
3x+1 \overline{\big) \, 3x^2 + 0x + 3} \\
\underline{3x^2 + 1x \quad\;} \\
-1x + 3 \\
\underline{-1x \;\; \vdots \;}
\end{array}
$$

EXAMPLE 2. Find the domain, any asymptotes, and hole of the rational function.

Key☞ Factor out and reduce to the lowest term
EXCEPT for the domain.

① $R(x) = \dfrac{3x}{x+4}$

② $R(x) = \dfrac{3x+5}{2x-6}$

③ $R(x) = \dfrac{2x^2 - 5x - 3}{x^3 - 2x^2 - 3x}$

④ $R(x) = \dfrac{2x^2 - 5x - 12}{12x - 3x^2}$

Blank : ① long division

⑤ $R(x)=\dfrac{x^4-16}{x^2-2x}$

⑥ $R(x)=\dfrac{x-3}{3x^2+1}$

⑦ $R(x)=\dfrac{6x^2+7x-5}{3x^4+5}$

⑧ $R(x)=\dfrac{x^4-2x^2+1}{x^2-x}$

⑨ $R(x)=\dfrac{x^3-4x^2+4}{x^2-2x+1}$

⑩ $R(x)=\dfrac{x^3}{x^2-5x-14}$

⑪ $R(x)=\dfrac{x^3-8}{x^2-5x+6}$

2) Finding intercepts

Remember, make sure the rational expression is in ①________ terms!(Reduce!)

x-intercepts : we let ②____________.

y-intercepts: we let ③____________.

EXAMPLE 3. Find the x-intercept and y-intercept

① $R(x)=\dfrac{3x+5}{x-6}$

② $R(x)=\dfrac{x^2-4x+4}{x^2-2x+1}$

③ $R(x)=\dfrac{x^3}{x^2-5x-14}$

④ $R(x)=\dfrac{x^3-8}{x^2-5x+6}$

⑤ $R(x)=\dfrac{x^4-16}{x^2-2x}$

⑥ $R(x)=\dfrac{2x^2-5x-3}{x^3-2x^2-3x}$

Blank : ① lowest ② y = 0 ③ x = 0

3) Behavior near vertical asymptotes

$y = \dfrac{x(2x-1)(x+4)}{x(x-1)(x+2)}$ has a vertical asymptotes at ①__________,_____________.

We need to know whether $y \to \infty$ or $y \to -\infty$ on each side of each vertical asymptote. To determine the sign of y near the vertical asymptotes, we use test values.

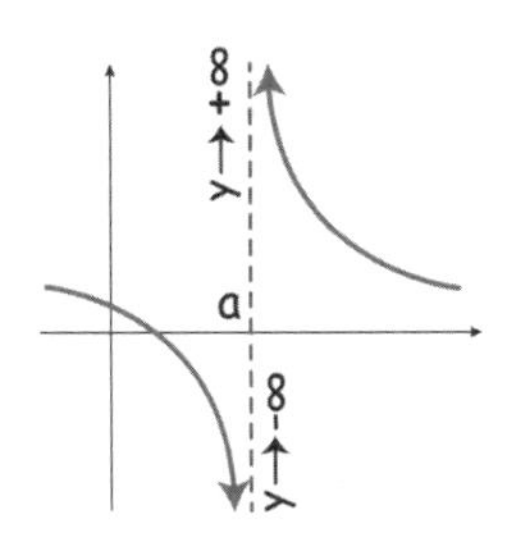

ex) around vertical asymptote x = 1

for right side of x = 1, let's use ②______(a test value close to the right side of 1)

$$y = \frac{(2(1.1)-1)((1.1)+4)}{((1.1)-1)((1.1)+2)}$$ whose sign is ③ $\dfrac{\square\,\square}{\square\,\square}$ = ④ $\square$

for left side of x = 1, let's use ⑤______(a test value close to the left side of 1)

$$y = \frac{(2(0.9)-1)((0.9)+4)}{((0.9)-1)((0.9)+2)}$$ whose sign is ⑥ $\dfrac{\square\,\square}{\square\,\square}$ = ⑦ $\square$

ex) around vertical asymptote x = -2

So we can conclude that;

	x=-2		x=1	
the sign of $\dfrac{(2x-1)(x+4)}{(x-1)(x+2)}$	$\dfrac{(-)(+)}{(-)(-)}$ $-\infty$	$\dfrac{(-)(+)}{(-)(+)}$ $+\infty$	$\dfrac{(+)(+)}{(-)(+)}$ $-\infty$	$\dfrac{(+)(+)}{(+)(+)}$ $+\infty$

Blank : ① x = 1 , x = -2 ② 1.1 ③ $\dfrac{++}{++}$ ④ $+\infty$ ⑤ 0.9 ⑥ $\dfrac{++}{-+}$ ⑦ $-\infty$

4) Finding Hole

$y = \dfrac{x(2x-1)(x+4)}{x(x-1)(x+2)}$ has a hole at ①__________.

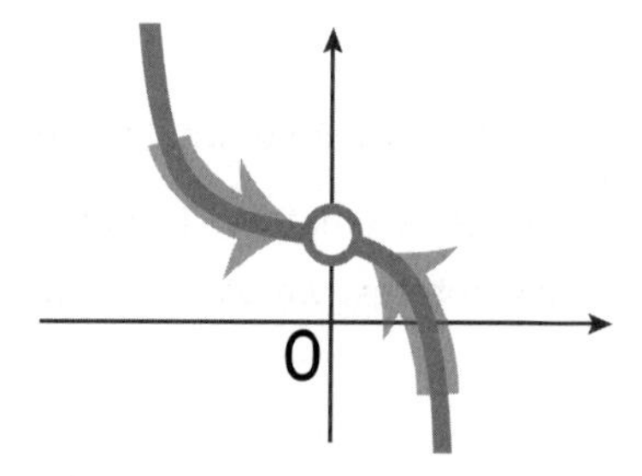

To find the exact location of the hole at x = 0,

we need to see where the graph f(x) ②____________________

when x approaches to x = 0.

So we need to set up ③_____________________________.

$$\lim_{x\to 0}\frac{x(2x-1)(x+4)}{x(x-1)(x+2)} = \lim_{x\to 0}\frac{\boxed{④\quad}}{(x-1)(x+2)} = \boxed{⑤\quad}$$

This means when x approaches at 0, f(x) approaches ⑥_______.

So the hole is at ⑦________________.

Blank : ① x = 0 ② approaches ③ $\lim_{x\to 0}\frac{x(2x-1)(x+4)}{x(x-1)(x+2)}$ ④ (2x − 1)(x + 4) ⑤ 2 ⑥ 2 ⑦ (0, 2)

5. Sketch the Graph (follow the guideline!)

BEFORE reduce	Find the domain	$y = \dfrac{2x^3+7x^2-4x}{x^3+x^2-2x} = \dfrac{x(2x-1)(x+4)}{x(x-1)(x+2)}$ ($\neq 0$, $\neq 0$, $\neq 0$) $x \neq 0, 1, -2$
AFTER reduce	Factor and Reduce	$y = \dfrac{2x^3+7x^2-4x}{x^3+x^2-2x} = \dfrac{\cancel{x}(2x-1)(x+4)}{\cancel{x}(x-1)(x+2)} = \dfrac{(2x-1)(x+4)}{(x-1)(x+2)}$
	Vertical asymptote	$y = \dfrac{(2x-1)(x+4)}{(x-1)(x+2)}$ ($=0$, $=0$) $x=1$, $x=-2$
	Horizontal asymptote	Degree! $y = \dfrac{2x^2+7x-4}{x^2+x-2}$ $y=2$
	y-intercept	$y = \dfrac{(2x-1)(x+4)}{(x-1)(x+2)}$ ($x \to 0$) $\rightarrow \dfrac{-4}{-2}$ $(0, 2)$
	x-intercept	$y = \dfrac{(2x-1)(x+4)}{(x-1)(x+2)}$ ($=0$, $=0$) $\left(\frac{1}{2}, 0\right)$ $(-4, 0)$
	Behavior near vertical asymptote	(see sign table below)
	holes	Where $x = 0$
	Graph	

the sign of $\dfrac{(2x-1)(x+4)}{(x-1)(x+2)}$	$x=-2$		$x=1$	
	$\dfrac{(-)(+)}{(-)(-)}$ $-\infty$	$\dfrac{(-)(+)}{(-)(+)}$ $+\infty$	$\dfrac{(+)(+)}{(-)(+)}$ $-\infty$	$\dfrac{(+)(+)}{(+)(+)}$ $+\infty$

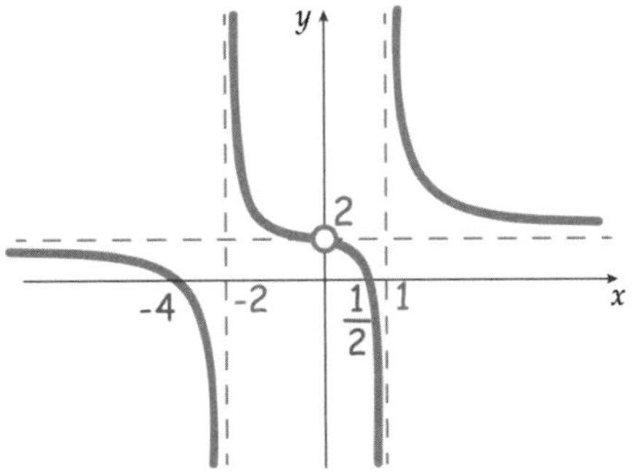

EXAMPLE 4. Analyze the given rational function. Graph the function.

① $f(x)=\dfrac{2x}{(x-3)(x+4)}$

Domain

Vertical asymptote

Horizontal asymptote

x-intercept

y-intercept

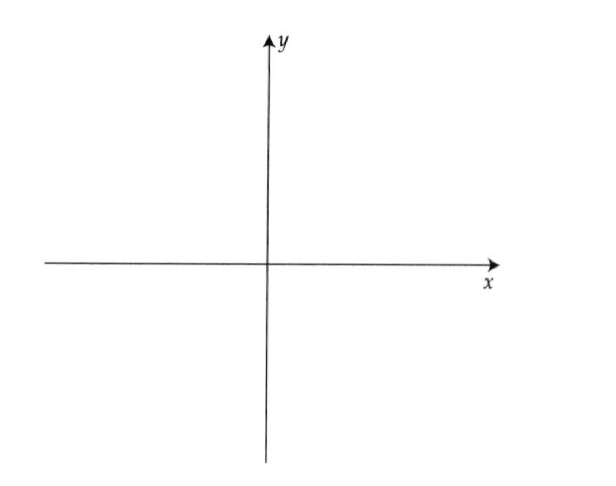

② $f(x)=\dfrac{x-3}{x^2-6x-16}$

Domain

Vertical asymptote

Horizontal asymptote

x-intercept

y-intercept

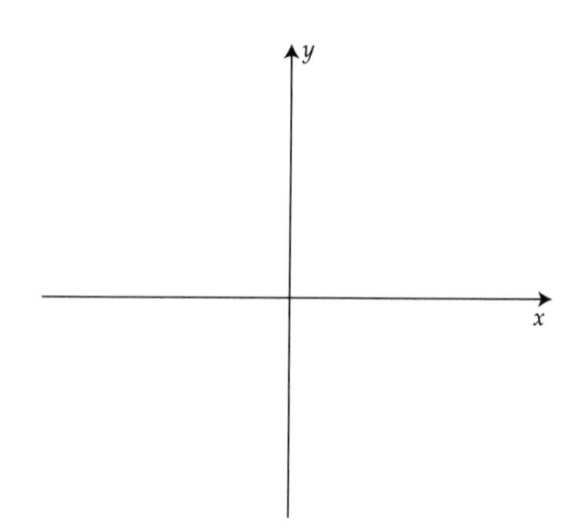

③ $f(x)=\dfrac{4}{x^2-4}$

Domain

Vertical asymptote

Horizontal asymptote

x-intercept

y-intercept

④ $f(x)=\dfrac{x-3}{(x-2)(x+3)}$

Domain

Vertical asymptote

Horizontal asymptote

x-intercept

y-intercept

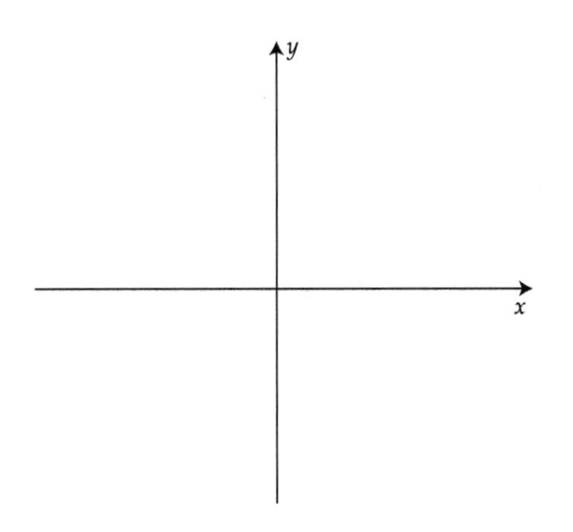

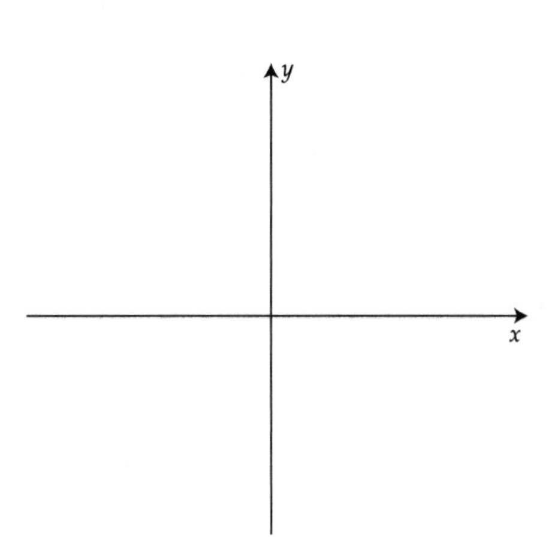

⑤ $f(x)=\dfrac{x^2+x-12}{x^2-x-6}$

Domain

Vertical asymptote

Horizontal asymptote

x-intercept

y-intercept

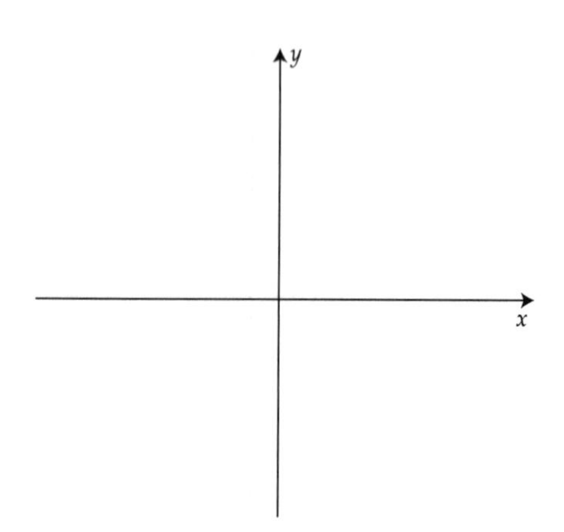

⑥ $f(x)=\dfrac{x^2+3x-4}{x^2+2x-8}$

Domain

Vertical asymptote

Horizontal asymptote

x-intercept

y-intercept

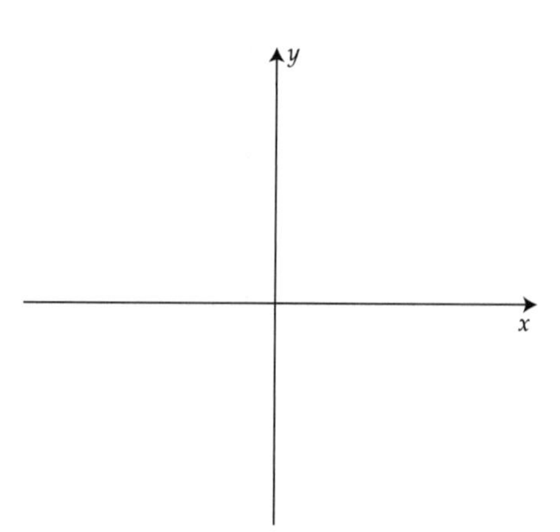

⑦ $f(x)=\dfrac{x^3+3x^2-18x}{x^2-36}$

Domain

Vertical asymptote

Horizontal asymptote

x-intercept

y-intercept

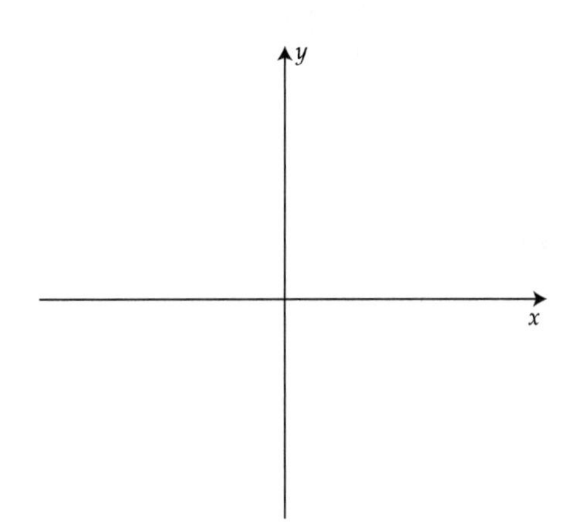

⑧ $f(x)=\dfrac{x^2-16}{x+2}$

Domain

Vertical asymptote

Horizontal asymptote

x-intercept

y-intercept

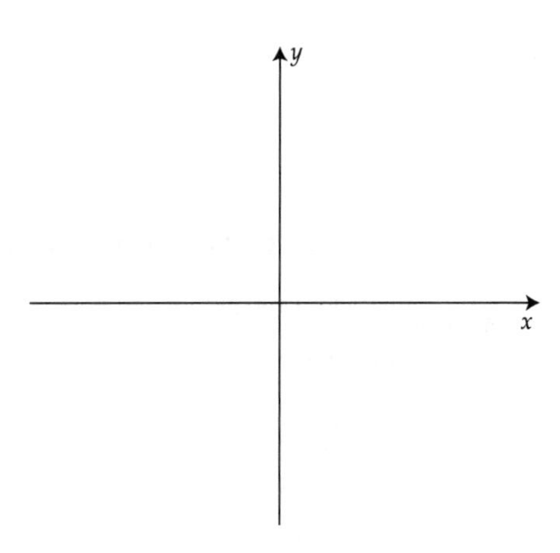

AP Style Problem

1. Which of the following could be the graph of $f(x)=\dfrac{x^2-6x-16}{(x-4)(x-8)}$?

A)

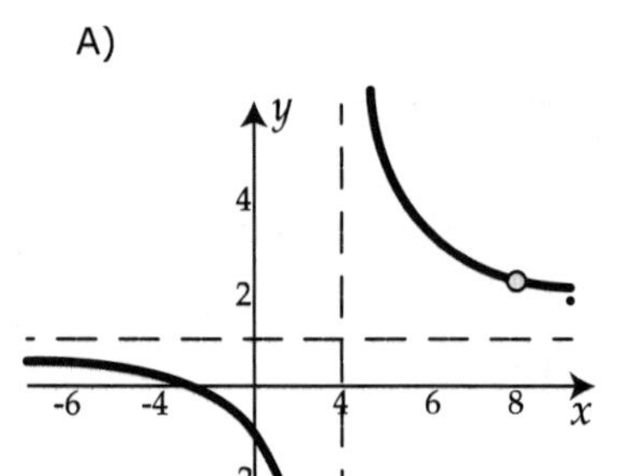

B)

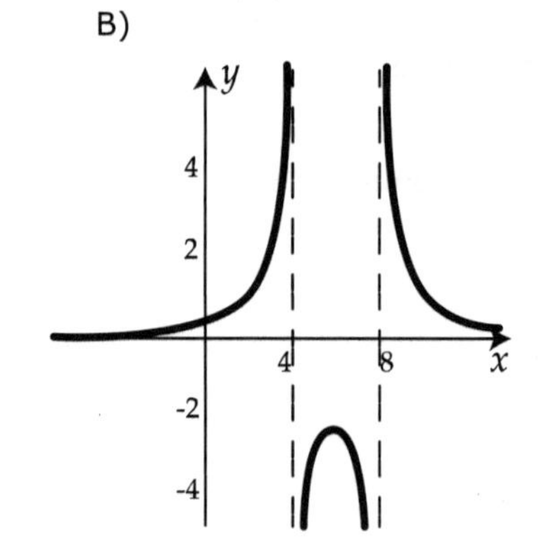

C)

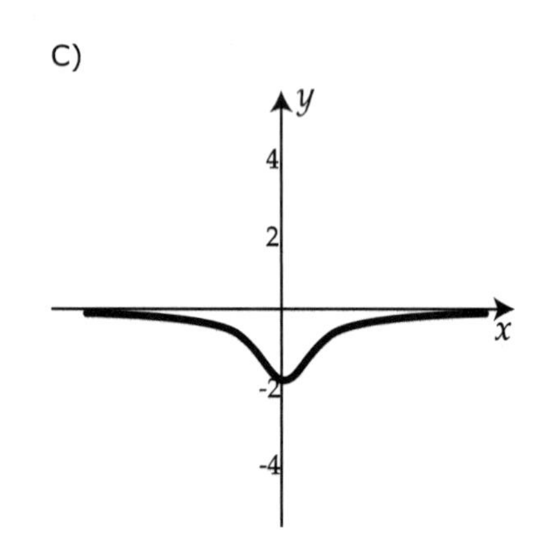

D)

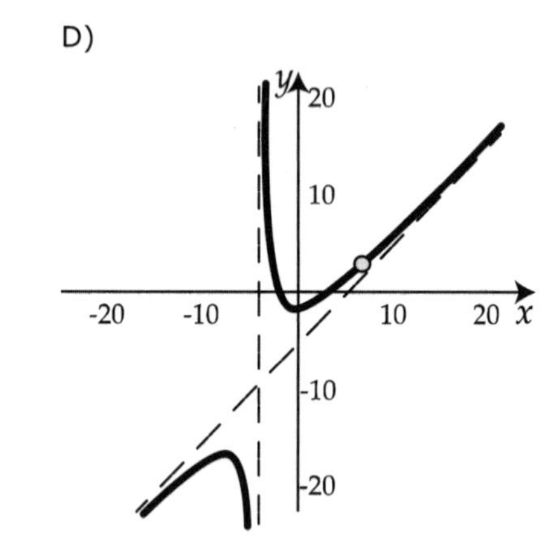

2. Which of the following function has exactly 3 asymptotes?

A) $y=\dfrac{(x-4)^2(x+1)^2}{(x+1)(x-4)^3}$

B) $y=\dfrac{(x-4)^3(x+1)^2}{(x+1)(x-4)}$

C) $y=\dfrac{(x-4)^2(x+1)}{(x+1)^2(x-4)^3}$

D) $y=\dfrac{(x-4)(x+1)^2}{(x+1)(x-4)^2}$

3.

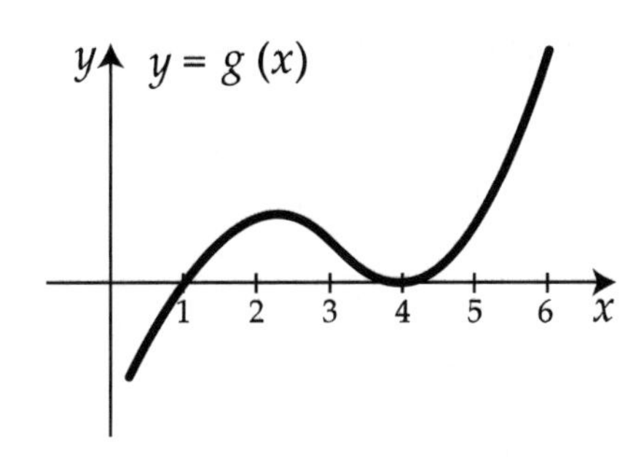

The function f is given by $f(x)=(x-1)(x-4)(x-7)$ and the graph of the cubic function g is given above.

1) Which of the following statement is true about $h(x)=\dfrac{g(x)}{f(x)}$?

A) h has holes at x = 1 and x = 4, h has no vertical asymptotes.

B) h has holes at x = 1, h has a vertical asymptote at x = 4 and x = 7.

C) h has holes at x = 4, h has a vertical asymptote at x = 7.

D) h has holes at x = 1 and x = 4, h has a vertical asymptote at x = 7.

2) What is the domain of $h(x)=\dfrac{g(x)}{f(x)}$?

A) All real numbers except 1

B) All real numbers except 7

C) All real numbers except 1 and 4

D) All real numbers except 1, 4, and 7

3) What is $\lim\limits_{x \to 7^-} h(x)$?

A) 0

B) 3

C) $-\infty$

D) ∞

4. Which of the following table could be rational function $f(x)=\dfrac{x^2}{x(x-1)^2}$?

A)

x	0	0.5	0.9	1	1.1	1.5
$f(x)$	und	4	900	und	1100	12

B)

x	0	0.5	0.9	1	1.1	1.5
$f(x)$	0	4	900	und	1100	12

C)

x	0	0.5	0.9	1	1.1	1.5
$f(x)$	0	-4	-900	und	1100	12

D)

x	0	0.5	0.9	1	1.1	1.5
$f(x)$	und	-4	-900	und	1100	12

5. If the vertical asymptote of the rational function $R(x)=\dfrac{x^4-5x^3+2x^2-bx+3}{(x-1)(x+1)}$ is $x=-1$,

1) What is the value of b?

A) 1

B) -1

C) 11

D) -11

2)* Which of the following is true about the hole of $R(x)$?

A) $R(x)$ has no holes

B) $R(x)$ has a hole at (1, -4)

C) $R(x)$ has a hole at (1, -3)

D) $R(x)$ has a hole at (-1, 2)

Mia's AP Precalculus

2.6 Polynomial and Rational Inequalities

1. Polynomial Inequalities Graphically

When we solve inequalities we try to find ①________or ②__________(s) of what the inequality sign says.

※ **Solving Inequalities**

$f(x) = g(x)$: When is the graph of **f(x)** ③______________ **the g(x)**?

$f(x) > g(x)$: When is the graph of **f(x)** ④______________ **the g(x)**?

$f(x) < g(x)$: When is the graph of **f(x)** ⑤______________ **the g(x)**?

$f(x) > 0$: When is the graph of **f(x)** ⑥______________ **the** ⑦ __________?

For example, using the graph of f(x) and g(x), we can find out

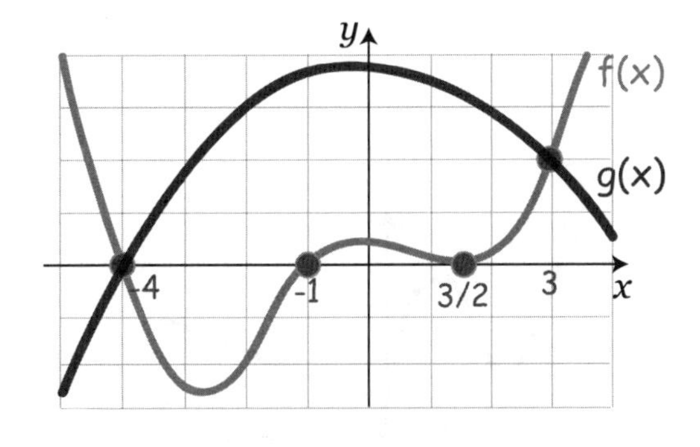

1) $f(x)>0$

2) $f(x)\geq0$

3) $f(x)\leq0$

4) $f(x) = g(x)$

5) $f(x) \leq g(x)$

Blank : ① intervals ② values ③ intersect ④ above ⑤ below ⑥ above ⑦ x axis

EXAMPLE 1. Solve the given inequality using the graph.

①

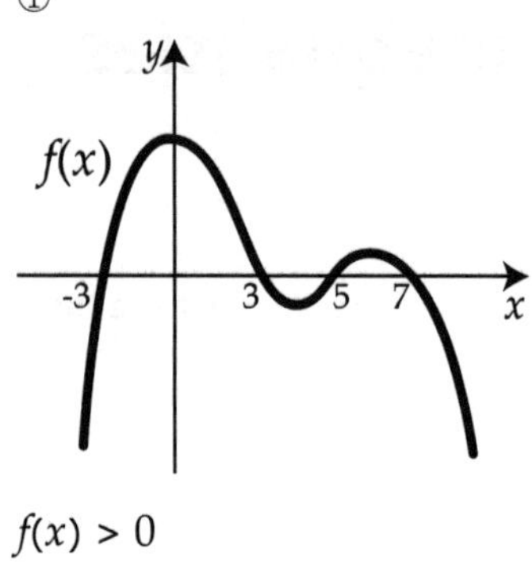

$f(x) > 0$

②

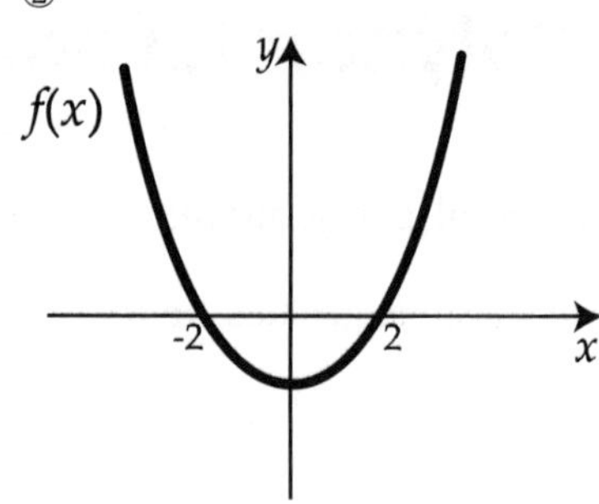

$f(x) \geq 0$

③

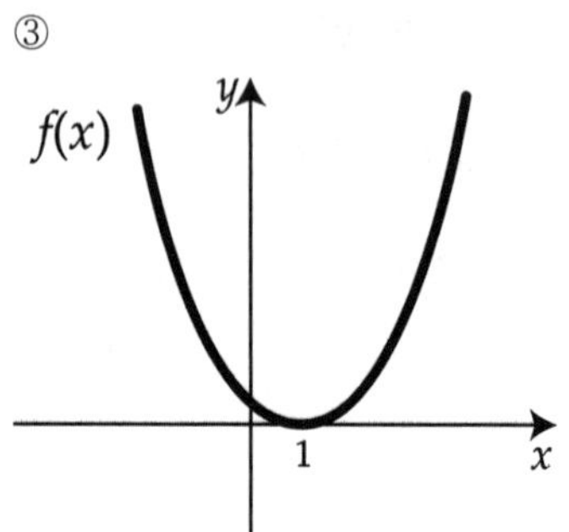

$f(x) \leq 0$

④

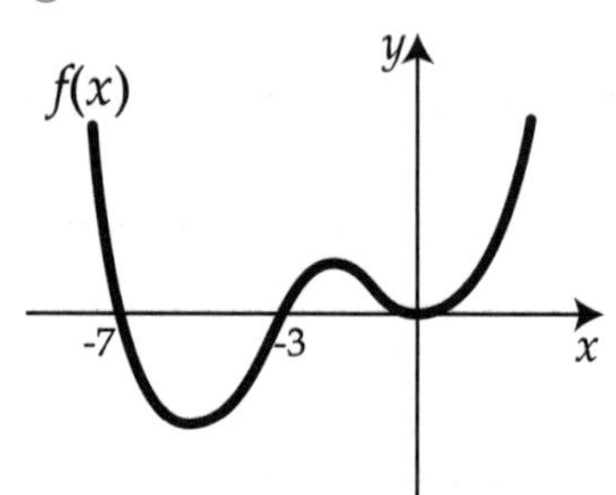

$f(x) \leq 0$

⑤

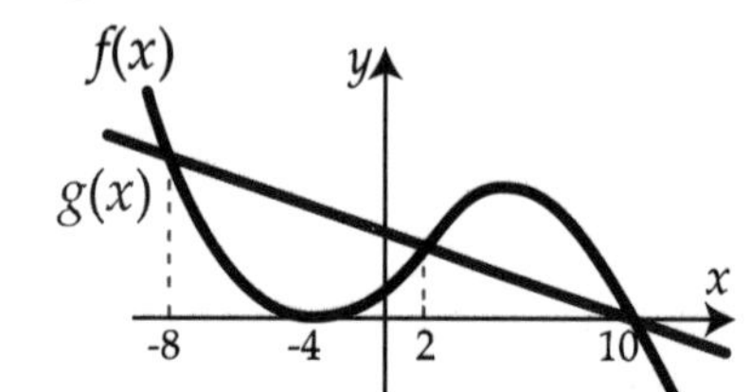

(i) $f(x) > 0$

(ii) $f(x) > g(x)$

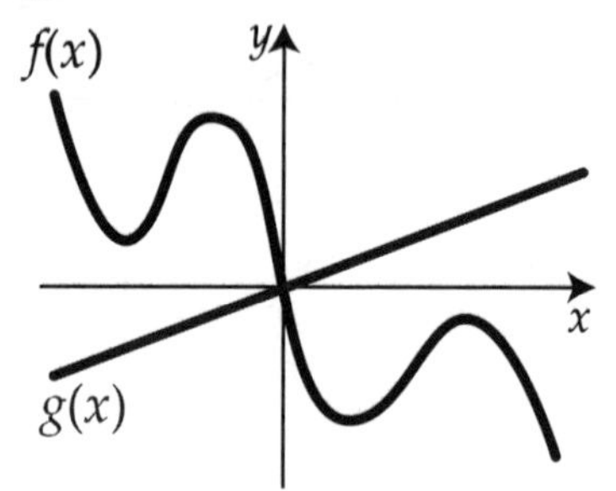

(i) $f(x) < 0$

(ii) $f(x) \geq g(x)$

⑦

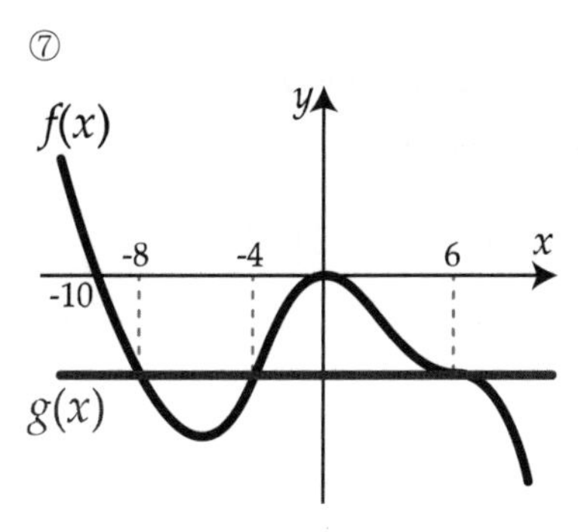

(i) $f(x) \geq 0$

(ii) $f(x) \leq g(x)$

⑧

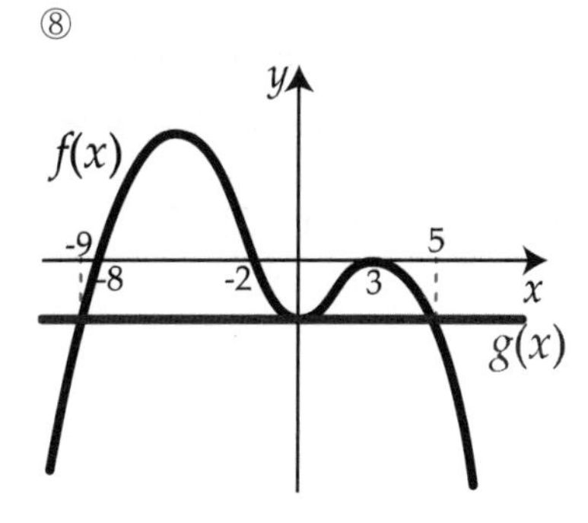

(i) $f(x) \leq 0$

(ii) $f(x) > g(x)$

2. Polynomial Inequalities Numerically

We want to solve $(x+4)(x+1)(2x-3)^2 \geq 0$

① Factor the polynomial.

② Find the **real zeros** (where f(x) = 0) and build up the intervals.

③ Pick a **test points** from each intervals

④ Determine if f is **positive**(f > 0)or **negative**(f < 0) on that interval by plugging in the test points.

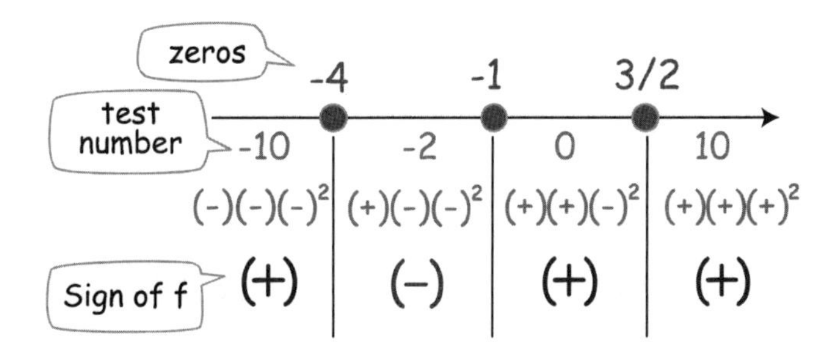

Determine whether the zeros include or not in my answer.

⑤ Read the inequality and find the appropriate interval(s).

Solution will be $(-\infty, -4] \cup [-1, \infty)$.

EXAMPLE 2. Solve the inequalities.

If the multiplicity is **Odd**, the sign will **Change** around zero.
If the multiplicity is **Even**, the sign will **stay the same** around zero.

① $(x+1)(x-3)(x-5)<0$

② $x(2x-1)(x+2)>0$

③ $x(x-1)(x+2)^2>0$

④ $x^2(3-x)(x-2)^3<0$

⑤ $x(1-x)(x+1)^2(x-3)^4\geq 0$

⑥ $x^3(x-2)^2(x+2)^3(x-3)^4\geq 0$

⑦ $x^3(x+1)^2(x^2+1)\geq 0$

⑧ $(x-1)(x+1)^2(x^2+3)\leq 0$

⑨ $x^4 - 13x^2 + 36 > 0$

⑩ $x^3 - 7x^2 - 18x > 0$

⑪ $x^4 < 36x^2$

⑫ $x^3 \geq 5x^2$

⑬ $2x^3 + 17x^2 - 2x - 80 > 0$

⑭ $x^3 + 5x^2 - 4x - 20 \geq 0$

3. Rational Function Inequalities Graphically

Same thing goes with rational functions.

For example, using the graph of $P(x) = \dfrac{x}{(x-4)(x+4)}$

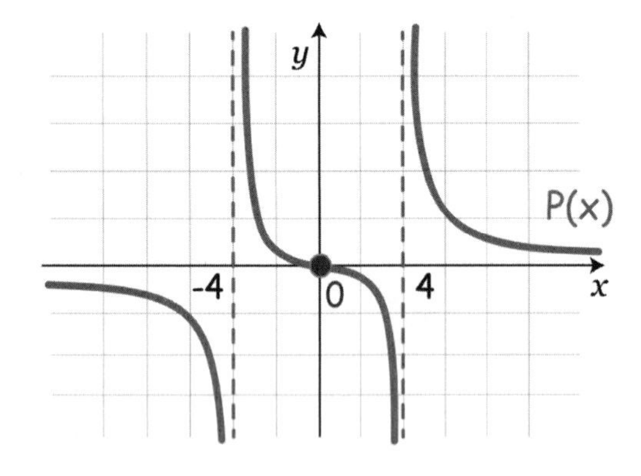

we can find out

1) P(x)>0

2) P(x)≥0

3) P(x)≤0

EXAMPLE 3. Solve the given inequality using the graph.

①

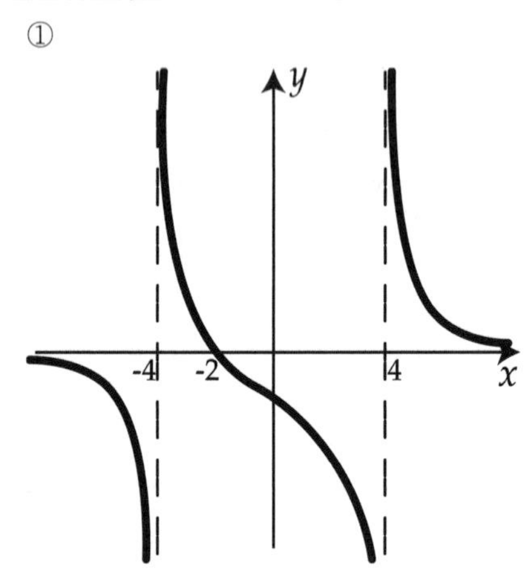

$f(x) > 0$

②

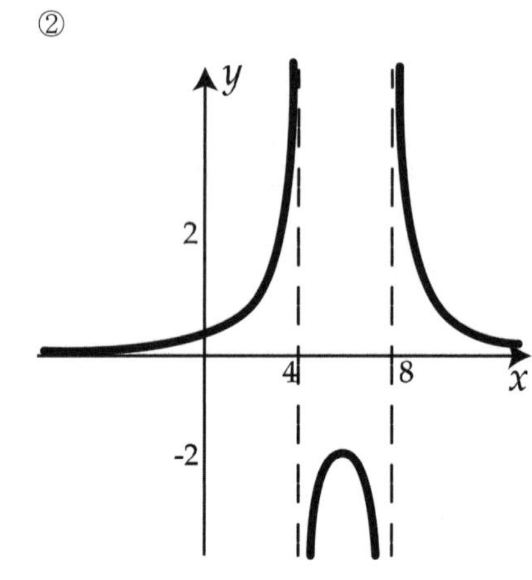

$f(x) \ge 0$

③

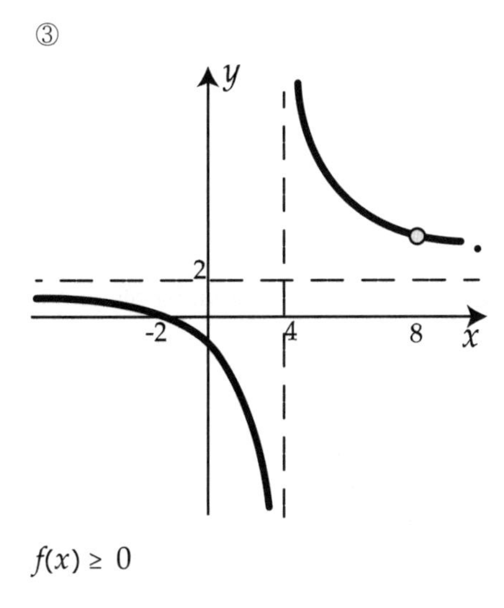

$f(x) \geq 0$

④

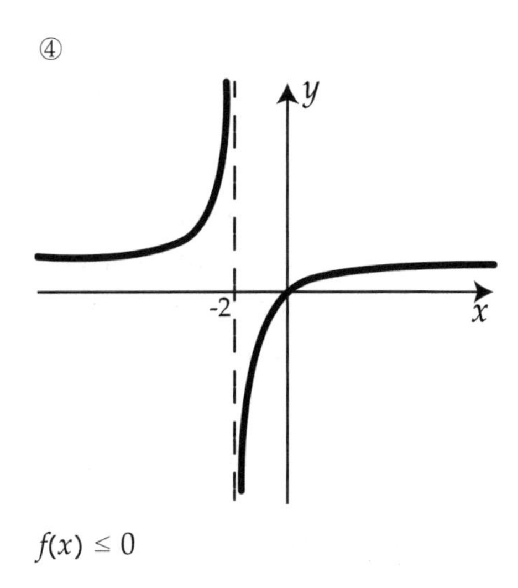

$f(x) \leq 0$

4. Rational Function Inequalities Numerically

※ Rational Function Inequalities Numerically

For rational inequality $\dfrac{f(x)}{g(x)} \geq 0$, then we multiply ① ________ on both sides.

(NOT $g(x)$, since $g(x)$ could be positive or negative).

multiply $g(x)^2$

$$\frac{f(x)}{g(x)} \geq 0 \rightarrow f(x)\,g(x) \geq 0,\ g(x) \neq 0$$

Do the same thing for $\dfrac{f(x)}{g(x)} \leq 0$, $\dfrac{f(x)}{g(x)} > 0$, $\dfrac{f(x)}{g(x)} < 0$.

We want to solve $\dfrac{x}{(x-4)(x+4)} \geq 0$

① Change it into ②____.

② Find the **real zeros** and build up the intervals.

③ Pick a **test points** from each intervals

④ Determine if f is **positive**(f > 0)or **negative**(f < 0) on that interval by plugging in the test points.

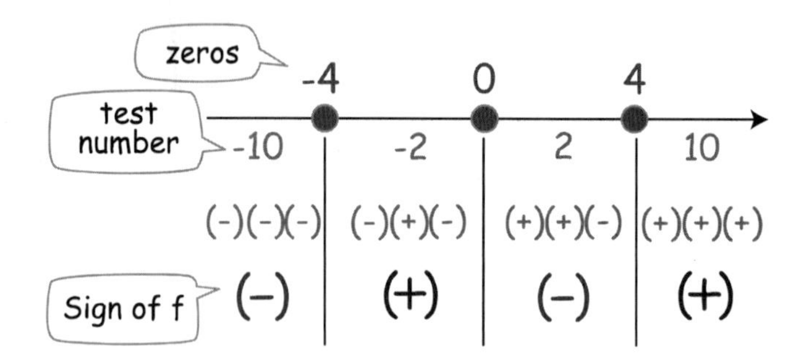

⑤ Read the inequality and find the appropriate interval(s).
Do not forget to **EXCLUDE the undefined x** in your answer.

Solution will be $(-4,0]\cup(4,\infty)$

EXAMPLE 4. Solve the inequality.

① $\dfrac{x-7}{x+1} \geq 0$

② $\dfrac{x-4}{x+5} > 0$

Blank : ① $g^2(x)$ ② $x(x-4)(x+4) \geq 0,\ x \neq 4, -4$

③ $\dfrac{x-9}{x+6}>1$

④ $\dfrac{x+28}{x+9}<2$

⑤ $\dfrac{(x+1)^2(x-3)^3}{(x+2)(x+3)}\ge 0$

⑥ $\dfrac{(x+1)^2(x-2)^4}{(1-x)^3}>0$

⑦ $\dfrac{(x-1)(3-x)}{(x-2)^2}<0$

⑧ $\dfrac{x^2(x-12)(x+2)^2}{x-7}\le 0$

⑨ $\dfrac{x^2-4}{x}<0$

⑩ $\dfrac{x^2-4}{x^2}\geq 0$

⑪ $\dfrac{3x}{5-x}\leq x$

⑫ $\dfrac{8x}{7-x}<x$

⑬ $\dfrac{12}{x-5}\leq\dfrac{10}{x-1}$

AP Style Problem

1. A rational function h is defined by $h(x)=\dfrac{x(x-2)^3(x-7)}{(x-1)}$. What are all intervals on which $h(x)\geq 0$?

A) $(-\infty,0]\cup[1,2]\cup[7,\infty)$

B) $(-\infty,0]\cup(1,2]\cup[7,\infty)$

C) $[0,1]\cup[2,7]$

D) $[0,1)\cup[2,7]$

2. Find the domain of $g(x)=\sqrt{(1-x)(x-5)^2(x+7)}$.

A) $[1,5]$

B) $(-\infty,7]\cup[1,5]$

C) $[-7,1]\cup x=5$

D) $[-7,1]$

2.7 Binomial Expansion

1. Binomial Theorem

$$(a + b)^n$$

The ***binomial theorem*** shows how to calculate a *power of a binomial*.

We already know....

$$(a + b)^0 = 1$$

$$(a + b)^1 = a + b$$

$$(a + b)^2 = a^2 + 2ab + b^2$$

$$(a + b)^3 = a^3 + 3a^2b + 3ab^2 + b^3$$

Let's talk about the patterns here.

1) Patterns of Exponents of a and b

Notice that the

$(a + b)^3 = a^3 + 3a^2b + 3ab^2 + b^3$ exponents of a start at ①__ and go *down*

$= a^3 + 3a^2b + 3ab^2 + b^3$ exponents of b start at ②__ and go *up*

Blank : ① 3 ② 0

2) Patterns of Coefficient

$$1$$
$$a + b$$
$$a^2 + 2ab + b^2$$
$$a^3 + 3a^2b + 3ab^2 + b^3$$

$$1$$
$$1a + 1b$$
$$1a^2 + 2ab + 1b^2$$
$$1a^3 + 3a^2b + 3ab^2 + 1b^3$$

Pascal's Triangle

If we look at just the coefficients, you'll notice that it makes a ①______________ ____________________!

For $(a + b)^4$, exponents of a starts at ②______ and go down.

exponents of b starts at ③______ and go up.

The coefficient goes ④____________________.

$(a + b)^4 =$ ⑤____

EXAMPLE 1. Use Pascal's triangle to Expand.

① $(x+2)^4$

② $\left(x-\frac{2}{x}\right)^4$

Blank : ① pascal's triangle ② 4 ③ 0 ④ 1, 4, 6, 4, 1 ⑤ $a^4 + 4a^3b + 6a^2b^2 + 4ab^3 + b^4$

③ $(2x-1)^5$

④ $(3+x)^5$

⑤ $\left(x^2+\dfrac{1}{x}\right)^5$

⑥ $(4x-3y)^3$

AP Style Problem

1. Which of the following is equivalent to $\left(x^2-3\right)^4$?

A) $x^4-12x^3+54x^2-108x+81$

B) $x^4-8x^3+24x^2-32x+16$

C) $x^8-12x^6+54x^4-108x^2+81$

D) $x^8-8x^6+24x^4-32x^2+16$

Free Response Questions (from ch1-ch2)

1. Analyzing Function

(Calculator) The function $f(x) = -2x^3 + 5.2x^2 + 3.5x + 6.1$ is given.

(a) Find the real zeros of the function f.

(b) Find the absolute maximum and minimum in the interval $1.5 \le x \le 2.7$.

(c) Find the number of imaginary zeros for $f(x)$. Give the reason for your answer.

(d) The graph of $f(x)$ is translated to the $g(x)$ by vertical translation by a. Find the value(s) a where $g(x)=0$ will have two solutions.

2. Rational Function

(Non-Calculator) The function f and g are given as

$$f(x) = x^3 + 2x^2 - 5x - 6 \text{ and } g(x) = 3(x-3)(x+2)(x+1).$$

A rational function h is given as $h(x) = \dfrac{f(x)}{g(x)}$.

(a) Factor $f(x)$ completely.

(b) Find the domain and the vertical asymptote of the rational function $h(x)$.

(c) Find the horizontal or slant asymptote of the rational function $h(x)$. Give the reason for your answer.

(d) The hole of the rational function $h(x)$ is at (a, b). Find the value of a and b.

Part 3

Exponential and Logarithmic Functions

3.1 Exponential Function

1. Properties of Exponents

If a and b are positive numbers,

$x^a \cdot x^b = x^{①\square}$ Multiplication Law

$\dfrac{x^a}{x^b} = x^{②\square}$ Division Law

$(x^a)^b = x^{③\square}$ Power Law

$(xy)^a = x^a y^a$ Power of a Product Law

$\left(\dfrac{x}{y}\right)^a = \dfrac{x^a}{y^a}$ Power of a Quotient Law

$x^0 = ④\square$ Zero Exponent ($x \neq 0$)

$x^{-a} = ⑤\square$, $\dfrac{1}{x^{-a}} = ⑥\square$ Negative Exponent ($x \neq 0$)

$\left(\dfrac{x}{y}\right)^{-a} = ⑦\square$

$x^{\frac{a}{b}}$ (root) $= ⑧\square$

Fractional Exponent

$x^{\frac{1}{2}} =$ ⑨_____ $x^{\frac{1}{3}} =$ ⑩_____

$(-1)^n \begin{cases} ⑪\square, & n \text{ is odd} \\ ⑫\square, & n \text{ is even} \end{cases}$

Blank : ① $a+b$ ② $a-b$ ③ ab ④ 1 ⑤ $\dfrac{1}{x^a}$ ⑥ x^a ⑦ $\dfrac{y^a}{x^a}$ ⑧ $\sqrt[b]{x^a}$ ⑨ $\sqrt{x}$ ⑩ $\sqrt[3]{x}$ ⑪ -1 ⑫ 1

EXAMPLE 1. Simplify.

① $\dfrac{3^{n-2} \cdot 9^{2-n}}{3^{2-n}}$

② $\dfrac{2^{n-3} \times 8^{n+1}}{2^{2n-1} \times 4^{2-n}}$

③ $\dfrac{x^{-1}+y^{-1}}{x^{-1}y^{-1}}$

④ $\dfrac{x^{-1}y^{-1}}{x^{-1}+y^{-1}}$

⑤ $\sqrt{x\sqrt{x}}$

⑥ $\sqrt{a\sqrt{a\sqrt{a}}}$

⑦ $\sqrt{\sqrt[3]{a} \times a^3}$

⑧ $\sqrt[3]{a\sqrt{a}}$

⑨ $\sqrt[4]{\dfrac{\sqrt[3]{x}}{\sqrt{x}}}$

⑩ $\sqrt{\dfrac{\sqrt[4]{x}}{\sqrt[3]{x}}}$

⑪ $(a^{\frac{1}{2}}+b^{\frac{1}{2}})(a^{\frac{1}{2}}-b^{\frac{1}{2}})$

⑫ $(a^{\frac{1}{3}}-b^{\frac{1}{3}})(a^{\frac{2}{3}}+a^{\frac{1}{3}}b^{\frac{1}{3}}+b^{\frac{2}{3}})$

EXAMPLE 2. Factor or simplify.

① $2^{n+3}+2^{n}$

② $2^{n+3}+8$

③ $2^{3+n}+2^{2+n}$

④ $\dfrac{2^{m+3}+2^{m}}{9}$

⑤ $36^{x}-11(6^{x})+18$

⑥ $4^{x}-(2^{x})-6$

⑦ $9^{x}+3^{x+1}-4$

⑧ $25^{x}-5^{x+1}-6$

⑨ $9^{x}-4^{x}$

⑩ $25-16^{x}$

⑪ $\dfrac{6^{n}}{3^{n}}$

⑫ $\dfrac{8^{n}}{2^{n}}$

2. Exponential Function

※ Exponential Function and Graphs

$$f(x) = b^x$$

Exponent (variable)

Base (constant)

The **exponential function with base b** is defined for all real numbers x where $b \neq 1$ and $b > 0$.

When $0 < b < 1$ and when $b > 1$

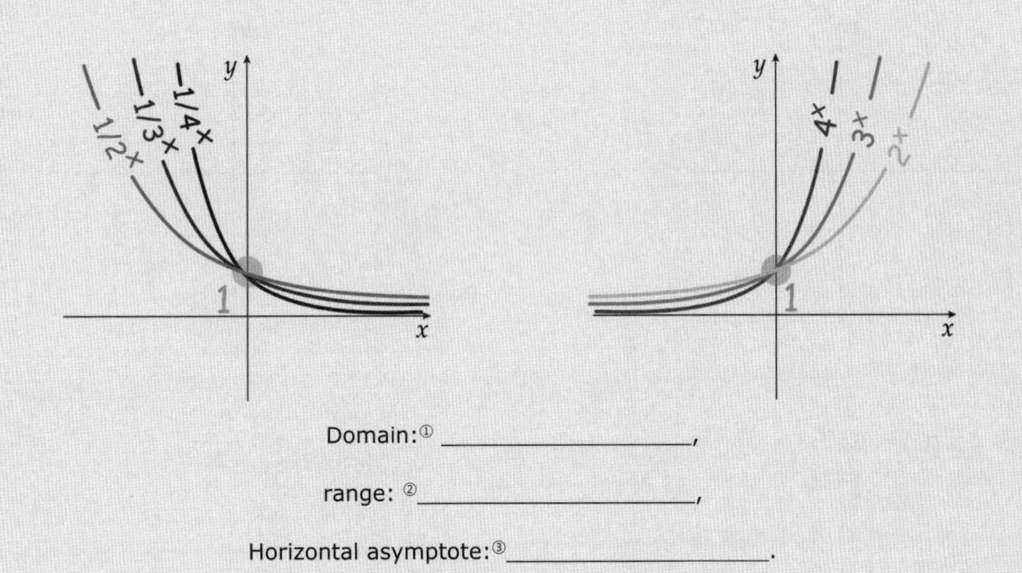

Domain: ① ________________,

range: ② ________________,

Horizontal asymptote: ③ ________________.

y intercept : ④ ____________

Blank : ① $(-\infty, \infty)$ ② $(0, \infty)$ ③ $y = 0$ ④ $(0, 1)$

EXAMPLE 3. Graph the function. State the domain, range, and asymptotes.

① $f(x) = 2^{x+2}$

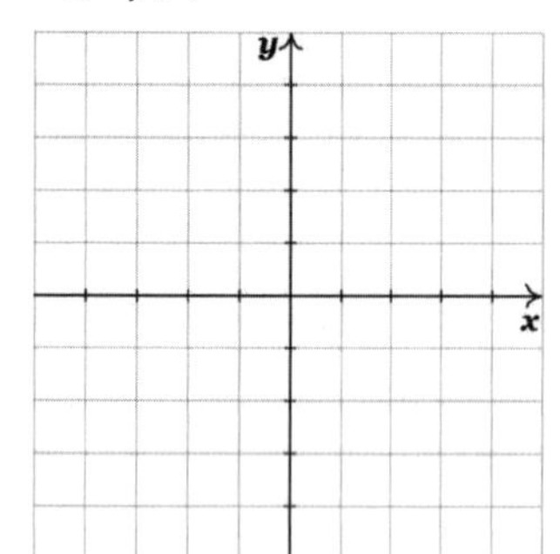

② $f(x) = (0.75)^x$

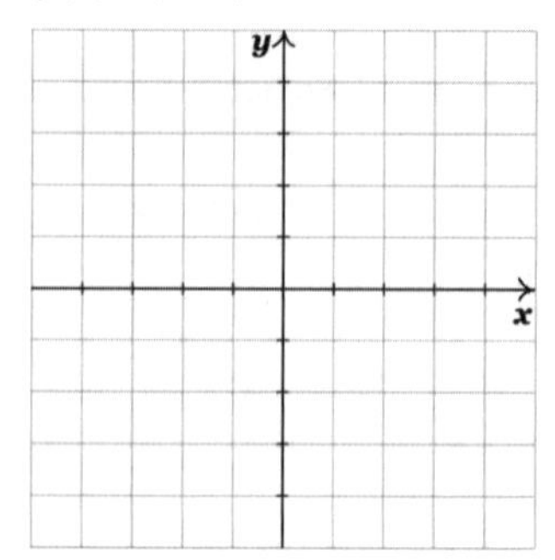

③ $f(x) = 4^{x-1} - 2$

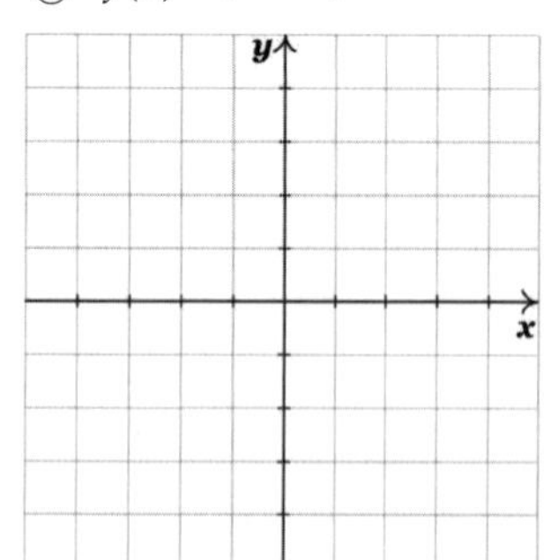

④ $f(x) = -2^{-x} + 1$

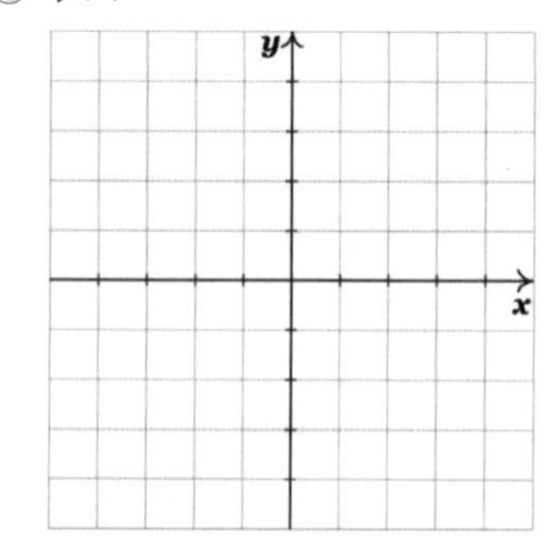

⑤ $f(x) = -0.5^x + 1$

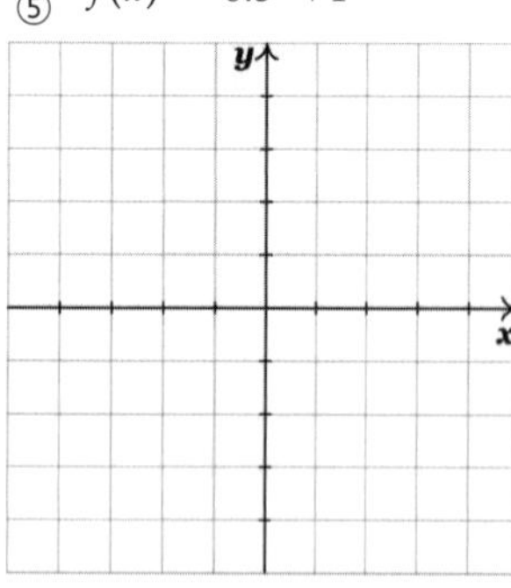

⑥ $f(x) = 2\left(\frac{1}{3}\right)^x - 2$

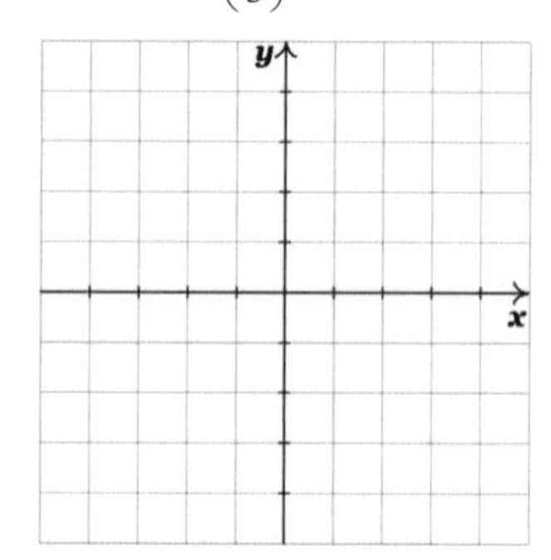

3. Euler Number

The number e is a famous (①rational/irrational) number, and is one of the most important numbers in mathematics.

$$e = 2.71828182845...$$

It is called ②____________ number (after Leonhard Euler).

e is found in many interesting areas (ex) compound interest, exponential growth...)
, so it is worth learning about.

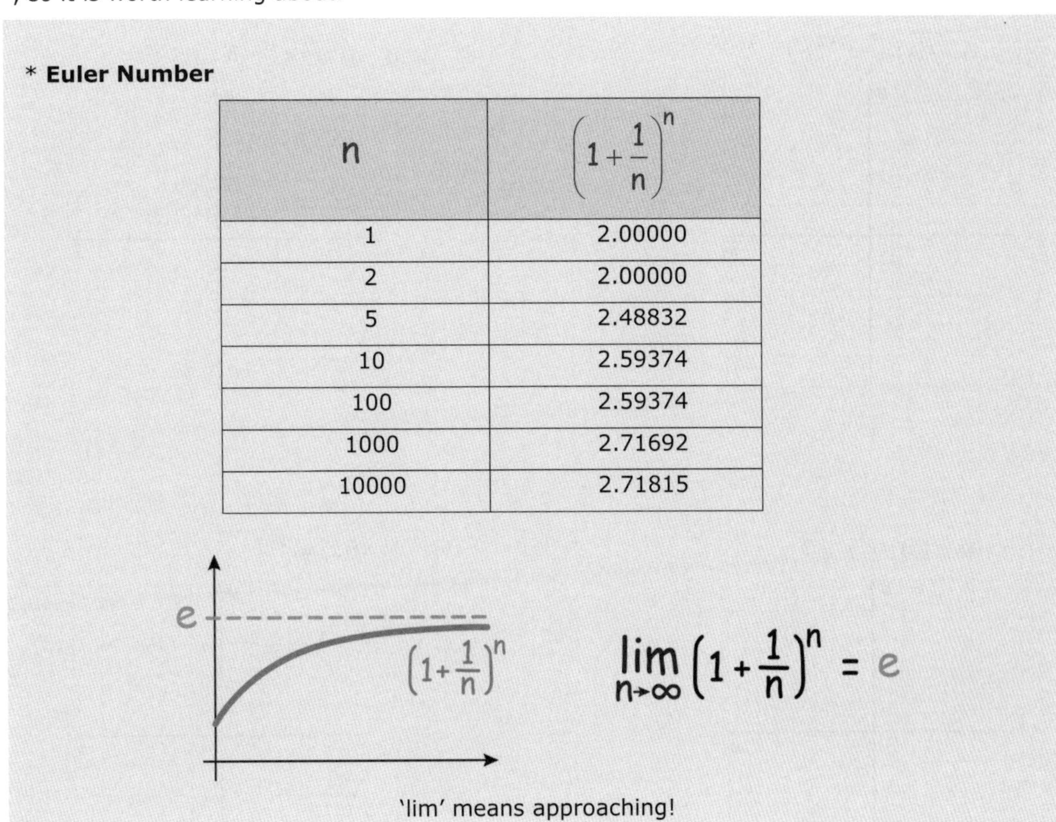

*** Euler Number**

n	$\left(1+\frac{1}{n}\right)^n$
1	2.00000
2	2.00000
5	2.48832
10	2.59374
100	2.59374
1000	2.71692
10000	2.71815

'lim' means approaching!

Blank : ① irrational ② Euler

EXAMPLE 4. Use transformation to graph the function. Determine the horizontal asymptote of the function.

① $f(x)=e^{-x}$

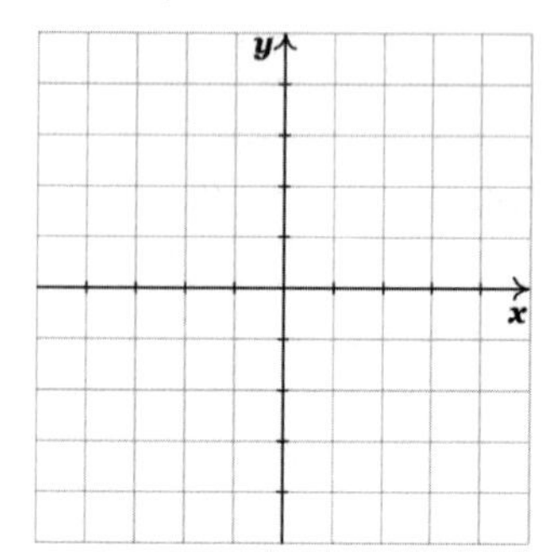

② $f(x)=1-e^{2+x}$

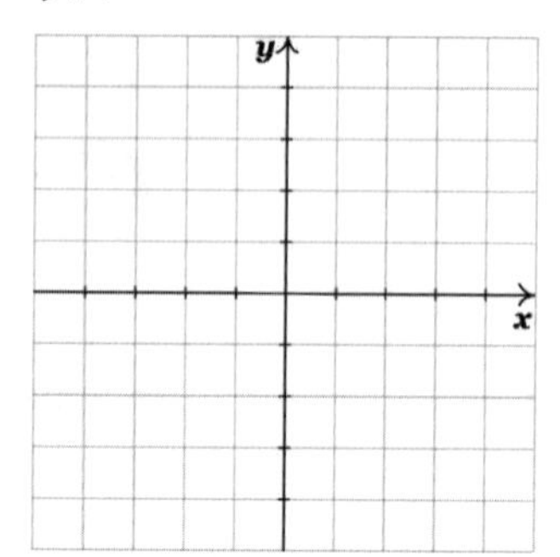

③ $f(x)=-e^{x+2}-1$

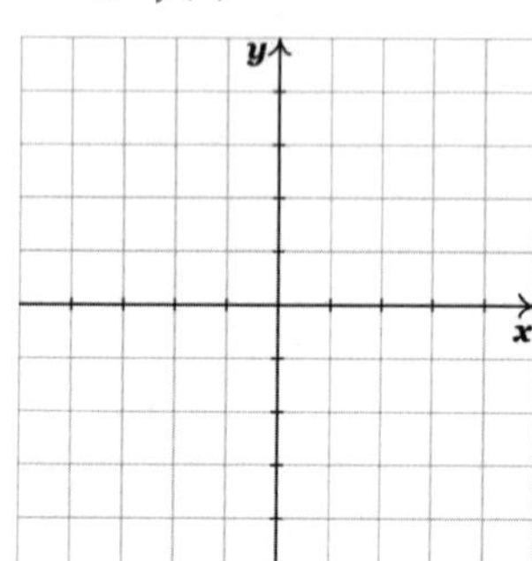

④ $f(x)=e^{-x}-3$

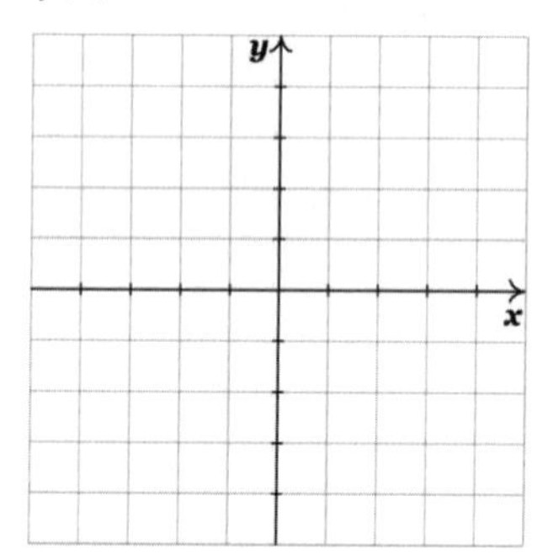

⑤ $f(x)=e^{1-x}-2$

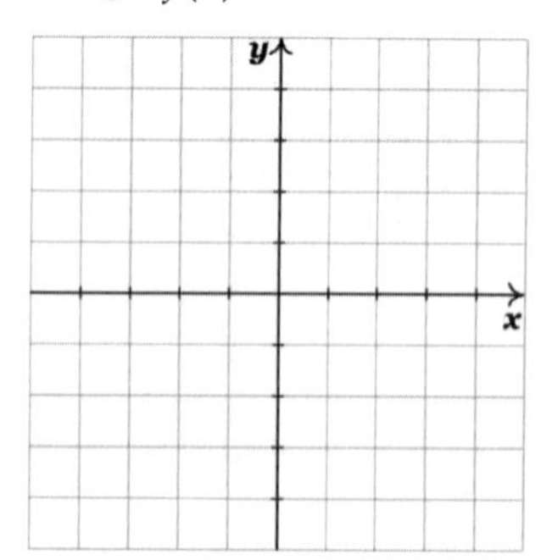

⑥ $f(x)=e^{2-x}$

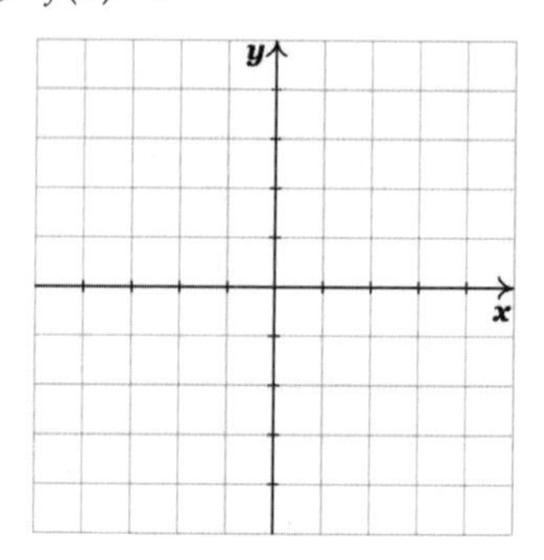

⑦ $f(x) = 2e^x - 3$

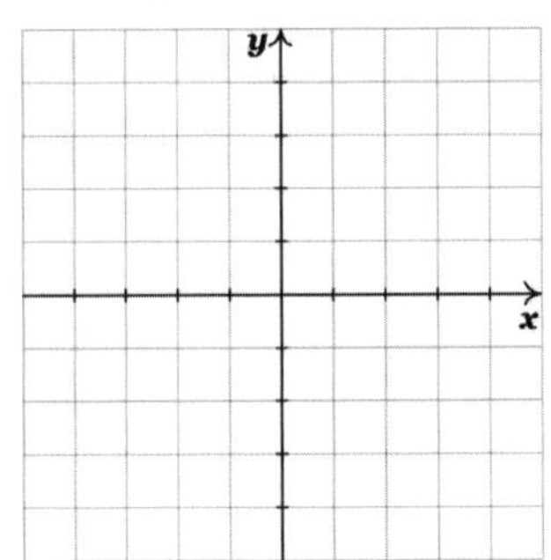

⑧ $f(x) = \frac{1}{2}e^{x-1}$

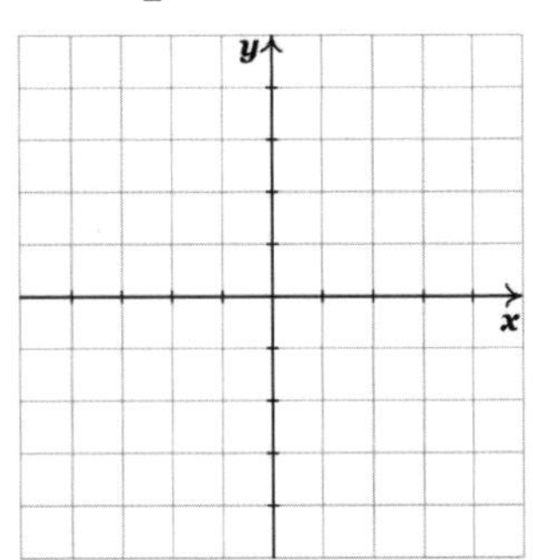

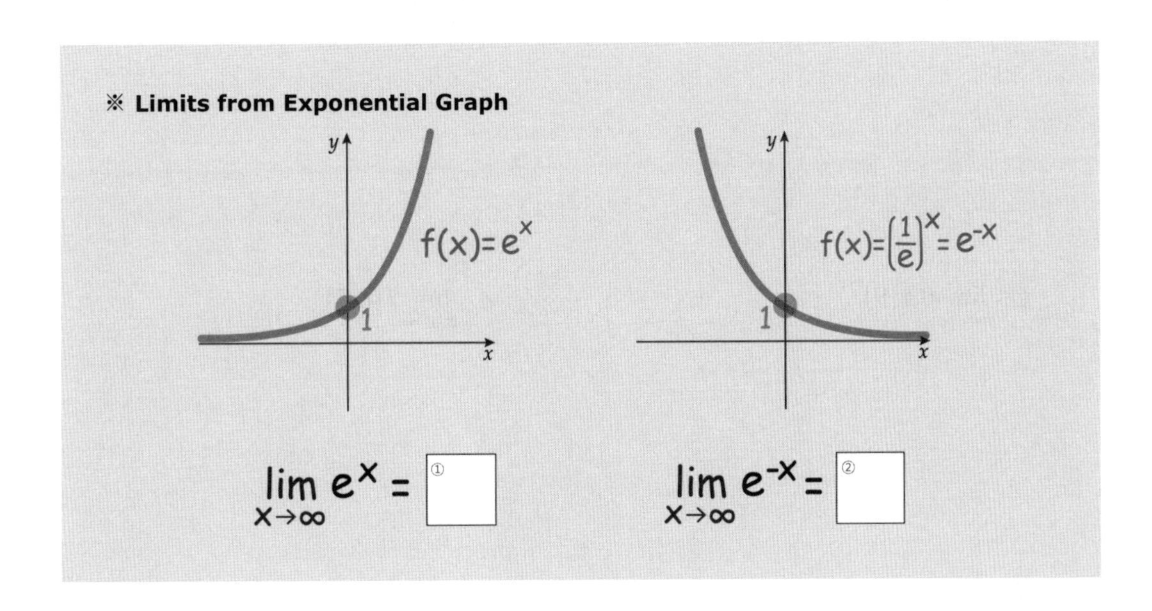

Blank : ① ∞ ② 0

EXAMPLE 5. Find the limit.

☺ Tip: Sketching the graph will help.

① $\lim_{x\to\infty} 3^x$

② $\lim_{x\to\infty} \pi^x$

③ $\lim_{x\to\infty} 5^{-x}$

④ $\lim_{x\to\infty} 0.3^x$

⑤ $\lim_{x\to-\infty} 3\left(\frac{1}{\pi}\right)^x$

⑥ $\lim_{x\to-\infty} 2(7)^{-x}$

⑦ $\lim_{x\to-\infty} \left(\frac{2}{3}\right)^x$

⑧ $\lim_{x\to-\infty} \left(\frac{5}{4}\right)^x$

⑨ $\lim_{x\to-\infty} 0.4(9)^{-x}$

⑩ $\lim_{x\to-\infty} 7(0.7)^x$

4. Exponential vs. Linear Function

	Linear Function $y=mx+b$	**Exponential Function** $y=a\cdot b^x$
Table looks like;	x: -1, 0, 1, 2 (+1, +1, +1) y: 2, 5, 8, 11 (+3, ①,)	x: -1, 0, 1, 2 (+1, +1, +1) y: 2, 4, 8, 16 (x2, ②,)
How to describe;	y increases by same ③__________ over the equal length of x = y values have equal ④__________ over the equal length of x	y increases by same ⑤__________ over the equal length of x = y values have equal ⑥__________ over the equal length of x = y values are *proportional* over the equal length of x
examples;	ex) increase 10 every year, grows 3cm each week	ex) increase 10% every year, grows 3% each week, doubling every month.

Blank : ① +3, +3 ② x2 (multiply 2), x2 ③ amount ④ difference ⑤ percentage ⑥ ratio

EXAMPLE 6. Determine a function for the given table.

☺ Tip: If it is exponential, write in $y = a \cdot b^x$ form
where base b is the $multiplier\ of\ y$ as x increase by 1.
where base b is the $\sqrt[n]{multiplier\ \ of\ \ y}$ as x increase by n.

①

x	$f(x)$
-2	5/9
-1	5/3
0	5
1	15
2	45

②

x	$f(x)$
-2	6
-1	2
0	-2
1	-6
2	-10

③

x	$f(x)$
-4	1
-2	4
0	7
2	10
4	13

④

x	$f(x)$
0	10
1	20
2	40
3	80
4	160

⑤

x	$f(x)$
-3	1
0	8
3	64
6	512
9	4096

⑥

x	$f(x)$
-4	1/3
-2	1
0	3
2	9
4	27

⑦

x	$f(x)$
0	64
2	16
4	4
6	1
8	1/4

⑧

x	$f(x)$
-3	135
0	45
3	15
6	5
9	5/3

AP Style Problem

1. For $f(x)=4(0.7)^x$, which of the following is true?

A) $f(x)$ has a same graph with 2.8^x

B) $\lim_{x\to-\infty} f(x)=0$

C) $\lim_{x\to\infty} f(x)=0$

D) The output of $f(x)$ decreases by 70% when x increases by 1.

2. The function $f(x)$ and $g(x)$ is given by $f(x)=3^x$ and $g(x)=3^{x-2}$. Which of the following steps of transformations maps the graph of f to the graph of g?

A) Vertical dilation by the factor of 9

B) Vertical dilation by the factor of $1/9$

C) Horizontal dilation by the factor of 2

D) Horizontal dilation by the factor of 0.5

3. How does $k(x)=3(0.6)^x$ change over the interval from $x=1$ to $x=3$?

A) $k(x)$ decreases by 36%

B) $k(x)$ decreases by 64%

C) $k(x)$ decreases by 60%

D) $k(x)$ decreases by 40%

4. The values of an exponential function $f(x)$ are given in the table.

x	2	6	10
$f(x)$	2500	?	1600

What is the value of $f(6)$?

A) 1620

B) 1800

C) 2000

D) 2050

5. Which of the following is true about $f(x)=2^x$ and $g(x)=2^x+1$?

A) $f(x)$ and $g(x)$ both increases by the same percentage over equal length of x.

B) $f(x)$ increases by the same percentage over equal length of x, but not $g(x)$.

C) $g(x)$ increases by the same percentage over equal length of x, but not $f(x)$.

D) Neither $f(x)$ nor $g(x)$ increases by the same percentage over equal length of x.

6. A linear function $f(x) = mx + n$, where $m>0$, $n>0$, and an exponential function $g(x) = a \cdot b^x$, where $a>0$, $b>0$ is given as shown.

x	1	2	4
$f(x)$	2	?	10
$g(x)$	2	?	10

Which of the following is true about $f(2)$ and $g(2)$?

A) $f(2) > g(2)$

B) $f(2) < g(2)$

C) $f(2) = g(2)$

D) $f(2) = g(2) = 6$

7. *Which of the following is NOT the property of $f(x) = 2^x$?

A) $f(x+y) = f(x)f(y)$

B) $f\left(\frac{x}{y}\right) = f(x) - f(y)$

C) $f(-x) = \frac{1}{f(x)}$

D) $f(nx) = [f(x)]^n$

3.2 Compound Interest

1. Compound Interest

Simple interest means the interest on principal (initial money) only.

$1,000 is invested at annual interest rate 10% with simple interest;

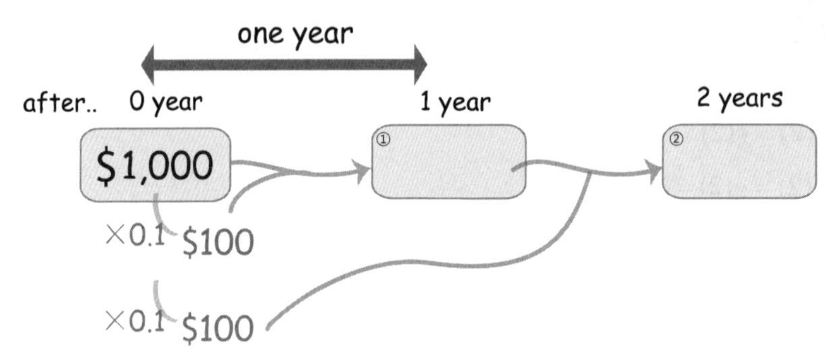

Compound interest means interest on principal (initial money) and interest previously earned.

$1,000 is invested at annual interest rate 10% compounded *annually*;

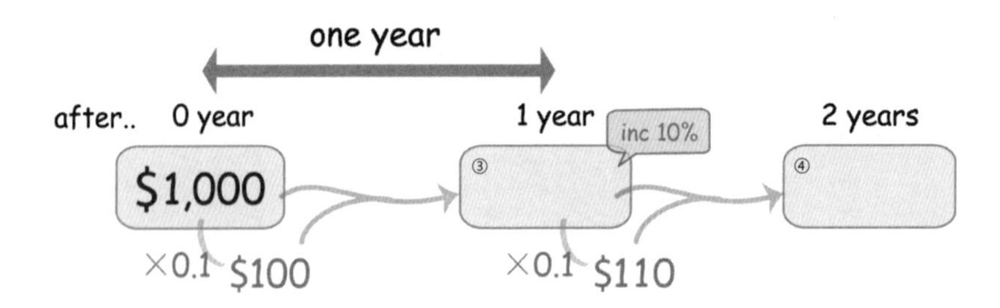

Blank : ① $1100 ② $1200 ③ $1100 ④ $1210

1) Compounded Annually

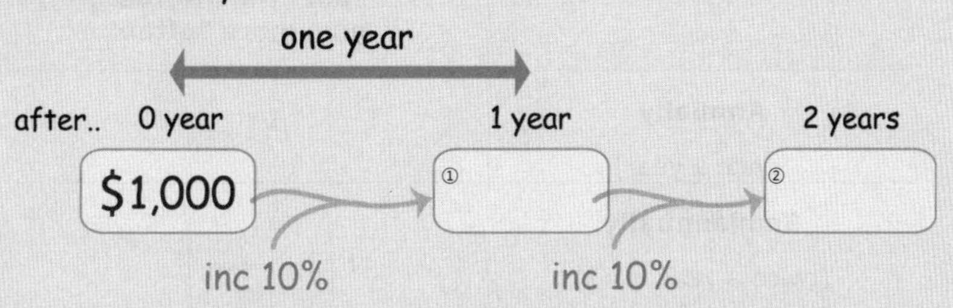

If a **principal P** is invested at a fixed annual **interest rate r** *compounded annually* calculated at the end of each year, then **after t years** the value of the investment is

$$\text{New} = P(1+r)^t$$

(rate: r, time(yr): t, principal: P)

2) Compounded n times a year

$1,000 is invested at annual interest rate 10% compounded *semiannually* *(2 times a year);*

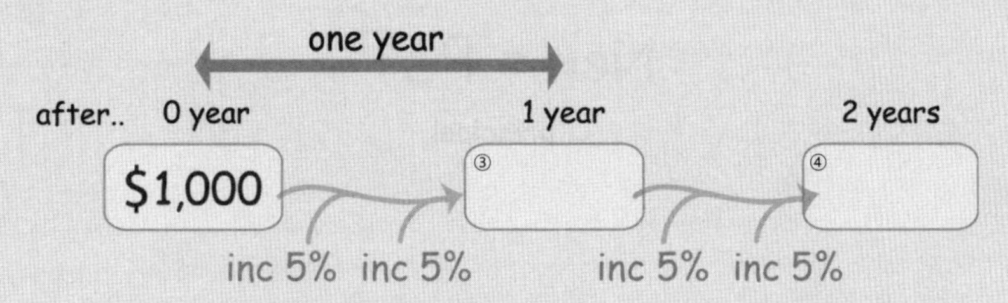

we **"split" the interest,** and give **more "often"!**

If a **principal P** is invested at a fixed annual **interest rate r** *compounded n times a year* calculated at the end of each year, then **after t years** the value of the investment is

$$\text{New} = P\left(1+\frac{r}{n}\right)^{tn}$$

(compound n times a year: n)

Blank : ① $1000(1+0.1)$ ② $1000(1+0.1)^2$ ③ $1000(1+0.05)^2$ ④ $1000(1+0.05)^4$

ex)

	"split" the interest , give more "often"
Annually (once a year)	$P(1+r)^t$
Semiannually (twice a year, n=2)	$P\left(1+\frac{r}{①\square}\right)^{t\times ②\square}$
Quarterly (4 times a year, n=4)	$P\left(1+\frac{r}{③\square}\right)^{t\times ④\square}$
Monthly (12 times a year, n=12)	$P\left(1+\frac{r}{⑤\square}\right)^{t\times ⑥\square}$

3) Compounded Continuously

If it is *compounded continuously* (when n is a large number) we use a special formula.

$$\text{New} = Pe^{rt}$$

rate
time(yr)
principal

☺ Proof:

$$P\left(1+\frac{r}{n}\right)^{nt} = P\left[\left(1+\frac{r}{n}\right)^{\frac{n}{⑦\square}}\right]^{t ⑧\square} = P\left[\left(1+\frac{1}{m}\right)^{⑨\square}\right]^{t ⑩\square}$$

When m is a big number the formula will be Pe^{rt} .

Blank : ① 2 ② 2 ③ 4 ④ 4 ⑤ 12 ⑥ 12 ⑦ r ⑧ r ⑨ m ⑩ r

☺ All in one

Compounded Annually	$New = P(1+r)^{t}$ (rate, time(yr), principal)
Compounded n times a year	$New = P\left(1+\frac{r}{n}\right)^{tn}$ (compound n times a year)
Compounded Continuously	$New = Pe^{rt}$ (rate, time(yr), principal)

EXAMPLE 1. Find the new amount that result from the investment.

① \$1,000 invested at 9% compounded annually after a period of 8 years

② \$1,000 invested at 12% compounded semiannually after a period of 3 years

③ \$14,000 invested at 14% compounded monthly after a period of 13 years

④ \$480 invested at 9% compounded quarterly after a period of 5 years

⑤ \$12,000 invested at 6% compounded continuously after a period of 4 years

⑥ \$4,000 invested at 0.5% compounded continuously after a period of 7 years

Increased by r% each year (or month, ...) : $P(1+r)^t$

Decreased by r% each year (or month, ...) : $P(1-r)^t$

EXAMPLE 2. The population of Smallville in the year 1890 was 2400. Assume the population increased at a rate of 0.2% per year. Estimate the population in 1925.

EXAMPLE 3. The 2000 population of Jacksonville was 32,000 and was decreasing at the rate of 0.01% each year. At that rate, estimate the population after 3 years.

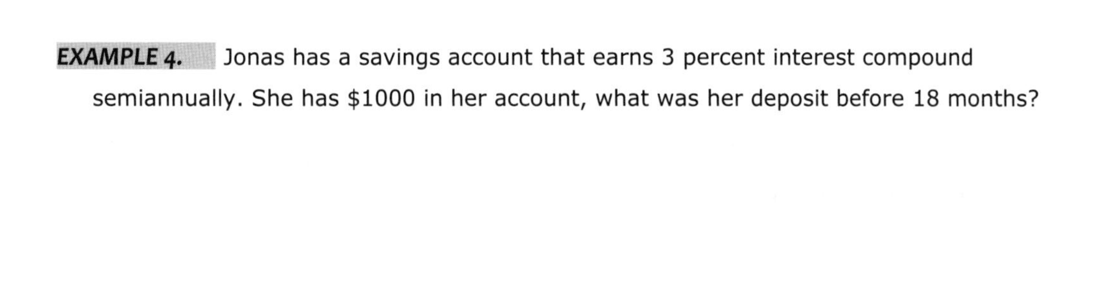

EXAMPLE 4. Jonas has a savings account that earns 3 percent interest compound semiannually. She has \$1000 in her account, what was her deposit before 18 months?

EXAMPLE 5. Jay will be buying a used car for \$20,000 in 3 years. How much money should he ask his parents for now so that, if he invests it at 3% compounded continuously, he will have enough to buy the car?

AP Style Problem

1. A bee population of 50000 is decreasing by 0.2% each month. Which of the following is the expression for the bee population in terms of t, where t is measured in years?

A) $50000(0.8)^{\frac{t}{12}}$

B) $50000(0.8)^{12t}$

C) $50000(0.998)^{\frac{t}{12}}$

D) $50000(0.998)^{12t}$

2. In 2000, there were approximately 200 million people living in Houston, Texas. The population was increasing by 0.92 percent per year. If the population continues to increase at the same rate, how many million people will be living in Houston in t quarter of a year?

A) $200(1.0092)^{\frac{t}{4}}$

B) $200(1.0092)^{4t}$

C) $200(1.92)^{\frac{t}{4}}$

D) $200(1.92)^{4t}$

3. What rate of interest compounded annually is required to double an investment in 3 years?

A) $\sqrt{2}+1$

B) $\sqrt[3]{2}+1$

C) $\sqrt[3]{2}-1$

D) $\sqrt{2}-1$

3.3 Logarithmic Function

1. Logarithm

A **logarithm** is another way of writing an ①____________.

A logarithm gives you the power(exponent) a base must be raised to produce a given number.

※ **Logarithmic function with base b** is defined by

$b^x = a \rightleftarrows \log_b a = x$

exponent
base
value
③
④
②

Remember, $b \neq 1$, $b > 0$, and a **has to be** ⑤____________.

We can read it as : 'log base b of a is equal to x'

※ **Special Logarithm**

⑥____________ log

Common Log

$\log_{10} x =$ ⑦

⑧____________ log

Natural Log

$\log_e x =$ ⑨

Blank : ① exponent ② base ③ value ④ exponent ⑤ positive (a>0) ⑥ Common ⑦ log x ⑧ Natural ⑨ ln x

EXAMPLE 1. Find x using the <u>definition</u> of log.

What power(exponent) a base must be raised to produce a given number?

① $\log_{10} 100{,}000 = x$

② $\log_2 32 = x$

③ $\log_{10} 0.01 = x$

④ $x = \log_5 \dfrac{1}{25}$

⑤ $\log_2 \dfrac{1}{4} = x$

⑥ $x = \log_{1/2} 2$

⑦ $\log_{16} 4 = x$

⑧ $\log_2 \sqrt{2} = x$

⑨ $4 = \log_x 625$

⑩ $\log_x 4 = -1$

⑪ $\log_5 x = 3$

⑫ $-2 = \log_{1/2} x$

⑬ $\log_{(x-1)} 5 = 1$

EXAMPLE 2. Find the domain.

Remember : If $y = \log_b a$, then $a > 0$.

① $\log(x+8)$

② $\ln(2-x)$

③ $\log_5(2x^2+9x-5)$

④ $\ln(x^2+3x+2)$

⑤ $\log_5\left(-\dfrac{2}{x+2}\right)$

⑥ $\log_5\left(\dfrac{1}{x+2}\right)$

⑦ $\log_5\left[\dfrac{(1+x)(3+x)^2}{1-x}\right]$

⑧ $\log_5\left[\dfrac{(x-1)(x+2)}{x-2}\right]$

2. Basic Properties of Logarithms

※ Basic Properties of Logarithms

$\log_b 1 =$ ①☐

$\log_b b =$ ②☐

$\log_b b^x = x$ ex) $\log_2 2^x =$ ③____, $\log 10^2 =$ ④____, $\ln e^y =$ ⑤____

$b^{\log_b x} = x$ ex) $2^{\log_2 3} =$ ⑥____, $10^{\log x} =$ ⑦____, $e^{\ln 4} =$ ⑧____

EXAMPLE 3. Evaluate the logarithm using basic log properties.

① $\log_2 8$

② $\log_3 \sqrt{3}$

③ $6^{\log_6 11}$

④ $\log_5 1$

⑤ $\log 100$

⑥ $\log \sqrt[5]{10}$

⑦ $\log \frac{1}{1000}$

⑧ $10^{\log 6}$

⑨ $10^{\log 5^2}$

⑩ $\log \frac{1}{\sqrt{10}}$

⑪ $e^{\ln 4}$

⑫ $\ln \sqrt{e}$

⑬ $\ln e$

⑭ $\ln e^5$

⑮ $\ln e^{kt}$

⑯ $\ln(\log 10^{e^2})$

Blank : ① 0 ② 1 ③ x ④ 2 ⑤ y ⑥ 3 ⑦ x ⑧ 4

EXAMPLE 4. Change into an equivalent expression.

To get rid of the LOG, you need ①________

To get rid of the BASE, you need ②________

$\log_2 8 = 3 \rightarrow$ ③[ ] $^{\log_2 8} = {}_2{}^3 \rightarrow$ ④[ ]

$2^3 = 8 \rightarrow$ ⑤[ ] $2^3 = \log_2 8 \rightarrow$ ⑥[ ]

	Exponential Form	Logarithmic Form
①		$\log 1000 = 3$
②	$5^x = 4000$	
③		$\ln \frac{1}{8} = x$
④		$\log_3 27 = 3$
⑤	$e^{1/2} = x$	
⑥	$10^x = 2$	
⑦		$\ln x^y = 2$
⑧	$10^{2x} = 5$	
⑨	$2^{x^2 - x} = 3$	
⑩	$e^{4x^2} = 7$	
⑪		$y = \log(x^2 - 4x)$

Blank : ① base ② log ③ 2 ④ $8 = 2^3$ ⑤ $\log_2$ ⑥ $3 = \log_2 8$

EXAMPLE 5. Find the inverse.

① $y=2^{x-1}$

② $y=5^{x}+3$

③ $y=3e^{x}-2$

④ $y=3(10^{x-2})$

⑤ $y=\log_5 x$

⑥ $y=\log_7 x+5$

⑦ $y=\ln(x+2)$

⑧ $y=2\log x-3$

⑨ $y=3\log x+2$

⑩ $y=\ln(x-1)+1$

⑪ $y=\frac{1}{2}\log_3(x-1)+2$

⑫ $y=\frac{\ln(x+2)-1}{2}$

3. Graph of Log

※ Graph of Log

Logarithm and exponential function are ①__________,
so we can draw the log graph
by taking the exponential function and reflecting across the ②________.

$y = \log_b x$ When $0 < b < 1$ $y = \log_b x$ and when $b > 1$

Domain: ③______________,
range:④ ______________,
Vertical asymptote:⑤______________.
x intercept : ⑥__________

EXAMPLE 6. Graph the function and find the asymptotes.

① $y = \log_3 x + 3$

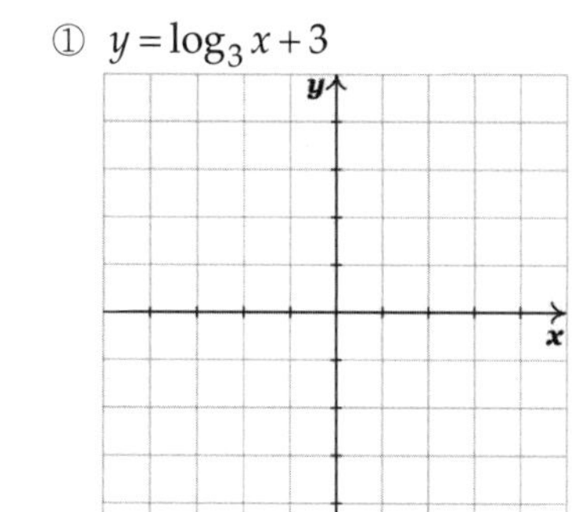

② $y = \ln(x + 3)$

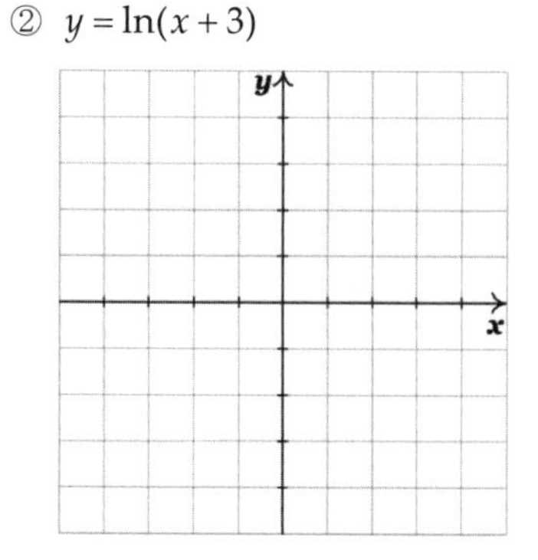

Blank : ① inverse ② y = x ③ (0, ∞) ④ (-∞, ∞) ⑤ x = 0 ⑥ (1, 0)

③ $y = \ln(-x+3)$

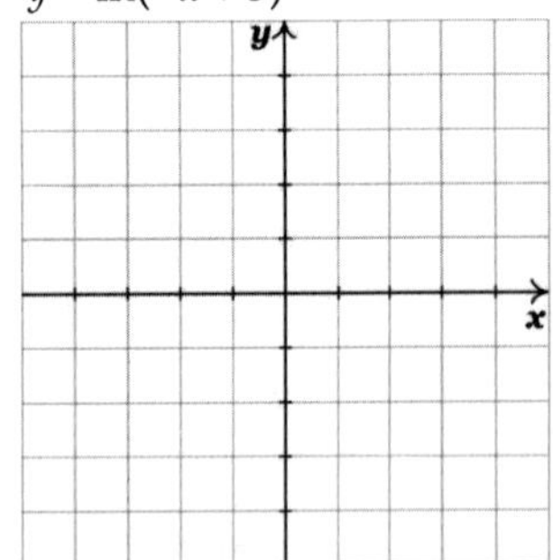

④ $y = \log_{0.3}(2-x)$

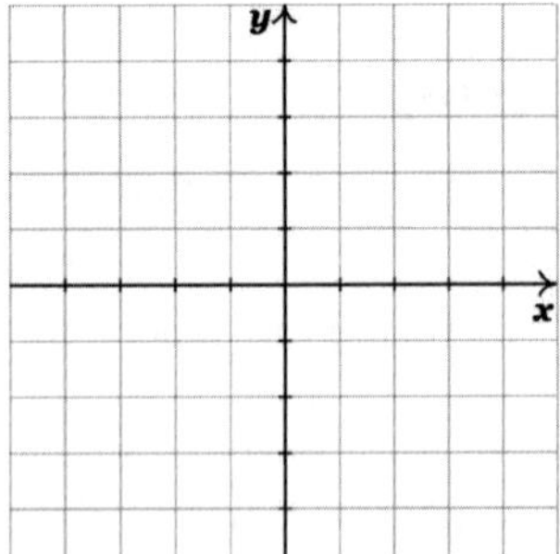

⑤ $y = \log_{0.2}(x-1)$

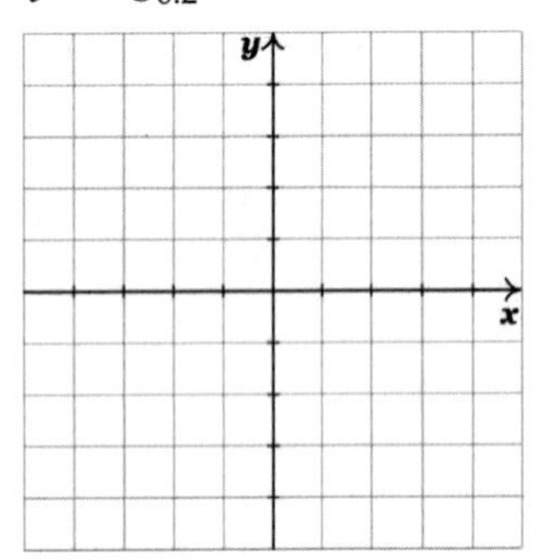

⑥ $y = -\log_3(-x)+2$

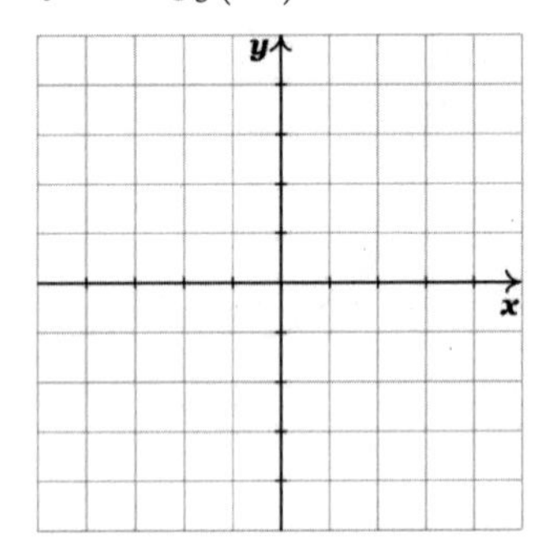

⑦ $y = 2\log x$

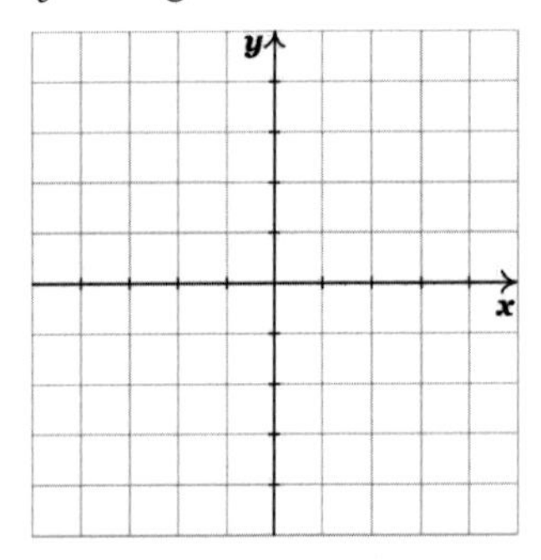

⑧ $y = \log(2x)$

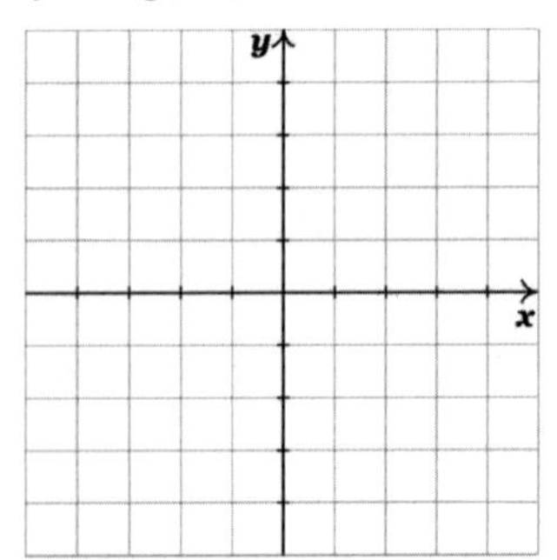

⑨ $y = \ln(0.5x)-1$

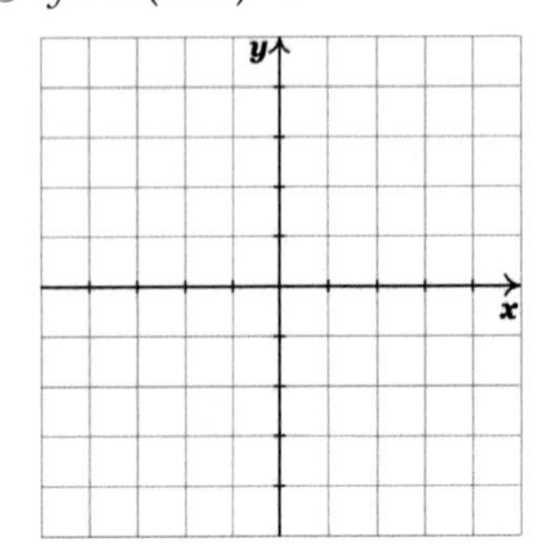

⑩ $y = 0.5\log x + 2$

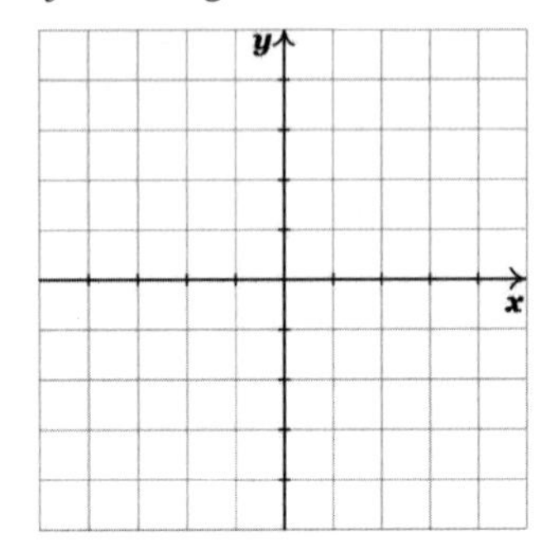

⑪ $y = |\log x|$

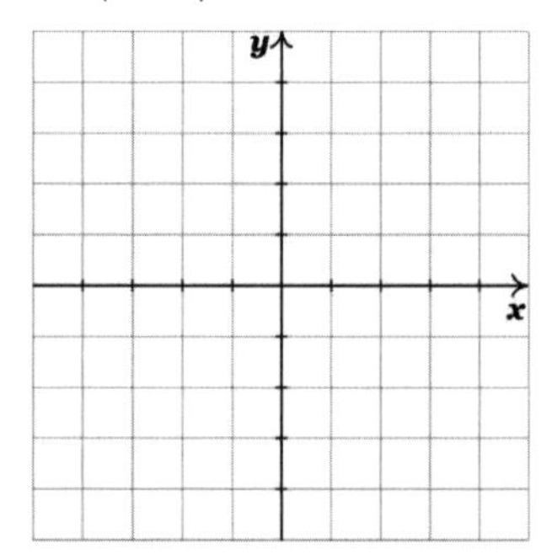

⑫ $y = \log|x|$

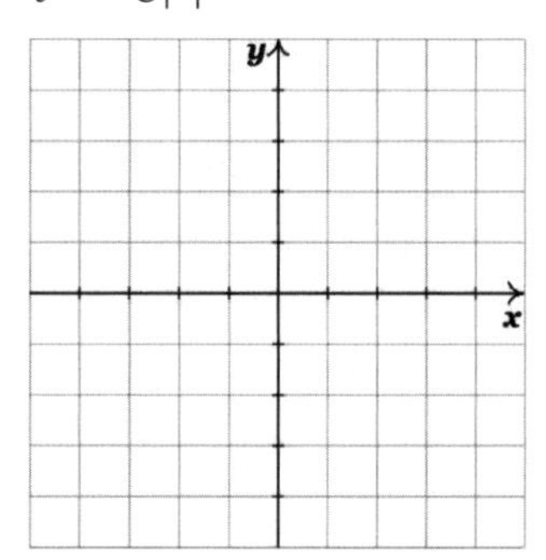

EXAMPLE 7. The values of $y = \log_b x$ are given. Determine a logarithmic function for the given table.

☺ Tip: If ①____ increases by same percentage (multiplied by same number)

over the equal length of ②____, it is logarithm function $y = \log_b x$

where base b is the *multiplier of* x as y increase by 1.

Blank : ① x ② y

①

x	$f(x)$
1	0
2	1
4	2
8	3
16	4

②

x	$f(x)$
3	-1
1	0
1/3	1
1/9	2
1/27	3

③

x	$f(x)$
64	-6
16	-4
4	-2
1	0
1/4	2

④

x	$f(x)$
1/81	-4
1/9	-2
1	0
9	2
81	4

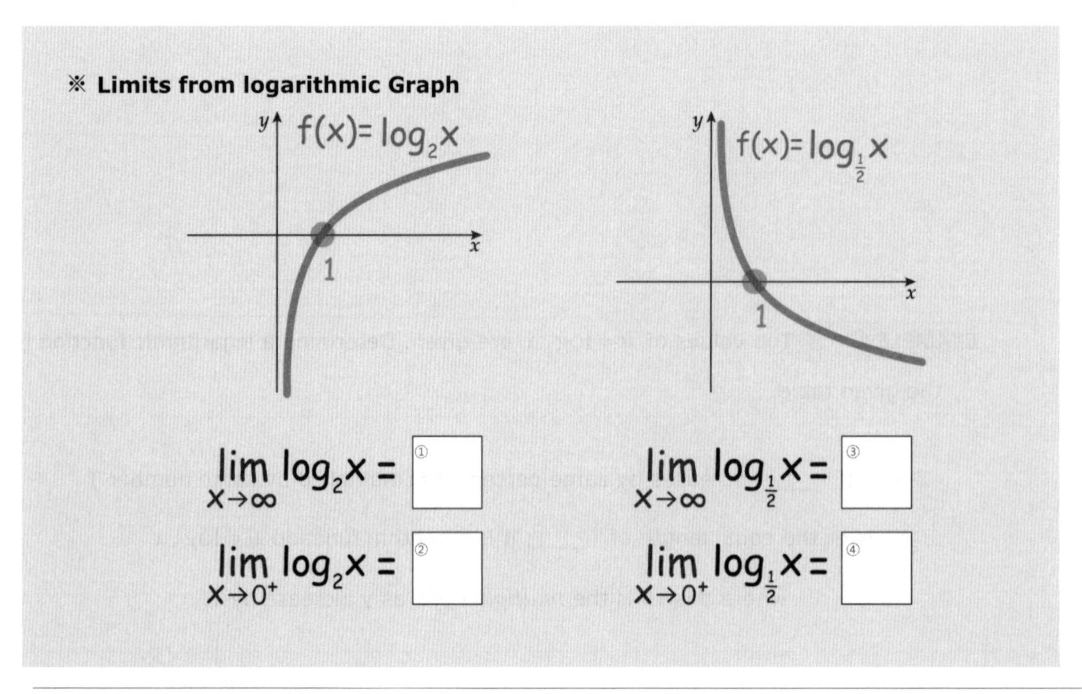

Blank : ① ∞ ② $-\infty$ ③ $-\infty$ ④ ∞

EXAMPLE 8. Find the limit.

☺ Tip: Sketching the graph will help.

① a) $\lim_{x\to\infty}\log_3 x$

b) $\lim_{x\to 0^+}\log_3 x$

② a) $\lim_{x\to\infty}\ln x$

b) $\lim_{x\to 0^+}\ln x$

③ a) $\lim_{x\to\infty}\log_{0.3}(x+1)$

b) $\lim_{x\to -1^+}\log_{0.3}(x+1)$

④ a) $\lim_{x\to\infty}\log_{\frac{1}{2}}(x-2)$

b) $\lim_{x\to 2^+}\log_{\frac{1}{2}}(x-2)$

⑤ a) $\lim_{x\to -\infty}\ln(2-x)$

b) $\lim_{x\to 2^-}\ln(2-x)$

⑥ a) $\lim_{x\to -\infty}\log_3(-x+1)$

b) $\lim_{x\to 1^-}\log_3(-x+1)$

AP Style Problem

1. For $g(x)$, x is proportional over the equal length of y. What could be the graph of $g(x)$?

A)

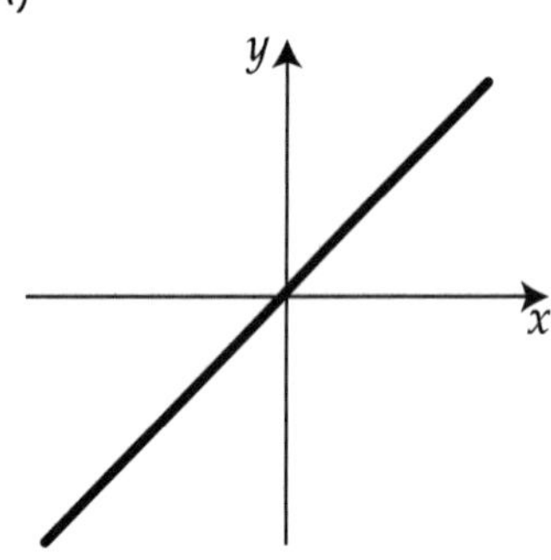

B)

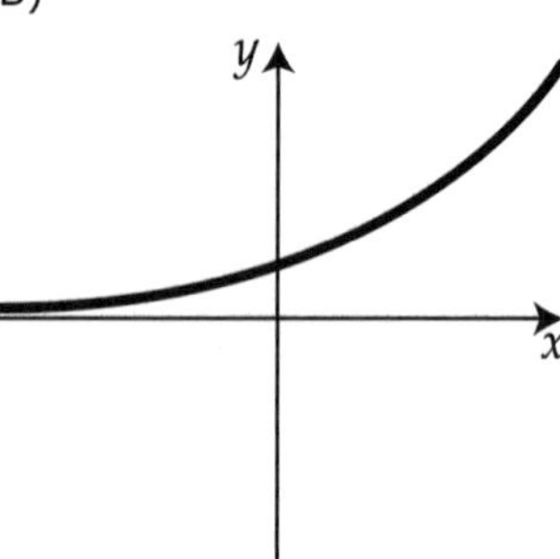

C)

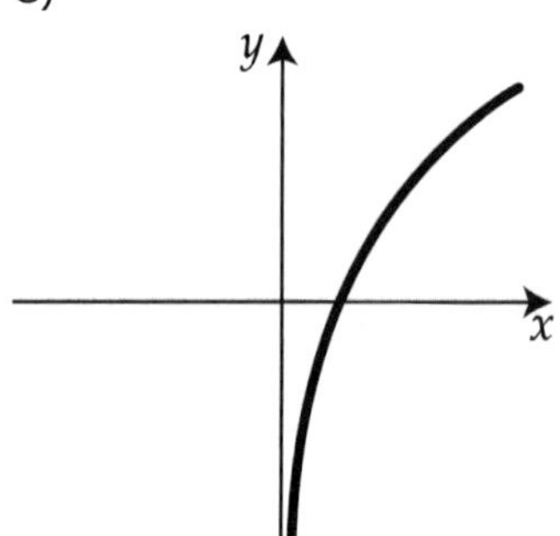

D)

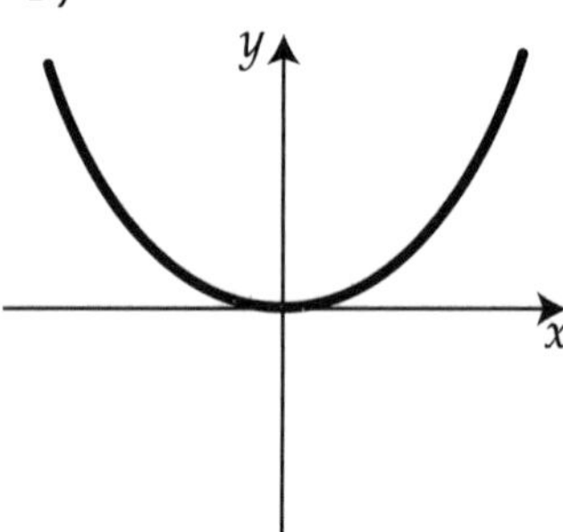

2. Which of the following is a function that decreases at a decreasing rate?

A) $y = e^{-x}$

B) $y = -e^{-x}$

C) $y = \ln(-x)$

D) $y = -\ln x$

3. For the function $f(x)=\log_7 x$, which of the following is NOT true?

A) f is increasing at a decreasing rate.

B) $\lim_{x\to\infty} f(x)=\infty$

C) $\lim_{x\to-\infty} f(x)=-\infty$

D) Every time the y increases by 1, the corresponding x value is multiplied by 7.

4. Find the inverse of the function $y=\ln(\ln x)$

A) $y=e$

B) $y=e^{x^2}$

C) $y=e^{x+1}$

D) $y=e^{e^x}$

5. Find the value of $\ln(\ln e^{e^4})-e^{\log_3 3^{\ln 4}}$.

A) 0

B) 2

C) 4

D) 6

6. The graph below shows $y = f(x)$, where $f(x) = x + \ln x$. Find the solution of $f(x) = f^{-1}(x)$.

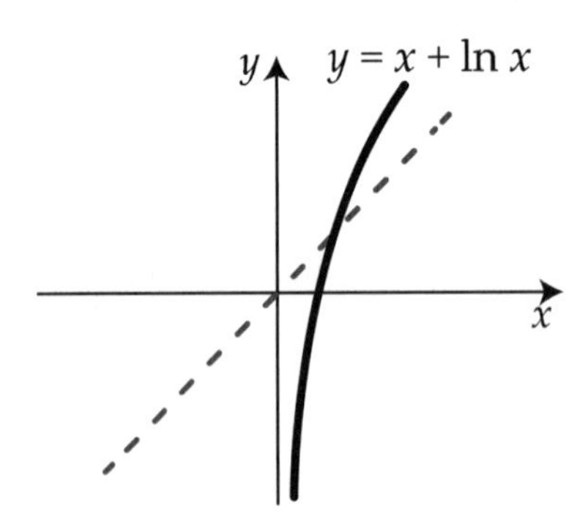

A) 1

B) 2

C) 3

D) 4

3.4 Properties of Logarithms

1. Properties of Logarithms

※ **Properties of Log** ($x, y > 0$, $b > 0$, $b \neq 1$)

$$\log_b xy = \log_b x + \log_b y$$ Product Property

$$\log_b \frac{x}{y} = \log_b x - \log_b y$$ Quotient Property

$$\log_b x^n = n \log_b x$$ Power Property

$$\log_{b^n} x = \frac{1}{n} \log_b x$$

$$\log_b x = \frac{\log_c x}{\log_c b}$$ Change of Base

EXAMPLE 1. Use the properties of logarithms to simplify the expression.

① $\log 3 + \log 10$

② $\log_3 24 - \log_3 8$

③ $3\ln 2+\ln 6-2\ln 3$

④ $\log 32-2\log 4-\frac{1}{3}\log 8$

⑤ $2\ln xy-3\ln y-\frac{1}{2}\ln z$

⑥ $2\ln(xy)-4\ln(xz)+3\ln(xz)$

⑦ $\log 2x+2(\log x-\log y)$

⑧ $2(\log_5 x+2\log_5 y-3\log_5 z)$

⑨ $2\log_6(x-1)-\log_6(x^2-1)$

⑩ $\log(x^2-2x-3)-\log(x+1)$

⑪ 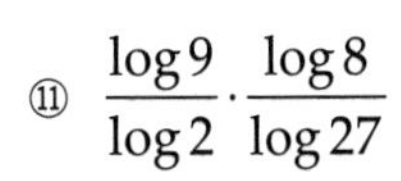 $\dfrac{\log 9}{\log 2} \cdot \dfrac{\log 8}{\log 27}$

⑫ $\dfrac{\log 8}{\log 4}$

⑬ $\log_4 23 \cdot \log_{23} 16$

⑭ $\log_2 3 \cdot \log_3 4 \cdot \log_4 5$

⑮ $\log_4 10 \div \log_4 5$

⑯ $\log_5 23 \div \log_8 23$

⑰ $e^{2\ln 5}$

⑱ $10^{3\log x}$

⑲ $10^{2(\log 30 - \log 5)}$

⑳ $5^{2\log_5 4 - 3\log_5 2}$

㉑ $\log_2 x + \log_4 x$

㉒ $\log_4 x + \log_8 x$

EXAMPLE 2. Expand the given log. Given $a=\log x, b=\log y$ and $c=\log z$, find an expression in terms of a, b, and c for

① $\log xy^3$

② $\log(xy)^3$

③ $\log x^2 y \sqrt[3]{z}$

④ $\log 100\sqrt{xy}$

⑤ $\log\left(\frac{10y^2}{\sqrt{z}}\right)$

⑥ $\log \frac{\sqrt{x}}{y^3 z}$

⑦ $\frac{\log x^3}{\log y^2}$

⑧ $\log\sqrt{x\sqrt{y}}$

⑨ $\log_{xy} \sqrt[3]{z\sqrt{x}}$

⑩ $\log_x \sqrt{yz}$

EXAMPLE 3. Write $\log_{13} \dfrac{10}{\sqrt[5]{x^2(4x+1)^3}}$ in expanded form.

EXAMPLE 4. Write $\log(x^2 - y^2)$ in expanded form.

EXAMPLE 5. What is $\log_2 3 \times \log_3 4 \times \log_4 5 \times . . . \times \log_{63} 64$?

EXAMPLE 6. True or false?

I. $\log(x+y)=\log x+\log y$

II. $\log x\cdot\log y=\log(x+y)$

III. $\log x+\log y=\log xy$

IV. $\dfrac{\log x}{\log y}=\log x-\log y$

V. $\log x-\log y=\log\dfrac{x}{y}$

VI. $\log(x-y)=\log\dfrac{x}{y}$

VII. $(\log x)^2=2\log x$

VIII. $\log\dfrac{1}{x}=-\log x$

EXAMPLE 7. Which one is different from others?

$$\log_3 14 \qquad \log_3 2+\log_3 7 \qquad \frac{\log 14}{\log 3} \qquad \frac{\ln 14}{\ln 3} \qquad \log_3 2\cdot\log_3 7 \qquad \frac{1}{2}\log_3 196$$

EXAMPLE 8. *Simplify the followings. (Hint: $n!=1\cdot2\cdot3\cdots n$)

① $2^{\log_2 1+\log_2 2+\log_2 3+...+\log_2 10}$

② $\log_2(2\cdot2^2\cdot2^3\cdot2^4\cdots2^{10})^2$

③ $(\log_2 2)(\log_2 2^2)(\log_2 2^3)\cdots(\log_2 2^{10})$

AP Style Problem

1. Let $x=\ln 3$ and $y=\ln 5$, then what is $\ln 45$ in terms of x and y?

A) $x+y^2$

B) $2x+y$

C) x^2+y

D) $x+2y$

2. What is $\log_5(\log_2 3)+\log_5(\log_3 4)+\log_5(\log_4 5)+\ldots+\log_5(\log_{31} 32)$?

A) 1

B) 5

C) 32

D) 64

3. Find the value of $16^{\log_2 3}$.

A) 2

B) 12

C) 27

D) 81

4. The function $f(x)$ and $g(x)$ is given by $f(x)=\ln x$ and $g(x)=\ln\sqrt{x}$. Which of the following steps of transformations maps the graph of f to the graph of g?

A) Vertical dilation by the factor of 2

B) Vertical dilation by the factor of 0.5

C) Horizontal dilation by the factor of 2

D) Horizontal dilation by the factor of 0.5

5. The function $f(x)$ and $g(x)$ is given by $f(x)=\ln x$ and $g(x)=\log x$. Which of the following steps of transformations maps the graph of f to the graph of g?

A) Vertical dilation by the factor of $\ln 10$

B) Vertical dilation by the factor of $\dfrac{1}{\ln 10}$

C) Horizontal dilation by the factor of e

D) Horizontal dilation by the factor of $\dfrac{1}{e}$

6. Let $f(x) = \ln(x+1) + \ln 4$. Find $f^{-1}(x)$.

A) e^{x-4}

B) $e^{x}-4$

C) $\dfrac{e^{x}}{4}-1$

D) $\dfrac{e^{x}}{4}-4$

7. * Which of the following describes the property of $f(x)=\log x$.

A) $f(x-y)=\dfrac{f(x)}{f(y)}$

B) $[f(x)]^{n}=nf(x)$

C) $f(xy)=f(x)+f(y)$

D) $f\left(\dfrac{1}{x}\right)=\dfrac{1}{f(x)}$

3.5 Exp and Log Equations and Inequalities

1. Logarithmic Equations

EXAMPLE 1. Solve the equation.

You must CHECK your Answer!!
Type1. If you want to get rid of the log,
raise both sides to be power of that base!

① $\log_5(x+1)=2$

② $2\ln(x-2)=3$

③ $\ln\left(\log_3\left(\frac{2}{x-1}\right)\right)=1$

④ $\log_2(\log_3 x)=2$

⑤ $\log_3 x+\log_3(x-24)=4$

⑥ $\log_2 x-\log_2(x-1)=3$

⑦ $\log(x^2+9)=1+2\log x$

⑧ $2\log(x-2)-\log x=0$

⑨ $\log_3 x+\log_9 x=2$

⑩ $\log_{16} x-\log_{32} x=0.5$

Type2. Use the one to one property.

If $\log_b x=\log_b y$, then $x = y$.

⑪ $\log_8 2+\log_8 x=\log_8 7$

⑫ $\log_2(x+5)=\log_2 x+\log_2 5$

⑬ $\ln(x-2)+\ln(2x-3)=2\ln x$

⑭ $\log_3 x+\log_3(x+3)=\log_3 4$

⑮ $\log x-\log(x+2)=\log(x+6)$

⑯ $\ln(5x^2+4)=2\ln 3x^2-\ln(2x^2-1)$

⑰ $\ln(3x^2-4)+\ln(x^2+1)=\ln(2+6x^2)$

⑱ $\log(4+x)-\log(x-4)=\log 5$

Type 3. Make a Substitution!

⑲ $\left(\log x\right)^2 = \log x^4 - 3$

⑳ $2(\log x)^2 = \log x^3 + 5$

㉑ $4\log_4 x = 9\log_x 4$

㉒ $4\log_x 10 + \log_{10} x = 5$

EXAMPLE 2. Solve the following simultaneous equations

$$xy = 3$$
$$2\log_3 x - \log_3 y = 2$$

2. Solving Exponential Equation

EXAMPLE 3. Solve the equation using log.

Type1. Make the base same if you can.

① $3(2^x) = 24$

② $5\left(\frac{1}{2}\right)^x = 20$

③ $32^{x-1} = 4^{x+2}$

④ $\left(\frac{5}{4}\right)^{4x} = \left(\frac{16}{25}\right)^{9-x}$

⑤ $\left(\frac{9}{16}\right)^{3x-2} = \left(\frac{4}{3}\right)^{x-4}$

⑥ $49^{\frac{x}{3}} = 7^{x-4}$

Type2. If you want to bring exponent x down,
take 'log' on both sides!

⑦ $2^x = 3$

⑧ $3^{x+1} = 8$

⑨ $3e^{2x} = 15$

⑩ $2e^{3x-1} = 14$

⑪ $5^{x-2} = 3^{3x+1}$

⑫ $11^{x-1} = 7^{2x+1}$

⑬ $7^{1-2x} = 2^x$

⑭ $4^{x-3} = 6^{2x}$

Type3. Make a substitution!

⑮ $2e^{2x} - e^{x} = 6$

⑯ $e^{2x} - 15e^{x} = -56$

⑰ $3^{2x+1} - 11 \times 3^{x} = 4$

⑱ $5^{2x} - 5^{x+1} + 4 = 0$

⑲ $e^{x} + 5e^{-x} = 6$

⑳ $e^{x} - 6e^{-x} = 1$

㉑ $3e^{5x}-10e^{3x}-8e^{x}=0$

㉒ $2e^{6x}+e^{3x}-1=0$

㉓ $3xe^{x}+x^{2}e^{x}=0$

3. Application

EXAMPLE 4. If you deposit $5000 into an account paying 6% annual interest compounded annually, how long until there is $8000 in the account?

EXAMPLE 5. If you deposit $8000 into an account paying 7% annual interest compounded quarterly, how long until there is $12400 in the account?

EXAMPLE 6. At 3% annual interest compounded continuously, how long will it take to double your money?

4. Exponential and Logarithmic Inequalities

※ **Property of Inequality for Exponential and Logarithmic Functions**

If b > 1, then $b^x > b^y$ if and only if $x > y$.

If b > 1, then $\log_b x > \log_b y$ if and only if $x > y$.

(But remember! When you have log A, A part should be ①____________.)

EXAMPLE 7. Solve the inequalities.

① $2^{x-1} > 3$

② $e^{2x+1} < 2$

③ $2 < 10^x \le 4$

④ $8 \le 3^x < 27$

⑤ $xe^x - 3e^x < 0$

⑥ $2^x x^2 - 2^x(2x+3) < 0$

Blank : ① positive

⑦ $\log_2 2x < 2$

⑧ $\log(2x-1) < 0$

⑨ $\log_2(3x+1) < 4$

⑩ $\ln x - 2 > 3$

⑪ $2\ln x - 1 \geq 5$

⑫ $\frac{1}{2} < \log x < 2$

⑬ $2 \leq \log_3 x \leq 3$

AP Style Problem

1. Find the x coordinates of the point of intersection of $f(x)=\log_2 x+\log_2(x+2)$ and $g(x)=\log_2(x+6)$.

A) x = 2, -3

B) x = 2

C) x = -3

D) x = 0

2. Solve $\ln x^2=(\ln x)^2$.

A) x = 0

B) x = e^2, 2

C) x = e^2, 1

D) x = e^2

3. Solve $\ln(x-3)+\ln(x+4)<3\ln 2$.

A) $-6<x<4$

B) $-5<x<4$

C) $-4<x<4$

D) $3<x<4$

4. The values of an exponential function $f(x)$ are given in the table.

x	1	3	5
$f(x)$	0.25	0.49	0.9604

What is the value of x when $f(x) = 10$?

A) $\dfrac{\ln 56}{\ln 1.4}$

B) $\dfrac{\ln 28}{\ln 1.4}$

C) $\dfrac{\ln 56}{\ln 14}$

D) $\dfrac{\ln 28}{\ln 14}$

5. Suppose the population of Minnesota grows 5% every decade. If the population was 40 million in 1990, how many *years* will it take for the population to reach 48 million?

A) $\dfrac{10\ln 1.2}{\ln 1.05}$ years

B) $\dfrac{10\ln 1.05}{\ln 1.2}$ years

C) $\dfrac{\ln 1.2}{\ln 1.05}$ years

D) $\dfrac{\ln 1.05}{\ln 1.2}$ years

6.* Solve $x^{\log_4 x} = x$.

A) $x = 0$

B) $x = 0, 1$

C) $x = 1, 4$

D) $x = 4$

7. * Solve $\log_8(2^x + 1) + \log_8(2^x + 3) = 1$.

A) $x = 0$

B) $x = 1/32$

C) $x = 0, 1/32$

D) no solution

3.6 Exponential Growth and Modeling

1. Exponential Growth and Decay Model (Halving or doubling)

A single cell amoeba *doubles* every 4 days. If there were 5 cells initially what is the population of the amoeba after t days?

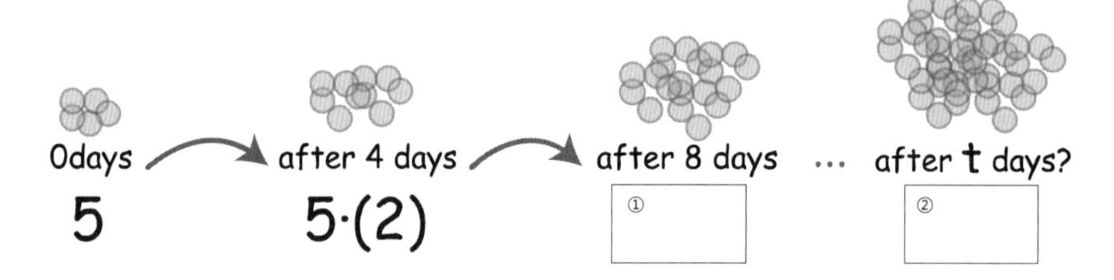

after t days, we will get : ③____________

※ **Exponential Growth Model** (Doubling or Halving)

Doubling every n days
(or other time units)

$$New = P(2)^{\frac{t}{n}}$$

(n: Doubling time)

Halving every n days:
= Half life is n days
(or other time units)

$$New = P\left(\frac{1}{2}\right)^{\frac{t}{n}}$$

(n: Halving time)

where P is the initial value, t is the time (with same time units with n).

Blank : ① $5 \cdot 2^2$ ② $5 \cdot 2^{t/4}$ ③ $5 \cdot 2^{t/4}$

EXAMPLE 1. Find the exponential function that satisfies the given conditions.

① Initial population = 1078, doubling every 8 hours

② Initial mass = 420 g, halving once every 26 years

③ Initial mass = 416 g, halving once every 23 days

④ Initial population = 1081, doubling every hour

EXAMPLE 2. Under ideal conditions a certain bacteria population doubles every three hours. Initially there are 1000 bacteria.

(a) Find a model for the bacteria population after t hours.

(b) How many bacteria are in the colony after 15 hours?

(c) When will the bacteria count reach 200,000?

EXAMPLE 3. A single cell amoeba doubles every 9 days. About how long will it take one amoeba to produce a population of 400?

EXAMPLE 4. The half-life of a certain radioactive substances is 20 days. There are 45g present initially.

(a) Express the amount of substance remaining as a function of time t.

(b) Find the amount of substance remaining after 26 days.

EXAMPLE 5. The half-life of a certain radioactive substances are 7 days. There are 23g present initially. When will there be 25% remaining?

EXAMPLE 6. A certain radioactive isotope has a half-life of approximately 1900 years. How many years to the nearest year would be required for a given amount of this isotope to decay to 30% of that amount?

2. Exponential Growth/decay

In our world, a lot of things grow (or decay) exponentially.
(not forever, but at least for a while)

So we have a generally useful formula:

※ **Exponential Growth/decay Model** (Relative Growth Rate)

A population that experiences exponential growth increases according to the model

$$\text{New} = Pe^{rt}$$

where P is the initial value, t is the time, r is the relative growth rate.

(If $r > 0$, then it is ① ____________________

If $r < 0$, then it is ② ____________________)

Blank : ① exponential growth ② exponential decay

EXAMPLE 7. The initial bacterium count in a culture is 600. A biologist later makes a sample count of bacteria in the culture and finds that the relative rate of growth is 30% per hour.

(a) Find a function that models the number of bacteria after t hours.

(b) What is the estimated count after 10 hours?

(c) When will the bacteria count reach 80,000?

(d) Sketch the graph of the function.

EXAMPLE 8. The population of rabbits is increasing according to the law of exponential growth. In an experiment, it was observed that there were 200 rabbits after the second day and 1000 rabbits after the fourth day. How many rabbits were there in the original population?

EXAMPLE 9. In a research experiment, the population of a city is growing according to exponential model. If the growth rate per year is 4% of the current population, how long will it take for the population to double?

3. Logistic Growth

That growth can't go on forever as they will soon run out of available food.

In many real life applications,
the growth is now always unlimited,
but **may have a limit** (= ①________________).

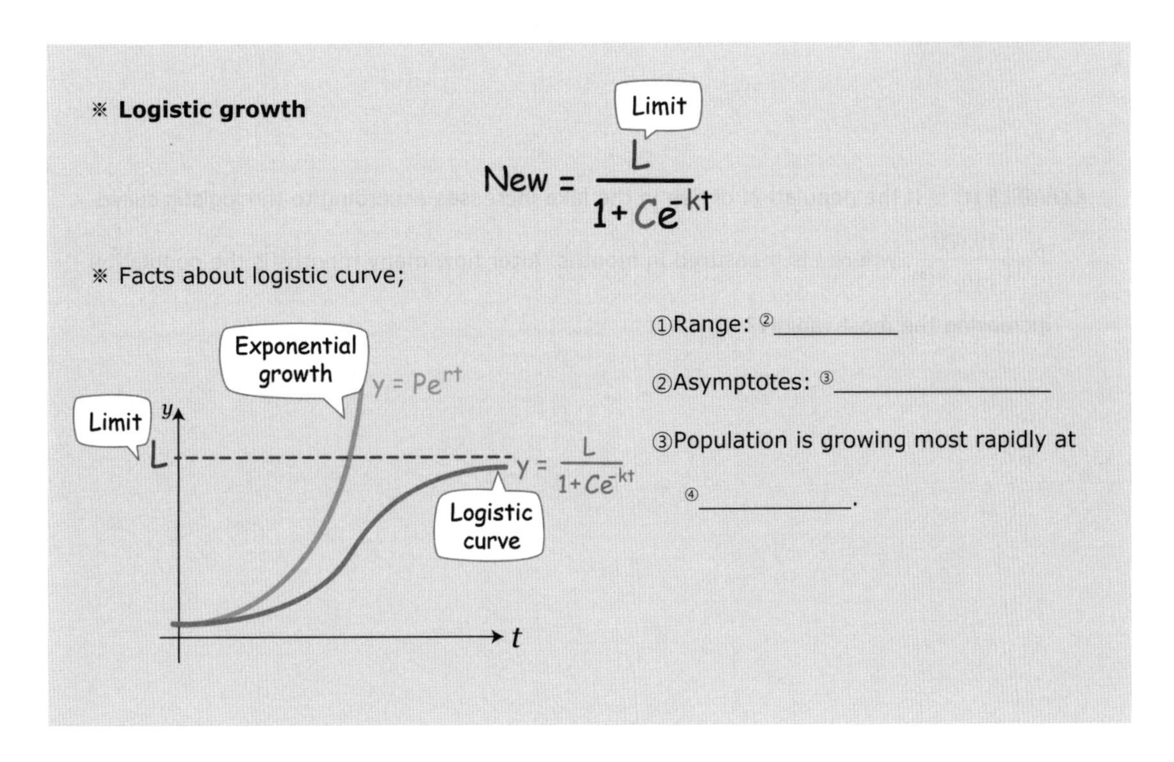

Blank : ① carrying capacity ② $0 < y < L$ ③ $y = 0,\ y = L$ ④ $y = \frac{L}{2}$

EXAMPLE 10. The spread of a disease through a college of 6000 students can be modeled with the logistic equation $y = \dfrac{6000}{1+590e^{-0.1t}}$, where y is the number of people infected after t days. The college will cancel class when 60% or more students are infected.

(a) How many people are infected when the disease is spreading the fastest?

(b) After how many days will the college cancel classes?

EXAMPLE 11. If the population of fish in the lake increases according to the logistic curve $P = \dfrac{10,000}{1+10e^{-t/3}}$, where t is measured in months. After how many months is the population increasing the most rapidly?

AP Style Problem

1. A certain population has 6 members at first and doubles every 20minutes. What is the size of the population after t hours?

A) $6(2^{\frac{20}{t}})^{60}$

B) $6(2^{\frac{t}{20}})^{60}$

C) $6(2^{\frac{20}{t}})^{\frac{1}{60}}$

D) $6(2^{\frac{t}{20}})^{\frac{1}{60}}$

2. Polonium-210 has a half-life of 140days. Calculate the time required for the Polonium-210 to decay to one fifth of its initial value.

A) $\dfrac{\ln 0.2}{140 \ln 0.5}$

B) $\dfrac{140 \ln 0.2}{\ln 0.5}$

C) $\dfrac{\ln 2}{140 \ln 5}$

D) $\dfrac{140 \ln 2}{\ln 5}$

$$h(t) = \frac{200}{1+120e^{-0.1t}}$$

3. The function h models the height, in feet, of the oak tree when it is t years old. Based on the model, how old is the oak tree if it is 150 feet tall?

A) $10\ln 36$

B) $0.1\ln 36$

C) $10\ln 360$

D) $0.1\ln 360$

4. The model of the population of rabbits is given by $R(t) = 100e^{kt}$, where k is a constant and t is in days. It was observed that there were 200 rabbits after the second day. How many rabbits will be after 10 days?

A) 100^{32}

B) 200^{5}

C) 3200

D) 1000

3.7 Sequences

1. Sequence

※ **Sequence**: list of numbers which follows a certain rule

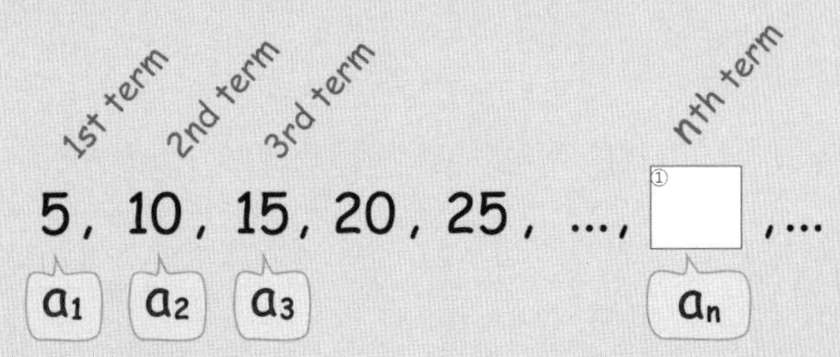

※ **The notation for sequences**

a_n represents the ②__________ of the sequence a.

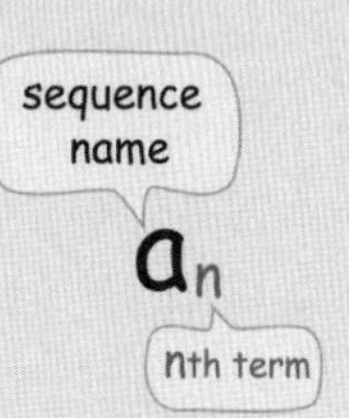

※ Sequences can be described in two ways:

① **Deductive rules** shows you the nth term formula

(defines how a_n depends on n, where n = 1, 2, 3, ...)

ex) $a_n = 5n$

② **Recursive rules** shows you the relationship between the terms

ex) $a_n = a_{n-1} + 5,\ a_1 = 5$

Blank : ① 5n ② nth term

EXAMPLE 1. Find the first 4 terms of the sequence.

① $a_n = 3^n + 2$

② $a_n = (-1)^{n+1}(n-2)$

③ $a_n = \dfrac{(-1)^n}{2^{n-1}}$

④ $a_n = \dfrac{n+1}{n}$

⑤ $a_n = 3(a_{n-1} + 2),\ a_1 = 1$

⑥ $b_n = 2b_{n-1} - 1,\ b_1 = 2$

⑦ $a_n = a_{n-1} - a_{n-2}$, $a_1 = 1,\ a_2 = -1$

⑧ $b_n = b_{n-1} + b_{n-2}$, $a_1 = 1,\ a_2 = 3$

⑨ $b_{n+1} = \frac{1}{b_n}$, $b_1 = 3$

EXAMPLE 2. Write a general term a_n for the sequence.

☺ Tip : Usually when you have;

$1,2,3,4,.... =$ ①	even numbers $2,4,6,.... =$ ④
$1,-1,1,-1,1,.... =$ ②	odd numbers $1,3,5,.... =$ ⑤
$-1,1,-1,1,-1.... =$ ③	odd numbers $3,5,7,.... =$ ⑥

① $\frac{1}{1}, \frac{1}{2}, \frac{1}{3}, \frac{1}{4}, \cdots$

② $\frac{2}{1}, \frac{3}{2}, \frac{4}{3}, \frac{5}{4}, \cdots$

③ $-1,1,-1,1,\ldots$

④ $2,-4,6,-8,10,\ldots$

Blank : ① n ② $(-1)^{n+1}$ ③ $(-1)^{n}$ ④ 2n ⑤ 2n–1 ⑥ 2n+1

⑤ $1, -\frac{1}{3}, \frac{1}{5}, -\frac{1}{7}, \frac{1}{9} \cdots$

⑥ $-7, 14, -21, 28, -35, \ldots$

⑦ $4^{1/3}, 4^{1/5}, 4^{1/7}, \ldots$

⑧ $x^{1/2}, x^{1/3}, x^{1/4}, \ldots.$

⑨ $\frac{a^2}{4}, \frac{a^3}{5}, \frac{a^4}{6}, \ldots$

⑩ $\frac{5}{6^2}, \frac{5^2}{6^3}, \frac{5^3}{6^4}, \frac{5^4}{6^5}, \ldots$

⑪ $\frac{1}{x}, \frac{2}{x}, \frac{3}{x}, \ldots.$

⑫ $x-1, x-2, x-3, \ldots.$

⑬ $1 \cdot 1,\ 2 \cdot 3,\ 3 \cdot 5,\ 4 \cdot 7, \ldots$

2. Arithmetic Sequences

※ **Arithmetic Sequence**

An arithmetic sequence has a constant difference, d, between two consecutive terms.

1st term, 2nd term, 3rd term, 4th term, nth term

a , $a+d$, $a+2d$, $a+$ ① ☐ d, ⋯ , ② ☐

+d +d +d

where a is the ③________ term, and d is a ④__________ __________.

General term (nth term): $a_n = a + (n-1)d$ (1st term; Common difference)

Recursive definitions: $a_{n+1} = a_n + d$

Notice that the general term a_n of arithmetic sequenc is in ⑤__________form;

where the common difference = ⑥______________.

EXAMPLE 3. Determine whether the sequence is arithmetic. If it is arithmetic, find the common difference.

① $a_n = 2 + 3(n-1)$ ② $a_n = 5 - 2n$

Blank : ① 3 ② a + (n–1)d ③ first ④ common difference ⑤ linear ⑥ slope d

③ $a_n = 2 - 7^{n-1}$

④ $a_n = \dfrac{1}{5n+6}$

⑤ $a_n = \dfrac{n-1}{3}$

⑥ $a_n = 3 + n^2$

EXAMPLE 4. Find the common difference d and the general term (nth term) of given sequence.

① 2, 5, 8, 11, . . .

② 7, 1, -5, -11, . . .

③ $2, \frac{5}{2}, 3, \frac{7}{2}, \ldots$

④ $8, 8\frac{1}{4}, 8\frac{1}{2}, 8\frac{3}{4}, 9, \ldots$

⑤ $x-1, x-3, x-5, \ldots$

⑥ $2t+1, 2t+7, 2t+13, \ldots$

⑦ $14\sqrt{3}, 19\sqrt{3}, 24\sqrt{3}, \ldots$

⑧ $\ln 3, \ln 6, \ln 12, \ln 24, \ldots$

EXAMPLE 5. Find the general term (nth term) of given arithmetic sequence.

① 6th term is 17, 12th term is 29.

② 3rd term is 13, 8th term is 38.

EXAMPLE 6. Find the number of terms in the sequence.

① 3, 5, 7, . . ., 37

② 7, 3, -1, . . . , -81

3. Geometric Sequences

※ **Geometric Sequence**

An geometric sequence has a constant ratio, r, between two consecutive terms:

1st term, 2nd term, 3rd term, 4th term

$a,\ ar,\ ar^2,\ ar^{①\square},\ \dots,\ ②\square$

×r ×r ×r

where a is the ③________ term, and r is a ④____________ ____________.

General term (nth term): $a_n = ar^{n-1}$ (exponential form)

(Common ratio: r; 1st term: a)

Recursive definitions: $a_{n+1} = a_n r$

Notice that the general term a_n of geometric sequenc is in ⑤___________form;

where the common ratio = ⑥_______________.

EXAMPLE 7. Find the common ratio r and the general term (nth term) of given sequence.

① $2, -1, \frac{1}{2}, -\frac{1}{4}, \dots$

② $-\frac{1}{4}, \frac{1}{2}, -1, \dots$

Blank : ① 3 ② ar^{n-1} ③ first ④ common ratio ⑤ exponential ⑥ base r

③ $\sqrt{2},\ 2,\ 2\sqrt{2},\ 4, . . .$

④ $5, -5, 5, -5, . . .$

⑤ $x-3,\ -3x+9,\ 9x-27, . . .$

⑥ $2x,\ 10x,\ 50x, . . .$

⑦ $x^{a+2}, x^{a+5}, x^{a+8}, . . .$

⑧ $x^{1/3},\ x^{2/3},\ x\ ,\ x^{4/3},\ . . .$

⑨ $\ln 2, \ln 2^3, \ln 2^9, \ln 2^{27}, . .$

⑩ $\sqrt{3}, \sqrt[3]{3}, \sqrt[6]{3}, 1, \cdots$

EXAMPLE 8. Find the general term (nth term) of given geometric sequence.

① 6th term is $\frac{1}{81}$, 9th term is $\frac{1}{3}$.

② 7th term is 4, 12th term is 128.

EXAMPLE 9. Find the number of terms in the sequence.

① $\frac{3}{4}, \frac{3}{2}, 3, 6, \ldots, 192$

② $-6, -12, -24, \ldots, -384$

4. Three Consecutive terms of Sequence

If we have 3 consecutive terms a, b, c;

arithmetic : ①____________________

geometric : ②____________________

EXAMPLE 10. Given the sequence $2, x, y, 9$. If the first three terms form an arithmetic sequence and the last three terms form a geometric sequence, find x and y. $(2 < x < y < 9)$

AP Style Problem

1. Consider the sequence;

$$-\frac{3}{2^3}, \frac{7}{2^6}, -\frac{11}{2^9}, \frac{15}{2^{12}}, \cdots$$

Which of the following is the general term a_n of the sequence?

A) $(-1)^{n+1}\dfrac{4n-1}{2^{3n}}$

B) $(-1)^{n}\dfrac{4n-1}{2^{3n}}$

C) $(-1)^{n+1}\dfrac{4n+1}{2^{3n}}$

D) $(-1)^{n}\dfrac{4n+1}{2^{3n}}$

Blank : ① $b - a = c - b$ ② $\frac{b}{a} = \frac{c}{b}$

2. For a geometric sequence a_n, the sixth term is $\frac{3}{2}$, and the ninth term is 12. What is the tenth term?

A) 8

B) 9

C) 24

D) 36

3. Which of the following is an arithmetic sequence?

A) $1,-1,1,-1,....$

B) $5x^2, 5x^4, 5x^6, ...$

C) $\ln x, \ln x^4, \ln x^7, ...$

D) $3^{a+1}, 3^{2a+1}, 3^{3a+1}, ...$

4. For an arithmetic sequence a_n, $a_8 - a_2 = 12$. Then what is $a_{20} - a_{15}$?

A) 10

B) 5

C) 2.5

D) 1

5. For a geometric sequence b_n, $\frac{b_6}{b_3} = 27$. What is $\frac{b_{21}}{b_{15}}$?

A) 1

B) 3^3

C) 3^6

D) 3^9

6. The first three terms of a geometric sequence are $2x+4$, $x+5$, $x+1$, where x is a real number. Find the possible values of x.

A) 3

B) 7

C) 3, -7

D) 7, -3

7. Which of the following graph corresponds to geometric sequence with common ratio $\frac{1}{2}$?

A)

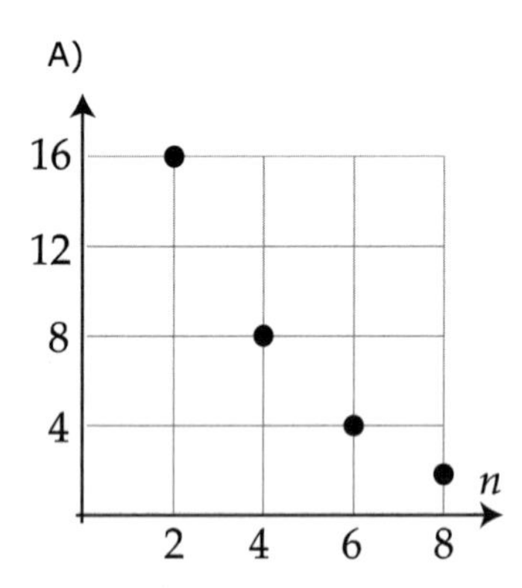

B)

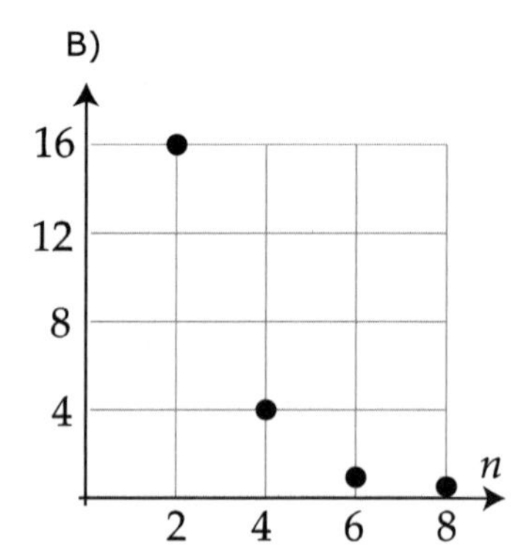

C)

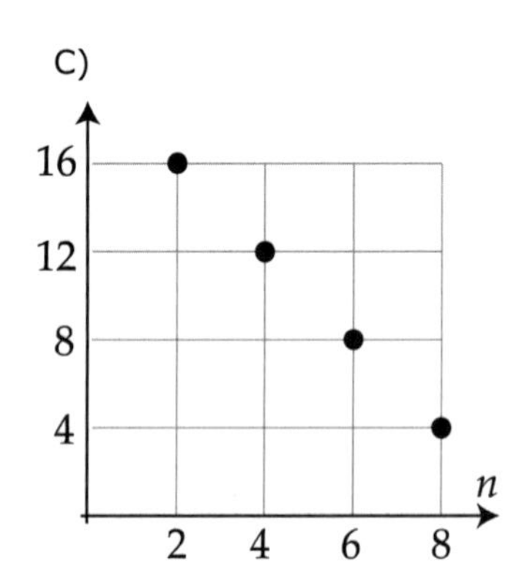

D)

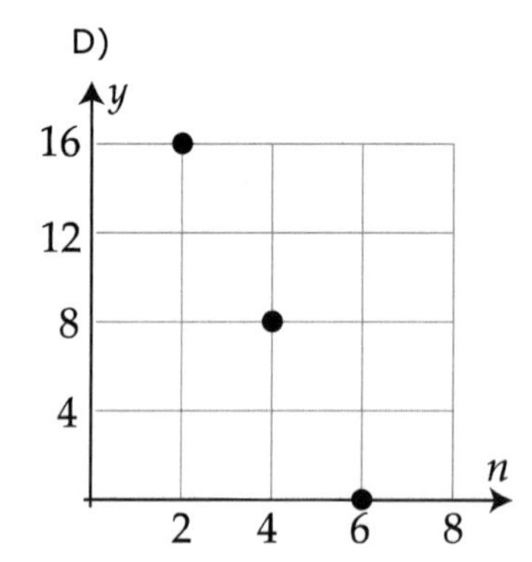

3.8 Regression

1. Regression

A Scatter Plot has points that show the relationship between two variables x and y.

Regression model gives a function that describes the relation between two sets of data.

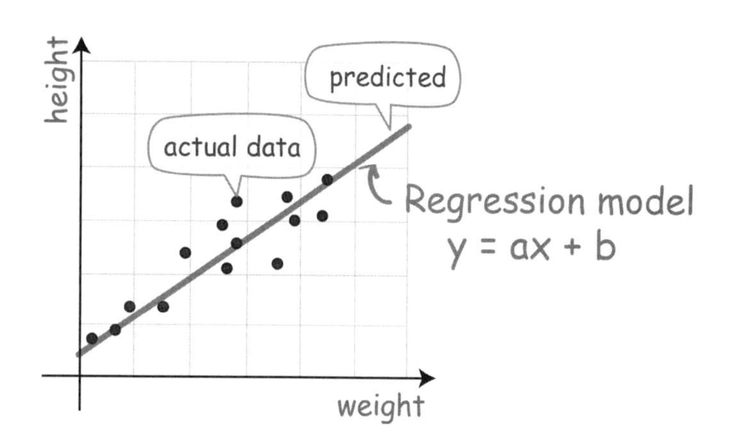

The data will often fit a linear, quadratic, or exponential type of model.

We use the regression model to ① ____________ the new data.

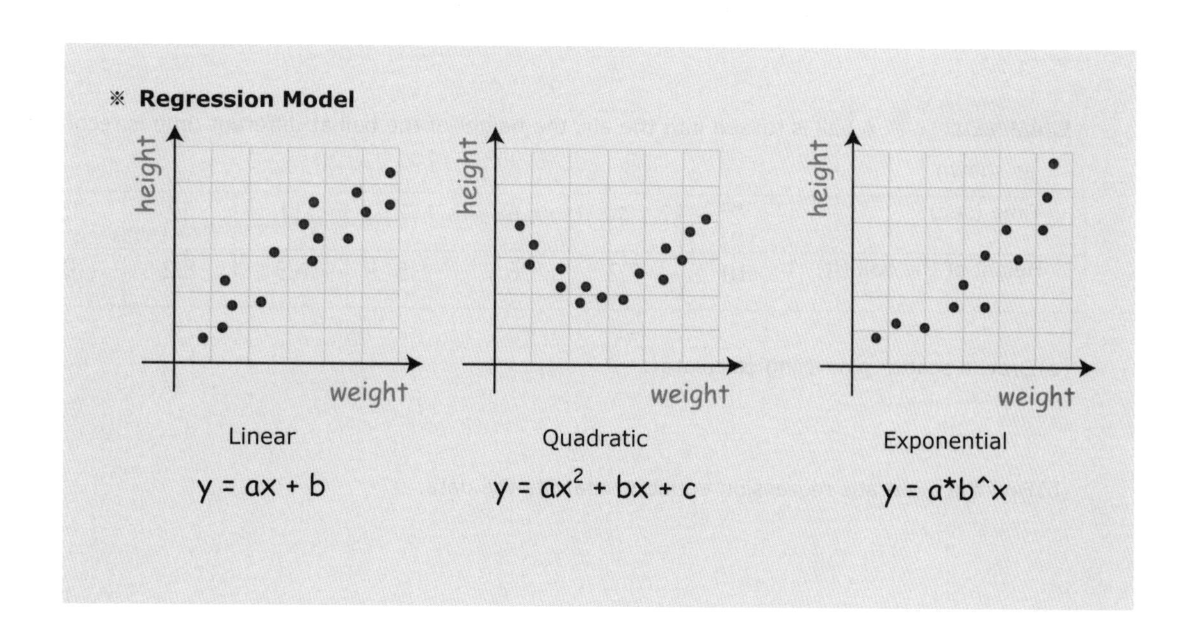

Blank : ① predict

EXAMPLE 1. The weight and height of six men are given below.

Weight(lb)	140	155	159	179	192	200
Height(in)	60	62	67	70	75	82

(a) Draw a scatter plot using calculator.

(b) Find the linear regression equation that fits the data.

(c) Interpret the slope.

(d) Predict the height when the weight is 162lb.

☺ Tip: Interpreting the slope

When x is increased by 1, y is inc/decreased by (the slope number).

EXAMPLE 2. A ball is tossed into the air, the height of the ball at different time is recorded as shown.

Time (sec)	2	4	6	7	11	12	15
Height of the ball (ft)	10	7.3	6	5	6.3	8.2	9

(a) Draw a scatter plot using calculator.

(b) Find the quadratic regression equation that fits the data.

(c) Predict the height of the ball when t = 9sec.

EXAMPLE 3. The table shows the population (in thousands) of a town after n years from 2000.

Year since 2000	1	2	4	5	6	8
population (in thousands)	3	8	26	84	230	580

(a) Draw a scatter plot using calculator.

(b) Find the exponential regression equation that fits the data.

(c) Predict the population in 2003.

EXAMPLE 4. The table shows the height of the pine tree and the age of the tree.

age	1	2	3	4	5	6
Height of the tree (ft)	6	13	16.5	18.5	19	19.7

(a) Draw a scatter plot using calculator.

(b) Find the logarithmic regression equation that fits the data.

2. Is the Regression model appropriate?

Let's say we want to perform a linear regression analysis.
Is that linear model appropriate for the data?

If the regression model we found is appropriate for explaining the data
, then the *residual plot* should show (①a pattern / no pattern).

If the *residual plot* shows a pattern, then other types of model might be more suitable.

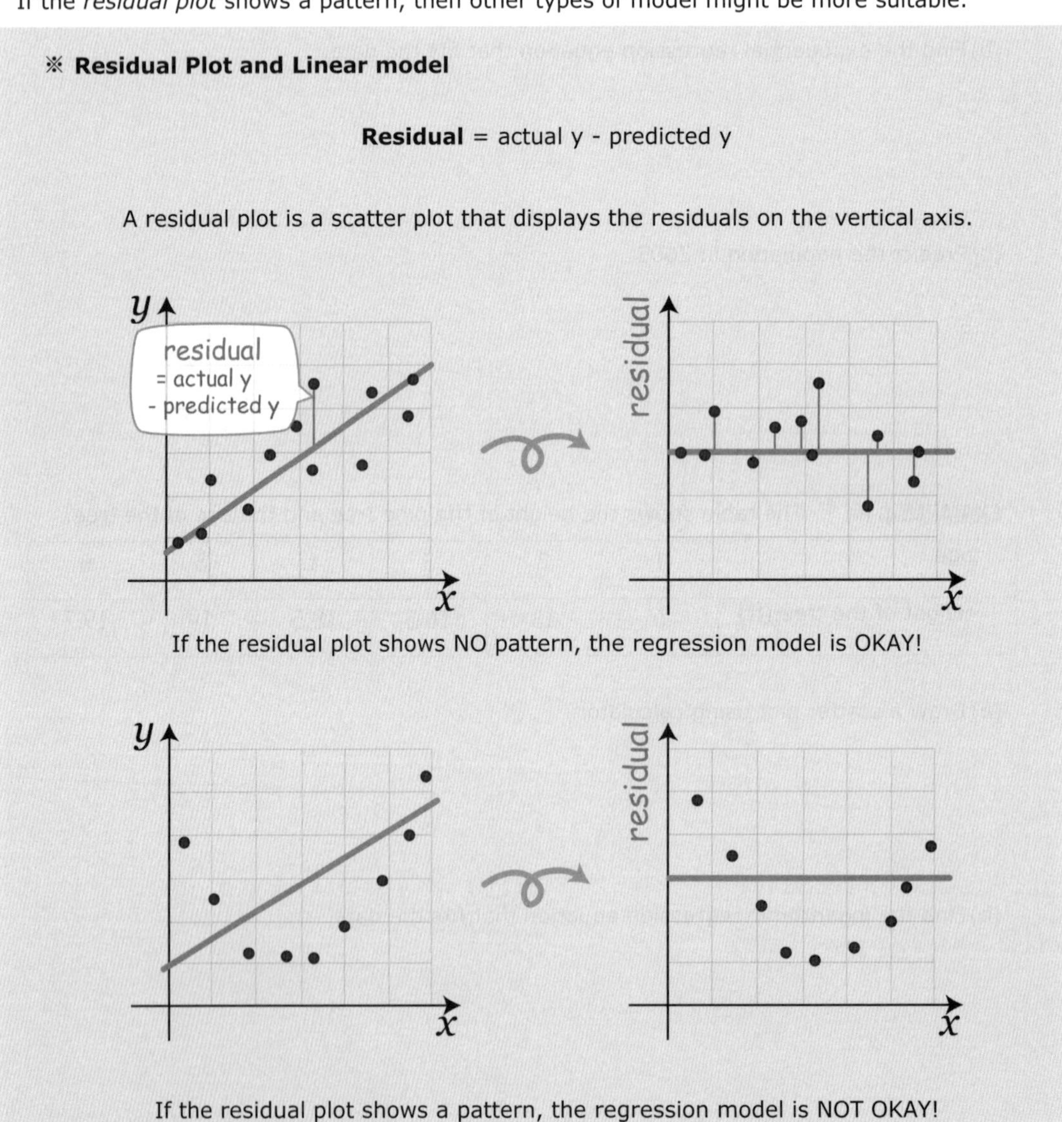

※ **Residual Plot and Linear model**

Residual = actual y - predicted y

A residual plot is a scatter plot that displays the residuals on the vertical axis.

If the residual plot shows NO pattern, the regression model is OKAY!

If the residual plot shows a pattern, the regression model is NOT OKAY!

Blank : ① no pattern

※ **Facts about residuals**

① Positive residuals mean the actual value is (①above/below) predicted value.

➔ (②Overestimated /Underestimated)

② Negative residuals mean the actual value is (③above/below) predicted value.

➔ (④Overestimated /Underestimated)

③ All of the residuals should add up to ⑤ ______.

; This is because some points are above the line and some are below the line.

④ If the residual plot for an exponential regression model shows no pattern, it suggests that the exponential model is appropriate for describing the relationship.

EXAMPLE 5. Find the linear regression model and the residual plot. Determine whether the residual plot suggest a linear relationship or not.

①

x	y
5	3
10	4
15	9
20	7
25	13
30	15

②

x	y
100	490
90	520
80	460
70	415
60	500
50	470

③

x	y
2	5
4	15
6	26
8	23
10	11
12	3

④

x	y
5	20
10	29
15	34
20	37
25	39
30	40

Blank : ① above ② underestimated ③ below ④ overestimated ⑤ 0

AP Style Problem

1. In recent years, the number of birth of twins in the United States has been declining. The number of twin births, in thousands, is given below where t is the number of years since 2010.

t	1	2	3	4	6
number of twin births	12.1	7.5	4.4	2.6	2.1

1) An exponential regression $R(t)$ is used to model these data. Find the regression model.

A) $-1.928x+11.911$

B) $-1.928+11.911x$

C) $0.696(14.817)^x$

D) $14.817(0.696)^x$

2) Use the exponential regression $R(t)$ to estimate the number of birth of twins in 2015 assuming the decline in births continues as predicted by the model.

A) 1305

B) 2425

C) 1.305

D) 2.425

3) Which of the following statement is true about the 2013 data (when t = 3)?

A) The exponential regression $R(t)$ underestimated the number of twin births.

B) The exponential regression $R(t)$ overestimated the number of twin births.

C) The predicted value is the same as the actual value of the number of twin births.

D) There is not enough information.

2. We perform a linear regression analysis. Which residual plot suggest the linear model would be appropriate than other model?

A)

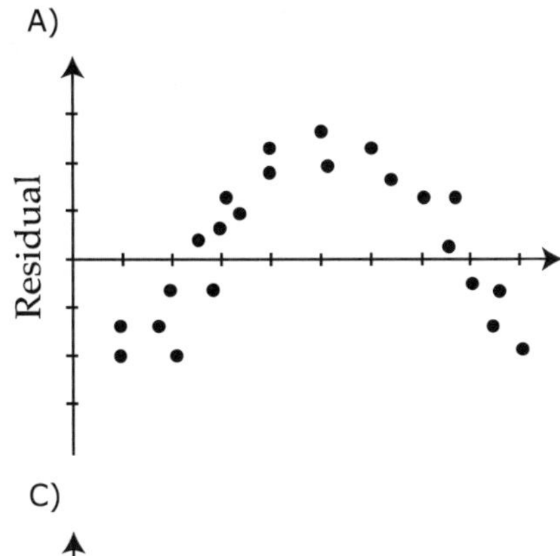

B)

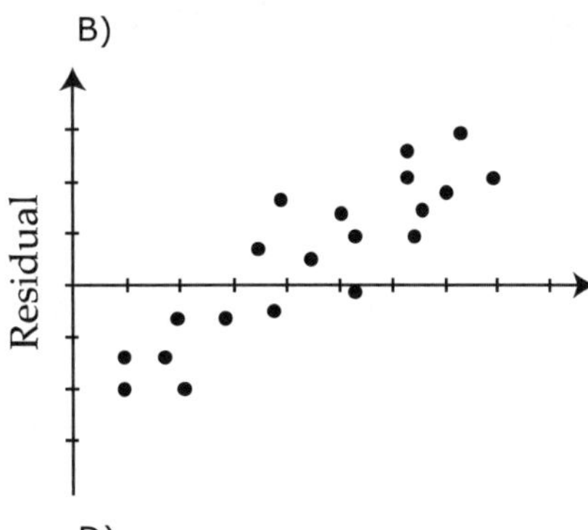

C)

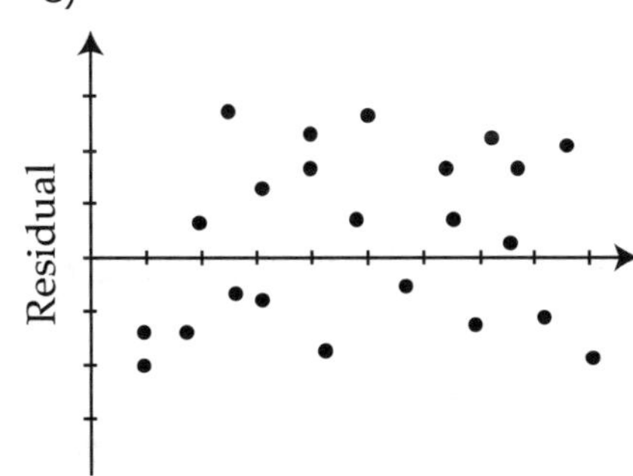

D)

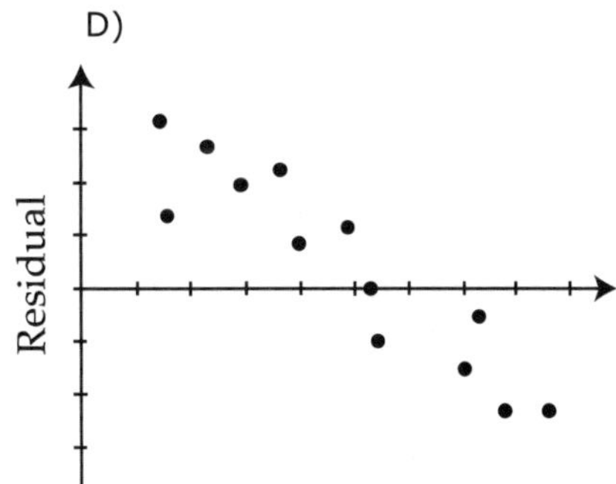

3. We perform a linear regression analysis. The following is a linear residual plot created from the data. Based on the residual plot, which of the following statements is(are) true?

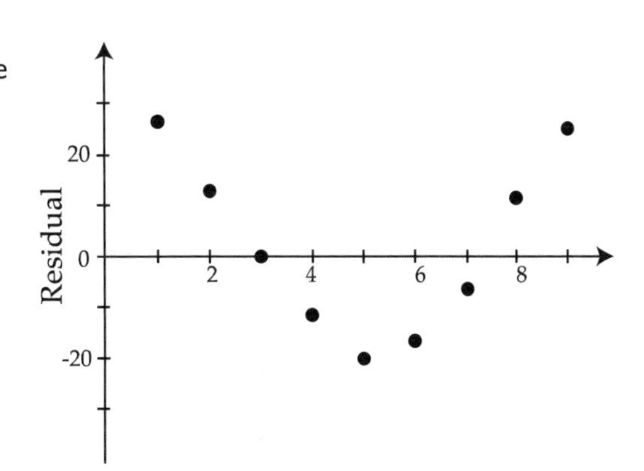

I. The relationship between x and y is linear.

II. The regression equation would underestimate at x = 6.

III. The predicted value is the same as the actual value at x = 3.

A) I

B) III

C) II, III

D) I, III

3.9 Semilog Plots

1. Log (y) and x

For the given table, take common log (= log) for the y value, and;

x	0	3	6	11	14
y	2	8	45	270	1000
$log\ y$	0.301	0.903	1.653	2.431	3

plot the points (x, y) on the x,y plane;

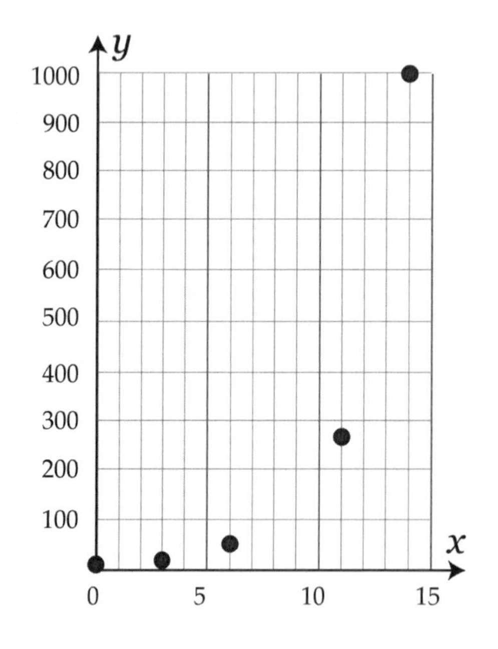

It shows an (①linear/exponential) graph in x,y plane.

pot the points $(x, log\ y)$ on the x,log y plane;

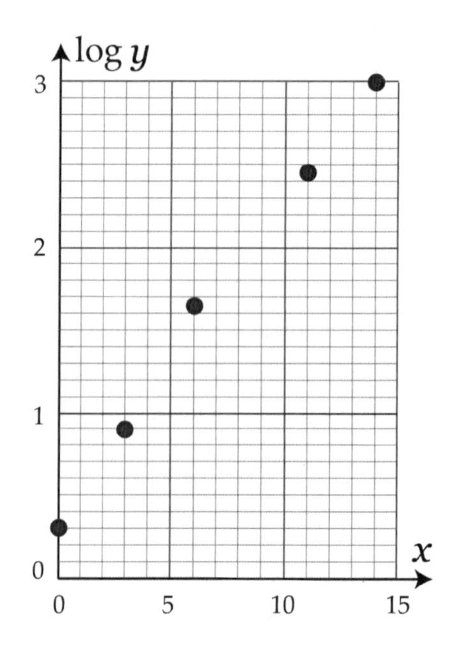

It shows an (②linear/exponential) graph in x, log y plane.

Blank : ① exponential ② linear

We can notice that;

When the graph is exponential in x, y plane, then it is ①____________ in x, log y plane.

☺ Proof : Let's see what happens when we take 'log' on an exponential model;

$$y = ab^x$$
$$\log y = \log ab^x$$
$$\log y = \log a + \boxed{②\quad}$$
$$\log y = \log a + x\boxed{③\quad}$$
$$\log y = A + xB$$

2. Log scale vs Linear scale

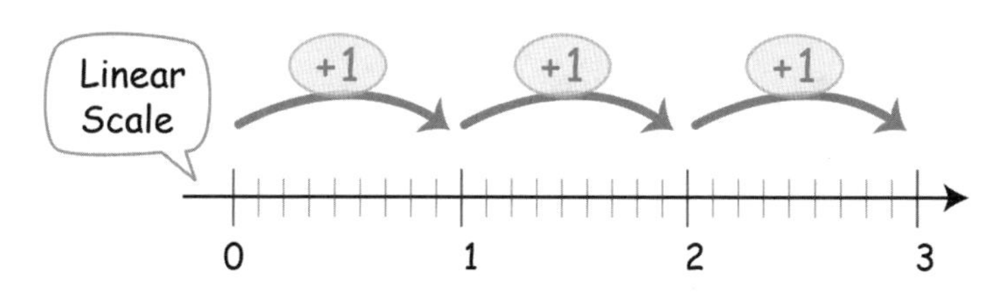

On linear scale, we add 1 to move up on the scale.

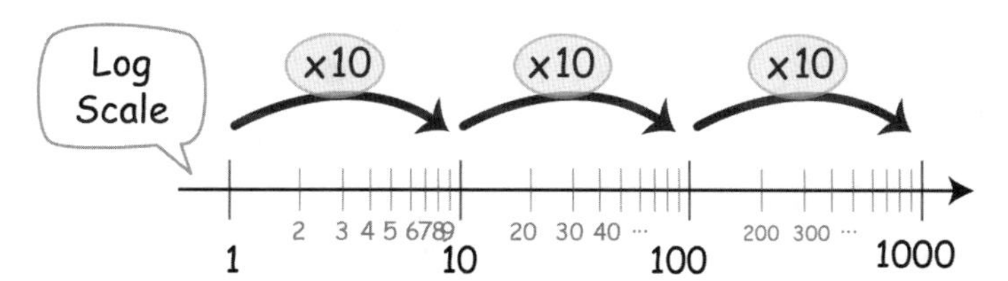

On logarithmic scale, we ④____________ to move up on the scale.

Using a logarithmic scale (or log scale) we can display a numerical data with very wide range in a compact way.

Blank : ① linear ② log b^x ③ log b ④ multiply 10

3. Semilog Plot

Semilog plot has one axis on a ①__________ scale, the other on a linear scale.

In other words, if the y-axis is in logarithmic scale then the x-axis must be in linear scale and vice versa.

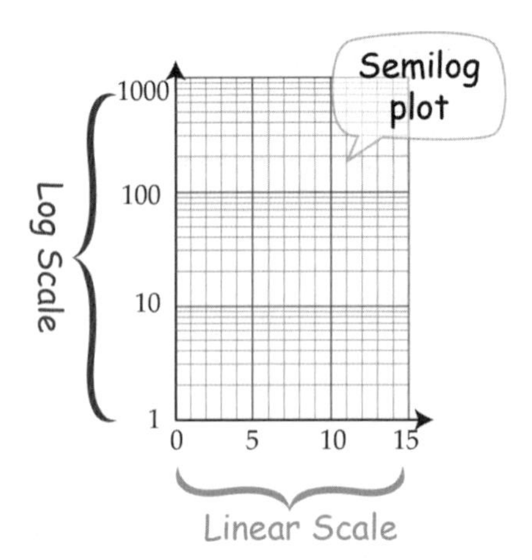

This time, plot the points in a semilog plot;

x	0	3	6	11	14
y	2	8	45	270	1000

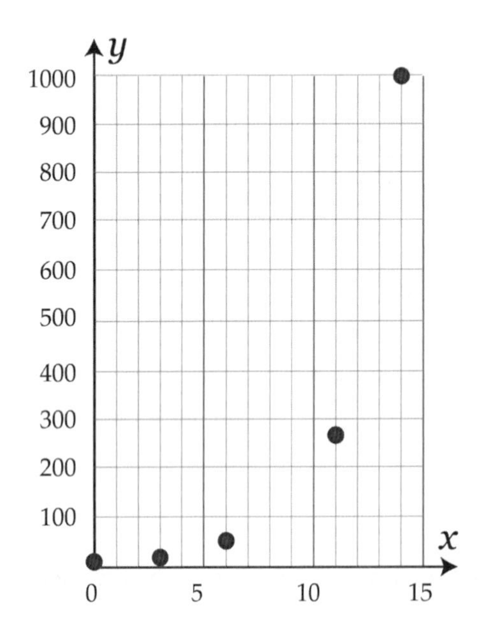

It shows an (②linear/exponential) graph in x,y plane.

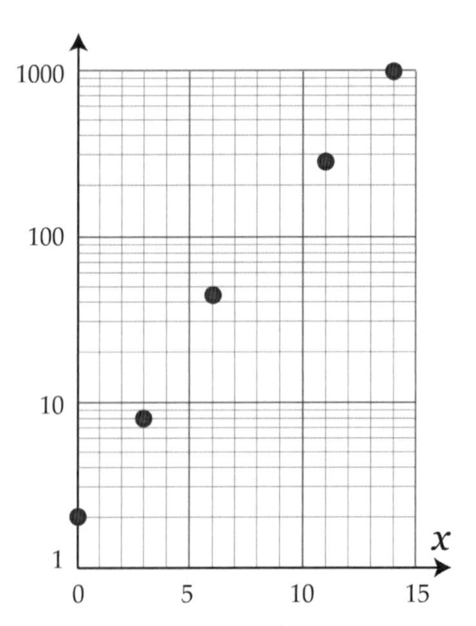

It shows an (③linear/exponential) graph in semi log plot.

Blank : ① log ② exponential ③ linear

※ **Facts about Semilog plot**

1) On logarithmic scale, we **multiply 10** to move up on the scale.
2) Log scales cannot show zero.
3) If it is **exponential** in xy plane, then it is **linear** in semilog plot and x,log y plane.

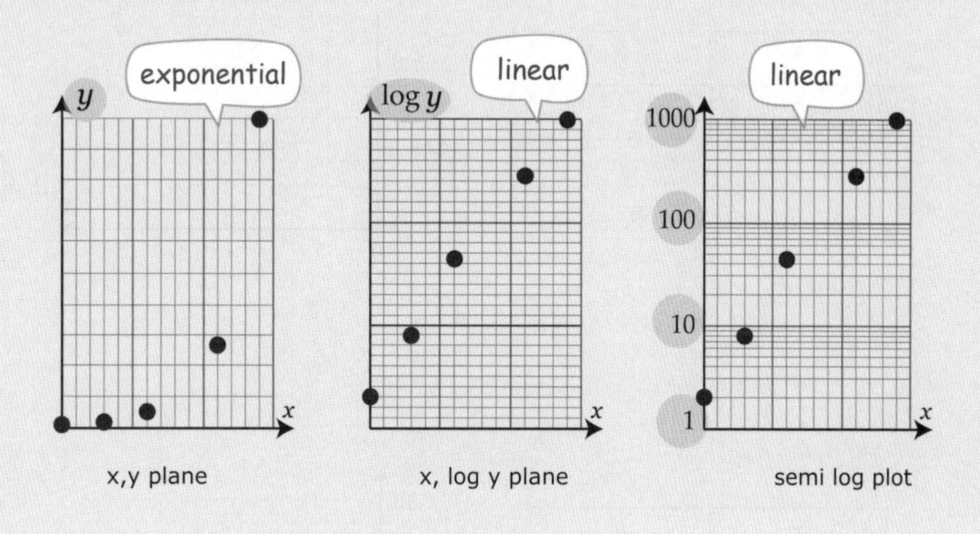

EXAMPLE 1. Plot the points on the semilog plot. Determine whether y = f(x) is an exponential or not.

①

x	y
-3	0.05
-2	0.14
-1	0.37
0	1
1	2.72
2	7.39

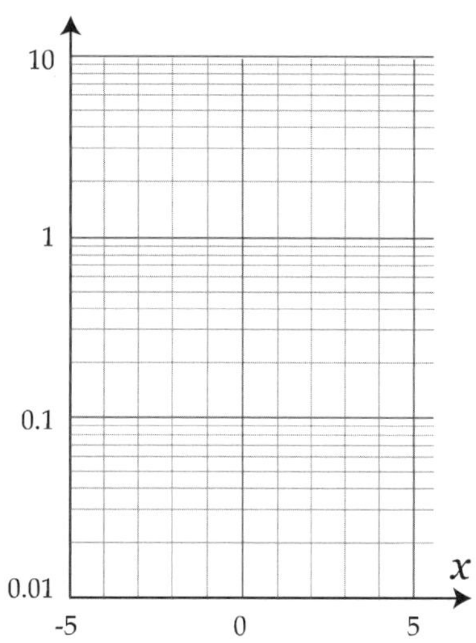

②

x	y
1	128
2	64
3	27
4	8
5	3

1000
100
10
1
0
5
x

③

x	y
2	0.05
3	0.1
7	0.8
10	3

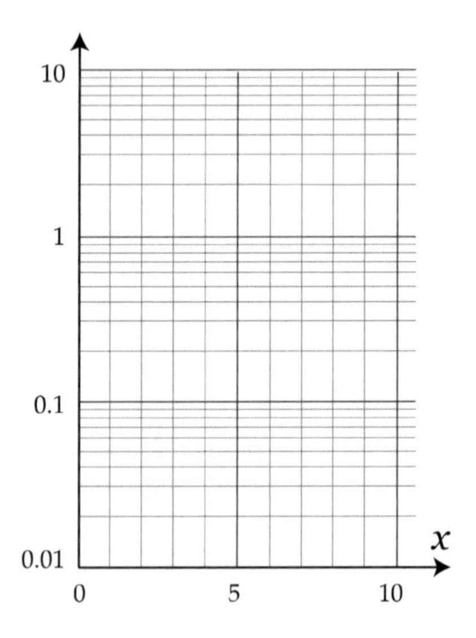

AP Style Problem

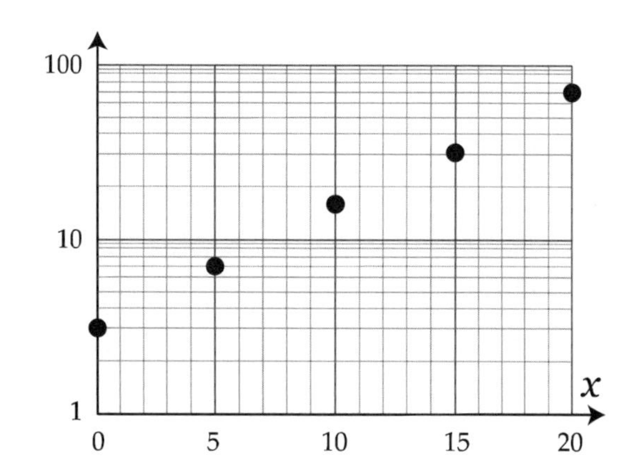

1. A semi lot plot of points from the function $f(x)$ is given. Which of the following could represent the table of $f(x)$?

A)

x	0	5	10	15
$f(x)$	5	7	16	30

B)

x	0	5	10	15
$f(x)$	3	7	10.6	12

C)

x	0	5	10	15
$f(x)$	5	7	10.6	12

D)

x	0	5	10	15
$f(x)$	3	7	16	30

2. Find the equivalent equation for the graph shown.

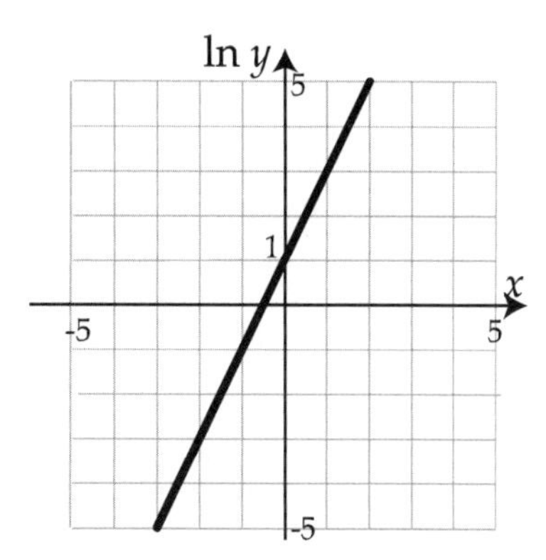

A) $y=2x+1$

B) $y=e^{2x}$

C) $y=\ln(2x+1)$

D) $y=e(e^2)^x$

3. Which of the following is NOT an exponential function $y = f(x)$?

A)

x	$\ln y$
1	2
2	7
3	12
4	17

B)

x	y
1	2
2	6
3	18
4	54

C)

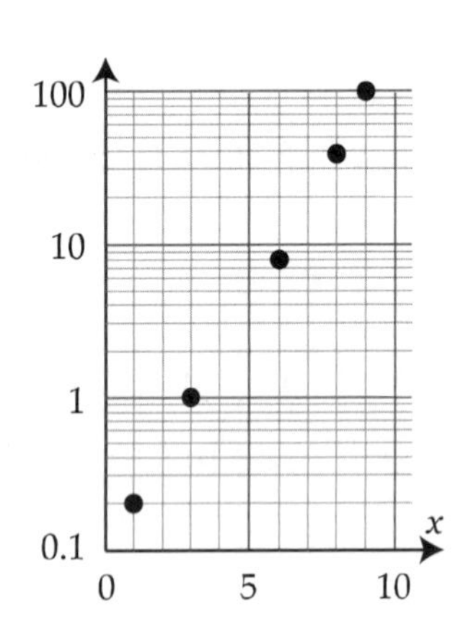

D)

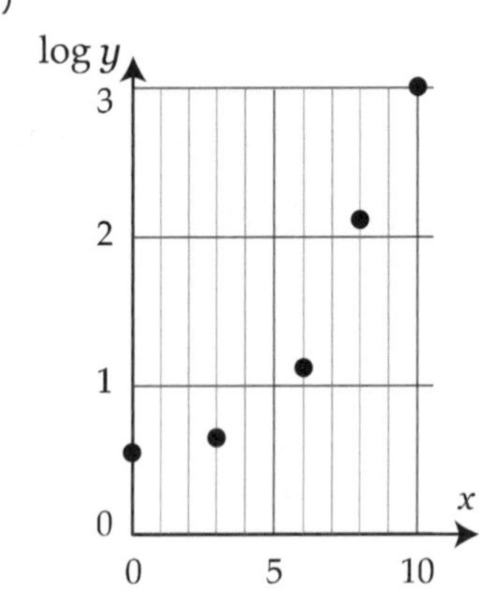

Free Response Questions (from ch3)

1. Exponential and Logarithmic Function

(Non-Calculator) The function f and g are given as

$$f(x)=1+e^{-x} \text{ and } g(x)=\ln(1-x).$$

(a) Find $\lim_{x\to\infty} f(x)$ and $\lim_{x\to-\infty} f(x)$.

(b) The graph of g is a transformation of the graph of $y=\ln x$. Give a full description of the transformation.

(c) Find $\lim_{x\to-\infty} g(x)$ and $\lim_{x\to1^-} g(x)$.

(d) If $(f\circ g)(x)=\dfrac{ax+b}{cx+d}$, then what are the values a, b, c, and d?

(e) The function h is given $h(x)=6e^{x}$. Find the x coordinate of the intersection point of f and h.

2. Exponential Growth Model

(Calculator) A dose of drug is injected into the body of patient. The initial dose was 3cc. And 40% of the drug remains in the body after 4 hours.

(a) The quantity of the drug remaining in the body after t hours can be modeled by

$Q(t) = Q_0 e^{kt}$. Find the constant k.

(b) How much of the initial dose of 3cc remains after 6 hours?

(c) After how long does half of the original amount remain?

(d) Use the exponential model $Q(t)$ to find the average rate of change from t = 4 to t = 6. Indicate units of measure, and interpret the meaning in the context.

Part 4

Trigonometry Definition and Graphs

4.1 Angles In Radian

1. Radian

We can measure *the size of the angle* by using *two units*,

① __________________ and ② ________________.

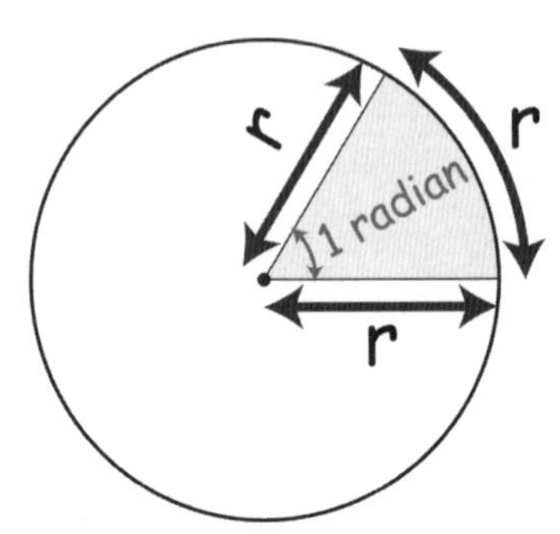

'Radian' is the angle made by taking the ③ ____________ and **wrapping it along the edge** of a circle

※ Facts about Radian and Degree

① *1 Radian* is about 57.2958°.

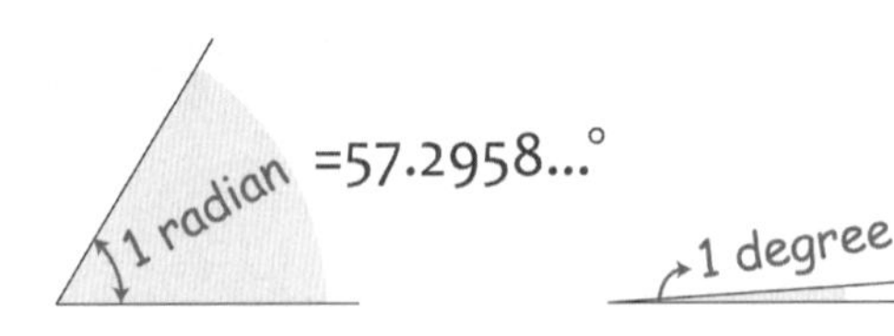

② The unit of *degree* is '°'.

The unit of *radian* is ④ ___________ or ⑤ ___________.

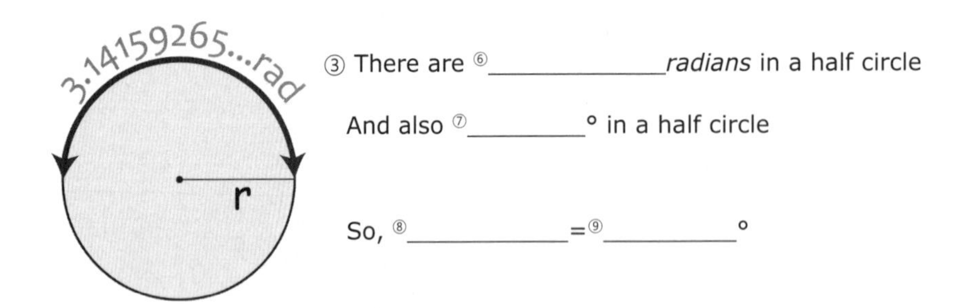

③ There are ⑥ _____________ *radians* in a half circle

And also ⑦ ________° in a half circle

So, ⑧ ___________ = ⑨ _________°

※ Radian and Degree

$$180° = \pi \text{ (rad)}$$

Blank : ① degree ②radian ③ radius ④ rad ⑤ none  ⑥ π  ⑦ 180  ⑧ π ⑨ 180

2. Converting Angles

Converting Degree → Radian

$$\text{Degree}^{\circ} \times \frac{\pi}{180^{\circ}}$$

Converting Radian → Degree

$$\text{Radian} \times \frac{180^{\circ}}{\pi}$$

EXAMPLE 1. Convert degrees into radians, and radians into degrees.

① 90°

② 100°

③ -135°

④ 150°

⑤ $\frac{\pi}{3}$

⑥ $\frac{5\pi}{3}$

⑦ $-\frac{4\pi}{3}$

⑧ $-\frac{\pi}{12}$

⑨ $\frac{2}{3}$

⑩ 7

☺ Memorize!

Degrees	0°	30°	45°	60°	90°	180°	270°	360°
Radians	①	②	③	④	⑤	⑥	⑦	⑧

Blank : ① 0 ② $\frac{\pi}{6}$ ③ $\frac{\pi}{4}$ ④ $\frac{\pi}{3}$ ⑤ $\frac{\pi}{2}$ ⑥ π ⑦ $\frac{3\pi}{2}$ ⑧ 2π

3. Angles in Standard position

When an angle is in "**standard position**", its vertex is at the Origin.

90° terminal side θ 180° 0° 360° initial side 270°

positive angle

-270° -180° 0° -360° θ -90°

negative angle

A **positive angle** goes *Counter Clock Wise* (CCW) from the X-axis (upward and to the left).

A **negative angle** goes *Clock Wise* (CW) from the X-axis (downward and to the left).

EXAMPLE 2. Draw a circle for each part and mark the points corresponding to the following angles.

① $\theta = -120^\circ$

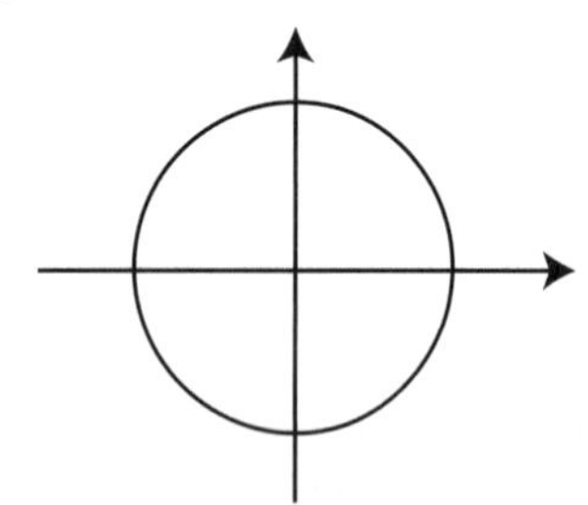

② $\theta = 720^\circ$

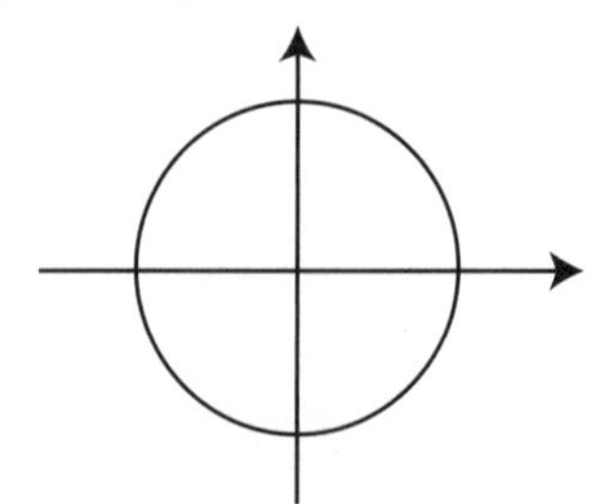

③ $\theta = \frac{5\pi}{4}$

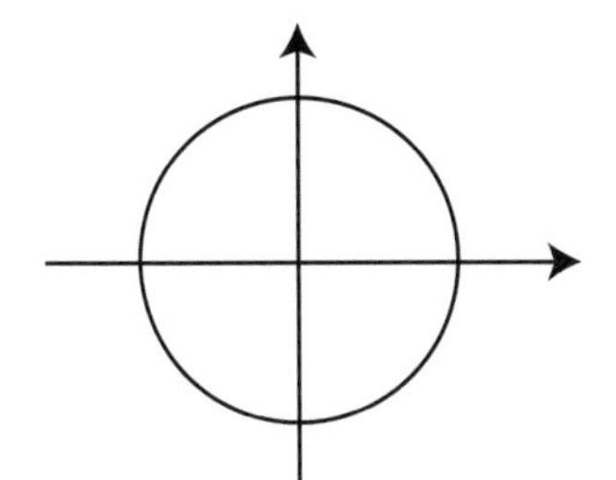

④ $\theta = \frac{5\pi}{2}$

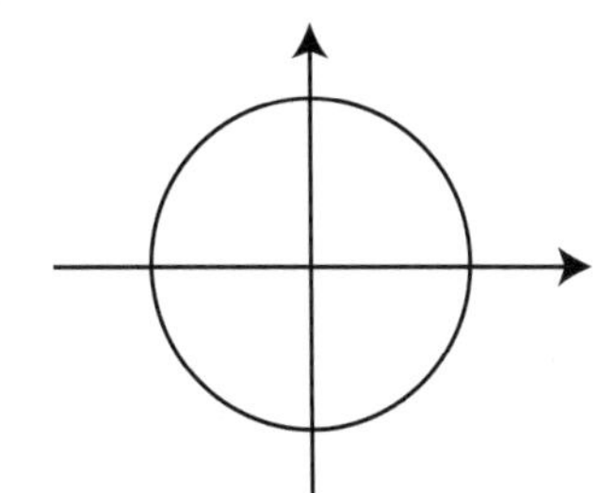

⑤ $\theta = -\frac{7\pi}{6}$

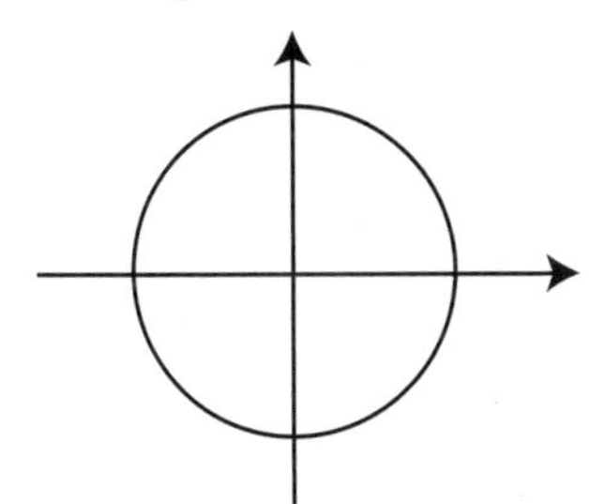

⑥ $\theta = \frac{11\pi}{6}$

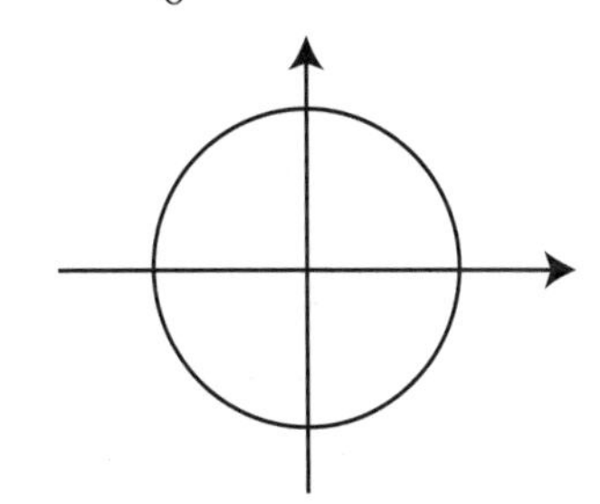

⑦ $\theta = \frac{17\pi}{6}$

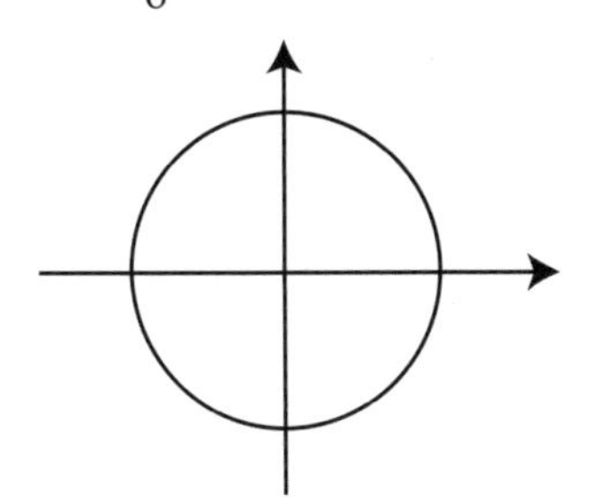

⑧ $\theta = -\frac{10\pi}{3}$

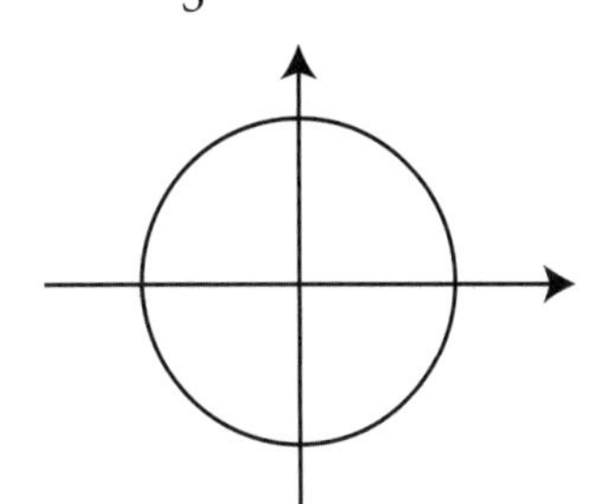

⑨ $\theta = -\frac{5\pi}{3}$

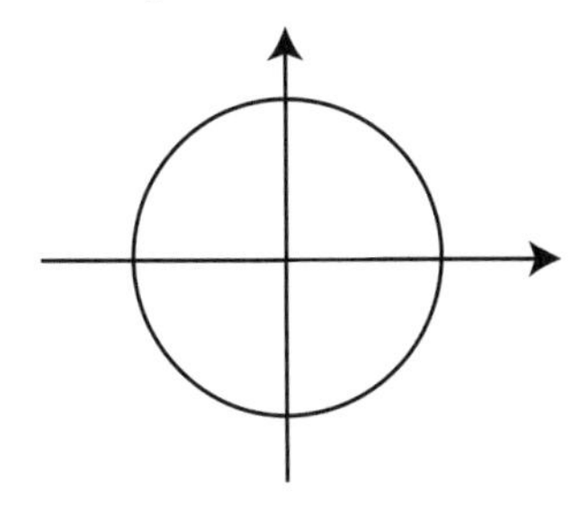

⑩ $\theta = -\frac{7\pi}{4}$

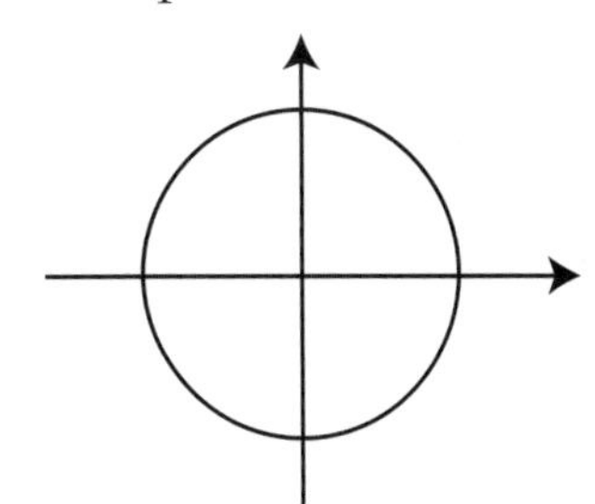

EXAMPLE 3. Fill in the blanks with appropriate angle.

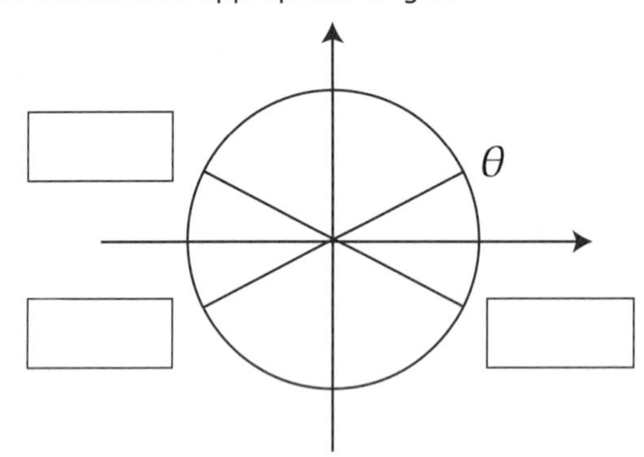

EXAMPLE 4. The measure of an angle in standard position is given. Find two positive angles and two negative angles that are coterminal with the given angle.

Coterminal angles are the angles that have the same terminal sides.

We can find the coterminal angle by adding or subtracting 360° or 2π to it.

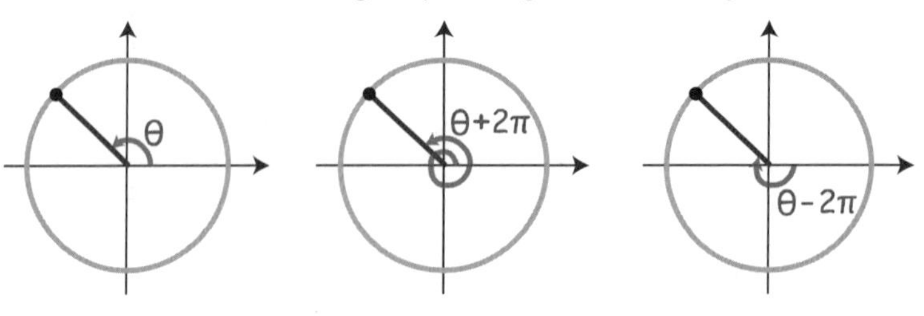

① 120°

② -145°

③ $\dfrac{4\pi}{3}$

④ $-\dfrac{3\pi}{4}$

4. Arc Length and Area of Sector

※ **Formulas for Arc Length** and **Sector Area**

If θ is a central angle (in *RADIAN*) in a circle of radius r, the arc length S and the area of sector A is;

arc length

$$S = r\theta$$

in Radians

sector area

$$A = \frac{1}{2}r^2\theta$$

in Radians

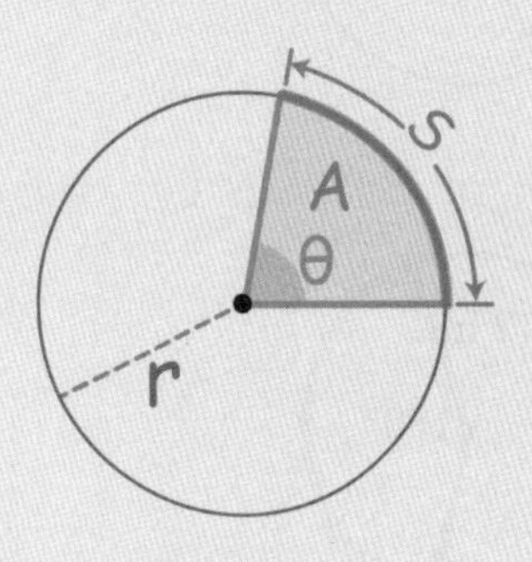

In geometry, we have learned that "the ratio of the measures of the angles equals the ratio of the corresponding lengths of the arcs subtended by these angles".

So, we can set;

$$\frac{S}{①\square} = \frac{②\square}{③\square}$$

$$S = ④\square$$

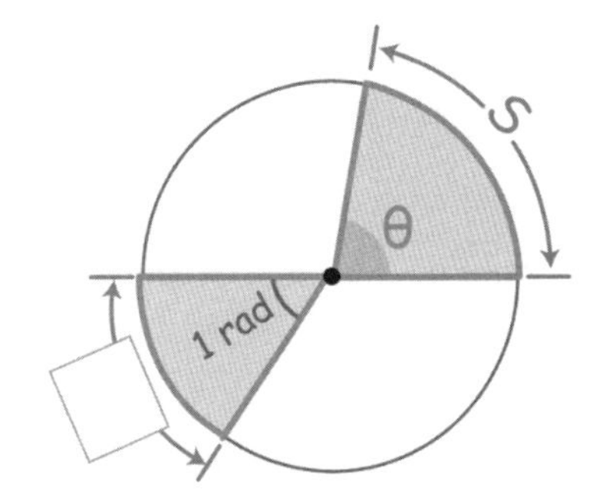

The ratio of the area of the sector to the area of the whole circle is the same as the ratio of angle θ to a full turn.

So, we can set;

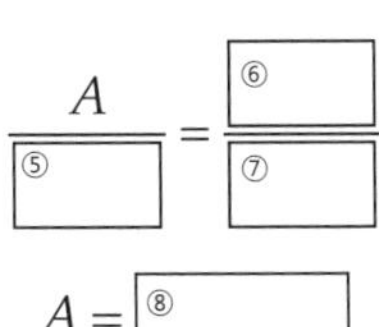

Blank : ① θ ② r ③ 1 ④ rθ ⑤ θ ⑥ πr^2 ⑦ 2π ⑧ $\frac{1}{2}r^2\theta$

EXAMPLE 5. Find the arc length and the area of the sector.

①

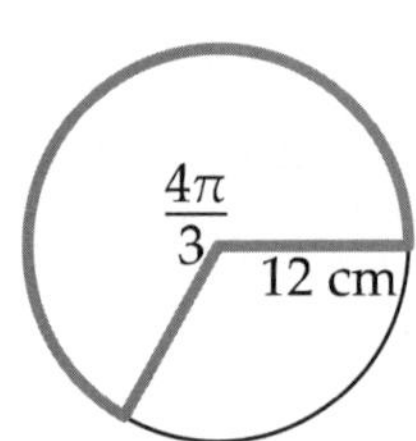

②

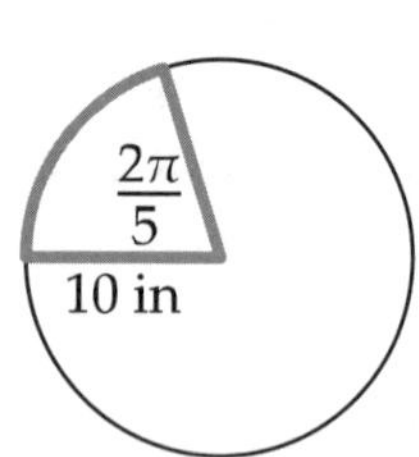

③

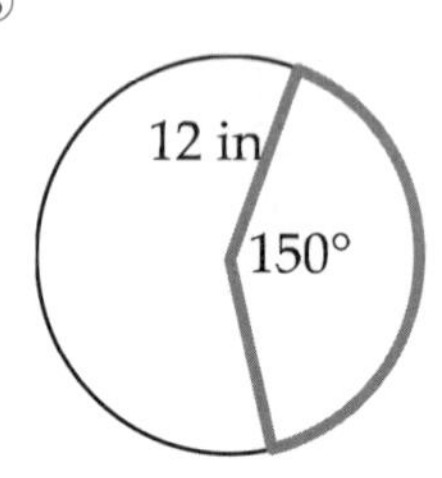

④

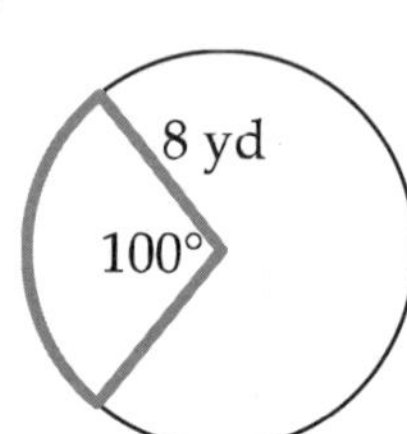

⑤

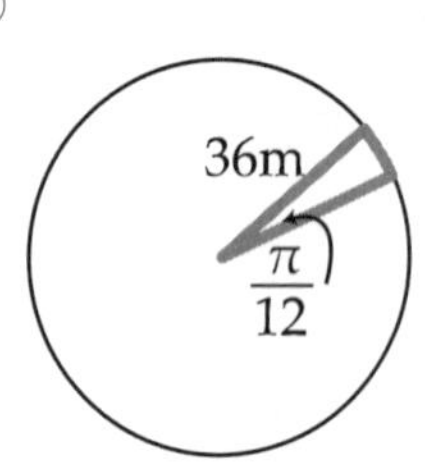

⑥

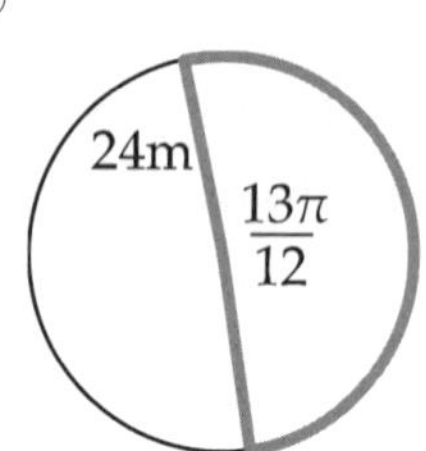

EXAMPLE 6. Find the measure of angle θ in radians

①

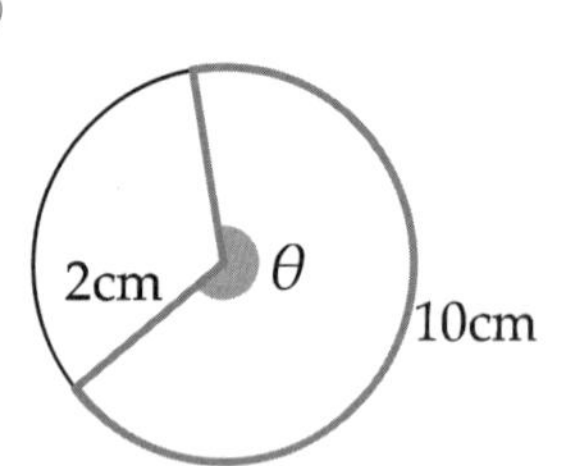

②

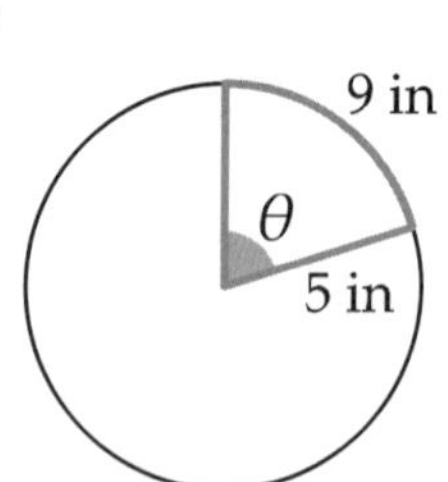

③

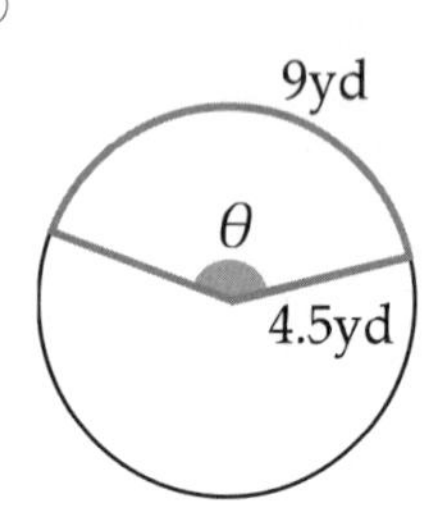

④

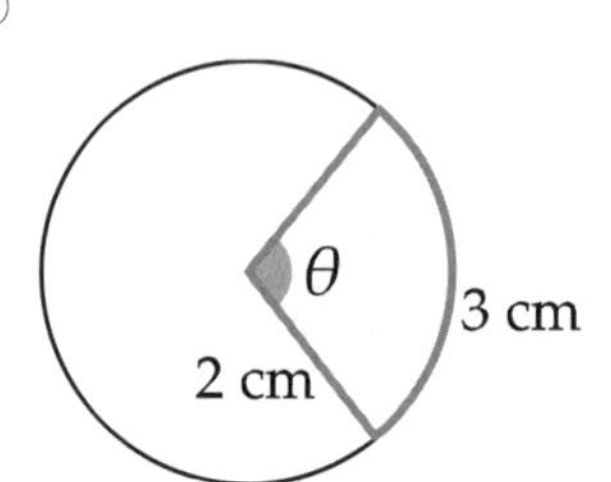

EXAMPLE 7. Two points, A and B, lie on the circumference of a circle of radius r cm. The minor arc AB has length 10 cm and subtends an angle of 2 at the center of the circle.

(a) Find the value of r.

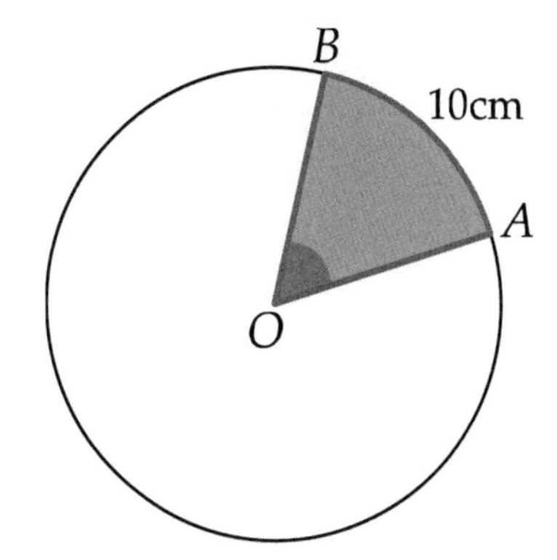

(b) Find the length of the major arc AB.

(c) Calculate the perimeter of the shaded region.

EXAMPLE 8. A sector of a circle has perimeter 12 cm and angle at the centre $\theta = 50°$. Find the radius of the sector.

AP Style Problem

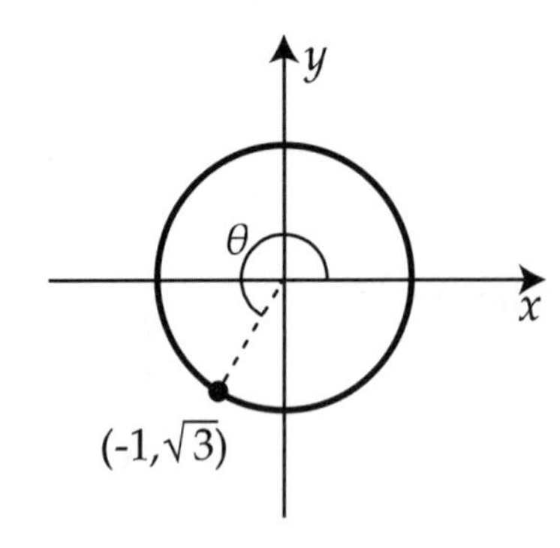

1. In the xy-plane above, angle θ is formed by the x-axis and the line segment shown. What is the measure, in radians, of angle θ?

A) $\frac{2\pi}{3}$

B) $\frac{4\pi}{3}$

C) $\frac{5\pi}{6}$

D) $\frac{7\pi}{6}$

2. The figure below shows an equilateral triangle ABC with side 6 in, and three arcs with centers at the vertices of the triangle. Calculate the perimeter of the figure.

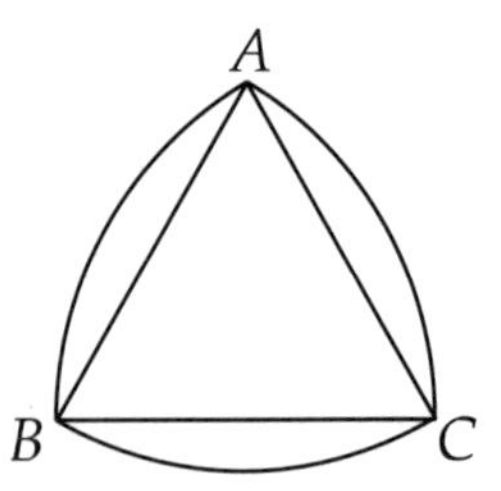

A) 3π inches

B) 6π inches

C) 9π inches

D) 12π inches

3. A cone in made by rolling a piece of paper as shown in the figure.

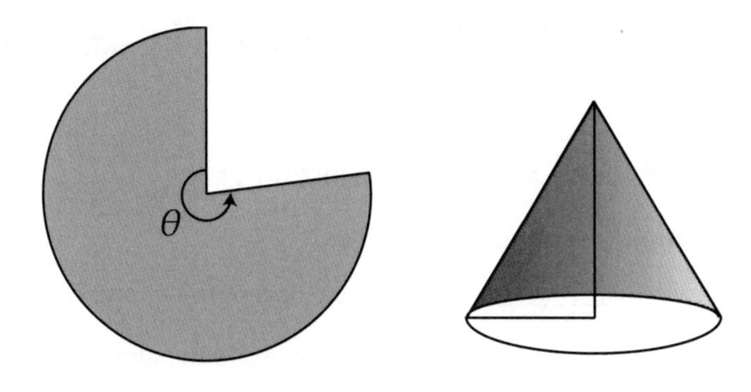

If the cone has a height 4 cm and base diameter 6 cm, find the size of the angle marked θ.

A) $\dfrac{4\pi}{5}$

B) $\dfrac{4\pi}{3}$

C) $\dfrac{6\pi}{5}$

D) $\dfrac{2\pi}{3}$

4. A sheep is tied to a pole which is 3m from a long fence. The length of the rope is 6m. Find the area, in square meters, which the sheep can feed on.

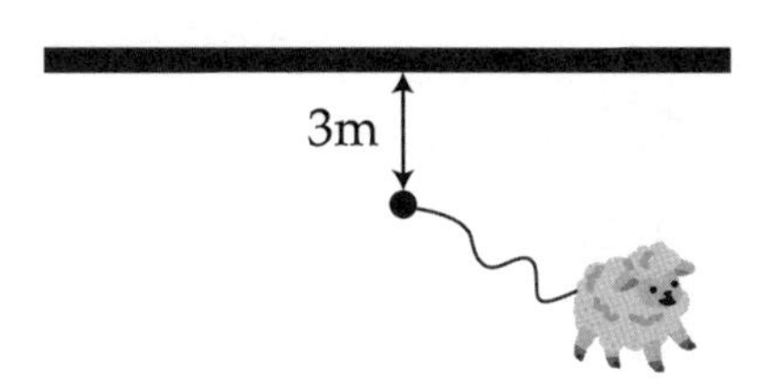

A) $6\sqrt{3}+8\pi$

B) $6\sqrt{3}+24\pi$

C) $9\sqrt{3}+8\pi$

D) $9\sqrt{3}+24\pi$

4.2 Trigonometry of Right Triangles

1. Trigonometry of Right Triangle (Definition 1)

A **right-angled triangle** has names for each side:

① ______________ side is adjacent to the angle,

② ______________ side is opposite the angle,

and the longest side is the ③ ______________.

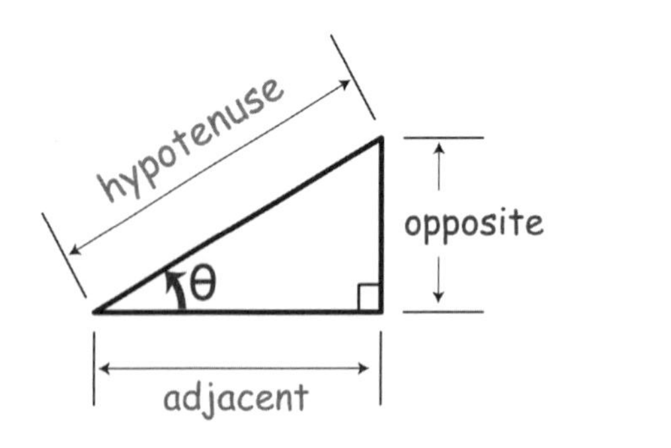

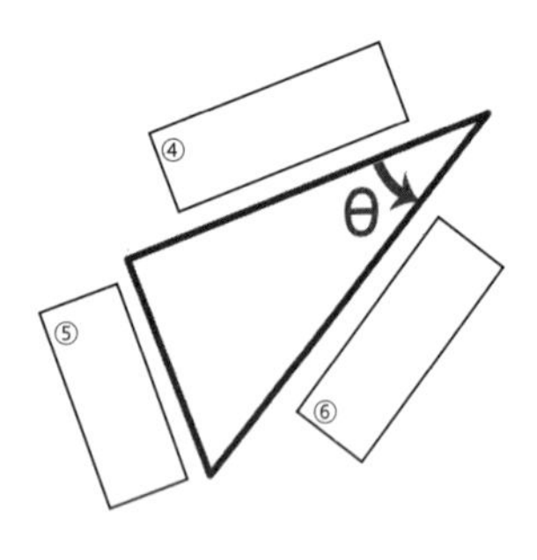

There are **Six Trigonometric Functions**

which tells you the ⑦ ____________ of the ⑧ ______________ of a right triangle.

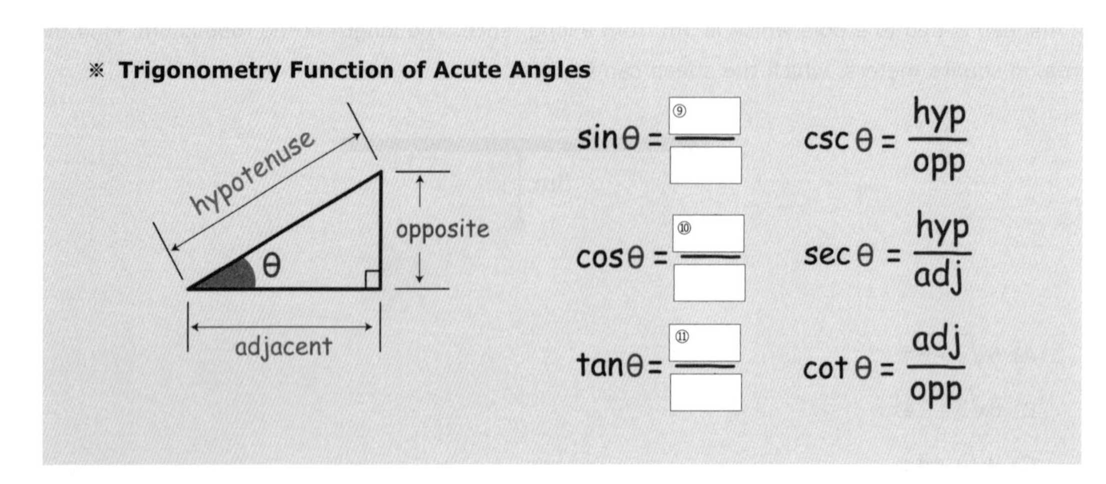

Blank : ① adjacent ② opposite ③ hypotenuse ④ adjacent ⑤ opposite ⑥ hypotenuse ⑦ratio ⑧ two sides

⑨ $\frac{opp}{hyp}$ ⑩ $\frac{adj}{hyp}$ ⑪ $\frac{opp}{adj}$ ⑫ reciprocal

How to remember? Think

"Soh Cah Toa"

csc ("cosecant"), sec("secant"), cot ("cotangent") are the ⑫______________of sin("sine"), cos("cosine"), and tan("tangent") respectively.

EXAMPLE 1. Find the exact values of the given trigonometric ratios of the angle θ in the triangle.

①

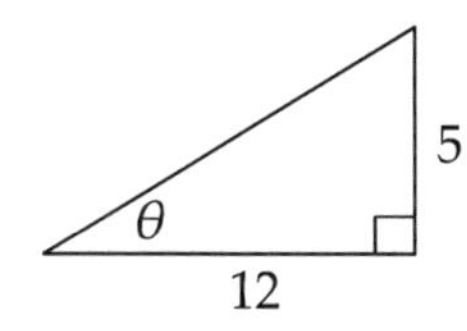

All trig ratios;

②

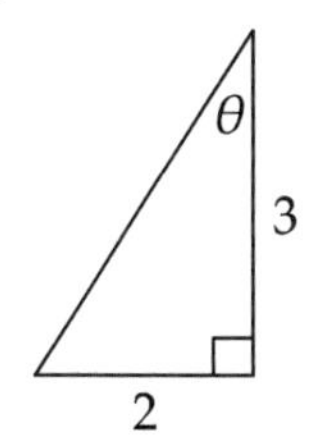

All trig ratios;

③

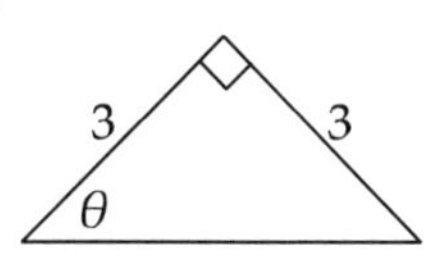

$\sin\theta =$

$\sec\theta =$

$\tan\theta =$

④

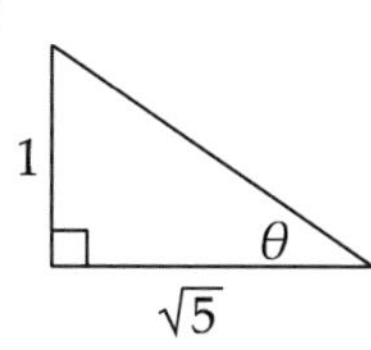

$\csc\theta =$

$\cos\theta =$

$\cot\theta =$

⑤

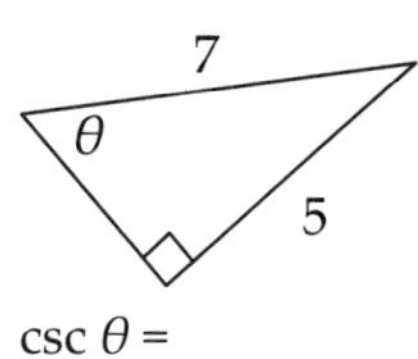

$\csc\theta =$

$\cos\theta =$

$\cot\theta =$

⑥

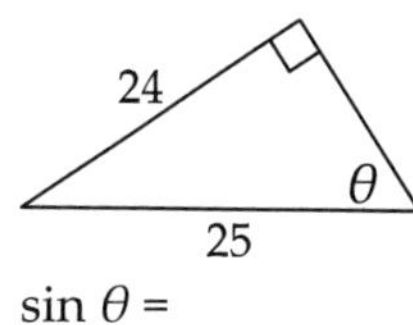

$\sin\theta =$

$\sec\theta =$

$\tan\theta =$

EXAMPLE 2. Assume that $0 < \theta < \frac{\pi}{2}$ and a, b > 0 satisfying the given conditions. Evaluate the indicated trigonometric function.

① $\sin\theta = \frac{3}{4}$;　$\cos\theta$

② $\sin\theta = \frac{1}{3}$;　$\cot\theta$

③ $\csc\theta = \frac{a}{b}$;　$\cos\theta$

④ $\sec\theta = a$;　$\cot\theta$

⑤ $\tan\theta = a$;　$\sec\theta$

EXAMPLE 3. Express x and y in terms of trigonometric ratios of θ.

①

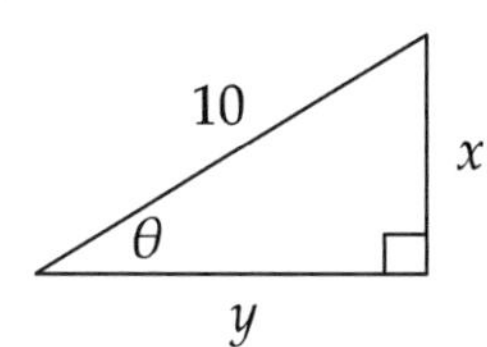

②

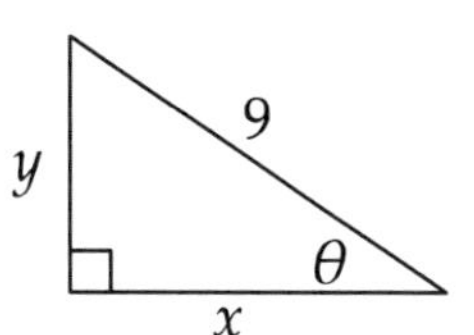

③

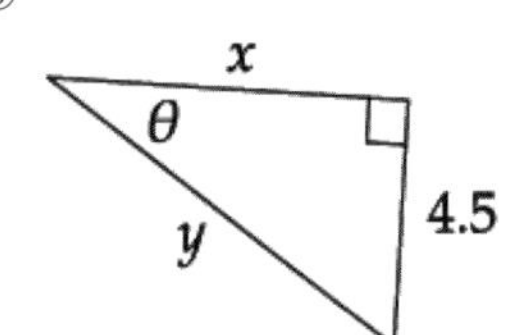

④

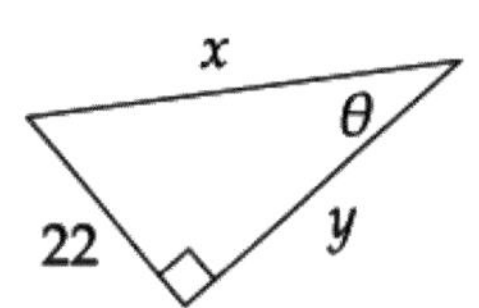

2. Special values of the trigonometric functions

※ **Special Values of Trigonometry Function**

	0°	30°	45°	60°	90°
	0	π/6	π/4	π/3	π/2
sin	0	$\frac{\sqrt{1}}{2}$	$\frac{\sqrt{2}}{2}$	$\frac{\sqrt{3}}{2}$	1
cos	1	$\frac{\sqrt{3}}{2}$	$\frac{\sqrt{2}}{2}$	$\frac{\sqrt{1}}{2}$	0
tan	0	$\frac{1}{\sqrt{3}}$	1	$\sqrt{3}$	und

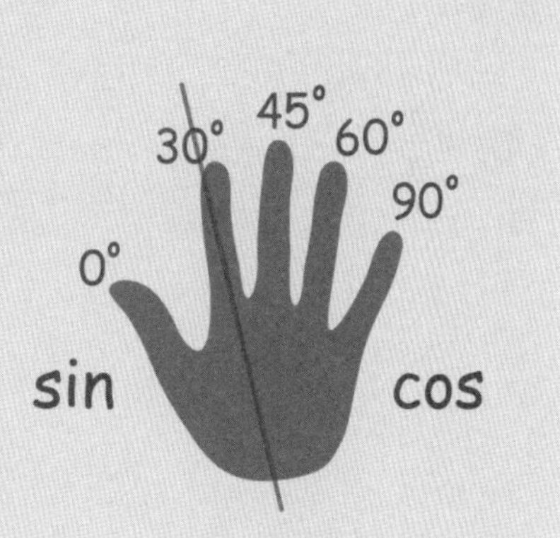

EXAMPLE 4. Memorize the table and then find the trig ratio.

① $\sin 60°$

② $\sin 45°$

③ $\cos 45°$

④ $\cos 30°$

⑤ $\tan\frac{\pi}{3}$

⑥ $\cos 0$

⑦ $\sin 0$

⑧ $\cos\frac{\pi}{3}$

⑨ $\tan\frac{\pi}{4}$

⑩ $\sin\frac{\pi}{6}$

⑪ $\tan\frac{\pi}{6}$

⑫ $\sin\frac{\pi}{2}$

⑬ $\sec\frac{\pi}{6}$

⑭ $\sec\frac{\pi}{4}$

⑮ $\csc\frac{\pi}{4}$

⑯ $\cot\frac{\pi}{6}$

⑰ $\cot\frac{\pi}{3}$

⑱ $\sec\frac{\pi}{3}$

3. Angle of Elevation and Depression

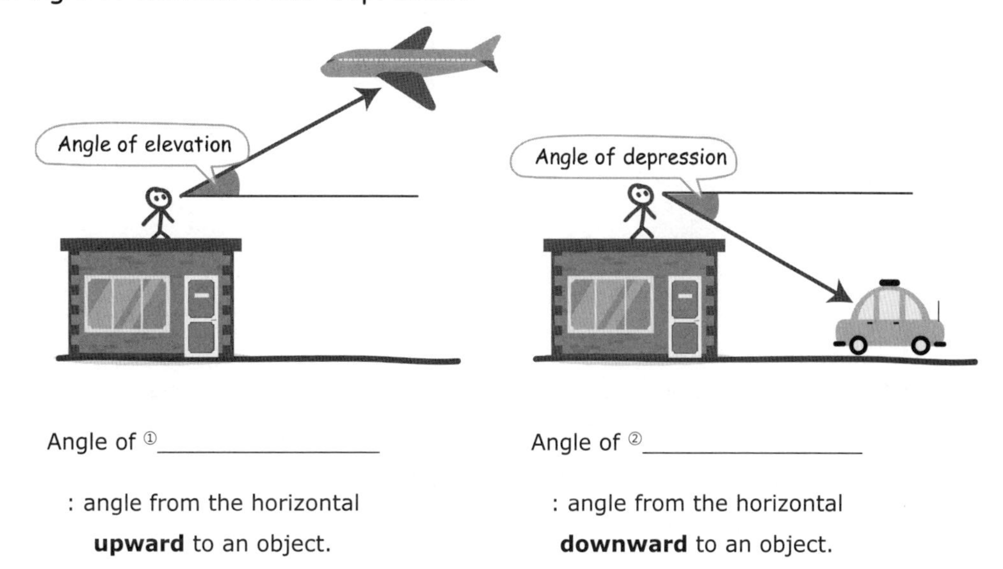

Angle of ①________________

: angle from the horizontal **upward** to an object.

Angle of ②________________

: angle from the horizontal **downward** to an object.

EXAMPLE 5. A giant redwood tree casts a shadow 200 ft long. Find the height of the tree if the angle of elevation of the sun is 57.6°.

EXAMPLE 6. An airplane is flying at a height of 2 miles above level ground. The angle of depression from the plane to the foot of the tree is 15°. What is the distance the plane must fly to be directly above the tree?

Blank : ① elevation ② depression

EXAMPLE 7. A 25ft ladder is leaning against a building making 35° angle with the ground. How far up the building does the ladder touch?

EXAMPLE 8. From a point on the ground 500 ft from the base of a building, an observer finds that the angle of elevation to the top of the building is 32° and that the angle of elevation to the top of a flagpole atop the building is 46°. Find the height of the building and the length of the flagpole.

EXAMPLE 9. Jay is standing 100 feet from the base of a tree, as shown in the figure. He measures the angle of elevation from the top of his head to the top of the tree to be 37°. If Jay is 6 feet tall, how tall is the tree?

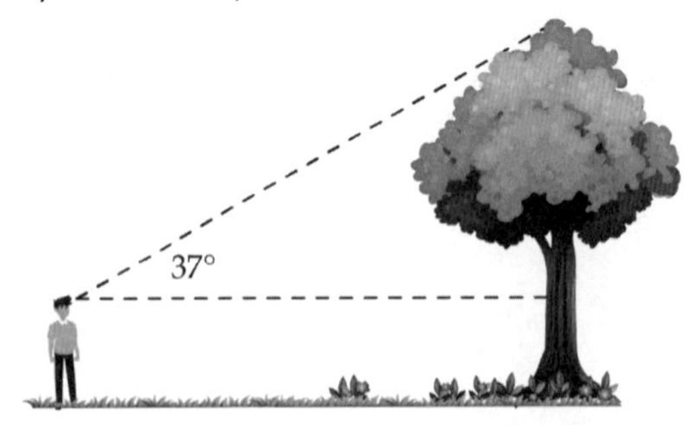

EXAMPLE 10. In traveling across flat land, you notice a mountain directly in front of you. Its angle of elevation (to the peak) is 3.5°. After you drive 13 miles closer to the mountain, the angle of elevation is 9°. Approximate the height of the mountain.

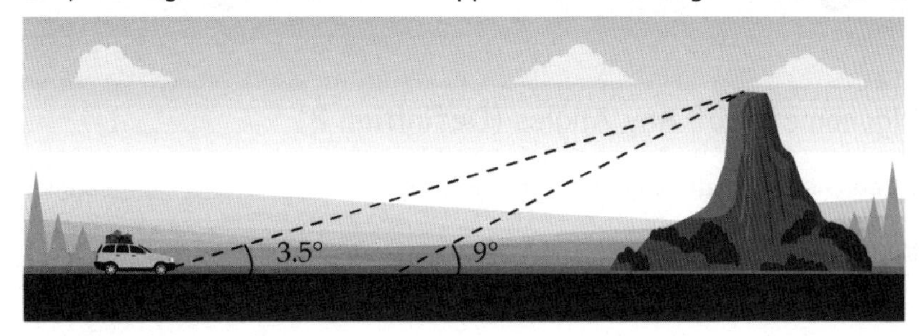

EXAMPLE 11. A Laser beam is to be directed toward the center of the circular disk in space, but the beam strays 0.05 degrees from its intended path. The disk has a radius of 100 feet and is situated 200 miles directly above the Earth where the laser is located. How many feet will the laser beam miss the edge of the disk? (5,280 feet = 1 mile)

4.3 Trigonometry of Any Angles

1. Trigonometry of Any Angles (Definition 2)

Soh Cah Toa is only for the trigonometric ratios of an ①______________ angles.

But what if we have an ②_______________ angle?

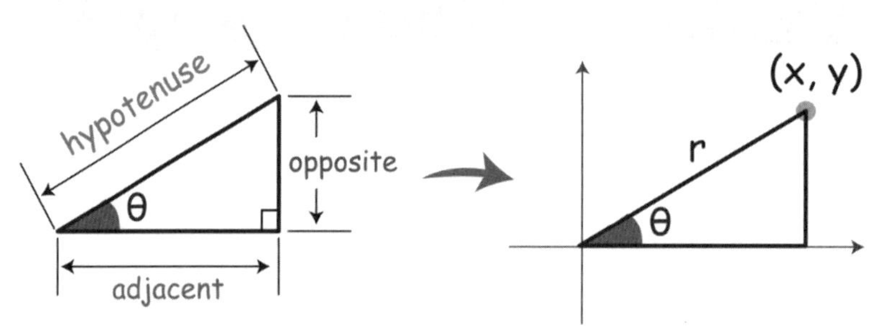

If we place the triangle in the coordinate
, we can find x, y, r instead of opp, adj , hyp.

※ Trigonometry Function of Any Angles

If θ is in standard position, (x, y) is on the terminal side of θ,

And r is the distance from (0, 0) to (x, y), then

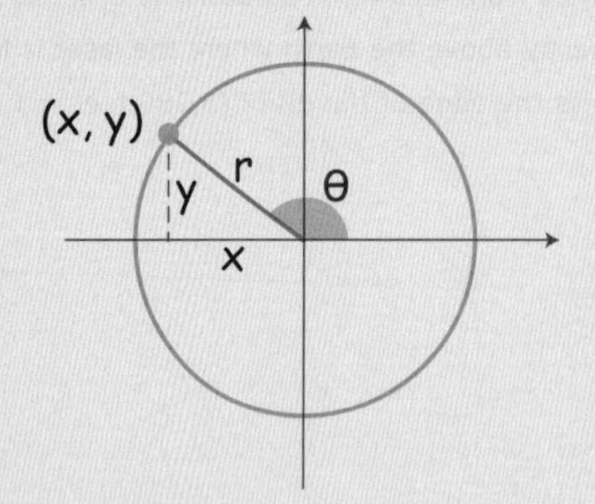

$\sin\theta = \dfrac{y}{r}$ $\csc\theta = \dfrac{r}{y}$

$\cos\theta = \dfrac{x}{r}$ $\sec\theta = \dfrac{r}{x}$

$\tan\theta = \dfrac{y}{x}$ $\cot\theta = \dfrac{x}{y}$

Remember! $r = ③\sqrt{\square^2 + \square^2}$ is always ④____________________.

Blank : ① acute ② other (obtuse or negative..) ③ $\sqrt{x^2 + y^2}$ ④ positive

EXAMPLE 1. Find the trigonometric functions of θ if the given point is on terminal side.

☺ Tip: You can use reference triangle (Drop a perpendicular line from the terminal side to the x-axis) and SOH CAH TOA!

①

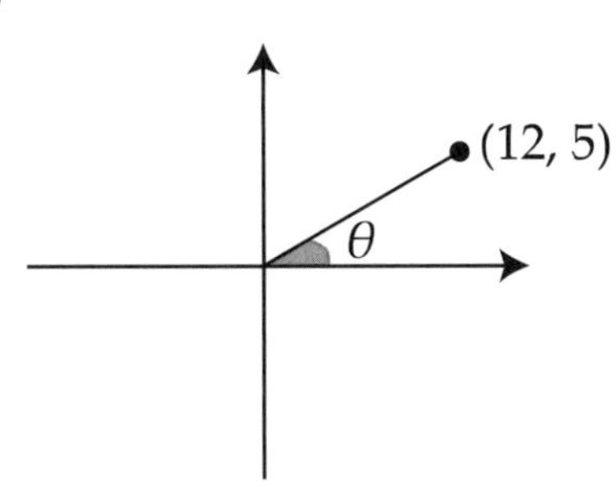

$\sin\theta =$

$\cos\theta =$

$\tan\theta =$

②

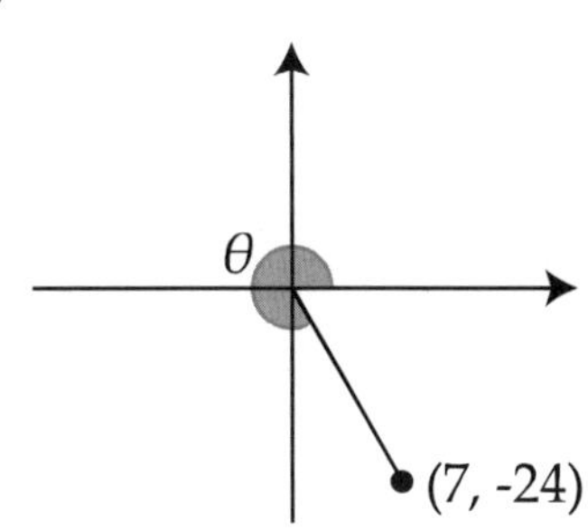

$\sin\theta =$

$\cos\theta =$

$\tan\theta =$

③

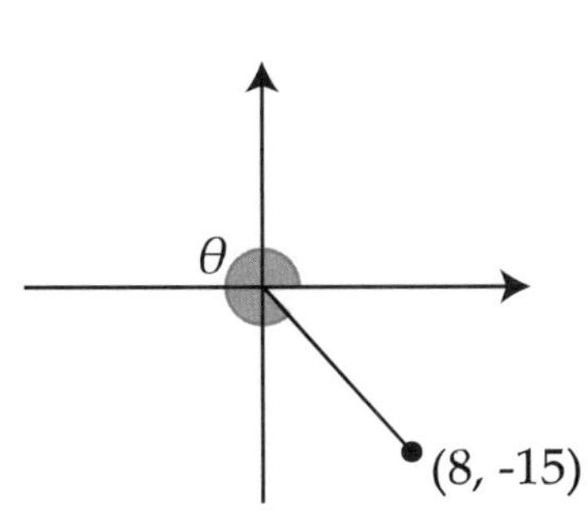

$\csc\theta =$

$\cos\theta =$

$\cot\theta =$

④

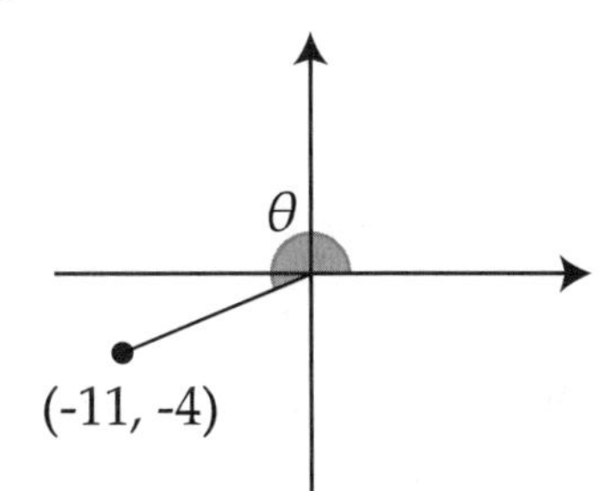

$\csc\theta =$

$\cos\theta =$

$\cot\theta =$

⑤

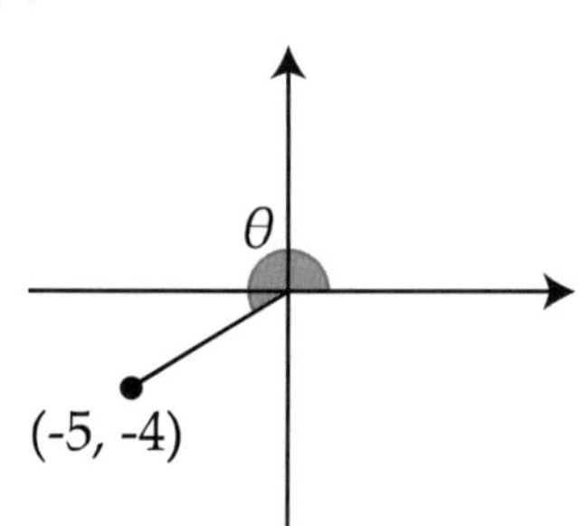

$\sin\theta =$

$\sec\theta =$

$\tan\theta =$

⑥

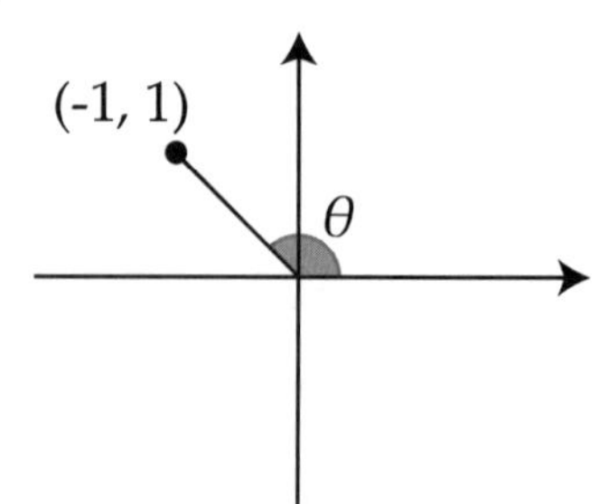

$\sin\theta =$

$\sec\theta =$

$\tan\theta =$

2. Trigonometry of Any Angles (Definition 3)

※ Signs of Trig Functions

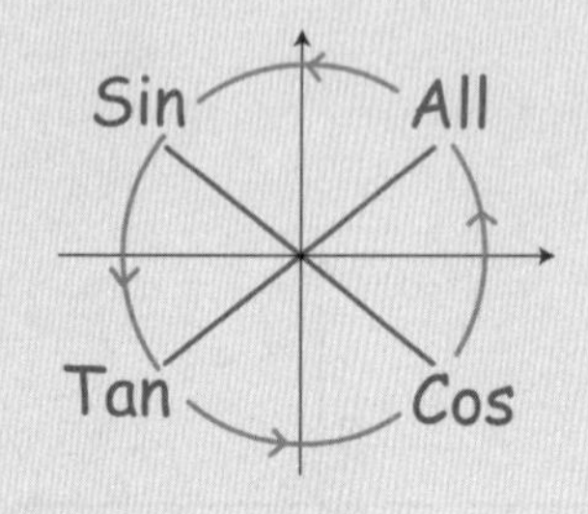

ARE ***positive****!!*

"Remember: **A**ll **S**tudents **T**ake **C**alculus"

EXAMPLE 2. Find the quadrant in which θ lies from the information given.

① $\sin\theta > 0,\ \cos\theta < 0$

② $\tan\theta > 0, \sin\theta < 0$

③ $\tan\theta < 0,\ \sec\theta > 0$

④ $\sec\theta < 0,\ \csc\theta > 0$

⑤ $\sin\theta > 0,\ \cot\theta > 0$

⑥ $\cot\theta > 0,\ \csc\theta > 0$

⑦ $\csc\theta < 0, \sec\theta < 0$

⑧ $\csc\theta < 0, \cot\theta < 0$

※ **Reference Angle** $\overline{\theta}$

: the **acute angle** between the terminal side of θ and x axis.

※ **Reference triangle**

: Drop a perpendicular line from the terminal side to the x-axis

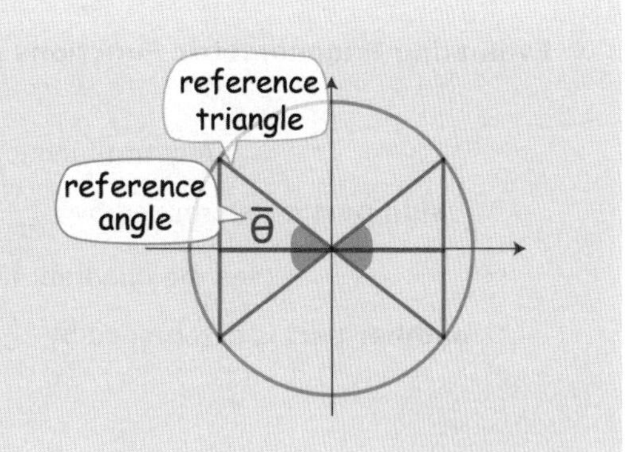

ex)

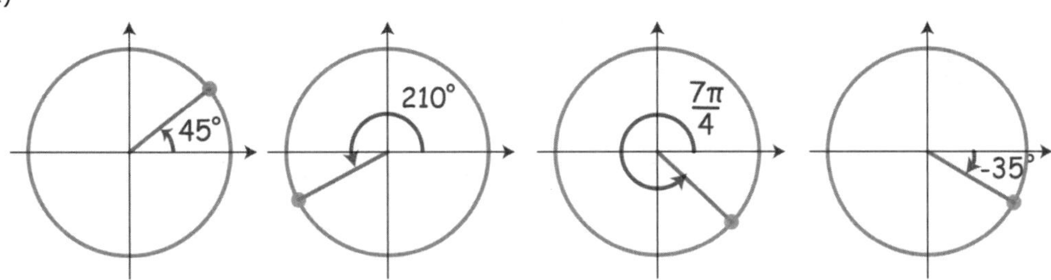

EXAMPLE 3. Find the reference number for each value of θ.

① $\theta = -300°$

② $\theta = 315°$

③ $\theta = \frac{7\pi}{3}$

④ $\theta = -\frac{\pi}{3}$

⑤ $\theta = -\frac{5\pi}{6}$

⑥ $\theta = \frac{7\pi}{4}$

※ **Evaluating Trigonometric Functions of Any Angles**

Any trig function of θ can be written as ;

sign part is determined by ① ____________________

(see the quadrant in which the terminal side lies)

number part is determined by ② ______________________

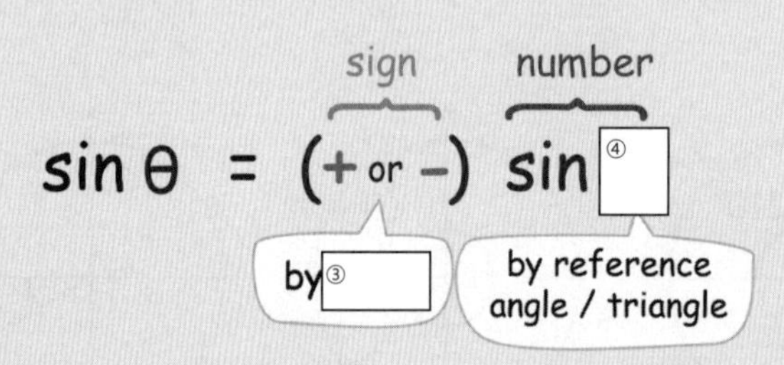

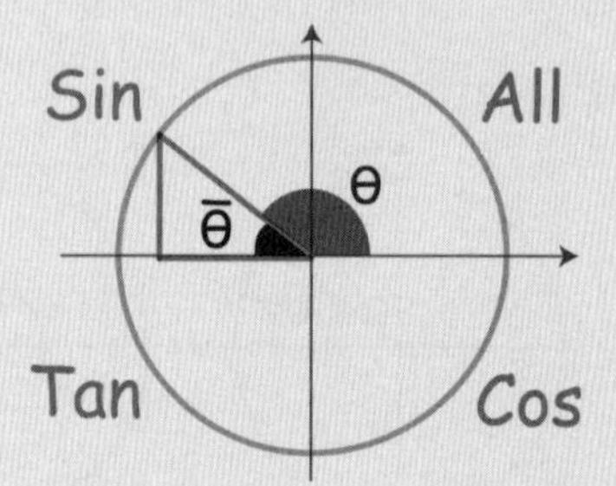

ex) tan 120° = ⑤___ tan ____   tan 210° = ⑥___ tan _____

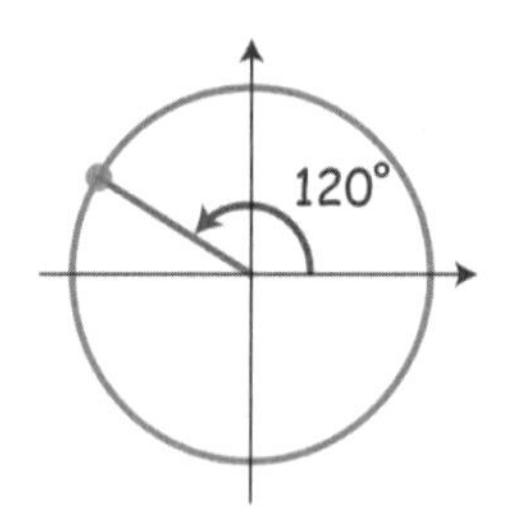

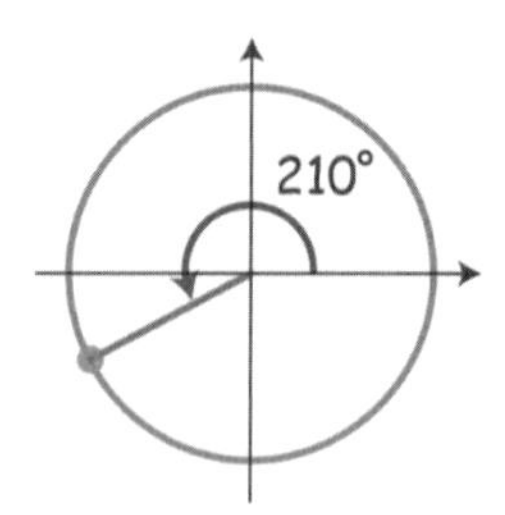

Blank : ① ASTC ② reference angle $\bar{\theta}$ ③ ASTC ④ $\bar{\theta}$ ⑤ −tan60° ⑥ +tan 30 °

EXAMPLE 4. Express the given trigonometric ratio as a function of its reference angle. Find the values of given trigonometric ratios.

① $\sin 225^{\circ}$

② $\cos 150^{\circ}$

③ $\tan(-210^{\circ})$

④ $\sec 300^{\circ}$

⑤ $\sin\dfrac{3\pi}{4}$

⑥ $\cos\dfrac{11\pi}{3}$

⑦ $\cos\dfrac{5\pi}{4}$

⑧ $\tan\dfrac{11\pi}{6}$

⑨ $\cos\left(-\dfrac{4\pi}{3}\right)$

⑩ $\sin\dfrac{4\pi}{3}$

⑪ $\csc\dfrac{9\pi}{4}$

⑫ $\sec\left(-\dfrac{5\pi}{4}\right)$

⑬ $\cot\left(-\dfrac{17\pi}{6}\right)$

⑭ $\csc\left(-\dfrac{8\pi}{3}\right)$

EXAMPLE 5. Find the exact values of the given trigonometric functions of θ from the given information.

① $\sin\theta = -\dfrac{4}{5}$, θ is in quadrant Ⅲ

Find $\cos\theta$, $\tan\theta$.

② $\tan\theta = -\dfrac{2}{7}$, θ is in quadrant Ⅱ

Find $\sin\theta$, $\cos\theta$.

③ $\csc\theta = \dfrac{4}{3}$, $\dfrac{\pi}{2} \le \theta \le \pi$

Find $\sec\theta$, $\tan\theta$.

④ $\cot\theta = 2$, $\pi \le \theta \le \dfrac{3\pi}{2}$

Find $\sin\theta$, $\sec\theta$.

⑤ $\cos\theta = \dfrac{5}{13}$, $\sin\theta < 0$

Find $\sin\theta$, $\cot\theta$.

⑥ $\sec\theta = -\dfrac{7}{5}$, $\csc\theta < 0$

Find $\sin\theta$, $\cot\theta$.

AP Style Problem

1. In the figure, the terminal ray of the angel θ passes through the point (-2, 3). What is cos θ ?

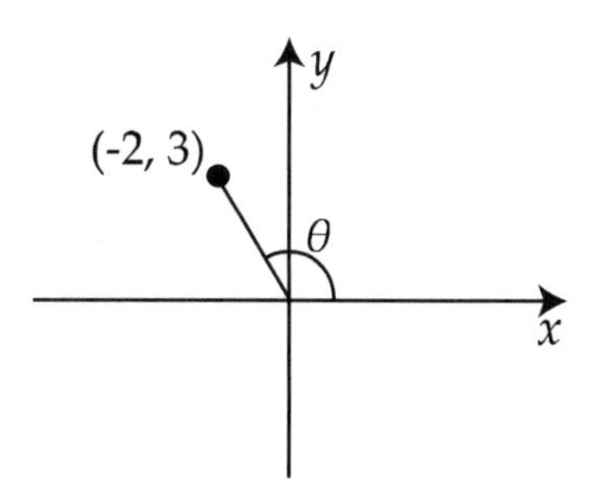

A) 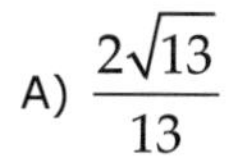 $\frac{2\sqrt{13}}{13}$

B) $-\frac{2\sqrt{13}}{13}$

C) $\frac{3\sqrt{13}}{13}$

D) $-\frac{3\sqrt{13}}{13}$

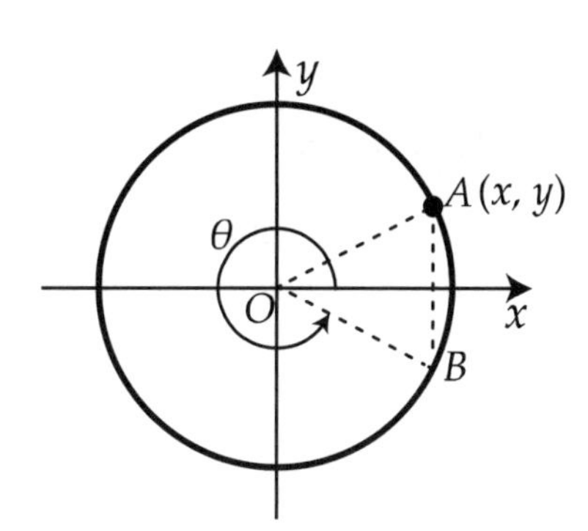

2. In the figure, $\overline{OA}$ and $\overline{OB}$ are the radii of a circle. Point A has a coordinate (x, y) and the terminal side of the angle θ is point B. Which of the following is true about angle θ?

A) $\tan\theta = \frac{x}{y}$

B) $\tan\theta = \frac{y}{x}$

C) $\tan\theta = -\frac{x}{y}$

D) $\tan\theta = -\frac{y}{x}$

3. Find the quadrant in which θ lies given that $\sin\theta\sec\theta<0$ and $\cos\theta\cot\theta<0$.

A) quadrant I

B) quadrant II

C) quadrant III

D) quadrant IV

4. If $\frac{3\pi}{2}<\theta<2\pi$, which of the following is equivalent to

$$\sin\theta+\cos\theta+\tan\theta+|\sin\theta|+|\cos\theta|+|\tan\theta|.$$

A) 0

B) $2\sin\theta$

C) $2\cos\theta$

D) $2\tan\theta$

4.4 Trigonometry in Unit Circle

1. Trigonometry of Unit Circle (Definition 4)

Reminder ☺

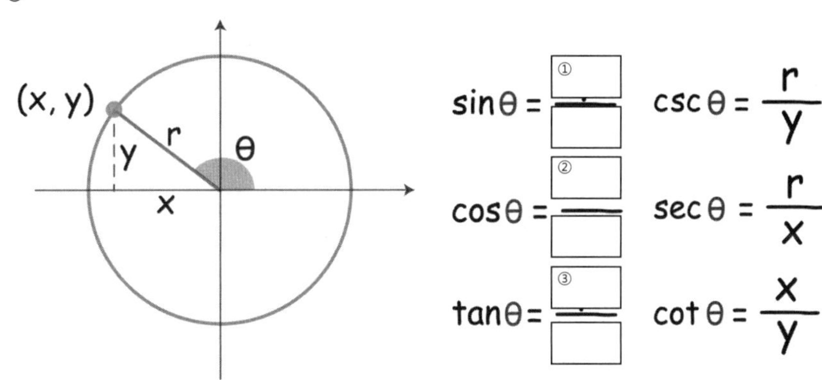

A '④______________________' is a circle with radius ⑤_____.

※ **Trig ratios in Unit Circle**

When θ is in standard position,

Let point (x, y) represent the point where the terminal side of the angle intersects the unit circle.

(r = ⑥______).

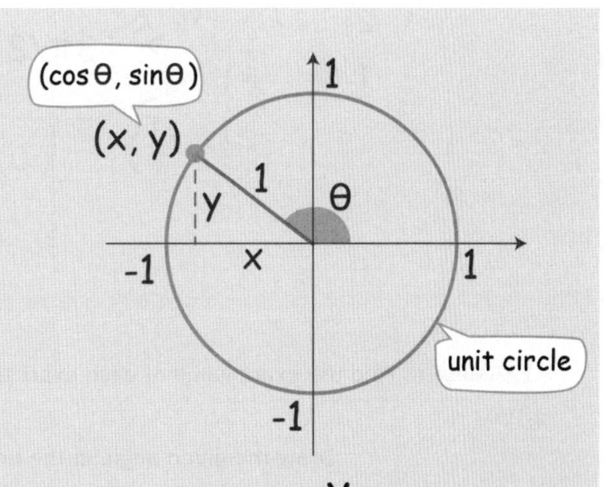

Then;

$$\sin\theta = y \quad \cos\theta = x \quad \tan\theta = \frac{y}{x}$$

(sin θ is the ⑦___ coordinate, cos θ is the ⑧___ coordinate of the point on the unit circle,

tan θ is the ⑨_______ of the terminal ray)

Blank : ① $\frac{y}{r}$ ② $\frac{x}{r}$ ③ $\frac{y}{x}$ ④ unit circle ⑤ 1 ⑥ 1 ⑦ y ⑧ x ⑨ slope

Therefore, the point on the unit circle will be;

$$(x, y) = (\cos\theta, \sin\theta)$$

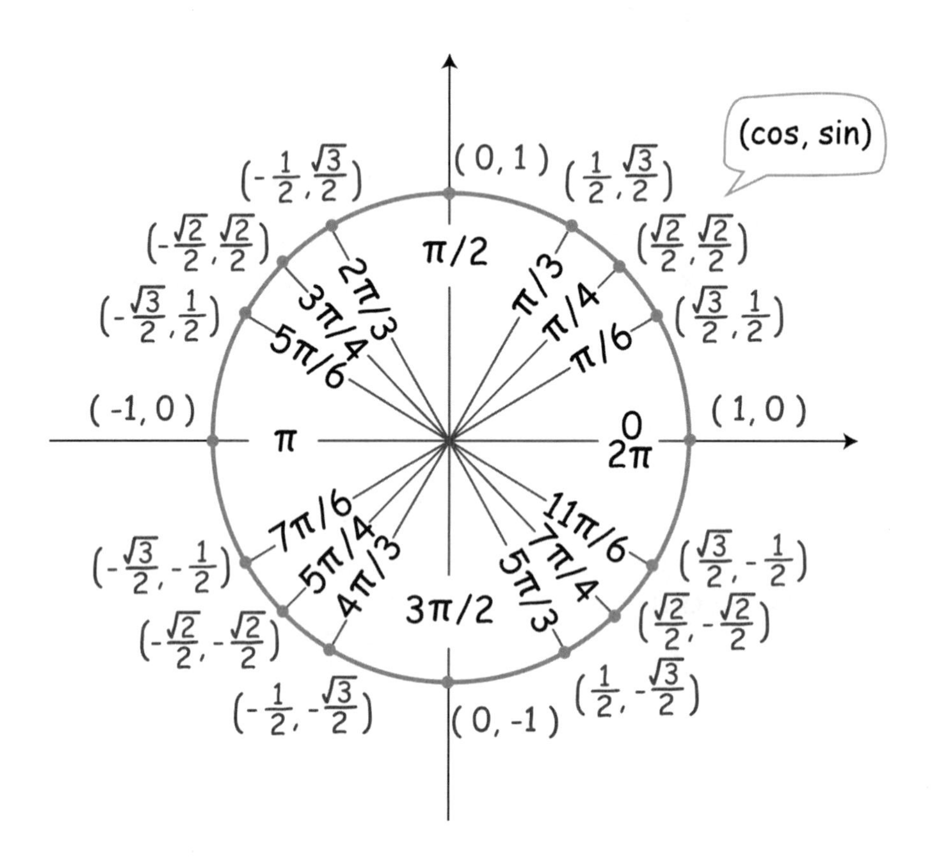

EXAMPLE 1. Find the exact value of each expression using unit circle.

☺ Tip:

Draw the given angle in the unit circle, then read the;

sin ratio: vertical displacement

cos ratio: horizontal displacement

tan ratio: slope of the terminal ray

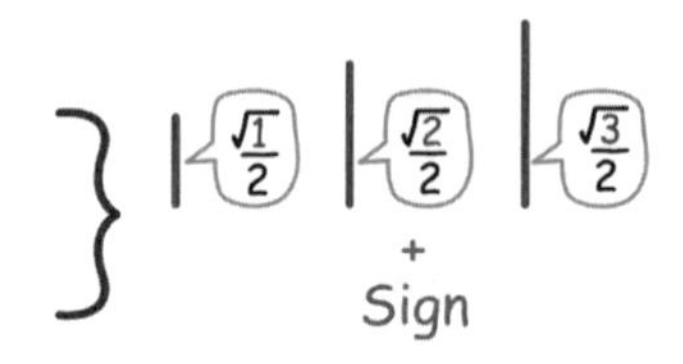

displacement: distance with plus and minus sign

$\sin\frac{2\pi}{3}=$ ①____ $\sin\frac{7\pi}{6}=$ ②____ $\cos\frac{7\pi}{4}=$ ③____ $\tan\frac{5\pi}{6}=$ ④____

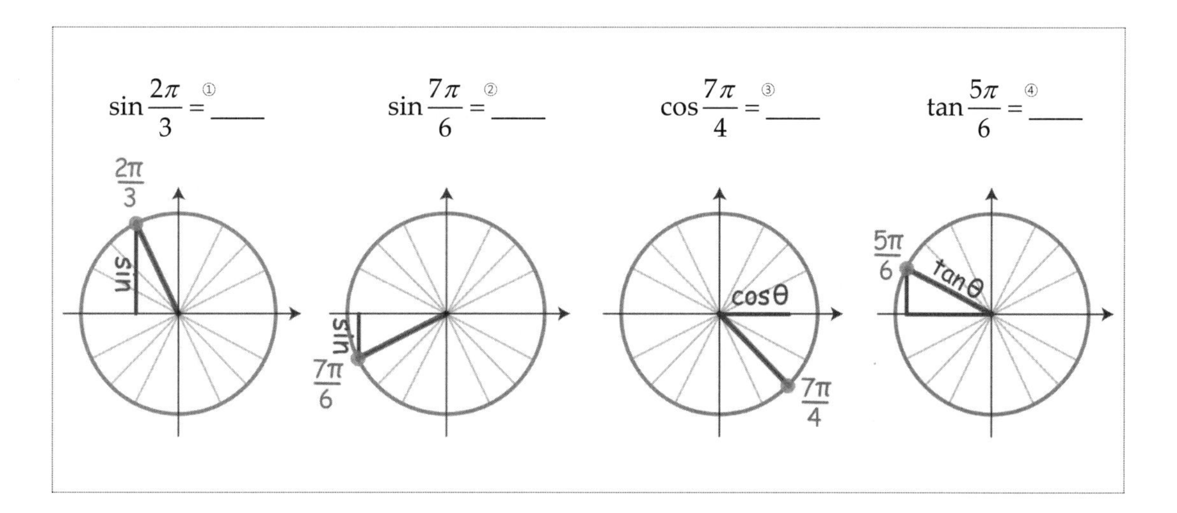

① $\sin\frac{11\pi}{6}$

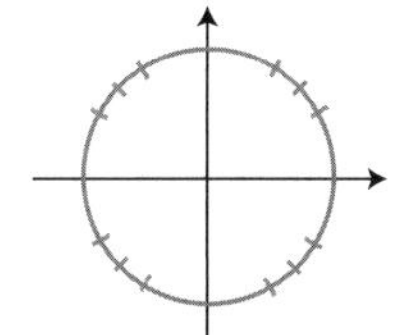

② $\sin\frac{5\pi}{6}$

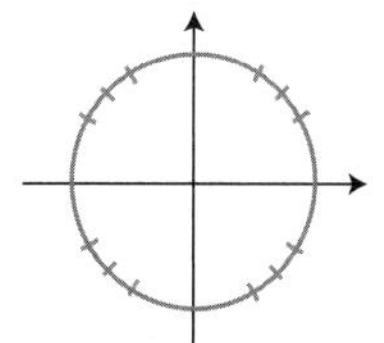

③ $\cos\frac{5\pi}{6}$

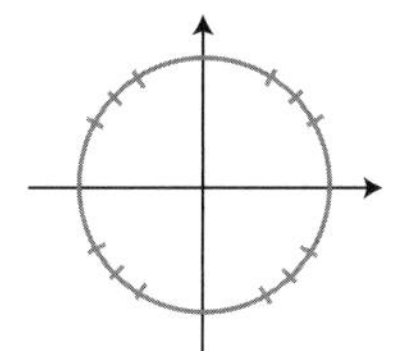

④ $\cos\frac{7\pi}{4}$

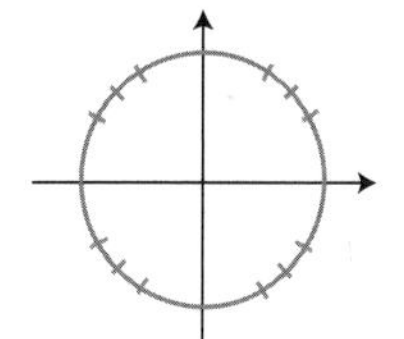

⑤ $\tan\frac{7\pi}{4}$

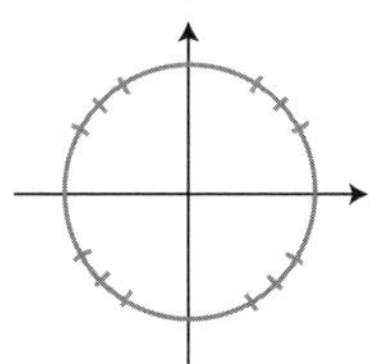

⑥ $\tan\frac{11\pi}{6}$

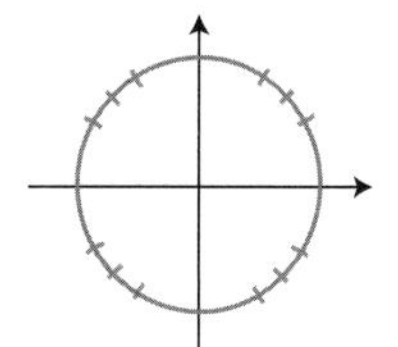

⑦ $\cos\frac{4\pi}{3}$

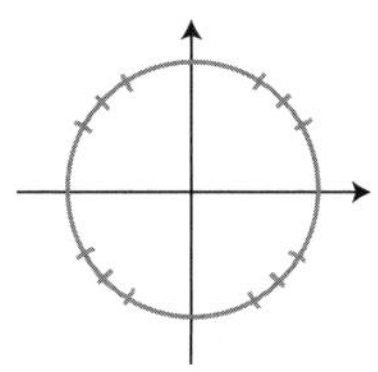

⑧ $\sin\frac{4\pi}{3}$

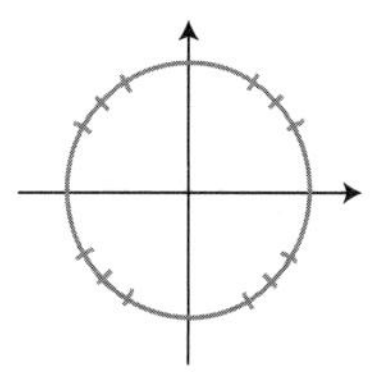

⑨ $\tan\frac{7\pi}{6}$

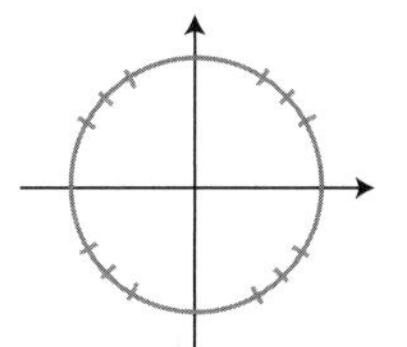

⑩ $\tan\frac{5\pi}{3}$

Blank : ① $\frac{\sqrt{3}}{2}$ ② $-\frac{1}{2}$ ③ $\frac{\sqrt{2}}{2}$ ④ $-\frac{1}{\sqrt{3}}=-\frac{\sqrt{3}}{3}$

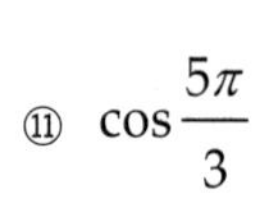
⑪ $\cos\frac{5\pi}{3}$

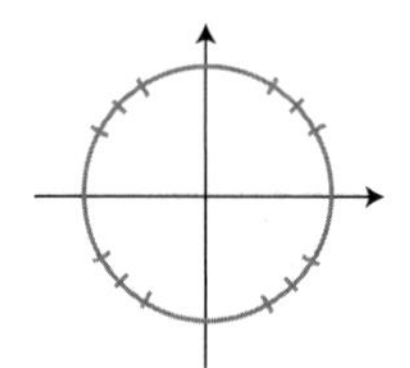

⑫ $\cos\frac{7\pi}{6}$

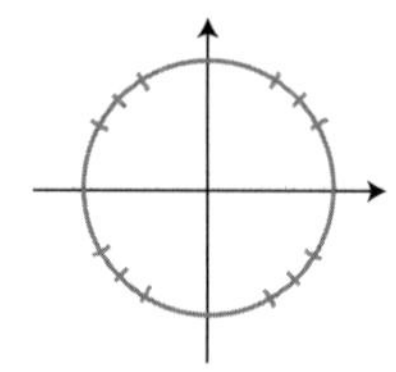

⑬ $\sec\frac{4\pi}{3}$

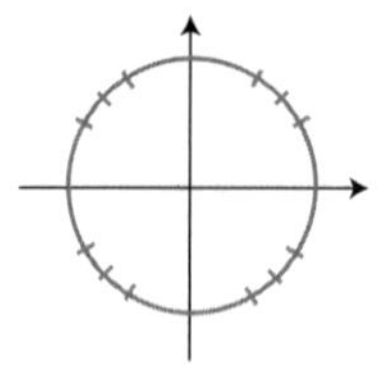

⑭ $\csc\frac{5\pi}{4}$

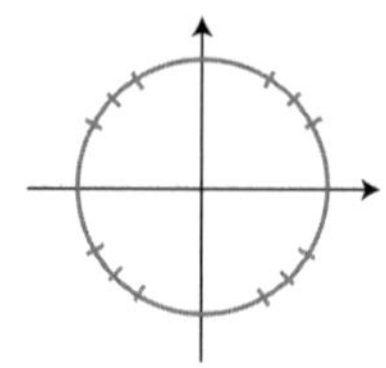

⑮ $\sin 2\pi$

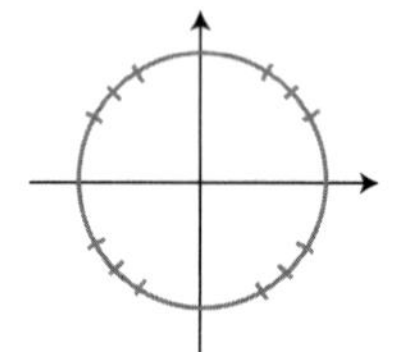

⑯ $\sin\frac{5\pi}{2}$

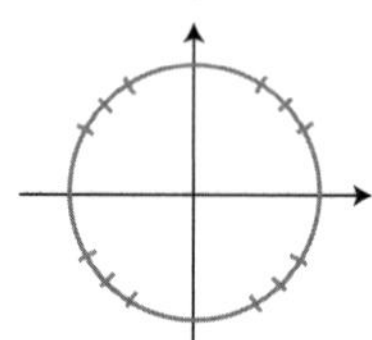

⑰ $\cos\frac{\pi}{2}$

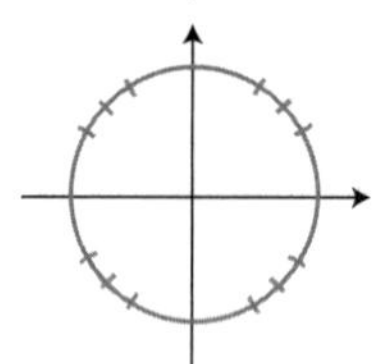

⑱ $\cos 5\pi$

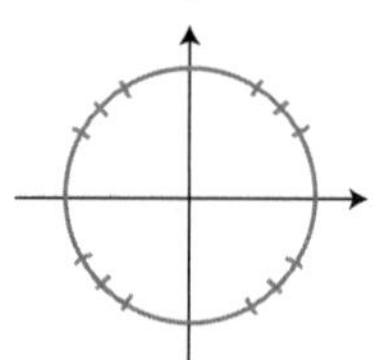

⑲ $\sin\frac{3\pi}{2}$

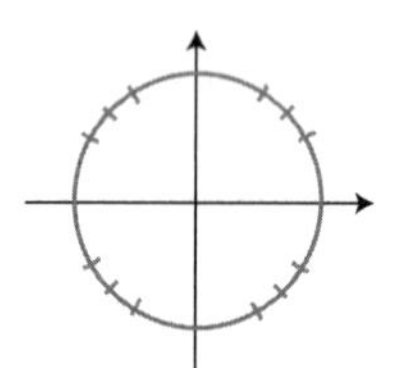

⑳ $\sin 3\pi$

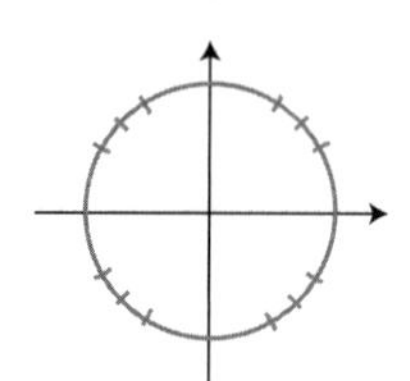

㉑ $\tan 5\pi$

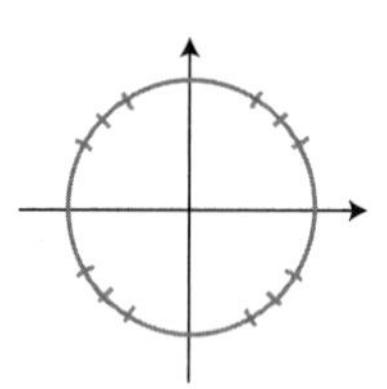

㉒ $\tan\frac{3\pi}{2}$

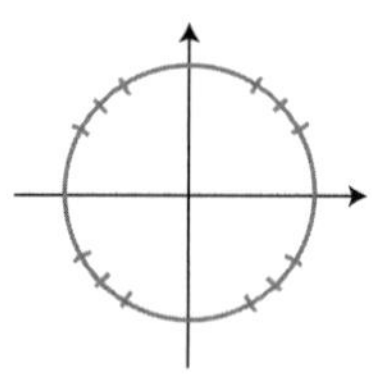

㉓ $\tan\frac{\pi}{2}$

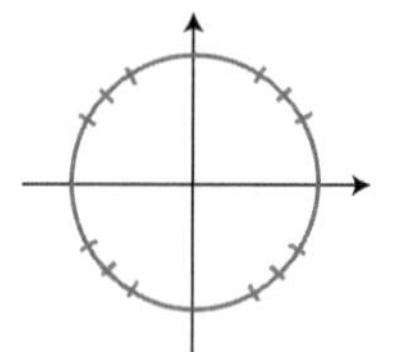

㉔ $\tan(-4\pi)$

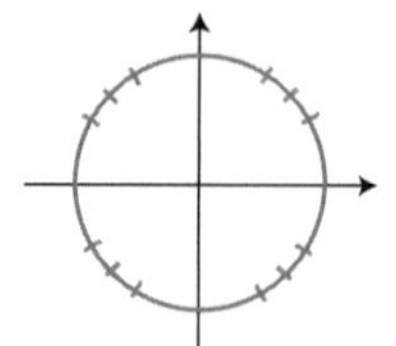

※ **Facts from Unit circle**

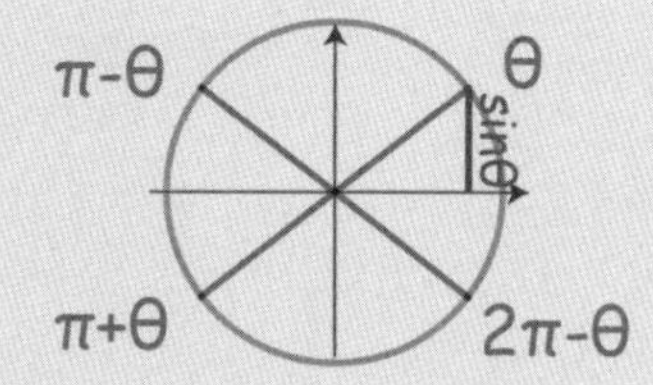

$\sin(\pi-\theta) = \sin\theta$

$\sin(\pi+\theta) =$ ① ☐

$\sin(2\pi-\theta) =$ ② ☐

$\sin(2\pi+\theta) =$ ③ ☐

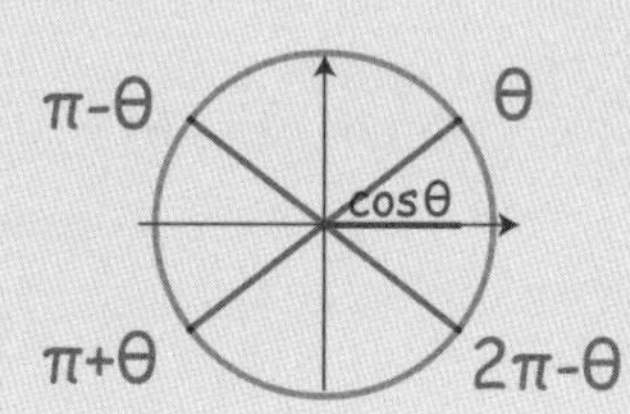

$\cos(\pi-\theta) = -\cos\theta$

$\cos(\pi+\theta) =$ ④ ☐

$\cos(2\pi-\theta) =$ ⑤ ☐

$\cos(2\pi+\theta) =$ ⑥ ☐

EXAMPLE 2. Given the $\sin\theta = 0.6$, find the value of

① $\sin(\pi - \theta)$

② $\sin(\theta + \pi)$

③ $\sin(2\pi - \theta)$

④ $\sin(-\theta)$

Blank : ① $-\sin\theta$ ② $-\sin\theta$ ③ $\sin\theta$ ④ $-\cos\theta$ ⑤ $\cos\theta$ ⑥ $\cos\theta$

EXAMPLE 3. Given the $\cos\theta = 0.4$, find the value of

① $\cos(\pi + \theta)$

② $\cos(\pi - \theta)$

③ $\cos(-\theta)$

④ $\cos(2\pi - \theta)$

※ **Trig ratios in Circle with radius r**

When θ is in standard position,

Let point (x, y) represent the point where the terminal side of the angle intersects the circle with radius r ;

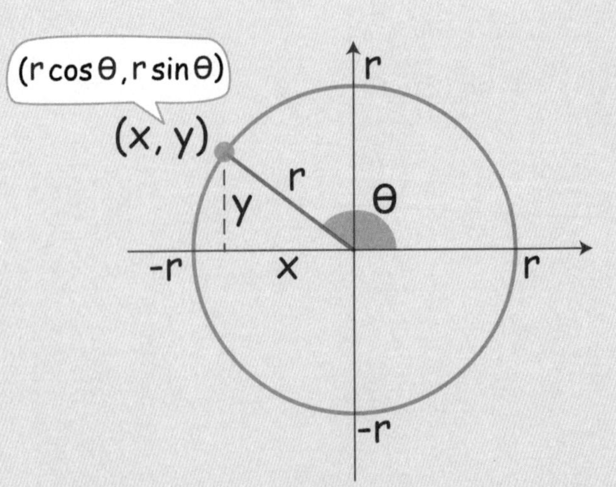

Then ;

$$(x, y) = (r\cos\theta, r\sin\theta)$$

AP Style Problem

1. What is the slope of the given line l?

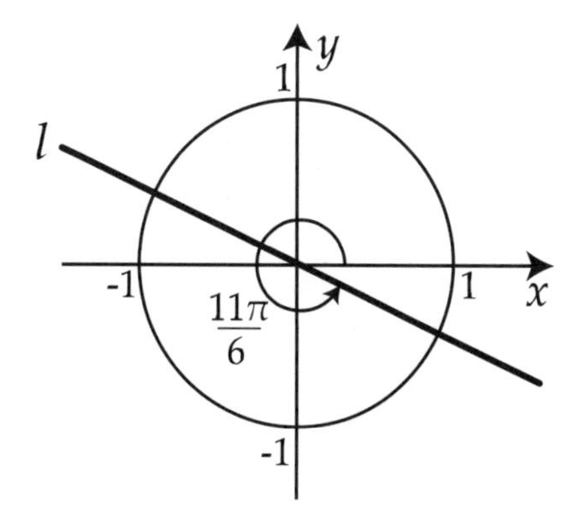

A) $\cos\frac{\pi}{6}$

B) $\cos\frac{11\pi}{6}$

C) $\tan\frac{\pi}{6}$

D) $\tan\frac{11\pi}{6}$

2. In the figure, what is the length of CD?

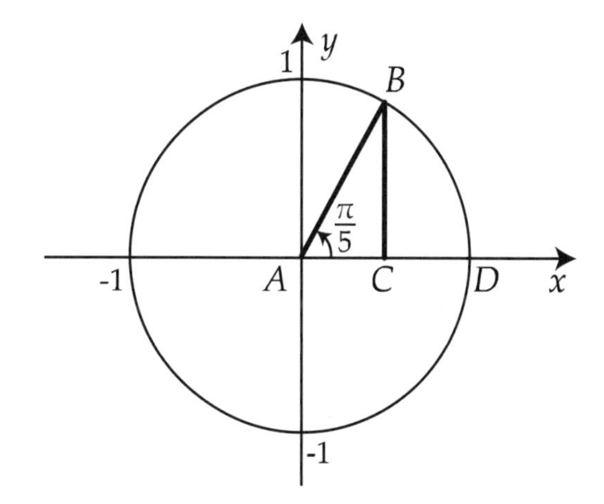

A) $1-\sin\frac{\pi}{5}$

B) $1-\cos\frac{\pi}{5}$

C) $\sin\frac{\pi}{5}$

D) $\cos\frac{\pi}{5}$

3. Point A is on the circle with radius 6. What is the coordinate of point A?

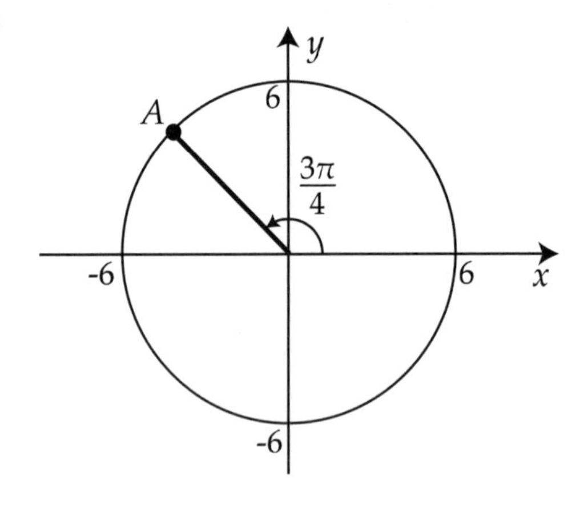

A) 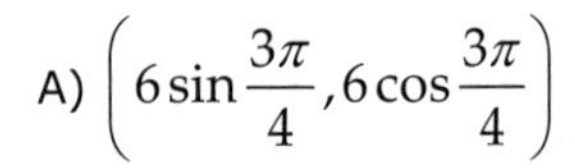 $\left(6\sin\frac{3\pi}{4}, 6\cos\frac{3\pi}{4}\right)$

B) $\left(3\cos\frac{3\pi}{4}, 3\sin\frac{3\pi}{4}\right)$

C) $\left(6\cos\frac{11\pi}{4}, 6\sin\frac{11\pi}{4}\right)$

D) $\left(6\cos\frac{\pi}{4}, 6\sin\frac{\pi}{4}\right)$

4. Which of the following is equivalent to

$$\cos\theta + \cos(\pi - \theta) + \cos(\pi + \theta) + \cos(2\pi + \theta)?$$

A) 0

B) $-\cos\theta$

C) $\cos\theta$

D) $2\cos\theta$

5. Angel A and B are in standard position is given as shown, where $B > A$. Which of the following is true?

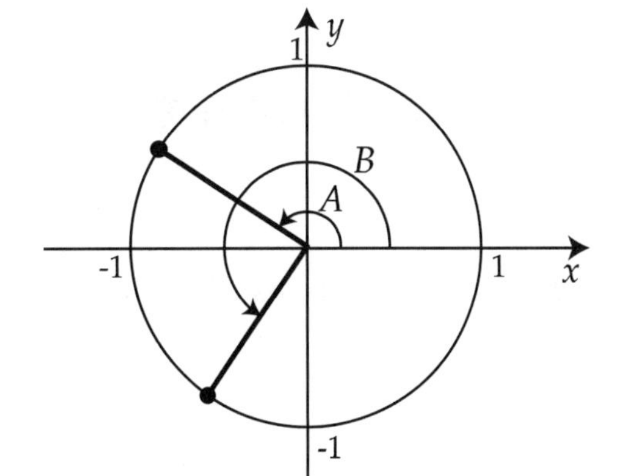

A) $\sin A < \sin B$

B) $\cos A < \cos B$

C) $\tan A > \tan B$

D) none of these

6. A unit circle is divided into 10 equal angles. Let the ten terminal points are $A_1, A_2, \ldots, A_{10}$ as shown. If $\angle A_1OA_2 = \theta$, what is $\sin\theta + \sin 2\theta + \ldots + \sin 10\theta$?

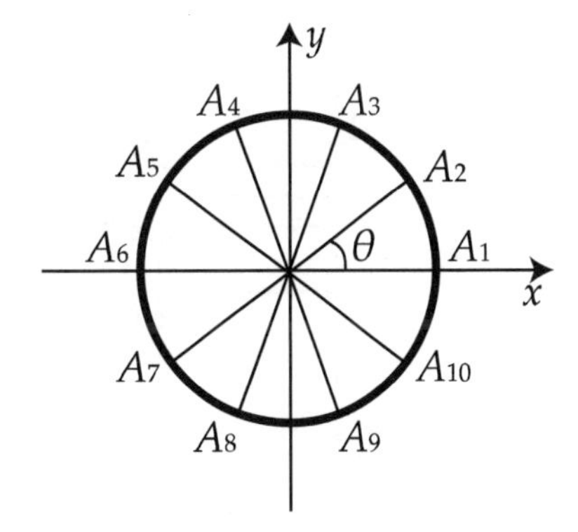

A) 0

B) 1

C) $\sin\theta$

D) $\sin 55\theta$

4.5 Trigonometric Graphs for Sin, Cos

1. Transformations of trigonometric graphs

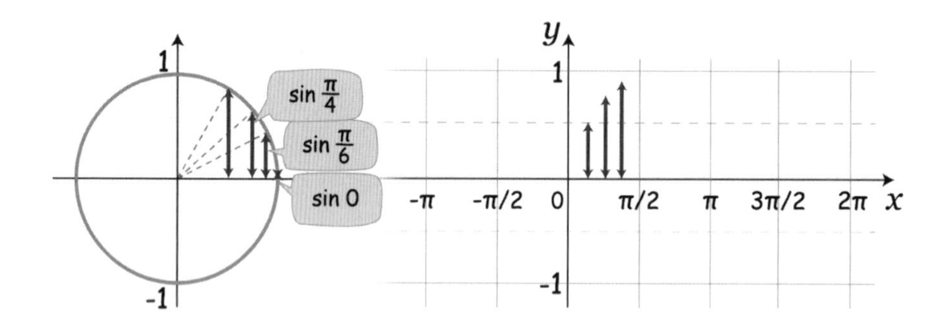

The **Sine Function** y = ①__________ has this 'wavy' up-down curve

which repeats every②________ radians and starts at ③______ (when x = 0).

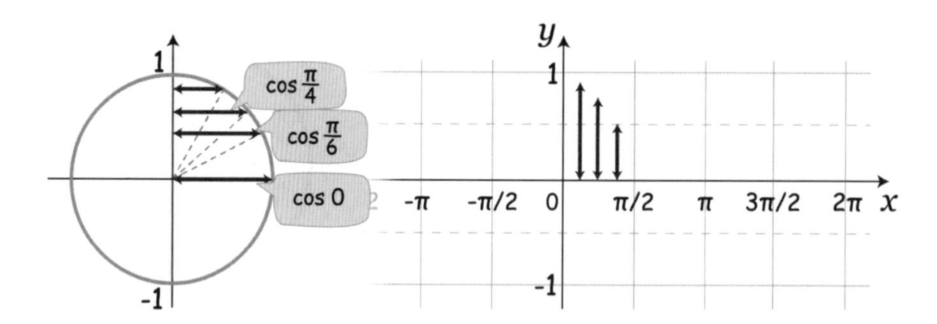

The **Cosine Function** y = ④__________ has this 'wavy' up-down curve

which repeats every ⑤________ radians and starts at ⑥______ (when x = 0).

Blank : ① sin x ② 2π ③ 0 ④ cos x ⑤2π ⑥ 1

⑦ periodic ⑧ repeats ⑨ $\frac{\text{max} - \text{min}}{2}$ ⑩ [-1,1] , 1, 2π , origin , odd ⑪[-1,1] , 1, 2π , y axis , even

Trig functions are ⑦ ________________ functions,

because all possible y values ⑧_________ **in the same sequence** over a given set of x values.

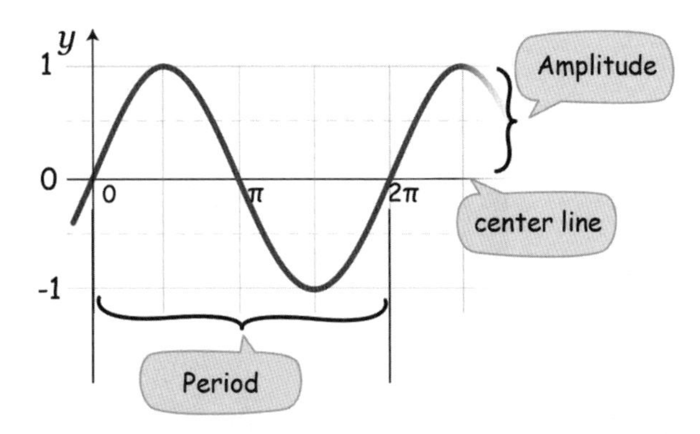

☺ Words

Amplitude: distance between the center line and the maximum point

(=half the distance from the max point to the min point=⑨ ___________)

Period: length of x for which the graph repeats.

※ **Graphs of Trigonometric Functions**

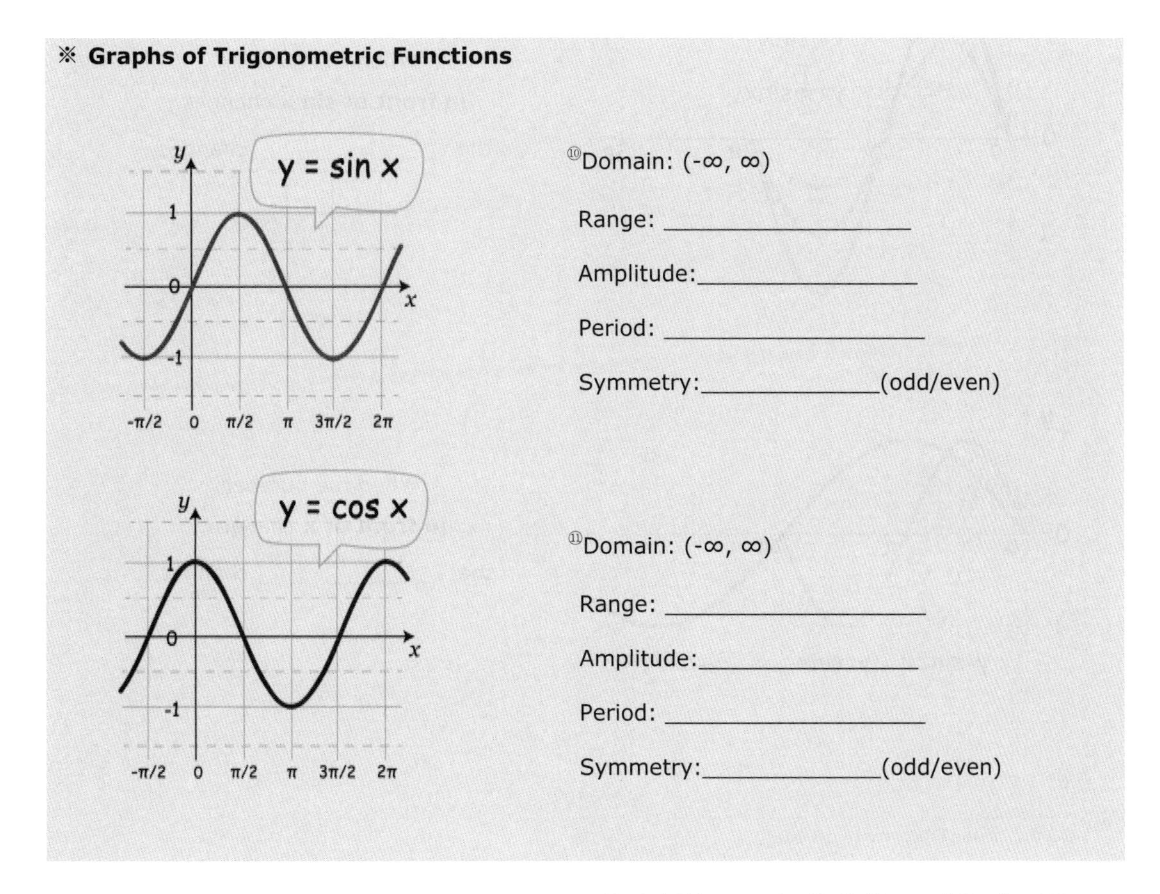

⑩Domain: (-∞, ∞)

Range: ___________________

Amplitude:_________________

Period: ____________________

Symmetry:_____________(odd/even)

⑪Domain: (-∞, ∞)

Range: ____________________

Amplitude:_________________

Period: ____________________

Symmetry:_____________(odd/even)

EXAMPLE 1. Evaluate using the graph of sin and cos.

① $\sin\frac{3\pi}{2}+\sin 2\pi$　　② $\sin 5\pi+\sin\frac{\pi}{2}$

③ $\cos\frac{\pi}{2}+\cos 2\pi-\cos\pi$　　④ $\cos\frac{3\pi}{2}-\cos\pi$

2. Transforming the Graph of Sine and Cosine

Let's look at how the ***amplitude*** and ***period changes*** as numbers of the function *change.*

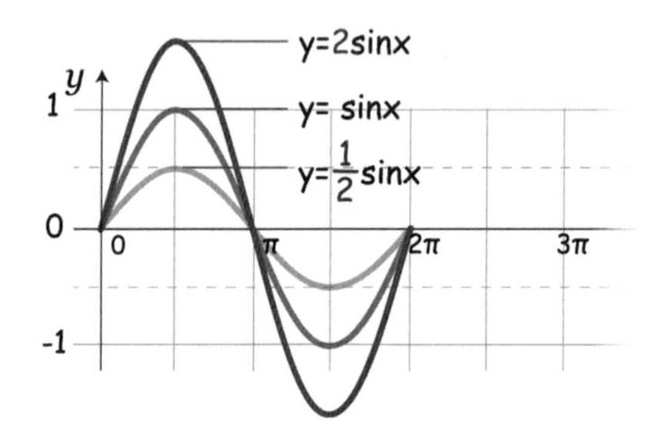

When the number

in front of sin x changes.

the ①____________ changes,

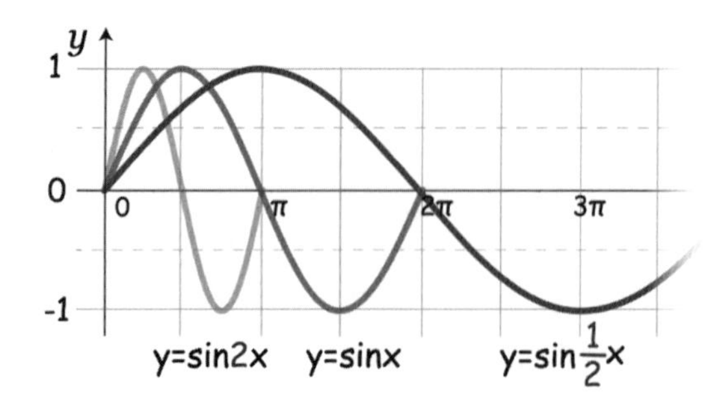

When the number

in front of x changes.

the ②____________ changes,

Blank : ① amplitude　② period

※ **Transformation of Sine and Cosine**

$$y = A \begin{matrix} \sin \\ \cos \end{matrix} (B(x-C)) + D$$

Amplitude:① ____________ Period:② ____________

Translation : ③ ________________

EXAMPLE 2. Find the amplitude, period, translations and range of the function.

① $y = 4\sin\left(5x + \frac{\pi}{8}\right) + 3$

② $y = -\sin\left(\frac{x}{3} - \pi\right) - 1$

③ $y = -\sin\left(\frac{1}{5}x + 3\right) + 2$

④ $y = 4\cos(2x + 3\pi) - 2$

Blank : ① $|A|$ ② $\frac{2\pi}{B}$ ③ Left or right C units, up or down D units

⑤ $y=-\frac{1}{2}\cos\left(\frac{\pi x}{3}-\frac{\pi}{5}\right)$

⑥ $y=-3\sin\left(\frac{3}{\pi}x+\pi\right)-3$

EXAMPLE 3. Graph the function over a one-period interval.

① $y=3\sin x$

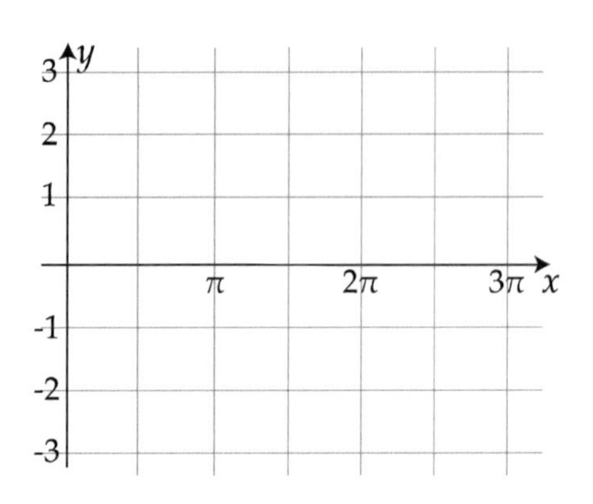

② $y=2\cos x$

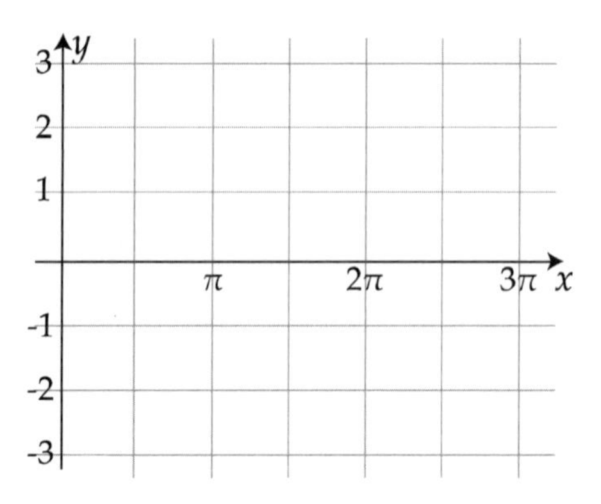

③ $y=-2\sin x+2$

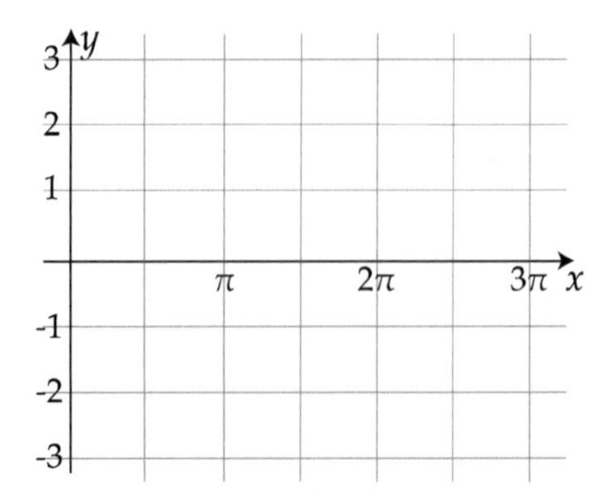

④ $y=3\sin x-1$

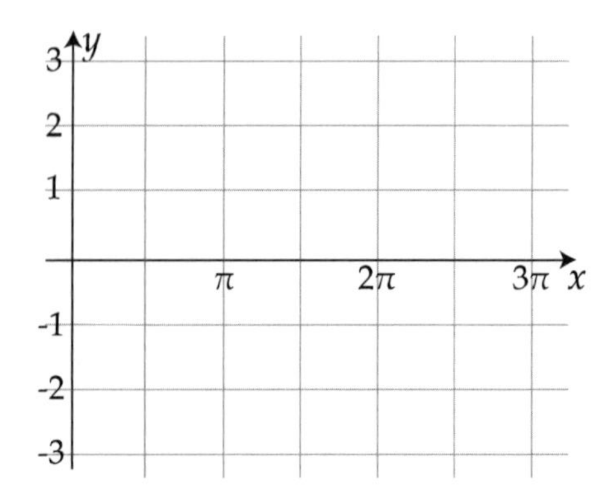

⑤ $y=\cos\left(x-\frac{\pi}{2}\right)+2$

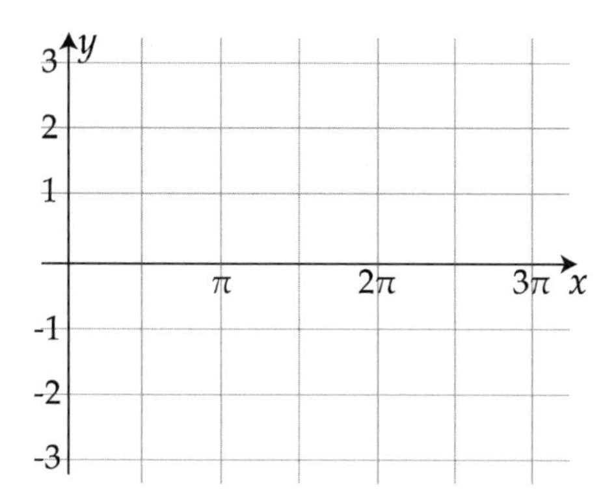

⑥ $y=3\sin\left(x+\frac{\pi}{4}\right)$

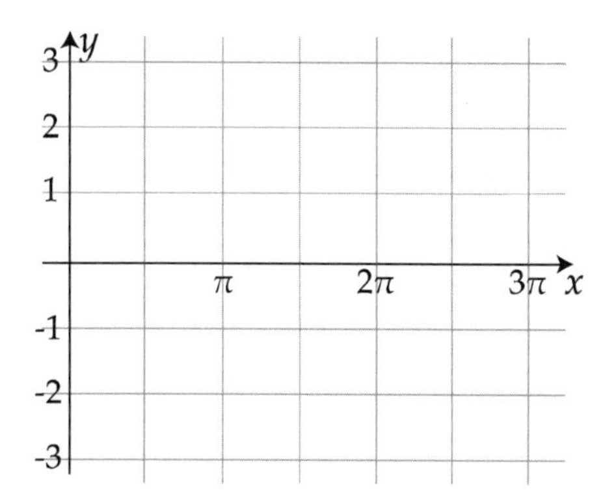

⑦ $y=3\cos(2x)$

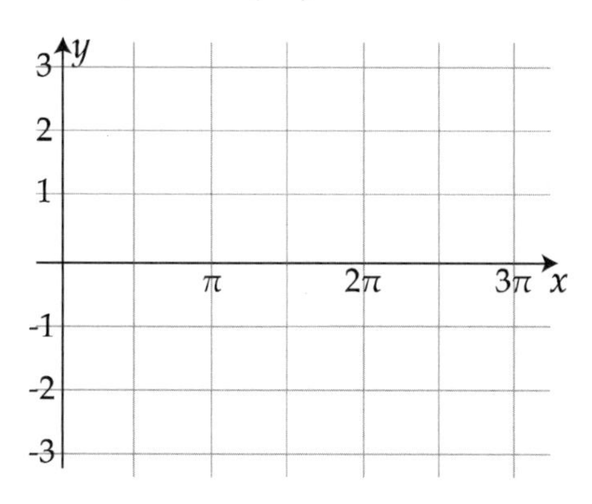

⑧ $y=2\sin(2x)+1$

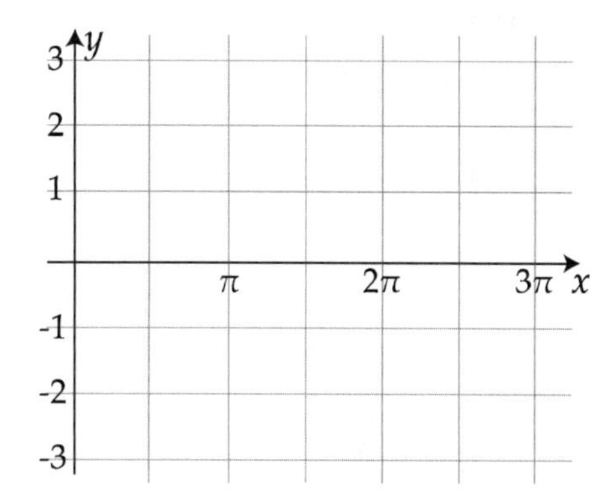

⑨ $y=2\sin\left(\frac{x}{3}\right)$

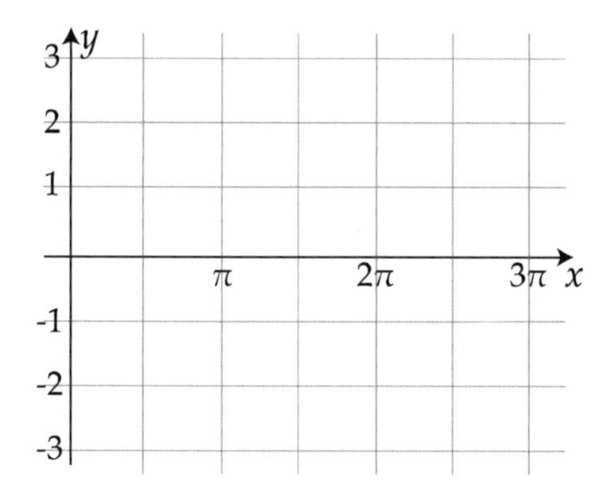

⑩ $y=-3\cos\left(\frac{x}{2}\right)-1$

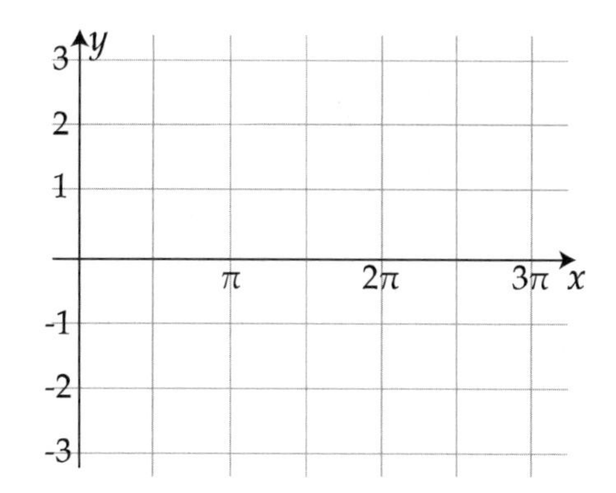

⑪ $y = -2\cos(\pi x)$

⑫ $y = -\sin\left(\dfrac{\pi x}{3}\right)$

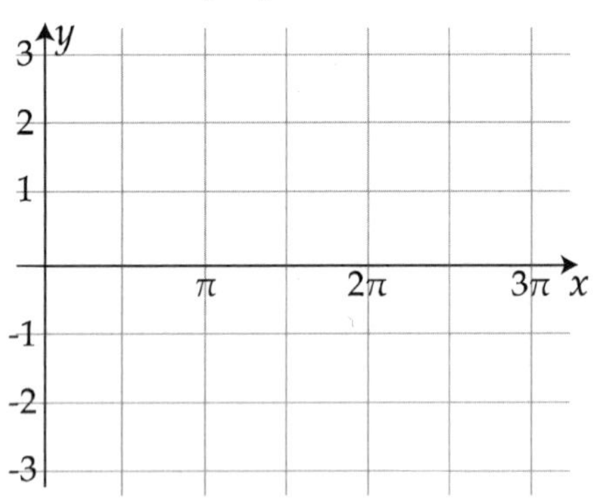

EXAMPLE 4. The graph of one complete period of a sine or cosine curve is given. Write an equation that represents the curve.

①

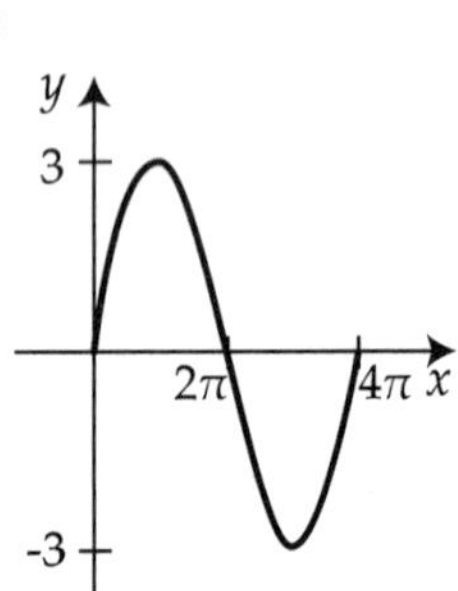

②

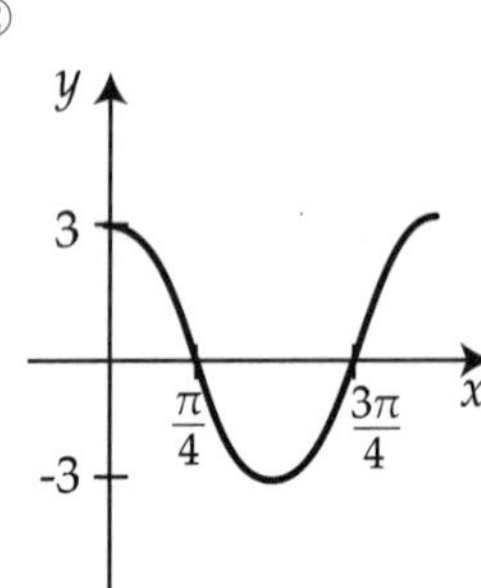

③

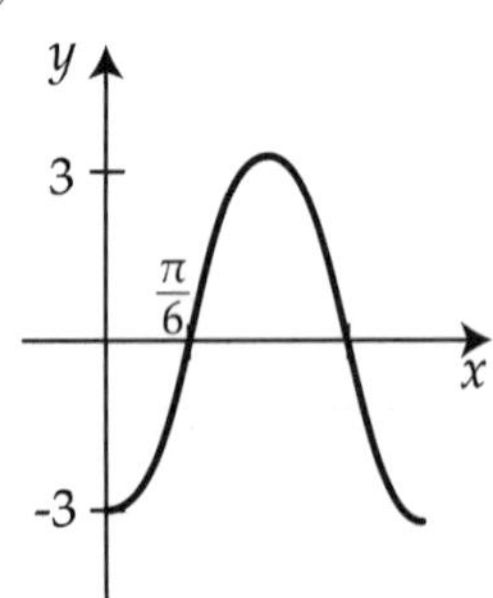

④

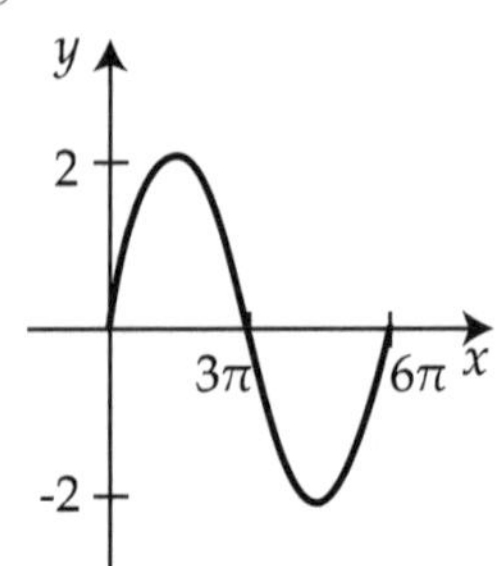

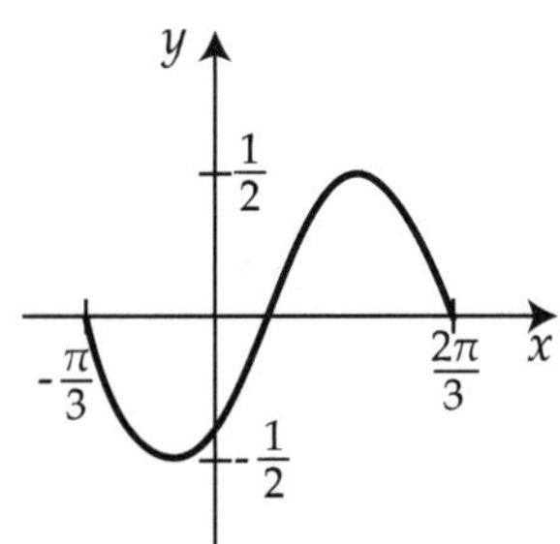

⑥

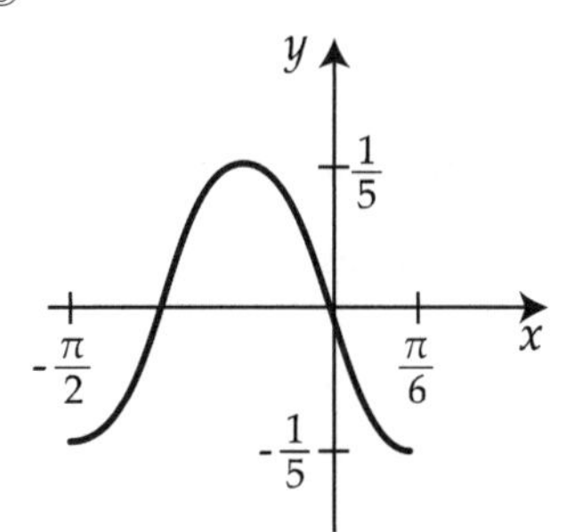

⑦

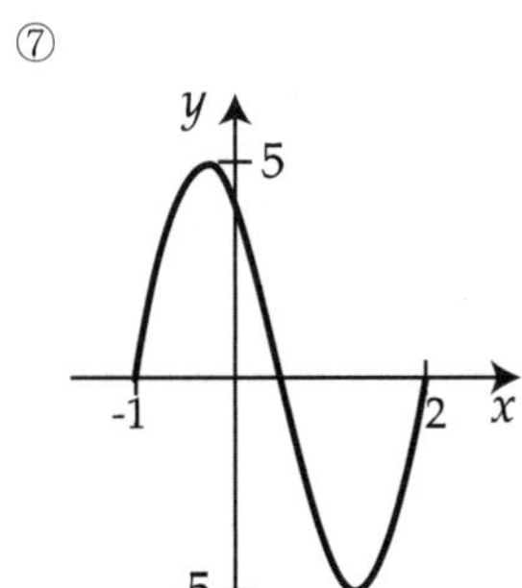

⑧

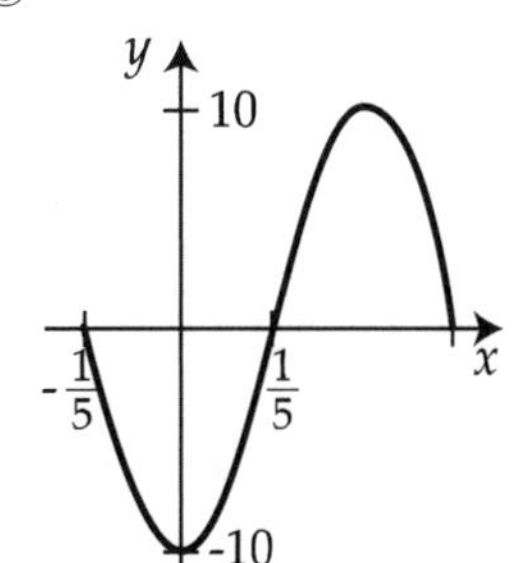

⑨

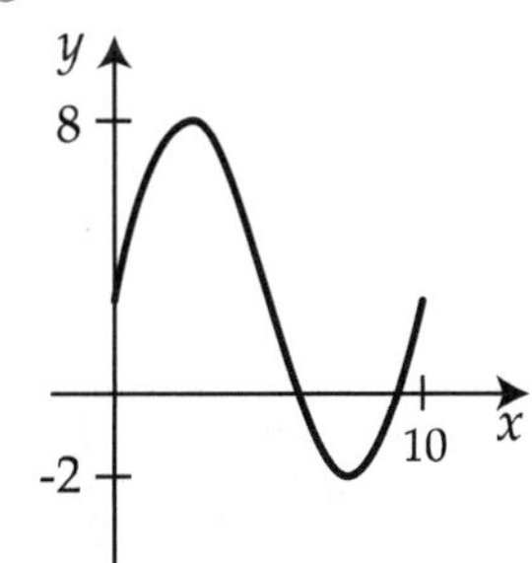

⑩

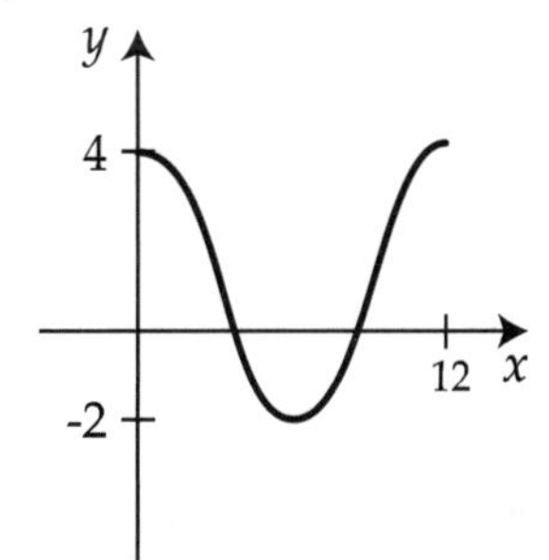

3. Periodic Function

※ Periodic Function

: all possible y values ①__________ in the same sequence over a given set of x values.

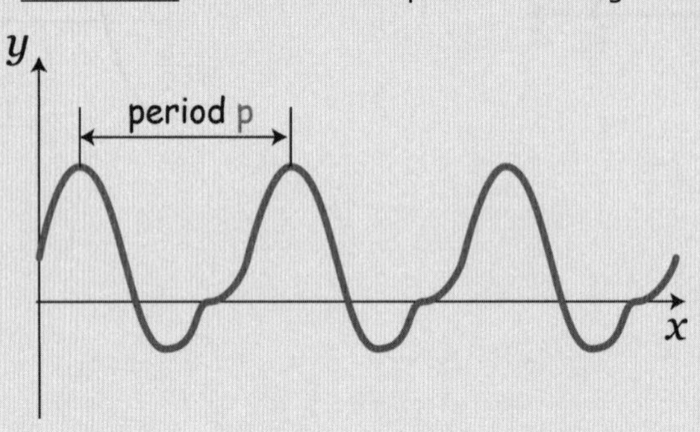

If a function f is periodic, where p is a positive number, called a period of f, then

$$f(x + p) = f(x)$$

EXAMPLE 5. What is the period of each function?

①

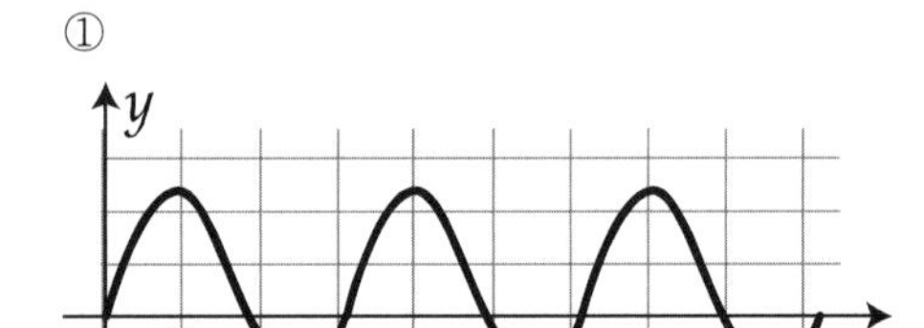

②

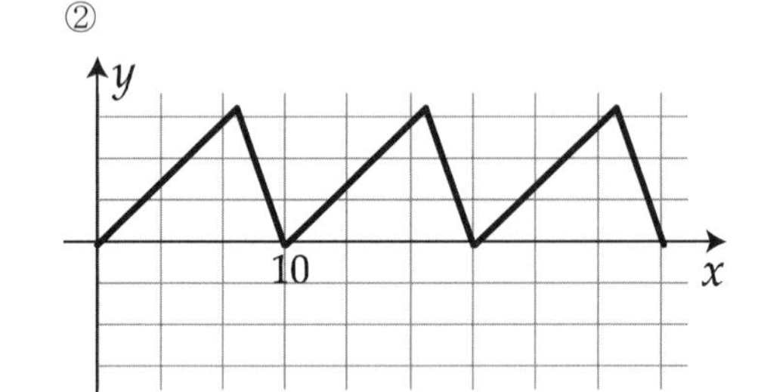

③

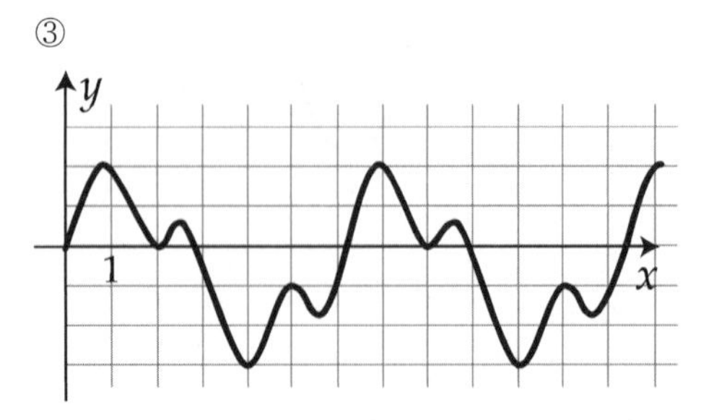

④

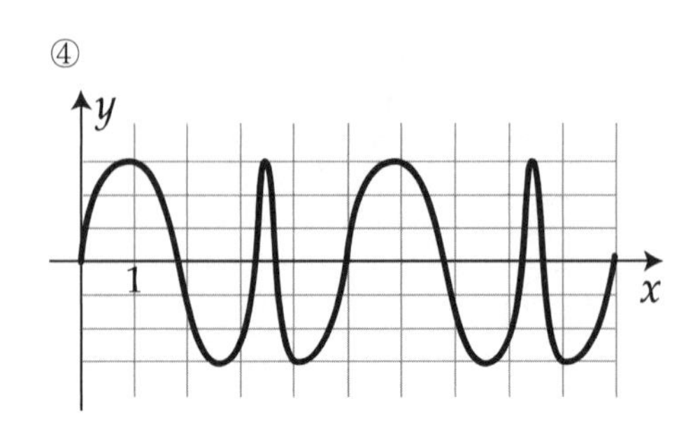

Blank : ① repeats

⑤

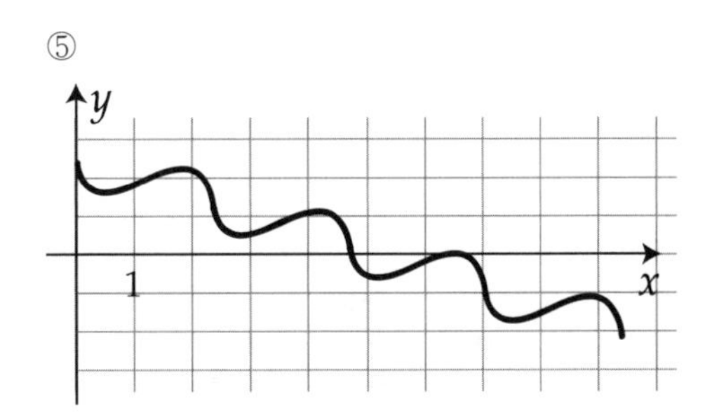

⑥

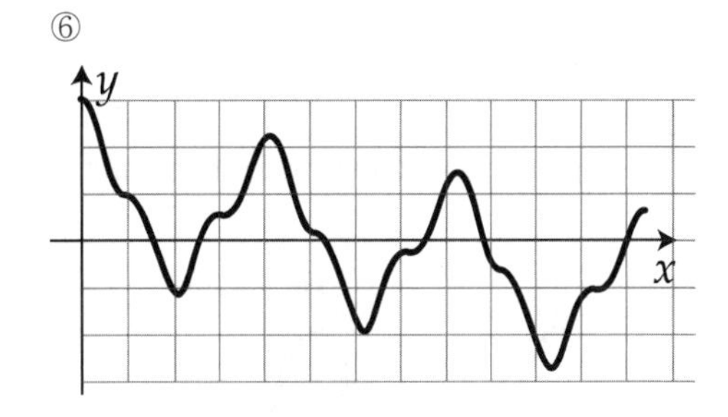

⑦ $f(x+5)=f(x)$

⑧ $f(x+10)=f(x)$

⑨ $f(x-2)=f(x+4)$

⑩ $f(x-7)=f(x+2)$

AP Style Problem

1. Which of the following is the graph of $y=-3\cos 2\pi x$?

A)

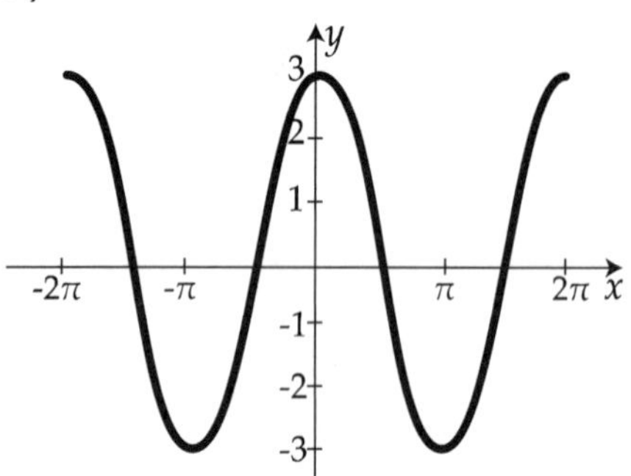

B)

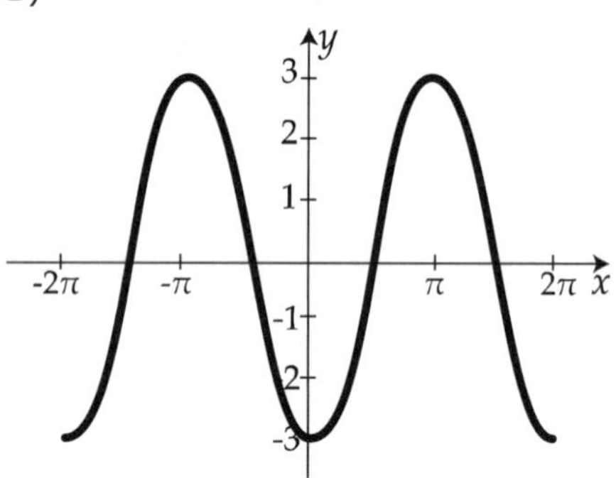

C)

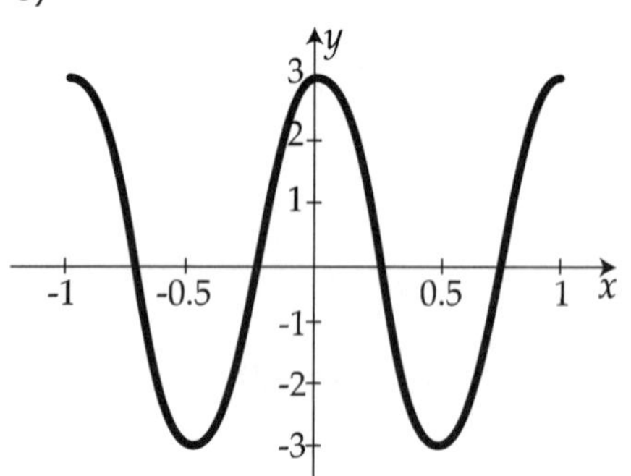

D)

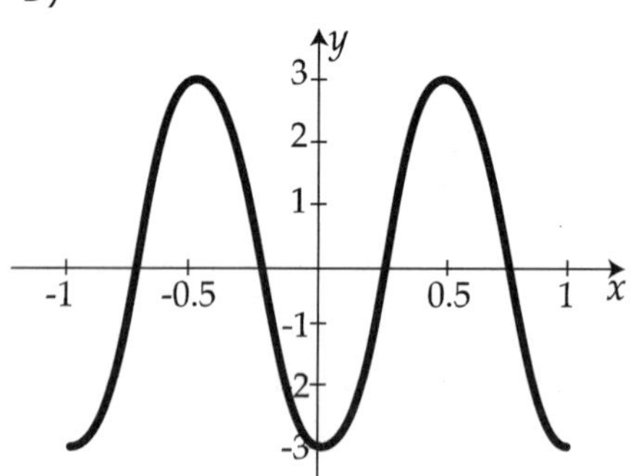

2. The values of a sinusoidal function $g(x)$ are given in the table. Determine the equation of the function $g(x)$.

x	0	4π	8π	12π	16π
$g(x)$	-2	4	-2	-8	-2

A) $y=6\sin\frac{1}{8}x-2$

B) $y=6\sin\frac{1}{8}x-8$

C) $y=6\sin 8x-2$

D) $y=6\sin 8x-8$

3. The range of trigonometric function $y = a\sin 2\pi x + b$ is $-2 \le y \le 10$. What is the value of b?

A) 2

B) 4

C) 6

D) 8

4. What is the minimum value of $G(t) = 440 - 220\cos\left(\frac{\pi t}{12}\right)$?

A) 220

B) 440

C) 660

D) 880

5. A portion of the sinusoidal function $f(x)$ is given as shown. Of the following, on which interval is f increasing and concave down?

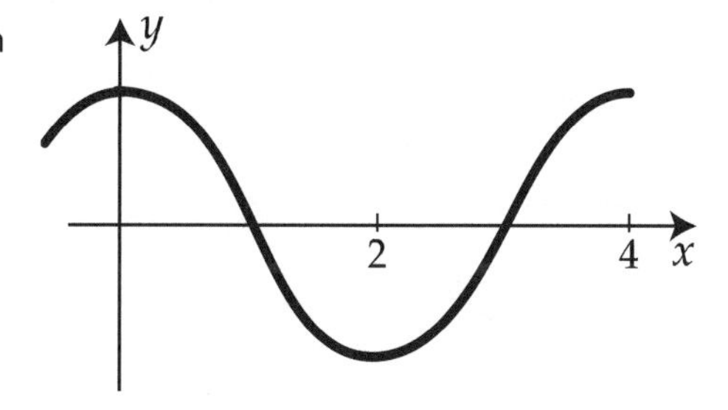

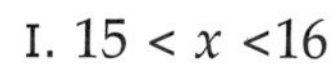 I. $15 < x < 16$

II. $40 < x < 41$

III. $22 < x < 24$

A) I

B) III

C) II, III

D) I, III

6. A sinusoidal function $f(x)$ has a period of 3π and has maximum value at $x=\frac{\pi}{4}$. Then which of the following statement is true?

I. In the interval $\frac{\pi}{4}\le x\le\pi$, f is decreasing at a decreasing rate.

II. In the interval $\frac{7\pi}{4}\le x\le\frac{5\pi}{2}$, f is increasing at a increasing rate.

III. In the interval $-\frac{\pi}{2}\le x\le\frac{\pi}{4}$, f is decreasing at a decreasing rate.

A) I

B) III

C) I, II

D) I, III

7. * Find the range of the function $f(x)=\frac{3}{5+2\sin x}$.

A) $3\le y\le 7$

B) $-1\le y\le\frac{7}{3}$

C) $1\le y\le\frac{7}{3}$

D) $\frac{3}{7}\le y\le 1$

4.6 Modeling using Sin,Cos Functions

1. Modeling using trigonometric functions

If we have a maximum point and a minimum point of a sinusoidal function,
then it is easy to find the model of the function.
A sinusoidal function can be modeled by either a sine function or a cosine function, depending on which period I choose.

※ **Modeling with Sine Function**

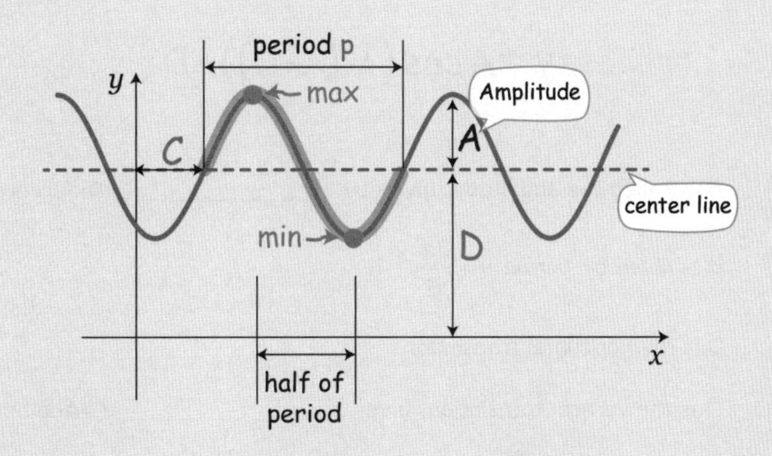

$$y = A\sin(B(x - C)) + D$$

A is called the amplitude, given by $\dfrac{\max y - \min y}{2}$ (=**D**ifference/2)

B is given by $period\ p = \dfrac{2\pi}{B}$

C is the horizontal translation

D is the vertical translation, given by $\dfrac{\max y + \min y}{2}$ (=**A**ddition/2)

※ **Modeling with Cosine Function**

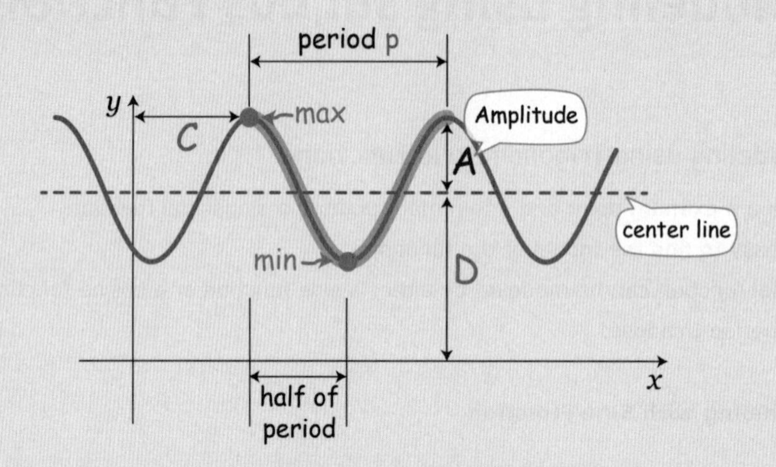

$$y = A\cos(B(x - C)) + D$$

A is called the amplitude, given by ①__________ (=**D**ifference/2)

B is given by $period\ p = \dfrac{2\pi}{B}$

C is the horizontal translation

D is the vertical translation, given by ②__________ (=**A**ddition/2)

Blank : ① $\dfrac{\max y - \min y}{2}$ ② $\dfrac{\max y + \min y}{2}$

EXAMPLE 1. Write a function for a sinusoid model $y = a\sin(bx) + d$ or $y = a\cos(bx) + d$.

①

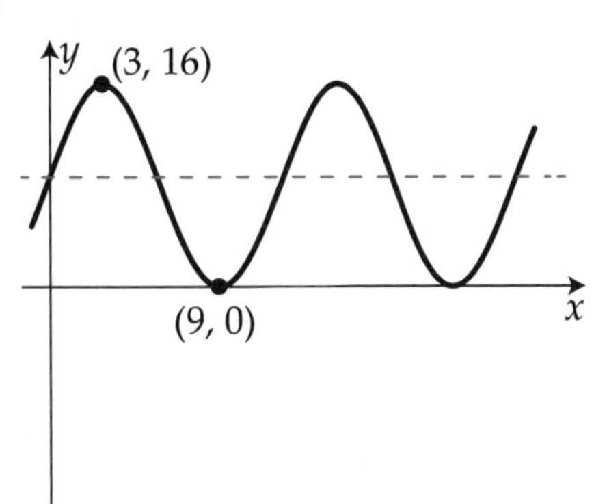

②

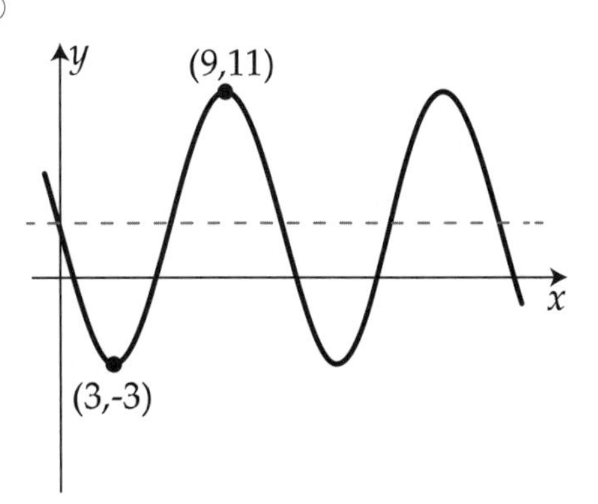

③

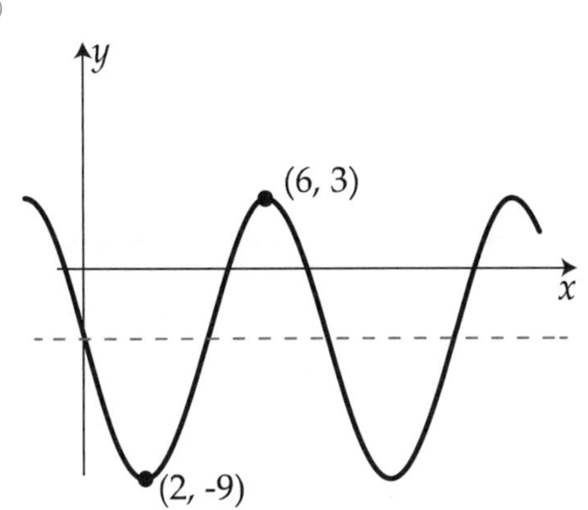

④

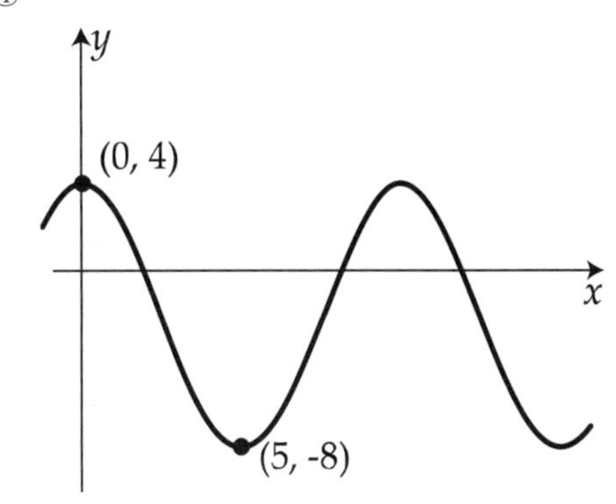

⑤

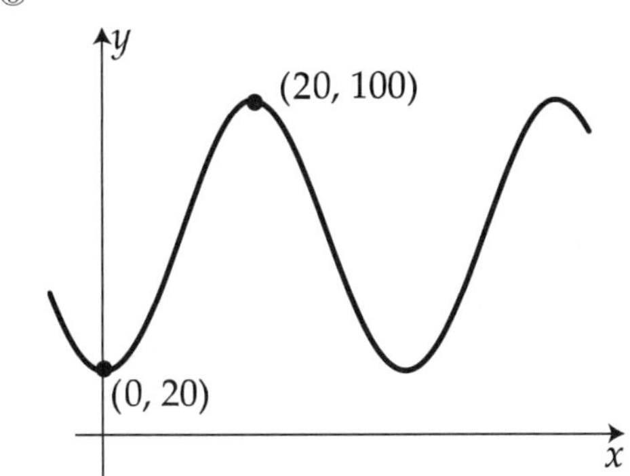

⑥

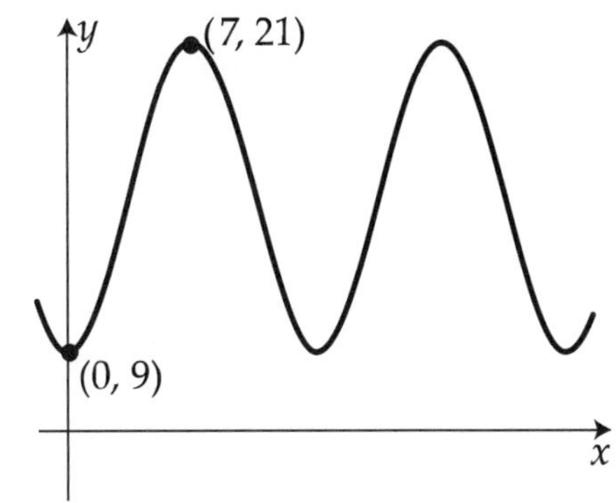

EXAMPLE 2. The height of water in the harbor is 18 m at high tide, and 12 hours later at low tide, it is 10 m. The graph below shows how the height of water changes with time over 24 hours.

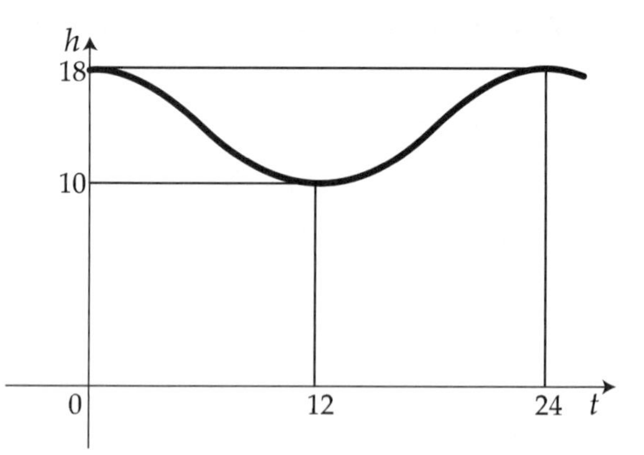

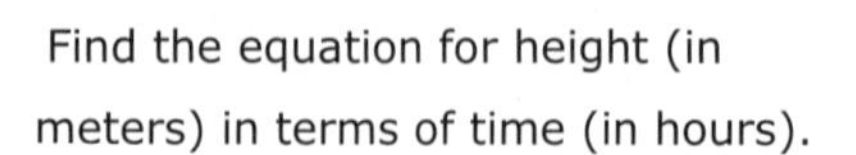

Find the equation for height (in meters) in terms of time (in hours).

EXAMPLE 3. The Ferris wheel with diameter 50ft makes one complete turn every 200 sec. The bottom of the wheel is 8ft above the ground and a seat A starts at the bottom of the wheel.

1) Graph the model of the height of seat A with respect to time.

2) Find an equation that describes the height of seat A with respect to time.

EXAMPLE 4. The following diagram shows a waterwheel with a bucket. The wheel rotates at a constant rate in an anticlockwise (counterclockwise) direction.

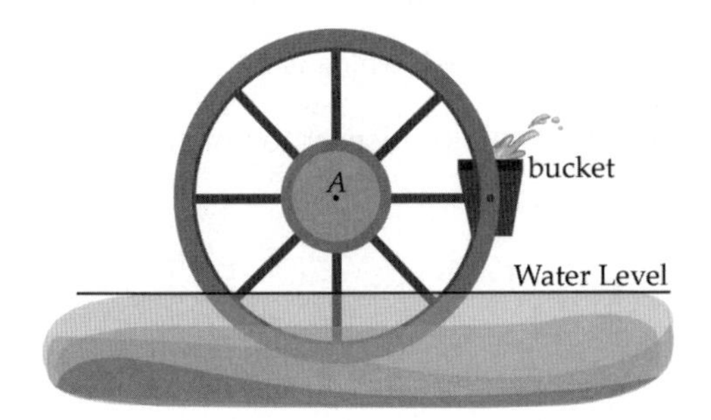

The diameter of the wheel is 8 m. The center of the wheel, A, is 2 m above the water level. The water bucket turns at a rate of one rotation every 30 seconds. Find an equation that describes the height of the water bucket with respect to time.

AP Style Problem

1. The Ferris wheel with diameter of 20 feet is rotating at the rate of 3 revolutions per hour. When t = 0, a chair starts at the *highest point* on the wheel, which is 22 feet above the ground. Find an equation that describes the height of chair $h(t)$ with respect to time t, where t is in minutes.

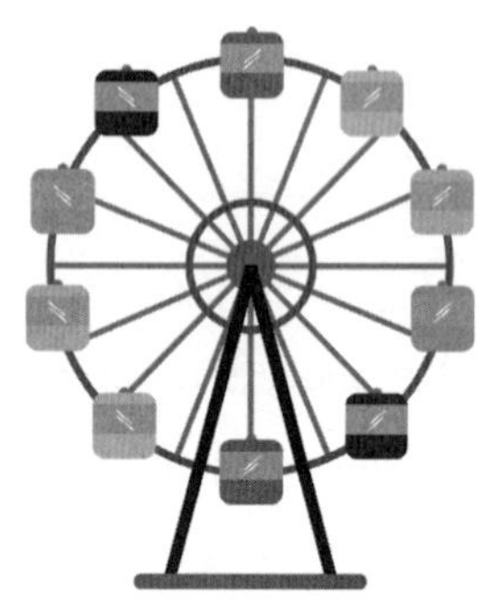

A) $h(t)=10\cos 6\pi t+12$

B) $h(t)=12\cos 6\pi t+10$

C) $h(t)=12\cos\frac{\pi t}{10}+10$

D) $h(t)=10\cos\frac{\pi t}{10}+12$

2. A wright is attached to a spring suspended from a beam. At time t = 0, it is pulled down to a point 7cm above the ground and released. After than, it bounces up and down between its minimum height of 7 cm and a maximum height of 21 cm. It first reaches a maximum height 0.6 seconds after starting. Find an equation that describes its motion.

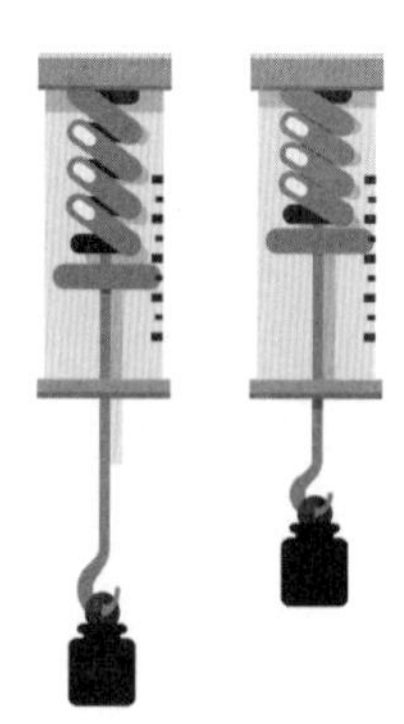

A) $h(t)=-14\cos\frac{5\pi t}{3}+7$

B) $h(t)=14\cos\frac{5\pi t}{3}+7$

C) $h(t)=-7\cos\frac{5\pi t}{3}+14$

D) $h(t)=7\cos\frac{5\pi t}{3}+14$

3. Table below shows the high termperature T for each month in Evanston, Illinois. Time t is measured in months, with t = 1 represents January.

t	1	2	3	4	5	6	7	8	9	10	11	12
T	20	25	37	50	61	71	80	72	62	51	38	25

Which of the following could model the temperature T as fucntion of t?

A) $T = 30\cos\left[\dfrac{\pi}{6}(t-1)\right] + 50$

B) $T = -30\cos\left[\dfrac{\pi}{6}(t-1)\right] + 50$

C) $T = 50\cos\left[\dfrac{\pi}{6}(t-1)\right] + 30$

D) $T = -50\cos\left[\dfrac{\pi}{6}(t-1)\right] + 30$

4. * A pendulum is pulled back $20°$ off center and released to swing. It takes the pendulum a quarter of a second to reach other side. When we take t = 0 to be the time when the pendulum is at very right, which of the following is a function that describes the angle of the pendulum with respect to time?

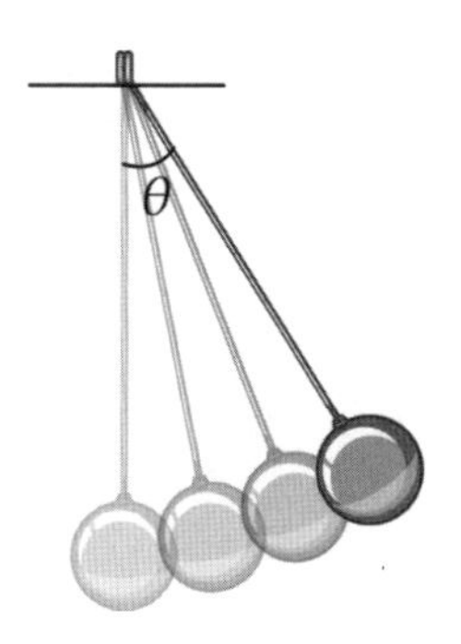

A) $20\cos 4\pi t$

B) $20\cos 2\pi t$

C) $10\cos 4\pi t + 10$

D) $10\cos 2\pi t + 10$

4.7 Trigonometric Graphs for Others

1. The Graph of Tangent

※ **Graph of Tangent**

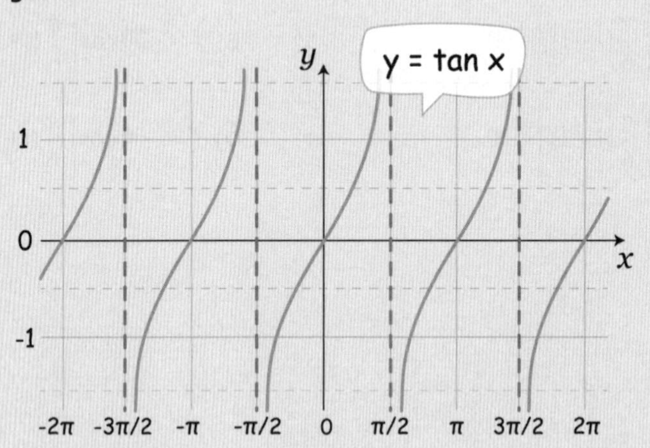

Domain: ① ____________________

Range: ② ____________________

Asymptotes: ③ ____________________

Period: ④ ____________________

Symmetry: ⑤ ______________ (odd/even)

※ **Transformation of Tangent graph**

$$y = A\tan(B(x-C)) + D$$

Period: ⑥ ____________________ translation: ⑦ ____________________

Blank : ① R, $x \neq \frac{(2n-1)\pi}{2}$ (where n is integer) ② R(all real numbers) ③ $x = \frac{(2n-1)\pi}{2}$ (where n is integer)

④ π ⑤ origin (odd) ⑥ $\frac{\pi}{B}$ ⑦ Left or right C units, up or down D units

EXAMPLE 1. Find the period and translation of given function.

① $y=-2\tan\left(2x-\dfrac{\pi}{4}\right)+1$

② $y=3\tan\left(\dfrac{x}{4}+\pi\right)+3$

③ $y=3\tan\left(\dfrac{1}{2}x-2\right)-5$

④ $y=4\tan(3\pi x-\pi)$

EXAMPLE 2. Graph the function.

① $y=3\tan x$

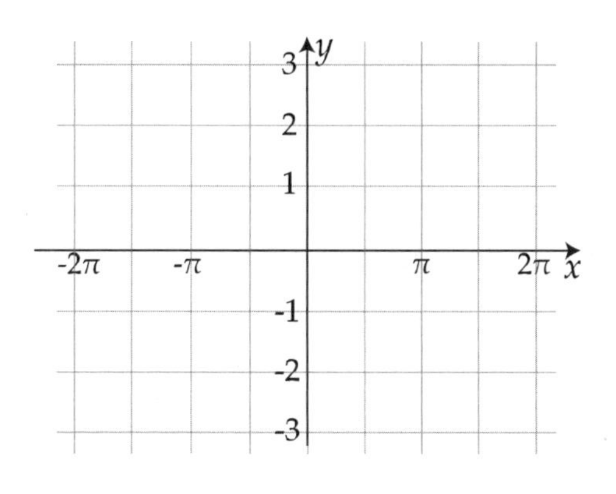

② $y=\tan 2x$

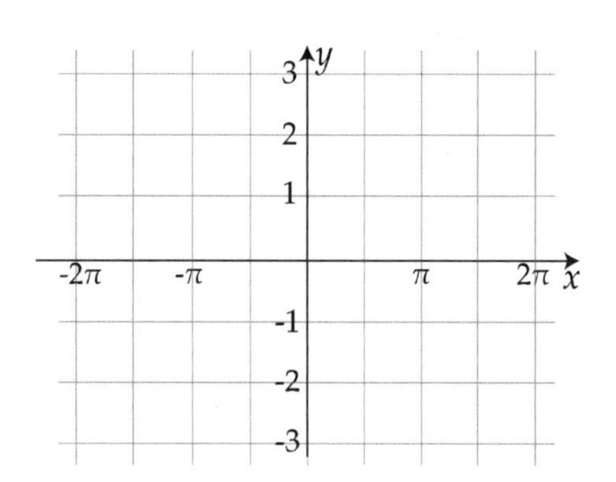

③ $y=-\tan\dfrac{x}{2}$

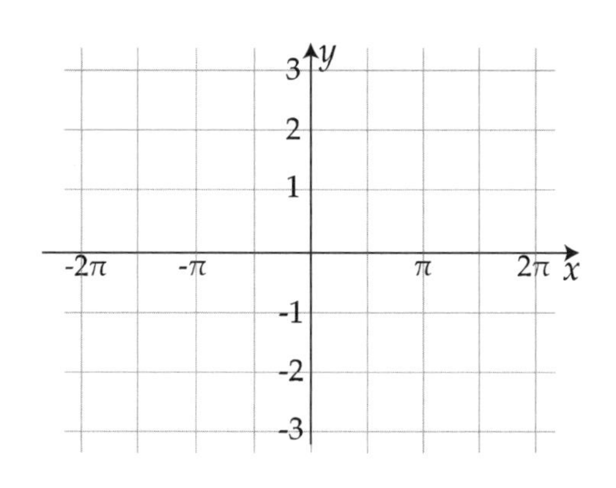

④ $y=\tan(x-\pi)$

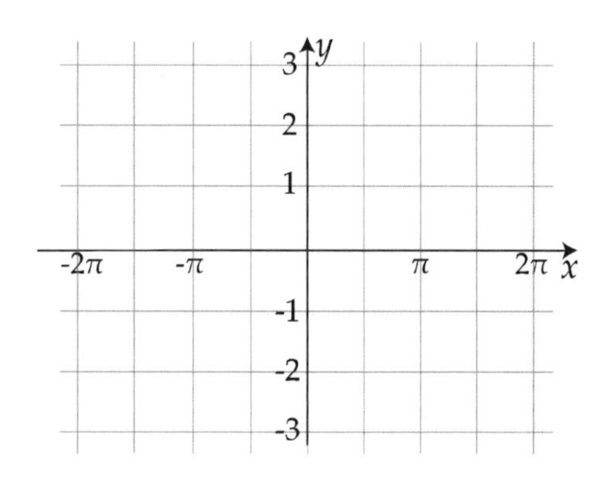

⑤ $y = \tan\left(x - \frac{\pi}{2}\right)$

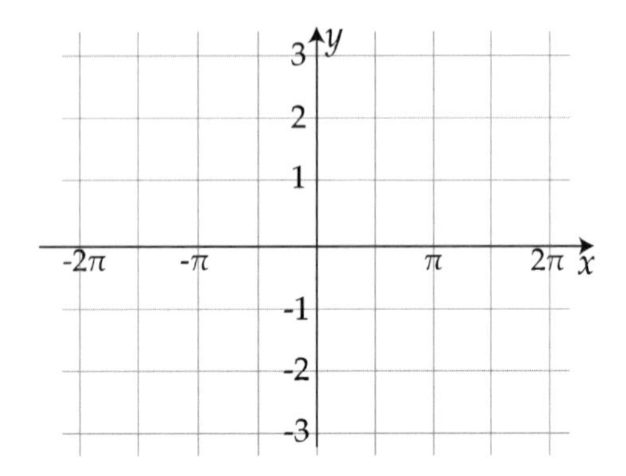

2. The Graph of Csc, Sec, and Cot

To sketch a reciprocal of function f(x) ;

$y = f(x)$	$y = \frac{1}{f(x)}$
0	①
vertical asymptote	
1	
-1	
Large and positive	
small and positive	
Large and negative	
small and negative	
Maximum point	
Minimum point	

Blank : ① und , 0 , 1 , -1 , small and positive , large and positive , small and negative , large and negative , min , max

EXAMPLE 3. Graph the function .

① Try to graph csc x (the reciprocal function of sin x).

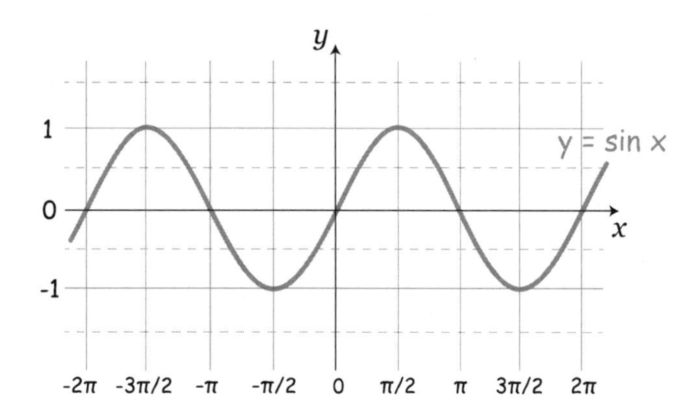

② Try to graph sec x (the reciprocal function of cos x).

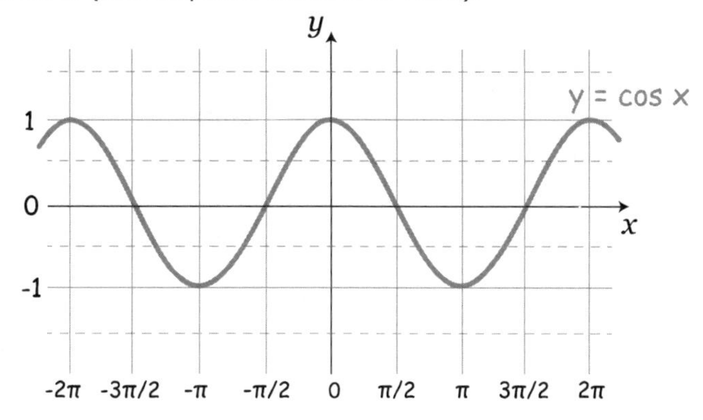

③ Try to graph cot x (the reciprocal function of tan x).

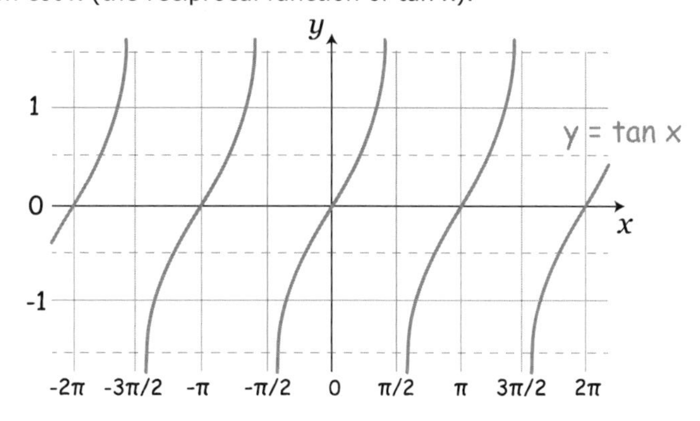

※ **Graph of Csc x, Sec x, Cot x**

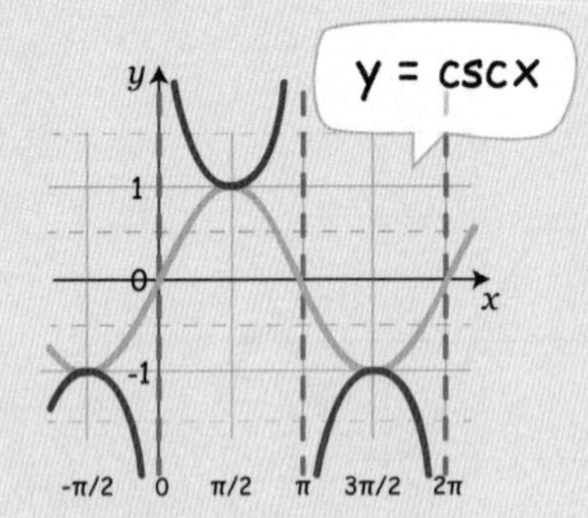

Domain: ① ____________________

Range: ____________________

Asymptotes: ________________

Period: ____________________

Symmetry: ______________(odd/even)

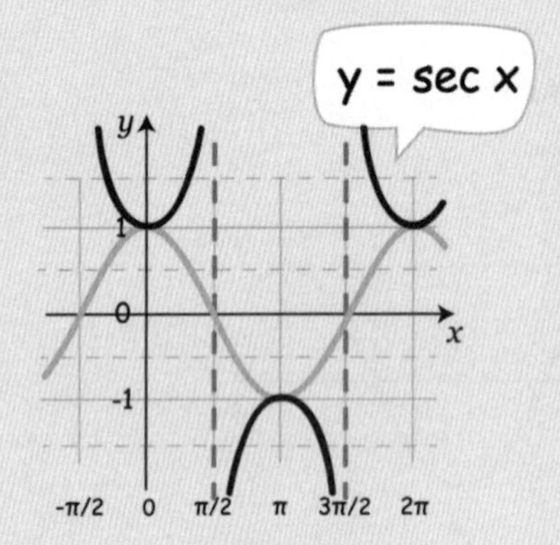

Domain: ② ____________________

Range: ____________________

Asymptotes: ________________

Period: ____________________

Symmetry: ______________(odd/even)

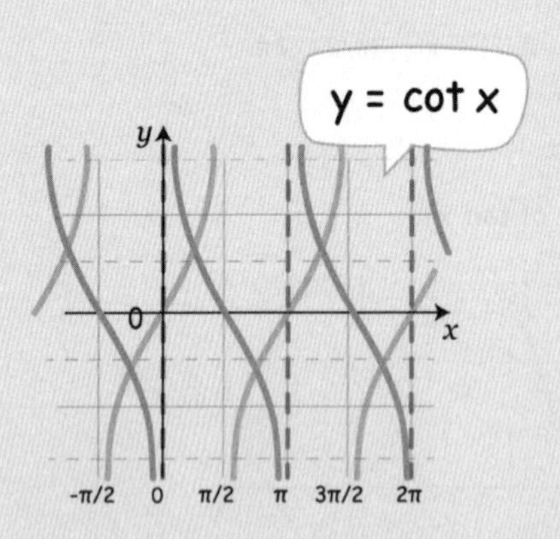

Domain: ③ ____________________

Range: ____________________

Asymptotes: ________________

Period: ____________________

Symmetry: ______________(odd/even)

Blank : ① All real numbers (R) where $x \neq n\pi$, $(-\infty,-1] \cup [1,\infty)$, $x = n\pi$, 2π, origin, odd

② All real numbers (R) where $x \neq \frac{(2n-1)\pi}{2}$, $(-\infty,-1] \cup [1,\infty)$, $x = \frac{(2n-1)\pi}{2}$, 2π, y axis, even

③ All real numbers (R) where $x \neq n\pi$, R , $x = n\pi$, π, origin, odd (n is integer)

EXAMPLE 4. Graph the function .

① $y = 2\csc x$

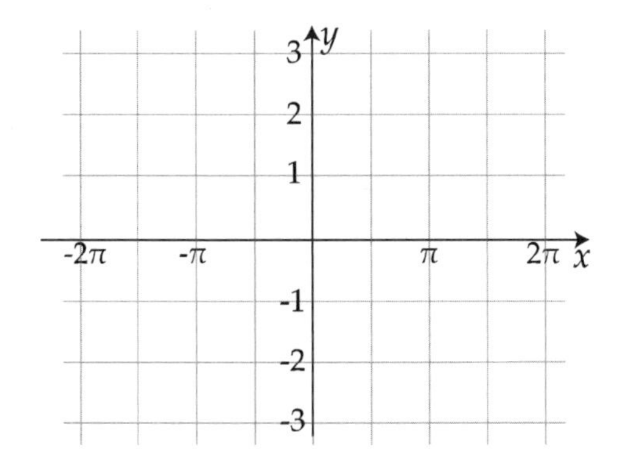

② $y = 0.5\sec x$

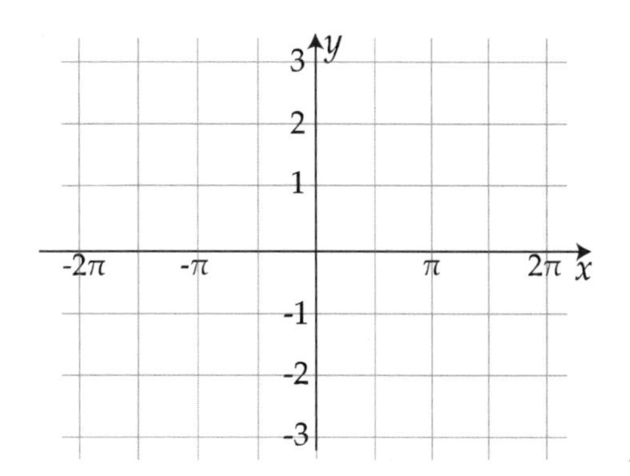

③ $y = \sec(0.5x)$

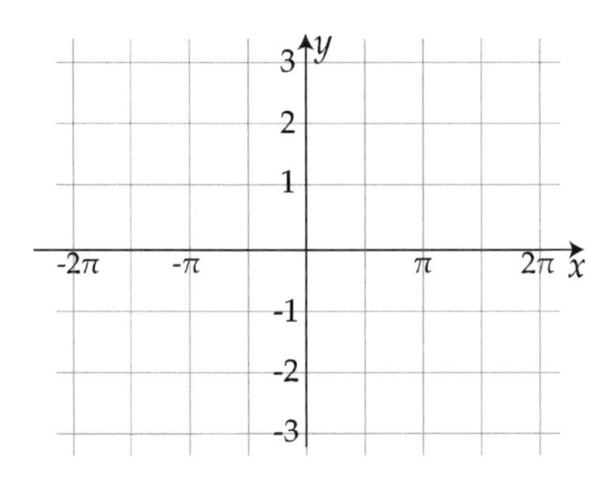

④ $y = \csc 2x$

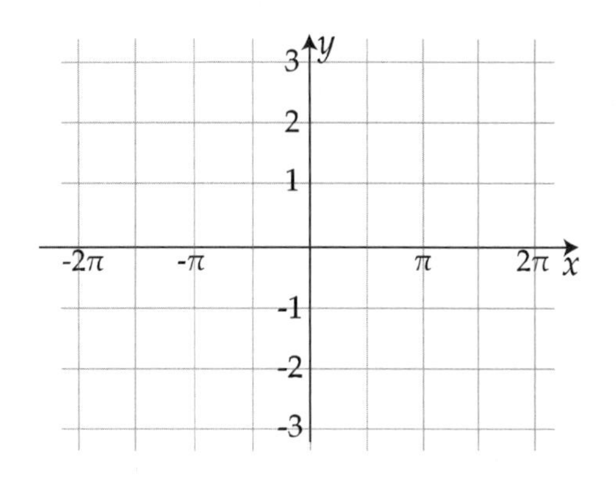

⑤ $y = \cot(x - \pi)$

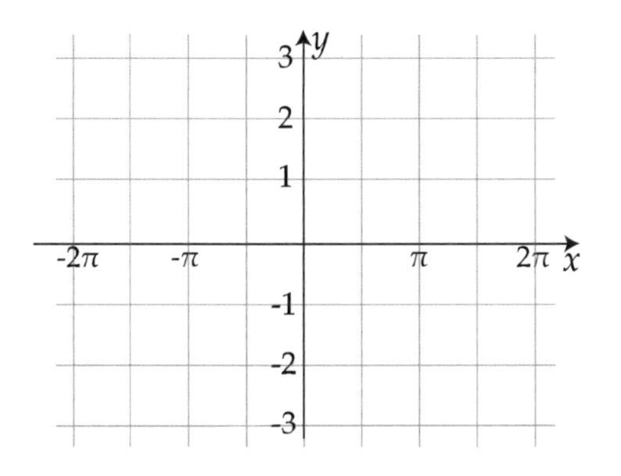

⑥ $y = \cot x + 1$

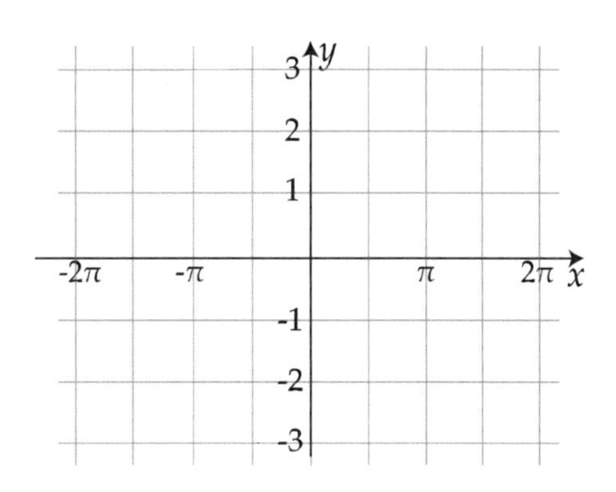

☺ ALL in one _ Graphs of trig functions

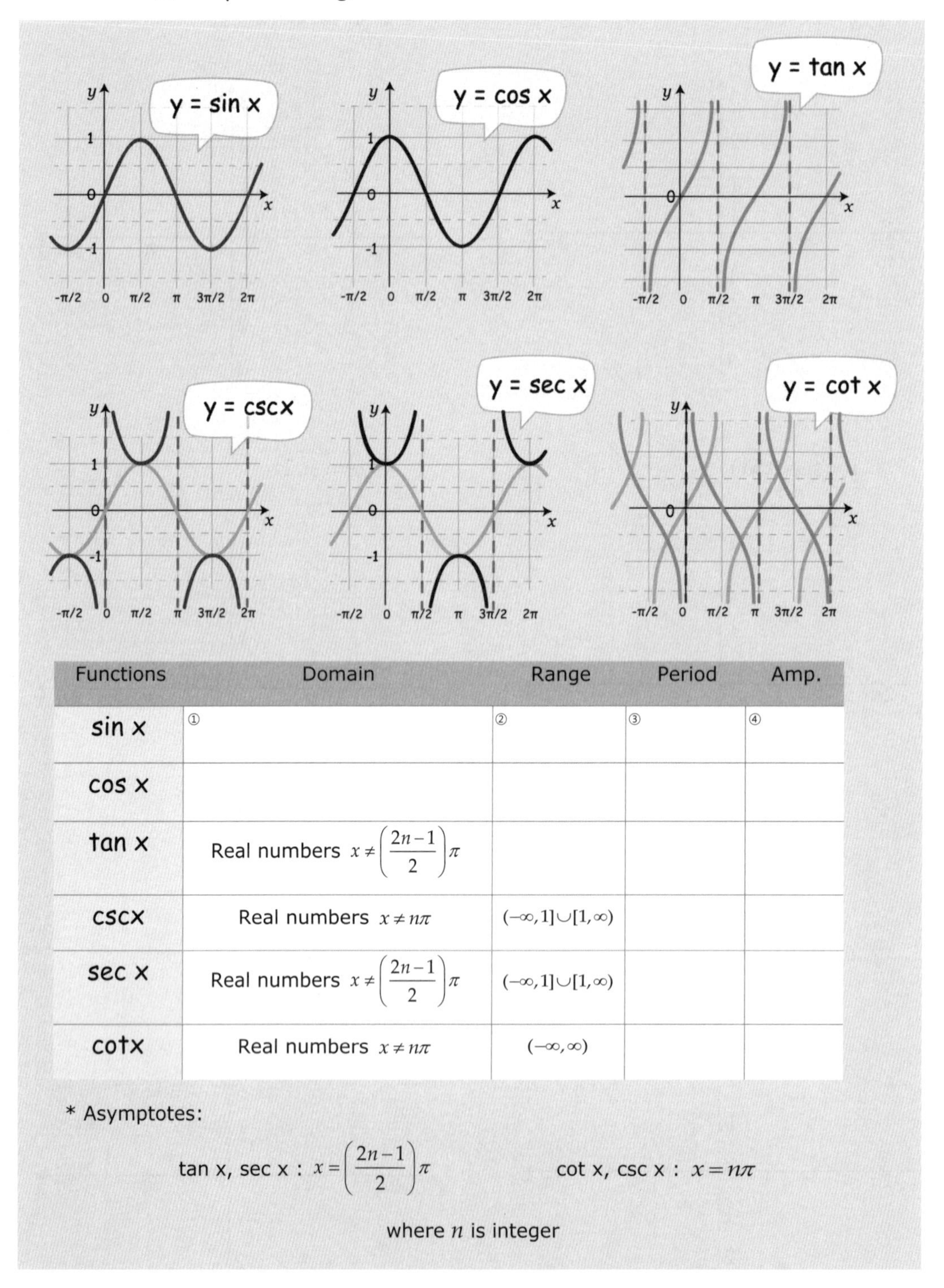

Functions	Domain	Range	Period	Amp.
sin x	①	②	③	④
cos x				
tan x	Real numbers $x \neq \left(\frac{2n-1}{2}\right)\pi$			
cscx	Real numbers $x \neq n\pi$	$(-\infty, 1] \cup [1, \infty)$		
sec x	Real numbers $x \neq \left(\frac{2n-1}{2}\right)\pi$	$(-\infty, 1] \cup [1, \infty)$		
cotx	Real numbers $x \neq n\pi$	$(-\infty, \infty)$		

* Asymptotes:

tan x, sec x : $x = \left(\frac{2n-1}{2}\right)\pi$ cot x, csc x : $x = n\pi$

where n is integer

Blank : ① R, R ② [-1, 1] , [-1, 1] , R ③ 2π , 2π , π , 2π , 2π , π ④ 1 , 1 , none , none , none , none

AP Style Problem

1. A graph of a trigonometric function f is given as shown. Which of the following could define $f(x)$?

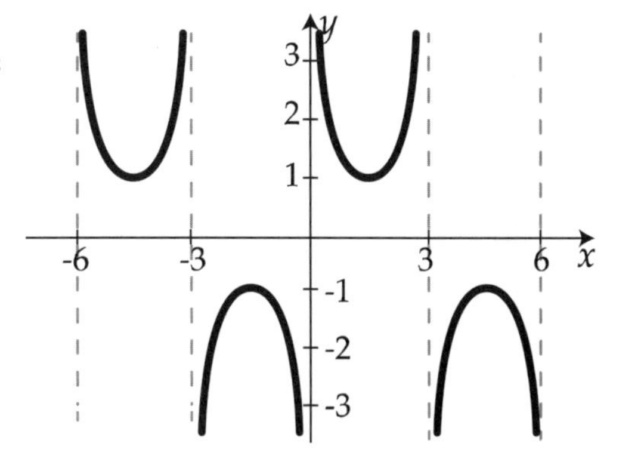

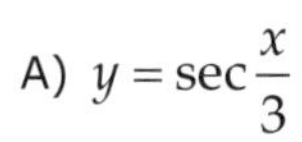

A) $y = \sec\dfrac{x}{3}$

B) $y = \csc\dfrac{x}{3}$

C) $y = \sec\dfrac{\pi x}{3}$

D) $y = \csc\dfrac{\pi x}{3}$

2. A graph of a trigonometric function g is given as shown. Which of the following could define $g(x)$?

I. $y = \tan x$

II. $y = -\cot x$

III. $y = \tan\left(x - \dfrac{\pi}{2}\right)$

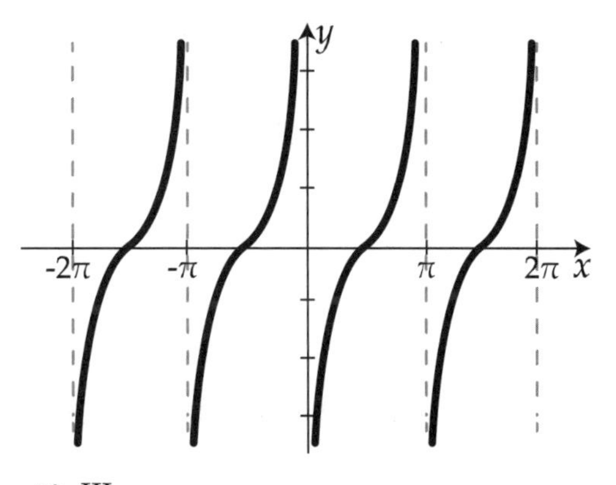

A) I

B) III

C) II, III

D) I, III

3. What is $\lim_{x \to \frac{7\pi}{2}^{+}} \tan x$?

A) $-\infty$

B) ∞

C) 1

D) 0

4.* Find the domain of $y = 2\csc 3x + 1$.

A) All real numbers except $x = \frac{n\pi}{3}$ where n is an integer.

B) All real numbers except $x = 3n\pi$ where n is an integer.

C) All real numbers except $x = \frac{n\pi}{2}$ where n is an integer.

D) All real numbers except $x = 2n\pi$ where n is an integer.

5.* Which of the following is true about $y = \tan\frac{x}{2}$?

A) Vertical asymptote is at $x = \frac{n\pi}{2}$ where n is an integer.

B) Vertical asymptote is at $x = n\pi$ where n is an integer.

C) Vertical asymptote is at $x = (2n-1)\pi$ where n is an integer.

D) Vertical asymptote is at $x = \frac{(2n-1)\pi}{2}$ where n is an integer.

6.* What is the range of $y = 3\sec\frac{x}{2} + 1$

A) $y \le -2 \ or \ y \ge 4$

B) $y \le -1 \ or \ y \ge 3$

C) $y \le 0 \ or \ y \ge 3$

D) $y \le -1 \ or \ y \ge 1$

7.* Write each value in increasing order.

$$A = \sin(1) \ , \ B = \cos(1), \ C = \tan(1)$$

A) $A < B < C$

B) $B < A < C$

C) $B < C < A$

D) $C < B < A$

Part 5

Trigonometry Identities

5.1 Inverse Trigonometry Function

1. Inverse Trig Function

※ **Inverse Trigonometry function**

The Sine function takes an ①__________ and gives us the ②______________.

Inverse Sine $\sin^{-1}$(= **arcsin**) takes the ③________ and gives us the ④__________.

Inverse Trig

$$\sin(30^\circ) = \frac{1}{2} \quad \rightleftarrows \quad 30^\circ = \sin^{-1}\left(\frac{1}{2}\right), \quad 30^\circ = \arcsin\left(\frac{1}{2}\right)$$

EXAMPLE 1. Find the measure of the indicated angle to the nearest degree.

①

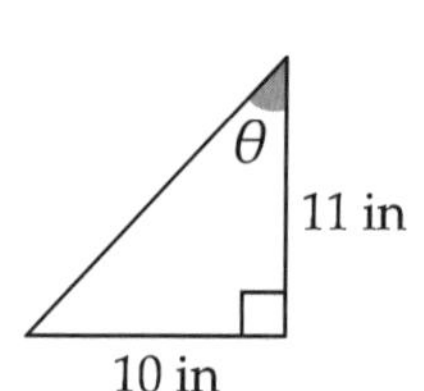

②

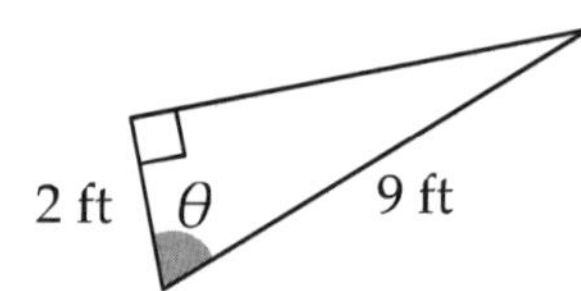

③

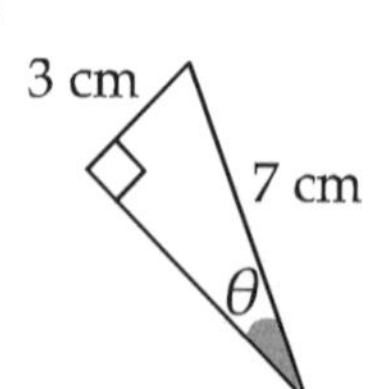

④

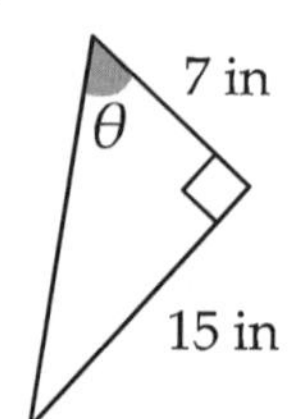

Blank : ① angle ② ratio ③ ratio ④ angle

EXAMPLE 2. A 40-ft ladder leans against a building. If the base of the ladder is 12 ft from the base of the building, what is the angle formed by the ladder and the building?

EXAMPLE 3. Joshua observes a boat in the sea below him from a point 6 ft above a 45 ft cliff. He has been told that the distance from the boat to the base of the cliff is 60 ft. What is the angle of depression, in degrees, from Joshua to the boat?

2. Graph of Inverse Trig Function

☺ remind:

① Only **ONE-TO-ONE function** has an inverse function!

② Functions and their inverses are always symmetric about the line ①________.

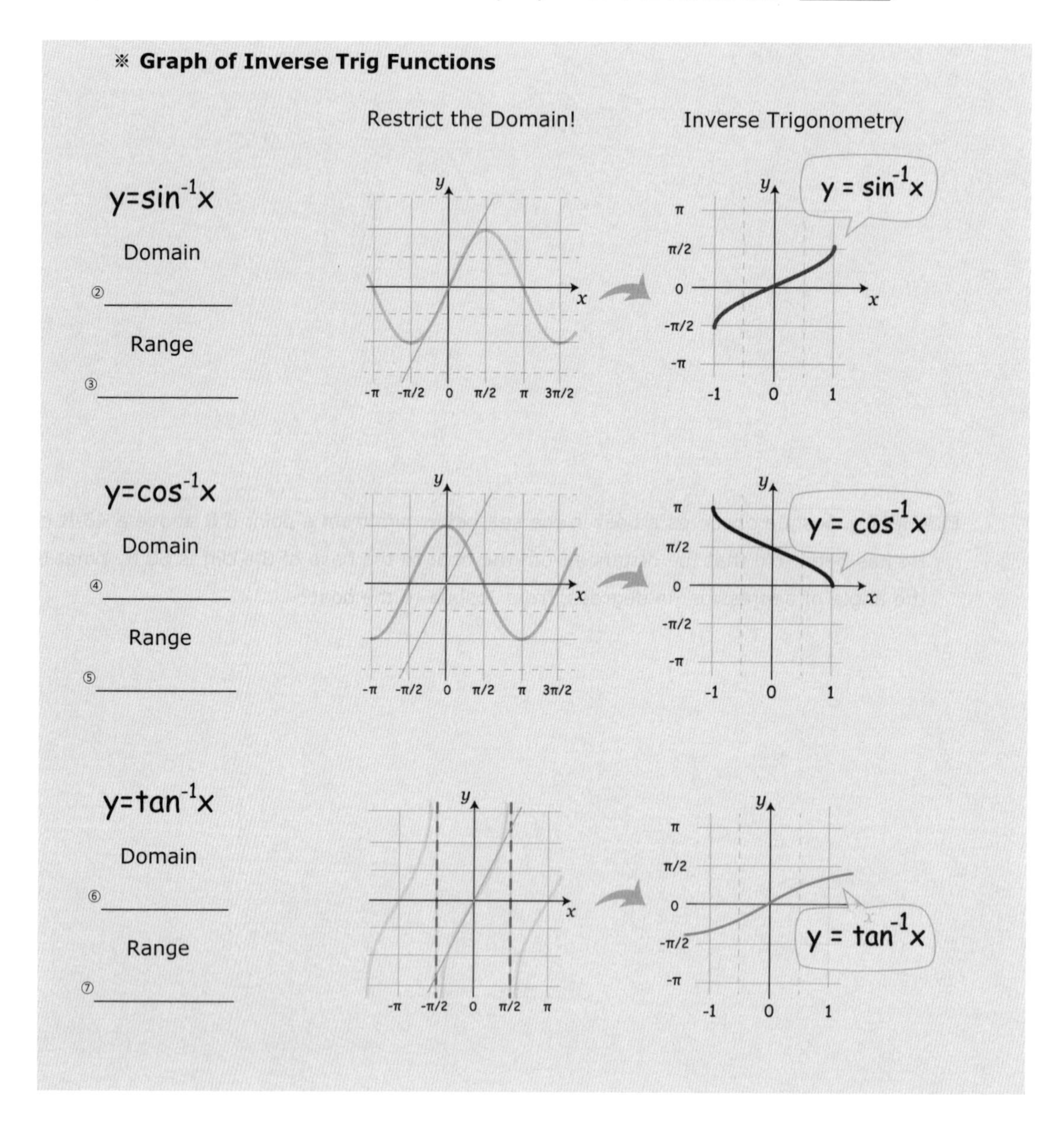

Blank : ① y=x ② [-1,1] ③ $\left[-\frac{\pi}{2}, \frac{\pi}{2}\right]$ ④ [-1,1] ⑤ [0,π] ⑥R ⑦ $\left(-\frac{\pi}{2}, \frac{\pi}{2}\right)$

In unit circle:

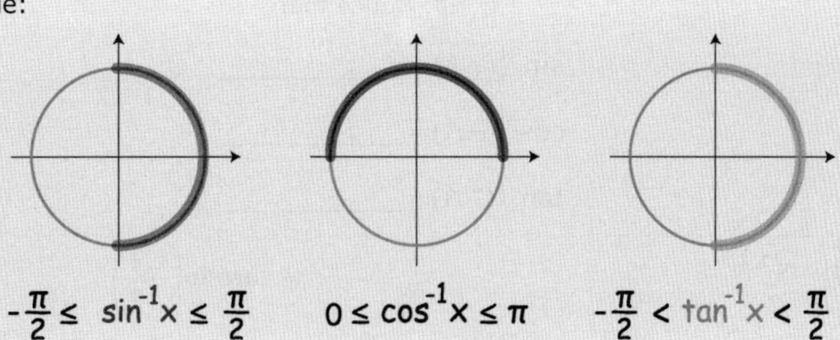

EXAMPLE 4. Evaluate in radians.

Inverse Trig gives you the ANGLE!

① $\sin^{-1}\left(\frac{1}{2}\right)$

② $\arccos\left(\frac{\sqrt{3}}{2}\right)$

③ $\tan^{-1}(1)$

④ $\arctan(0)$

⑤ $\arccos\left(\frac{\sqrt{2}}{2}\right)$

⑥ $\sin^{-1}(1)$

☺ Tip: Notice that $\csc^{-1} A = \sin^{-1}\left(\frac{1}{A}\right)$

⑦ $\text{arccsc}(2)$

⑧ $\text{arcsec}(1)$

⑨ $\cot^{-1}(\sqrt{3})$

⑩ $\text{arccsc}\left(\frac{2\sqrt{3}}{3}\right)$

⑪ $\text{arcsec}(\sqrt{2})$

⑫ $\text{arccot}(1)$

☺ Tip: Notice that

$\sin^{-1}(-A) =$ ①__________

$\cos^{-1}(-A) =$ ②__________

$\tan^{-1}(-A) =$ ③__________

⑬ $\cos^{-1}\left(-\frac{\sqrt{3}}{2}\right)$

⑭ $\arcsin\left(-\frac{1}{2}\right)$

⑮ $\arctan\left(-\sqrt{3}\right)$

⑯ $\tan^{-1}(-1)$

⑰ $\sin^{-1}\left(-\frac{\sqrt{2}}{2}\right)$

⑱ $\cos^{-1}\left(-\frac{1}{2}\right)$

3. Evaluating Inverse Trigs (Type1)

☺ remind: $f^{-1}(f(x)) = f(f^{-1}(x)) =$ ④_____

But if there is inverse trig or trig in your outer function, be careful with the range!

⑤ $\square \le \sin(\sin^{-1}x) = x \le$ ⑥ $\square$

⑦ $\square \le \cos(\cos^{-1}x) = x \le$ ⑧ $\square$

⑨ $\square < \tan(\tan^{-1}x) = x <$ ⑩ $\square$

Blank : ① $-\sin^{-1}A$ ② $\pi - \cos^{-1}A$ ③ $-\tan^{-1}A$ ④ x ⑤ -1 ⑥1  ⑦-1  ⑧ 1 ⑨-∞  ⑩ ∞

EXAMPLE 5. Find the exact value.

① $\sin\left(\sin^{-1}0.5\right)$

② $\tan\left(\arctan 0.2\right)$

③ $\cos\left(\cos^{-1}(-0.4)\right)$

④ $\cos\left(\cos^{-1}\pi\right)$

⑤ $\sin\left(\arcsin e\right)$

⑥ $\sin\left(\sin^{-1}0.8\right)$

⑦ $\tan\left(\arctan 2\right)$

⑧ $\tan\left(\tan^{-1}3\right)$

4. Evaluating Inverse Trigs (Type2)

① [ ] $\le \sin^{-1}(\sin x) = x \le$ ② [ ]

③ [ ] $\le \cos^{-1}(\cos x) = x \le$ ④ [ ]

⑤ [ ] $< \tan^{-1}(\tan x) = x <$ ⑥ [ ]

If the answer is NOT in the range, use ASTC and reference angle to find the angle that is IN THE RANGE.

Blank : ① $-\frac{\pi}{2}$ ② $\frac{\pi}{2}$ ③ 0 ④ π ⑤ $-\frac{\pi}{2}$ ⑥ $\frac{\pi}{2}$

ex) Find the exact value.

a) $\sin^{-1}\left(\sin\dfrac{\pi}{3}\right)$

b) $\cos^{-1}\left(\cos\dfrac{\pi}{3}\right)$

c) $\tan^{-1}\left(\tan\dfrac{\pi}{3}\right)$

d) $\sin^{-1}\left(\sin\dfrac{2\pi}{3}\right)$

e) $\cos^{-1}\left(\cos\dfrac{2\pi}{3}\right)$

f) $\tan^{-1}\left(\tan\dfrac{2\pi}{3}\right)$

g) $\sin^{-1}\left(\sin\dfrac{5\pi}{4}\right)$

h) $\cos^{-1}\left(\cos\dfrac{5\pi}{4}\right)$

i) $\tan^{-1}\left(\tan\dfrac{5\pi}{4}\right)$

j) $\sin^{-1}\left(\sin\dfrac{11\pi}{6}\right)$

k) $\cos^{-1}\left(\cos\dfrac{11\pi}{6}\right)$

l) $\tan^{-1}\left(\tan\dfrac{11\pi}{6}\right)$

EXAMPLE 6. Find the exact value.

① $\sin^{-1}\left(\sin\dfrac{\pi}{6}\right)$

② $\tan^{-1}\left(\tan\dfrac{\pi}{3}\right)$

③ $\cos^{-1}\left(\cos\frac{3\pi}{4}\right)$

④ $\cos^{-1}\left(\cos\left(-\frac{\pi}{3}\right)\right)$

⑤ $\tan^{-1}\left(\tan\left(-\frac{2\pi}{3}\right)\right)$

⑥ $\cos^{-1}\left(\cos\frac{4\pi}{3}\right)$

⑦ $\cos^{-1}\left(\cos\frac{7\pi}{4}\right)$

⑧ $\tan^{-1}\left(\tan\frac{11\pi}{6}\right)$

⑨ $\tan^{-1}\left(\tan\frac{7\pi}{3}\right)$

⑩ $\sin^{-1}\left(\sin\frac{10\pi}{3}\right)$

⑪ $\cos^{-1}\left(\cos\frac{7\pi}{5}\right)$

⑫ $\sin^{-1}\left(\sin\frac{5\pi}{8}\right)$

5. Evaluating Inverse Trigs (Type3)

EXAMPLE 7. Use a sketch to find the exact value.

☺ Tip: Use a sketch!

① $\sin\left(\tan^{-1}\frac{1}{2}\right)$

② $\tan\left(\cos^{-1}-\frac{1}{3}\right)$

③ $\cos\left(\arctan\left(-\frac{3}{4}\right)\right)$

④ $\cos\left(\sin^{-1}\left(-\frac{3}{5}\right)\right)$

⑤ $\tan\left(\cos^{-1}\left(-\frac{5}{13}\right)\right)$

⑥ $\sec\left(\cos^{-1}\left(-\frac{8}{17}\right)\right)$

EXAMPLE 8. Write the following as an algebraic expression in x, x > 0.

① $\tan(\arcsin x)$

② $\cos(\sin^{-1} x)$

③ $\sin(\sin^{-1} x)$

④ $\cos(\arccos x)$

⑤ $\cot(\cos^{-1} x)$

⑥ $\sec(\text{arc}\cot x)$

⑦ $\sin\left(\tan^{-1}\dfrac{x}{\sqrt{5}}\right)$

⑧ $\cot\left(\cos^{-1}\dfrac{3}{\sqrt{x^2+9}}\right)$

⑨ $\sin\left(\sec^{-1}\dfrac{\sqrt{x^2+4}}{x}\right)$

AP Style Problem

1. When $x > 0$, then $\cos(\tan^{-1} x) =$

A) $\sqrt{1+x^2}$

B) $\dfrac{\sqrt{1+x^2}}{x}$

C) $\dfrac{1}{\sqrt{1+x^2}}$

D) $\dfrac{x}{\sqrt{1+x^2}}$

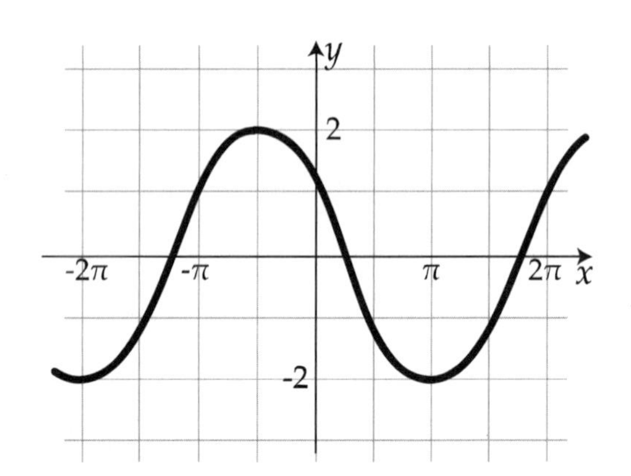

2. Graph of f(x) is given as shown. What is the possible restriction of the domain of f(x) so that its inverse f^{-1} will also be a function?

A) $\dfrac{\pi}{4} \le x \le \dfrac{7\pi}{4}$

B) $-\dfrac{\pi}{2} \le x \le \pi$

C) $-\pi \le x \le 0$

D) $-2\pi \le x \le 0$

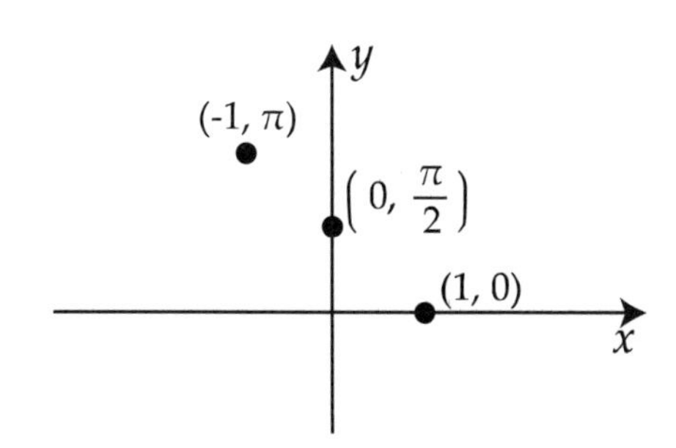

3. Three points on the function $f(x)$ are given as shown. Which of the following function could be $f(x)$?

A) $f(x) = \sin x$

B) $f(x) = \cos x$

C) $f(x) = \sin^{-1} x$

D) $f(x) = \cos^{-1} x$

4. What is the range of $y = \frac{1}{2}\tan^{-1}\frac{x}{2}$?

A) $0 < y < \frac{\pi}{2}$

B) $-\frac{\pi}{4} < y < \frac{\pi}{4}$

C) $0 \le y \le \frac{\pi}{2}$

D) $-\frac{\pi}{4} \le y \le \frac{\pi}{4}$

5. What is the range of $y = \arccos 2x + 3\pi$?

A) $0 \le y \le \pi$

B) $\frac{5\pi}{2} < y < \frac{7\pi}{2}$

C) $\frac{5\pi}{2} \le y \le \frac{7\pi}{2}$

D) $3\pi \le y \le 4\pi$

6. Find the inverse of the function $f(x) = 8\cos 2x,\ 0 \le x \le \frac{\pi}{2}$ and the domain of the f^{-1}.

A) $f^{-1}(x) = \frac{1}{2}\cos^{-1}\frac{x}{8}$, domain is $-8 \le x \le 8$

B) $f^{-1}(x) = \frac{1}{2}\cos^{-1}\frac{x}{8}$, domain is $0 \le x \le \frac{\pi}{2}$

C) $f^{-1}(x) = \frac{1}{8}\cos^{-1}\frac{x}{2}$, domain is $-8 \le x \le 8$

D) $f^{-1}(x) = \frac{1}{8}\cos^{-1}\frac{x}{2}$, domain is $0 \le x \le \frac{\pi}{2}$

5.2 Basic Trigonometric Identities

1. Fundamental Identities

※**Reciprocal Identities**

$$\sin\theta = \frac{1}{\csc\theta} \qquad \cos\theta = \frac{1}{\sec\theta} \qquad \tan\theta = \frac{1}{\cot\theta}$$

$$\csc\theta = \frac{1}{\sin\theta} \qquad \sec\theta = \frac{1}{\cos\theta} \qquad \cot\theta = \frac{1}{\tan\theta}$$

※ **Quotient Identities**

$$\tan\theta = \frac{\sin\theta}{\cos\theta} \qquad \cot\theta = \frac{\cos\theta}{\sin\theta}$$

☺Proof:

When $\sin\theta = \frac{y}{r}$ and $\cos\theta = \frac{x}{r}$,

then $\tan\theta = \frac{y}{x} = \dfrac{\frac{y}{①\ \square}}{\frac{x}{②\ \square}} = \dfrac{③\ \square}{\square}$

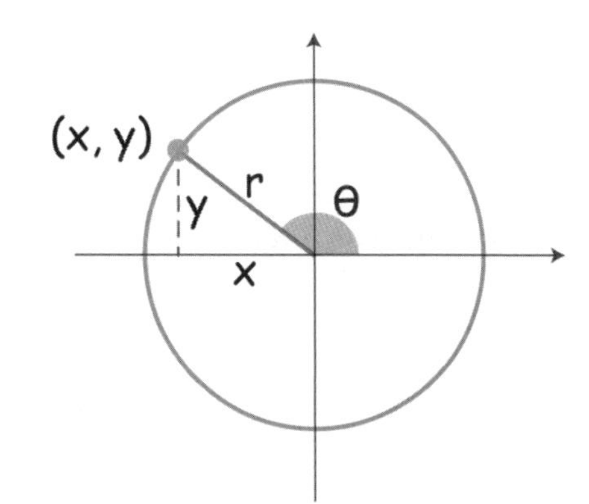

Blank : ① r ② r ③ $\frac{\sin\theta}{\cos\theta}$

EXAMPLE 1. Simplify the given expression.

① $\tan x \cos x$

② $\cot x \tan x$

③ $\cot u \sin u$

④ $\dfrac{\sec\theta}{\csc\theta}$

⑤ $\dfrac{\csc\theta}{\cot\theta}$

※ **Pythagorean Identities**

$$\sin^2\theta + \cos^2\theta = 1$$
$$1 + \tan^2\theta = \sec^2\theta$$
$$1 + \cot^2\theta = \csc^2\theta$$

☺ Careful! $\sin^2\theta = (\sin\theta)^2 \neq \sin\theta^2$

☺ Proof:

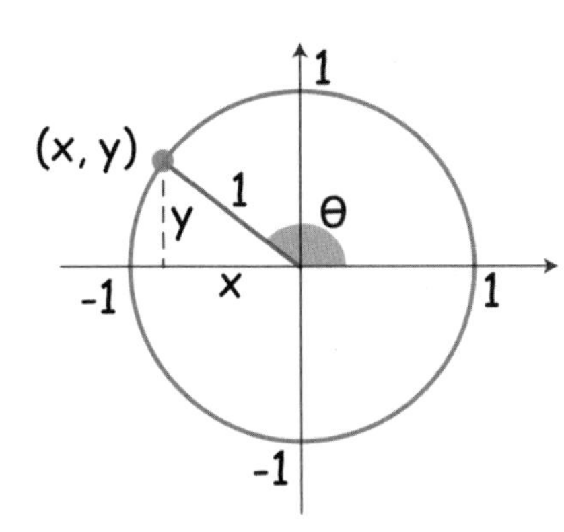

According to Pythagorean Theorem,

$$\sin^2\theta + \cos^2\theta = \boxed{①}$$

and if we divide everything by $\cos^2\theta$, then

$$\frac{\sin^2\theta}{\boxed{②}} + \frac{\cos^2\theta}{\boxed{③}} = \frac{1}{\boxed{④}}$$

$$\boxed{⑤} + 1 = \boxed{⑥}$$

EXAMPLE 2. Simplify the given expression.

① $\dfrac{1+\tan^2 x}{\csc^2 x}$

② $\dfrac{1-\cos^2\theta}{\sin\theta}$

③ $\dfrac{\cos x - \cos^3 x}{\sin^3 x}$

④ $\dfrac{\sin^2 x + \tan^2 x + \cos^2 x}{\sec x}$

⑤ $\dfrac{\sec\theta - \cos\theta}{\sin\theta}$

⑥ $\dfrac{1}{\cot^2\theta - \csc^2\theta}$

Blank : ① 1 ② $\cos^2\theta$ ③ $\cos^2\theta$ ④ $\cos^2\theta$ ⑤ $\tan^2\theta$ ⑥ $\sec^2\theta$

※ **Cofunction Identities**

Complementary

$$\sin\left(\frac{\pi}{2}-\theta\right) = \cos\theta \qquad \cos\left(\frac{\pi}{2}-\theta\right) = \sin\theta$$

$$\tan\left(\frac{\pi}{2}-\theta\right) = \cot\theta \qquad \cot\left(\frac{\pi}{2}-\theta\right) = \tan\theta$$

$$\csc\left(\frac{\pi}{2}-\theta\right) = \sec\theta \qquad \sec\left(\frac{\pi}{2}-\theta\right) = \csc\theta$$

※ **Odd-Even Identities**

$$\sin(-\theta) = -\sin\theta \qquad \csc(-\theta) = -\csc\theta$$

$$\cos(-\theta) = \cos\theta \qquad \sec(-\theta) = \sec\theta$$

$$\tan(-\theta) = -\tan\theta \qquad \cot(-\theta) = -\cot\theta$$

☺ Proof:

From the figure;

$\sin\theta = \frac{①\square}{\square}$ $\qquad$ $\cos\left(\frac{\pi}{2}-\theta\right) = \frac{②\square}{\square}$

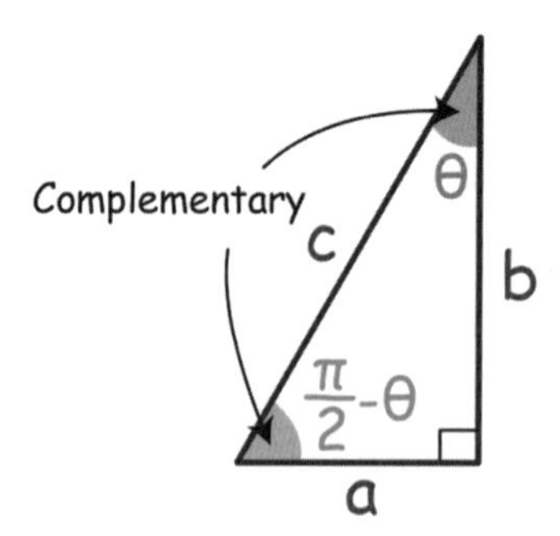

(③even/odd) function: $f(-x)=f(x)$

(④even/odd) function: $f(-x)=-f(x)$

Sin is (⑤even/odd) function, *cos* is (⑥even/odd) function, *tan* is (⑦even/odd) function.

Blank : ① $\frac{a}{c}$ ② $\frac{a}{c}$ ③ even ④ odd ⑤odd ⑥even ⑦odd

EXAMPLE 3. Simplify the given expression.

① $\sin x \csc(-x)$

② $\sec(-x)\sin\left(\frac{\pi}{2}-x\right)$

③ $\cos(-x)\csc\left(\frac{\pi}{2}-x\right)$

④ $\sec^2(-x)-\tan^2 x$

⑤ $\cot(-x)\tan(-x)$

⑥ $\sin^2(-x)+\sin^2\left(\frac{\pi}{2}-x\right)$

EXAMPLE 4. Use the even-odd properties to find the exact value of each expression.

① $\sin(-60^\circ)$

② $\cos(-270^\circ)$

③ $\tan(-\pi)$

④ $\csc\left(-\frac{\pi}{3}\right)$

⑤ $\cos\left(-\frac{\pi}{4}\right)$

⑥ $\tan\left(-\frac{\pi}{4}\right)$

⑦ $\cot\left(-\frac{\pi}{6}\right)$

EXAMPLE 5. Simplify the given expression.

① $\dfrac{\tan\left(\dfrac{\pi}{2}-x\right)\csc x}{\csc^2 x}$

② $\dfrac{1+\tan x}{1+\cot x}$

③ $\left(\sec^2 x+\csc^2 x\right)-\left(\tan^2 x+\cot^2 x\right)$

④ $\dfrac{\sec^2 u-\tan^2 u}{\cos^2 v+\sin^2 v}$

⑤ $\sin x\left(\tan x+\cot x\right)$

⑥ $\sin\theta-\tan\theta\cos\theta+\cos\left(\dfrac{\pi}{2}-\theta\right)$

⑦ $(\cot x-\csc x)(\cos x+1)$

⑧ $\dfrac{\left(\sec x-\tan x\right)\left(\sec x+\tan x\right)}{\sec x}$

⑨ $\dfrac{\tan x}{\csc^2 x}+\dfrac{\tan x}{\sec^2 x}$

⑩ $\sin x \cos x \tan x \sec x \csc x$

⑪ $\dfrac{\sec^2 x \csc x}{\sec^2 x+\csc^2 x}$

AP Style Problem

1. Which of the following is NOT equivalent to 1?

A) $\cot x \cdot \sec x \cdot \sin x$

B) $\cos^2 x(1+\tan^2 x)$

C) $(\cot x-\csc x)(\cot x+\csc x)$

D) $\dfrac{\sin x}{\csc x}+\dfrac{\cos x}{\sec x}$

2. If both the angles are acute and $\tan(3x+20^\circ)=\cot(2x-40^\circ)$, find x.

A) $22°$

B) $44°$

C) $60°$

D) $-60°$

3. What is $\sin^2 1°+\sin^2 2°+\sin^2 3°+\sin^2 87°+\sin^2 88°+\sin^2 89°$?

A) 0

B) 1

C) 2

D) 3

4.* If $\sin\theta+\cos\theta=\frac{1}{2}$, then $\sin\theta\cos\theta$ = ?

A) $\frac{1}{2}$

B) $\frac{3}{8}$

C) $-\frac{3}{8}$

D) 1

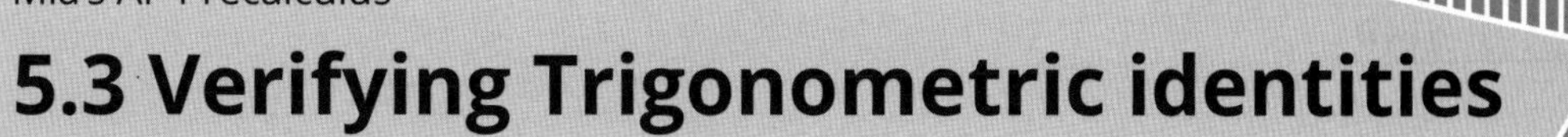

5.3 Verifying Trigonometric identities

1. Verifying Identities

EXAMPLE 1. Verify the identity.

Type1. Use Identities!
(Sometimes changing in terms of sin and cos will help.)

① $\cos^2 x - \sin^2 x = 1 - 2\sin^2 x$

② $\dfrac{\csc(-x)}{\sec(-x)} = -\cot x$

③ $(\tan x + \cot x)^4 = \csc^4 x \sec^4 x$

④ $\cot^2 x - \cos^2 x = \cot^2 x \cos^2 x$

⑤ $\dfrac{\cos\theta\cot\theta}{1-\sin\theta}-1=\csc\theta$

⑥ $\dfrac{\tan x-\sin x}{\tan x\sin x}=\dfrac{\tan x\sin x}{\tan x+\sin x}$

⑦ $\dfrac{\csc x+\sec x}{\sin x+\cos x}=\csc x\sec x$

Type2. Factoring could help.

⑧ $$\frac{\sin^3 x+\cos^3 x}{\sin x+\cos x}=1-\sin x\cos x$$

⑨ $$\frac{\cot u-\tan u}{\cot^2 u-\tan^2 u}=\sin u\cos u$$

⑩ $$\frac{\sin^2 x-\cos^2 x}{(\sin x+\cos x)^2}=\frac{(\sin x-\cos x)^2}{\sin^2 x-\cos^2 x}$$

Type3. Combine the fractions!

⑪ $\dfrac{1+\sin\theta}{\cos\theta}+\dfrac{\cos\theta}{1+\sin\theta}=2\sec\theta$

⑫ $\dfrac{1}{\csc\theta+\cot\theta}+\dfrac{1}{\csc\theta-\cot\theta}=2\csc\theta$

Type4. Multiply or divide the top and the bottom by same things.

⑬ $\dfrac{\sec x-1}{\tan x}=\dfrac{\tan x}{\sec x+1}$

⑭ $\dfrac{1+\sin x}{1-\sin x}=(\tan x+\sec x)^2$

⑮ $\dfrac{\tan x+\tan y}{1-\tan x\tan y}=\dfrac{\cot x+\cot y}{\cot x\cot y-1}$

Type5. Work each side separately!

⑯ $\dfrac{\csc x \sec x + 2}{\csc x \sec x} = (\sin x + \cos x)^2$

⑰ $\dfrac{\tan^2 x}{1 + \sec x} = \dfrac{1 - \cos x}{\cos x}$

5.4 Sum and Difference Identities

1. Sum and Difference Identities

※ **Sum and Difference Identities**

$$\sin(A \pm B) = \sin A \cos B \pm \cos A \sin B$$

$$\cos(A \pm B) = \cos A \cos B \mp \sin A \sin B$$

switch sign

$$\tan(A \pm B) = \frac{\tan A \pm \tan B}{1 \mp \tan A \tan B}$$

EXAMPLE 1. Find the exact value of the expression without using calculator.

① $\sin(75^\circ) = \sin(\boxed{})$

② $\cos(75^\circ) = \cos(\boxed{})$

③ $\cos(15^\circ) = \cos(\boxed{})$

④ $\tan(75^\circ) = \tan(\boxed{})$

⑤ $\tan(105^\circ) = \tan(\boxed{})$

EXAMPLE 2. Find the exact value of the expression without using a calculator.

① $\sin 20^\circ \cos 40^\circ + \cos 20^\circ \sin 40^\circ$

② $\sin 215^\circ \cos 95^\circ - \cos 215^\circ \sin 95^\circ$

③ $\cos\frac{5\pi}{12}\cos\frac{\pi}{4} - \sin\frac{5\pi}{12}\sin\frac{\pi}{4}$

④ $\cos\frac{5\pi}{18}\cos\frac{2\pi}{9} - \sin\frac{5\pi}{18}\sin\frac{2\pi}{9}$

⑤ $\dfrac{\tan 40^\circ + \tan 110^\circ}{1 - \tan 40^\circ \tan 110^\circ}$

⑥ $\dfrac{\tan 80^\circ - \tan(-40^\circ)}{1 + \tan 80^\circ \tan(-40^\circ)}$

EXAMPLE 3. Verify the following.

① $\cos\left(x+\frac{\pi}{2}\right)=-\sin x$

② $\sin(x+y)+\sin(x-y)=2\sin x\cos y$

③ $\frac{\cos(x+y)}{\cos(x-y)}=\frac{1-\tan x\tan y}{1+\tan x\tan y}$

EXAMPLE 4. Simplify $\tan\left(A+\dfrac{\pi}{4}\right)\tan\left(A-\dfrac{\pi}{4}\right)$.

EXAMPLE 5. Simplify $\dfrac{\tan(A+B)+\tan(A-B)}{1-\tan(A+B)\tan(A-B)}$.

EXAMPLE 6. Find the exact value of the expression.

① $\sin\left(\sin^{-1}\dfrac{3}{5}+\cos^{-1}\dfrac{1}{2}\right)$

② $\cos\left(\sin^{-1}\dfrac{3}{5}-\tan^{-1}\dfrac{1}{3}\right)$

③ $\cos\left(\tan^{-1}\frac{4}{3}+\sec^{-1}\frac{13}{5}\right)$

④ $\sin\left(\cos^{-1}\frac{2}{3}+\cot^{-1}2\right)$

⑤ $\sin(\arcsin x+\arccos x)$

⑥ $\cos(\arccos x-\arctan x)$

EXAMPLE 7.* Use the figure which shows two lines whose equations are $y=5x-5$ and $y=\frac{2}{3}x$. What is the acute angle between the two lines.

☺ Tip: If the angle between horizontal line and linear function $y=mx+b$ is θ,

then ① ____________________.

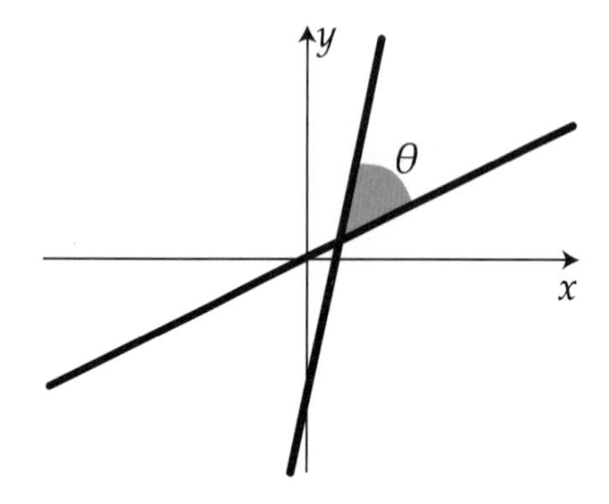

AP Style Problem

1. $\sin\left(\frac{\pi}{4}-x\right)$ is equivalent to

A) $\frac{\sqrt{2}}{2}(\sin x-\cos x)$

B) $\frac{\sqrt{2}}{2}(\cos x-\sin x)$

C) $\frac{\sqrt{2}}{2}(\sin x+\cos x)$

D) $\frac{1}{2}\sin x+\frac{\sqrt{3}}{2}\cos x$

Blank : ① slope m = tan θ

2. $\csc\left(\dfrac{\pi}{3}+\dfrac{\pi}{4}\right)$

A) $\dfrac{1}{\sqrt{6}+\sqrt{2}}$

B) $\dfrac{4}{\sqrt{6}+\sqrt{2}}$

C) $\dfrac{4}{\sqrt{6}-\sqrt{2}}$

D) $\dfrac{1}{\sqrt{6}-\sqrt{2}}$

3. Point $A\ (x, y)$ and $B\left(-\dfrac{1}{2}, \dfrac{\sqrt{3}}{2}\right)$ are in the unit circle as shown. If the origin is point O, then what is equivalent to cosine of $\angle AOB$?

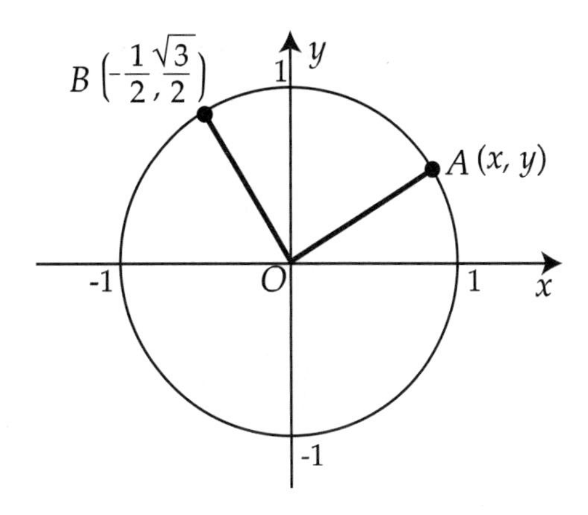

A) 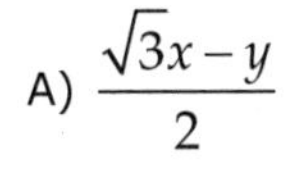 $\dfrac{\sqrt{3}x-y}{2}$

B) $\dfrac{x-\sqrt{3}y}{2}$

C) $\dfrac{\sqrt{3}y-x}{2}$

D) $\dfrac{y-\sqrt{3}x}{2}$

4. If $\cos^{-1}\left(\frac{4}{5}\right) = \alpha$ and $\cos^{-1}\left(\frac{2}{3}\right) = \beta$, what is $\cos(\alpha + \beta)$?

A) $\frac{8 - 3\sqrt{5}}{15}$

B) $\frac{3\sqrt{5} - 8}{15}$

C) $\frac{8 + 3\sqrt{5}}{15}$

D) $\frac{22}{15}$

5.* The figure shown consists of five squares and angle A and B are given as shown. Find the $\tan(A + B)$.

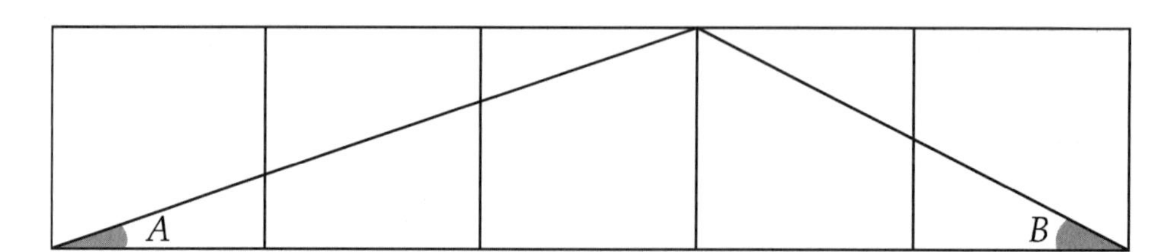

A) 1

B) $\frac{1}{2}$

C) $\sqrt{3}$

D) $\frac{\sqrt{3}}{3}$

5.5 Double-Angle Identity

1. Double Angle Identity

From sum identity we can find; $\sin(\theta+\theta) =$ ① ______________________.

※ **Double Angle Formulas**

$$\sin(2\theta) = 2\sin\theta\cos\theta$$

$$\cos(2\theta) = \cos^2\theta - \sin^2\theta$$
$$= 1 - 2\sin^2\theta$$
$$= 2\cos^2\theta - 1$$

$$\tan(2\theta) = \frac{2\tan\theta}{1-\tan^2\theta}$$

EXAMPLE 1. Find the exact value of the expression or simplify the expression using double identity.

① $\sin\left(120^\circ\right) = \sin\left(\boxed{}\right)$

② $\cos\left(120^\circ\right)$

③ $2\sin 6\alpha$

④ $\sin 10B$

⑤ $\cos 4A$

⑥ $2\cos 8\beta$

⑦ $\tan 4C$

Blank : ① sin θ cos θ + cos θ sin θ = 2 sin θ cos θ

EXAMPLE 2. Use a double-angle identity to simplify the expression.

① $4\sin A\cos A$

② $6\sin 2\beta\cos 2\beta$

③ $2\cos^2 3\alpha - 1$

④ $2\sin^2 2\beta - 1$

⑤ $\sin^2 2\theta - \cos^2 2\theta$

⑥ $2\cos^2 \frac{A}{2} - 1$

⑦ $4\sin^2 \frac{\beta}{2} - 2$

⑧ $\cos^2 3\theta - \sin^2 3\theta$

⑨ $\frac{2\tan 2\beta}{1-\tan^2 2\beta}$

⑩ $\frac{2\tan C}{1-\tan^2 C}$

EXAMPLE 3. Verify the following

① $\sin 4x = (4\sin x\cos x)(2\cos^2 x - 1)$

② $\cos 4x = 8\sin^4 x - 8\sin^2 x + 1$

③ $\dfrac{\sin 2x + \sin x}{1 + \cos 2x + \cos x} = \tan x$

④ $\tan 2x = \dfrac{2}{\cot x - \tan x}$

EXAMPLE 4. Find an expression for $\cos 4x$ in terms of $\cos x$.

EXAMPLE 5. Use the information given about the angle θ, $0 \le \theta \le 2\pi$, to find the exact value of the indicated trigonometric function.

① $\cos\theta = -\frac{8}{17}$, $\frac{\pi}{2} < \theta < \pi$,

Find $\sin(2\theta)$

② $\sin\theta = \frac{2\sqrt{6}}{5}$, $\tan\theta < 0$,

Find $\sin(2\theta)$

③ $\tan\theta = -\frac{11}{9}$, $\sin\theta < 0$,

Find $\cos(2\theta)$

④ $\cot\theta = \frac{3}{5}$, $\sin\theta < 0$,

Find $\cos(2\theta)$

EXAMPLE 6. Find the exact value of the expression. ($1 \le x \le 1$)

① $\sin\left(2\cos^{-1}\frac{5}{13}\right)$

② $\cos\left(2\tan^{-1}\frac{24}{7}\right)$

③ $\sec\left(2\tan^{-1}\frac{3}{4}\right)$

④ $\csc\left(2\sin^{-1}\frac{5}{6}\right)$

⑤ $\sin(2\arccos x)$

⑥ $\cos(2\arccos x)$

2. Power Reducing Formula

From $\cos(2\theta) =$ ① ______________________________,

we can find; $\sin^2\theta =$ ② ______________, $\cos^2\theta =$ ③ ______________.

※**Power-Reducing Formulas**

power reduce

$$\sin^2\theta = \frac{1}{2}[1 - \cos(2\theta)] \qquad \cos^2\theta = \frac{1}{2}[1 + \cos(2\theta)]$$

$$\tan^2\theta = \frac{1 - \cos(2\theta)}{1 + \cos(2\theta)}$$

EXAMPLE 7. Use the power-reducing formulas to rewrite the expression in terms of the first power of the cosine.

① $(\cos x + 1)^2$

② $(2 - \sin x)^2$

Blank : ① $\cos^2\theta - \sin^2\theta = 2\cos^2\theta - 1 = 1 - 2\sin^2\theta$ ② $\frac{1}{2}(1 - \cos 2\theta)$ ③ $\frac{1}{2}(1 + \cos 2\theta)$

③ $\sin^4 x$

④ $\cos^4 x$

⑤ $\sin^2 x \cos^2 x$

AP Style Problem

1. $\sin\left(2\sin^{-1}\frac{1}{4}\right) =$

A) $\frac{1}{2}$

B) $\frac{\sqrt{3}}{2}$

C) $\frac{\sqrt{15}}{8}$

D) $\frac{\sqrt{17}}{8}$

2. $6\sin\frac{5\pi}{12}\cos\frac{5\pi}{12} =$

A) 3

B) $3\sqrt{3}$

C) $\frac{3}{2}$

D) $\frac{3\sqrt{3}}{2}$

3. Which of the following is equivalent to $\cos 2x$?

I. $\cos^4 x - \sin^4 x$

II. $(\sin 2x)(\cot x) - 1$

III. $\frac{1-\tan^2 x}{1+\tan^2 x}$

A) I, II

B) I, III

C) II, III

D) I, II, III

4. What is the period of $y = 4\sin x\cos x - 1$?

A) π

B) $\dfrac{\pi}{2}$

C) 2π

D) 4π

5. What is the amplitude of $y = 6\cos^2 x + 5$?

A) 0.5

B) 1

C) 3

D) 6

6. Point $A\ (x, y)$ is in the unit circle as shown. If the line $\overline{AB}$ is perpendicular to the x axis, then what is equivalent to sine of $\angle AOB$?

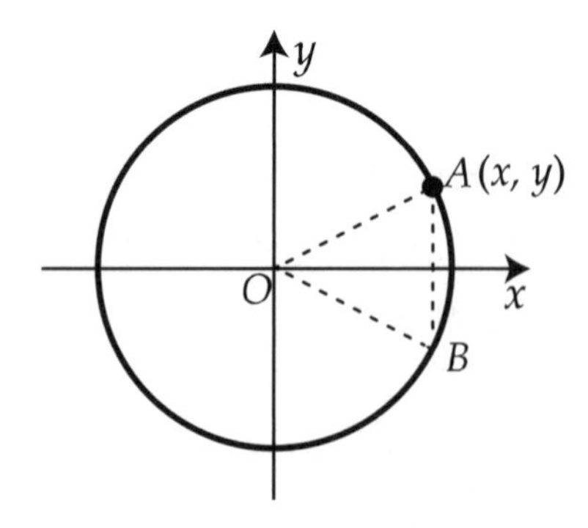

A) y

B) $2y$

C) $2x$

D) $2xy$

Part 6

Trig Equations and Inequalities

6.1 Basic Trigonometric Equations

1. Solving Trigonometric Equations

☺ Reminder: We need ①__________ and ②__________ to evaluate $\cos\frac{4\pi}{3}=$③______.

To solve the equation $\cos x=-\frac{1}{2}$, we need ④____________ and ⑤____________.

※ How to Solve a trig equation?

① Find the **reference angle**.

② Determine which **quadrants** your solutions lie in.

③ Express the solutions in standard form .

Case 1. i) Solve $\cos x=-\frac{1}{2}$ for $0\le x\le 2\pi$.

Sign part **tells you which quadrant** (ASTC)

Number part **tells you** reference angle

$$\cos x = -\frac{1}{2}$$

The reference angle is $\bar{x}=$ ⑥ .

$\cos x$ is ⑦__________ in ⑧ ___, ___ quadrant.

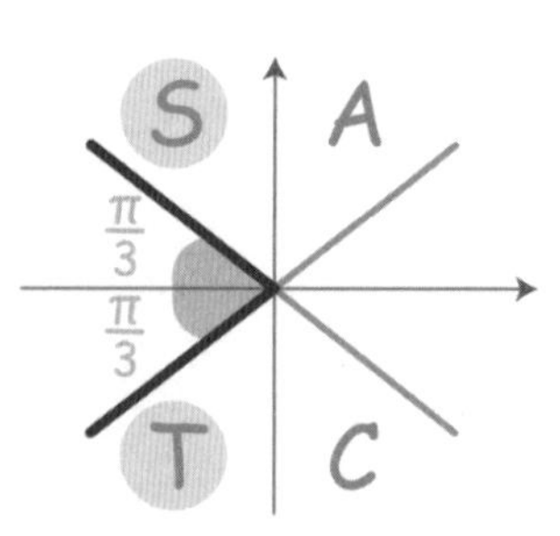

The solution is ⑨______________________________.

(Find all angles in the given interval is standard position)

Blank : ① ASTC ② reference angle ③ $-\frac{1}{2}$ ④ ASTC ⑤ reference angle ⑥ $\frac{\pi}{3}$ ⑦ negative ⑧ II, III ⑨ $\frac{2\pi}{3}$, $\frac{4\pi}{3}$

Case 2. ii) Solve $\cos x = -\frac{1}{2}$ for every x. (all solution)

Because the cosine function repeats its values every ① ______ units, we get

all solutions of the equation by adding integer multiples of ② ______ to these solutions:

The solution is ③ ____________________________.

EXAMPLE 1. Solve the equation i) when $0 \le x \le 2\pi$ and ii) for every x without calculator.

① $\sqrt{2}\sin x = -1$

② $2\cos x - \sqrt{3} = 0$

③ $\sqrt{3}\tan x + 1 = 0$

④ $\tan x = 1$

Blank : ① 2π ② 2π ($=2\pi k$) ③ $\frac{2\pi}{3} + 2\pi k$, $\frac{4\pi}{3} + 2\pi k$ where k is integer

⑤ $2\cos x - 1 = 0$

⑥ $2\sin x + \sqrt{2} = 0$

EXAMPLE 2. Solve the equation i) when $0 \le x \le 2\pi$ and ii) for every x without calculator.

When reference angle is 0 or 90°, using graph will be helpful.

① $\sin x = 1$

② $\sin x = 0$

③ $\cos x = 0$

④ $\cos x + 1 = 0$

EXAMPLE 3. Solve the equation when $0 \le x \le 2\pi$. You may answer in terms of inverse trig function.

☺ Tip: Inverse Trig only gives you ①________________________.

(DO NOT include the *negative sign* when finding the reference angle using inverse trig)

① $\sin x = 0.8$

② $\tan x = 2.5$

③ $\cos x = -0.3$

④ $\sin x = -\dfrac{2}{3}$

⑤ $\tan x = -2$

Blank : ① reference angle

EXAMPLE 4. Solve the equation for $0 \le x \le 2\pi$.

Type 1. Factor the quadratic

① $2\cos^2 x + 11\cos x = -5$

② $4\sin^2 x - \sqrt{3}\sin x = 2\sin^2 x$

③ $3\sin x\cos x = 2\sin x$

④ $\sin x - 2\sin x\cos x = 0$

Type 2. Take the square root

⑤ $\tan^2 x - 1 = 0$

⑥ $3\cot^2 x - 1 = 0$

EXAMPLE 5. Solve the equation $\sqrt{3}\tan x = -1$ when $0 \le x \le 4\pi$.

EXAMPLE 6. Solve the equation $\tan^2 x = 1$ when $-\pi \le x \le \pi$.

AP Style Problem

1. What are all values of x, for $0 \le x \le 2\pi$, where $\sin x \tan x = \sqrt{3}\sin x$?

A) $\frac{\pi}{3}, \frac{4\pi}{3}$

B) $\frac{\pi}{6}, \frac{7\pi}{6}$

C) $0, \frac{\pi}{3}, \pi, \frac{4\pi}{3}, 2\pi$

D) $0, \frac{\pi}{6}, \pi, \frac{7\pi}{6}, 2\pi$

2. What are all values of x where $3\tan x + 2 = 0$?

A) $k\pi - \tan^{-1}\frac{2}{3}$, where k is integer

B) $k\pi + \tan^{-1}\frac{2}{3}$, where k is integer

C) $2k\pi - \tan^{-1}\frac{2}{3}$, where k is integer

D) $2k\pi + \tan^{-1}\frac{2}{3}$, where k is integer

3. The function f is given by $f(x)=\sin^2 x$. What are all values of x where $f(x)=1$?

A) $x=n\pi$, where n is integer

B) $x=\frac{(2n-1)\pi}{2}$, where n is integer

C) $x=\frac{n\pi}{2}$, where n is integer

D) $x=n$, where n is integer

6.2 More Trigonometric Equations

1. More Trigonometric Equations

EXAMPLE 1. Solve the equations when $0 \le x \le 2\pi$.

Type1. Use reciprocal trig ratio.

① $3\csc x = 6$

② $\cot^2 x = 3$

③ $3\cot^2 x + 4 = 7$

④ $\sec^2 x = 4$

Type2. Change into single trig equation.

⑤ $2\cos^2 x + 9\sin x = 3\sin^2 x$

⑥ $1 = 2\sin^2 x + \cos x$

⑦ $\tan^2 x \sec^2 x + 2\sec^2 x - \tan^2 x = 2$

⑧ $1 - \csc x = \cot^2 x$

Type3. Divide both sides by cos.

⑨ $\sin x = \cos x$

⑩ $\sqrt{3}\sin x = \cos x$

Type4. Use double angle identities.

⑪ $\cos 2x = \cos x$

⑫ $\sin 2x - \sin x = 0$

Type5. Square both sides. You MUST check the answer!

⑬ $\tan x = 1 - \sec x$

⑭ $\csc x = \cot x + 1$

⑮ $\cos x = \sin x - 1$

2.Solving Trigonometric Equations Using Substitution

※ How to Solve a trig equation?

① Make a substitution for the angle part and adjust the INTERVAL!

② Find the **reference angle.**

③ Determine which **quadrants** your solutions lie in.

④ Express the solutions in standard form.

Solve $2\sin 2x + 1 = 0$ when $0 \leq x \leq 2\pi$.

Make a substitution $2x = A$.

Sign part **tells you** which quadrant (ASTC)

Number part **tells you** reference angle

$$\sin A = -\frac{1}{2}$$

When $0 \leq x \leq 2\pi$, ①_____ $\leq A \leq$ ②______.

The reference angle is $\overline{A} =$ ③ [ ].

$\sin A$ is ④_________ in ⑤___, ___ quadrant.

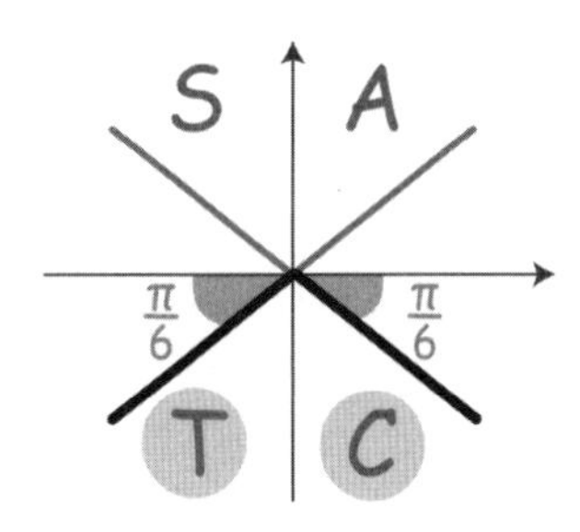

So A = ⑥________________________.

(Find all angles in the given interval is standard position)

The solution x is ⑦________________________.

Blank : ① 0 ② 4π ③ $\frac{\pi}{6}$ ④ negative ⑤ III, IV ⑥ $\frac{7\pi}{6}, \frac{11\pi}{6}, \frac{19\pi}{6}, \frac{23\pi}{6}$ ⑦ $\frac{7\pi}{12}, \frac{11\pi}{12}, \frac{19\pi}{12}, \frac{23\pi}{12}$

EXAMPLE 2. Solve the equation for given interval.

① $2\cos 2x = 1$, $0 \le x \le 2\pi$

② $\cos 3\theta = -\frac{\sqrt{3}}{2}$, $0 \le x \le 2\pi$

③ $\sin\left(\frac{x}{2}\right) = \cos\left(\frac{x}{2}\right)$, $0 \le x \le 2\pi$

④ $\sin(2x) = \cos(2x)$, $0 \le x \le \pi$

⑤ $4\tan 2x = 2\sec^2 2x$, $0 \le x \le \pi$

⑥ $2\sin^2 2x+1=3\sin 2x$, $0\le x\le 2\pi$

⑦ $\sqrt{2}\sin\left(x+\frac{\pi}{4}\right)=1$, $0\le x\le 2\pi$

AP Style Problem

1. What are all values of x, for $-\frac{\pi}{2} \le x \le \frac{\pi}{2}$, where $\sec 4x + 2 = 0$?

A) $-\frac{4\pi}{3}, -\frac{2\pi}{3}, \frac{2\pi}{3}, \frac{4\pi}{3}$

B) $-\frac{\pi}{3}, -\frac{\pi}{6}, \frac{\pi}{6}, \frac{\pi}{3}$

C) $-\frac{2\pi}{3}, \frac{2\pi}{3}$

D) $-\frac{\pi}{6}, \frac{\pi}{6}$

2. What are all values of x, for $0 \le x \le 2\pi$, where $6\cos^2 x + 7\sin x - 8 = 0$?

A) $\frac{\pi}{3}, \frac{2\pi}{3}, \sin^{-1}\frac{2}{3}, \pi - \sin^{-1}\frac{2}{3}$

B) $\frac{\pi}{6}, \frac{5\pi}{6}, \sin^{-1}\frac{2}{3}, \pi - \sin^{-1}\frac{2}{3}$

C) $\frac{\pi}{3}, \frac{2\pi}{3}, \pi + \sin^{-1}\frac{2}{3}, 2\pi - \sin^{-1}\frac{2}{3}$

D) $\frac{\pi}{6}, \frac{5\pi}{6}, \pi + \sin^{-1}\frac{2}{3}, 2\pi - \sin^{-1}\frac{2}{3}$

3. How many solutions are there to the equation $\cos^2 3x = \sin^2 3x$ if $0 \le x \le 2\pi$

A) 3

B) 6

C) 9

D) 12

6.3 Trigonometric Inequalities

1. Solving Trigonometric Inequalities using Unit Circle

☺ Reminder

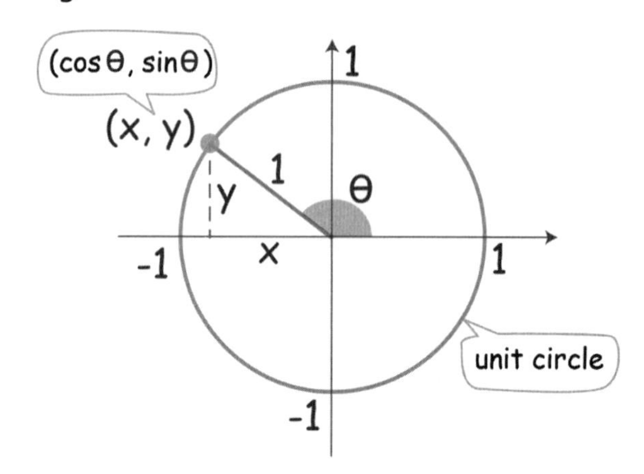

sin θ is the ①___ coordinate

cos θ is the ②___ coordinate

of the point on the unit circle,

tan θ is the ③________ of the terminal ray

※ How to Solve a trig Inequalities?

① Switch to equation and find two solutions of the equation.

② Use the unit circle definition to find the appropriate intervals.

Solve $\cos\theta < -\frac{1}{2}$ where $0 \le \theta \le 2\pi$.

Solutions to $\cos\theta = -\frac{1}{2}$ is ④______________.

Since $\cos\,\theta$ is x (unit circle definition)

, the question will change to ⑤__________.

The angles that satisfy $x < -\frac{1}{2}$ will be;

⑥____.

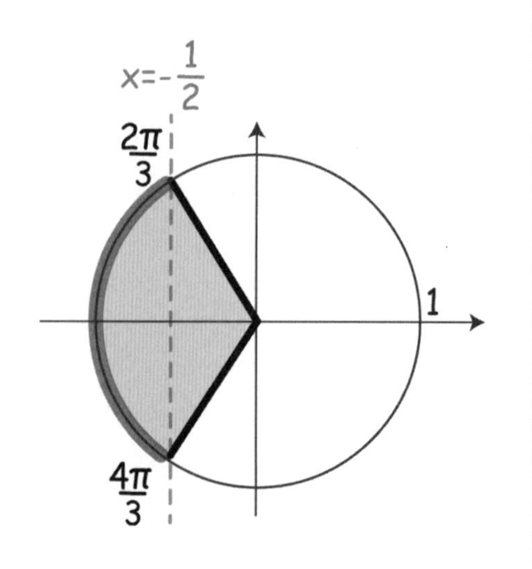

Solve $\sin\theta \geq -\frac{1}{2}$ where $-\frac{\pi}{2} \leq \theta \leq \frac{3\pi}{2}$.

Solutions to $\sin\theta = -\frac{1}{2}$ is ⑦________________.

Since $sin\ \theta$ is y(unit circle definition)

, the question will change to ⑧____________.

The angles that satisfy $y \geq -\frac{1}{2}$ will be;

⑨____.

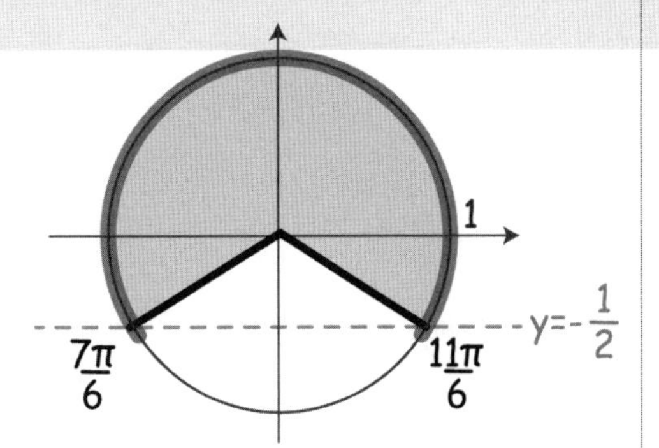

EXAMPLE 1. Solve the inequalities.

① $\sin\theta > \frac{\sqrt{3}}{2}$ where $0 \leq \theta \leq 2\pi$

② $\sin\theta \leq \frac{\sqrt{2}}{2}$ where $\frac{\pi}{2} \leq \theta \leq \frac{5\pi}{2}$

③ $\sin\theta \geq -\frac{\sqrt{2}}{2}$ where $-\frac{\pi}{2} \leq \theta \leq \frac{3\pi}{2}$

④ $\cos\theta < \frac{1}{2}$ where $0 \leq \theta \leq 2\pi$

Blank : ① y ② x ③ slope ④ $\frac{2\pi}{3}, \frac{4\pi}{3}$ ⑤ $x < -\frac{1}{2}$ ⑥ $\frac{2\pi}{3} < \theta < \frac{4\pi}{3}$ ⑦ $\frac{7\pi}{6}, \frac{11\pi}{6}$ ⑧ $y \geq -\frac{1}{2}$ ⑨ $-\frac{\pi}{6} \leq \theta \leq \frac{7\pi}{6}$

⑤ $\cos\theta > -\frac{\sqrt{2}}{2}$ where $0 \le \theta \le 2\pi$

⑥ $\sin\theta < -\frac{\sqrt{3}}{2}$ where $0 \le \theta \le 2\pi$

⑦ $\cos\theta > -\frac{1}{2}$ where $-\pi \le \theta \le \pi$

⑧ $\cos\theta < -\frac{\sqrt{3}}{2}$ where $0 \le \theta \le 2\pi$

Solve $\tan\theta > \sqrt{3}$ where $0 \le \theta \le 2\pi$.

Solutions to $\tan\theta = \sqrt{3}$ is ①______________.

Since *tan θ* is slope (unit circle definition)

, the question will change to ②__________.

The angles that satisfy $slope > \sqrt{3}$ will be;

③____________________________________.

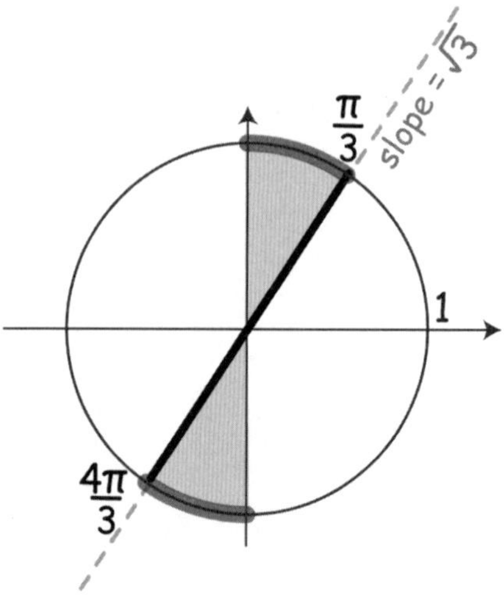

Solve $\tan x < \sqrt{3}$ where $-\frac{\pi}{2} \le \theta \le \frac{3\pi}{2}$.

Solutions to $\tan\theta = \sqrt{3}$ is ④______________.

Since *tan* θ is slope (unit circle definition)

, the question will change to ⑤___________.

The angles that satisfy $slope < \sqrt{3}$ will be;

⑥______________________________________.

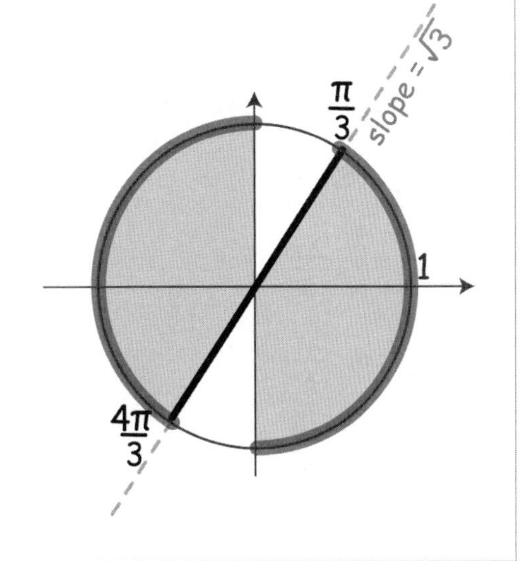

EXAMPLE 2. Solve the inequalities.

① $\tan\theta < 1$ where $-\frac{\pi}{2} \le \theta \le \frac{3\pi}{2}$

② $\tan\theta > 1$ where $0 \le \theta \le 2\pi$

③ $\tan\theta < -\frac{\sqrt{3}}{3}$ where $0 \le \theta \le 2\pi$

④ $\tan\theta \ge -\frac{\sqrt{3}}{3}$ where $-\frac{\pi}{2} \le \theta \le \frac{3\pi}{2}$

Blank : ① $\frac{\pi}{3}, \frac{4\pi}{3}$ ② slope $> \sqrt{3}$ ③ $\frac{\pi}{3} < \theta < \frac{\pi}{2}$ or $\frac{4\pi}{3} < \theta < \frac{3\pi}{2}$ ④ $\frac{\pi}{3}, \frac{4\pi}{3}$ ⑤ slope $< \sqrt{3}$ ⑥ $-\frac{\pi}{2} < \theta < \frac{\pi}{3}$ or $\frac{\pi}{2} < \theta < \frac{4\pi}{3}$

AP Style Problem

1. What are all values of θ that satisfy $2\cos\theta+1<0$ and $2\sin\theta-1>0$ where $0\le\theta\le 2\pi$?

A) $\dfrac{2\pi}{3}<\theta<\dfrac{5\pi}{6}$

B) $\dfrac{\pi}{6}<\theta<\dfrac{2\pi}{3}$

C) $\dfrac{\pi}{6}<\theta<\dfrac{5\pi}{6}$

D) $\dfrac{\pi}{6}<\theta<\dfrac{4\pi}{3}$

Part 7

Polar Curve

7.1 Polar Coordinates

1. Polar Coordinate

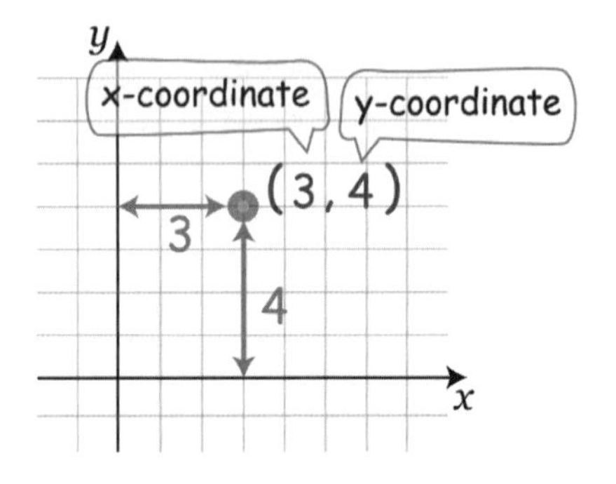

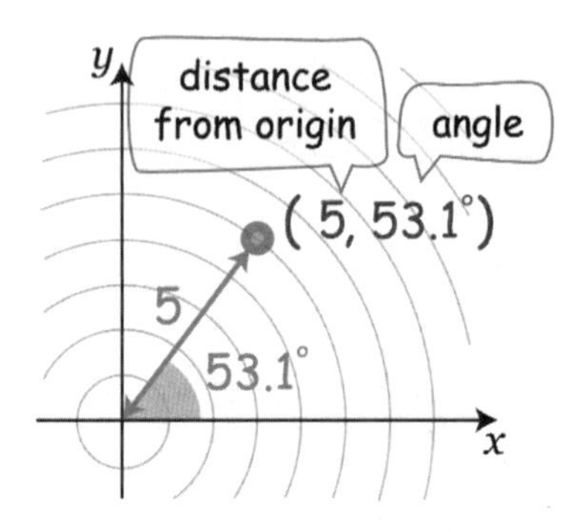

In ①______________ coordinate system, we mark a point by **how far along** and **how far up** it is.

In ②______________ coordinate system, we mark a point by **how far away**, and what **angle** it is.

※ **Polar Coordinates**

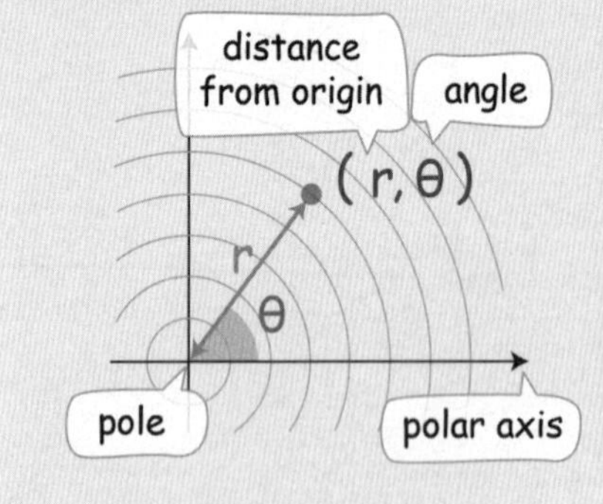

Polar Coordinate is a way to pinpoint where you are on a map or graph by how far away, and at what angle the point is.

$(-r, \theta)$ is the reflection over the pole (origin) of (r, θ)

☺ Vocabulary

Polar axis: The positive x axis

Pole: Origin in polar coordinate (0,0)

Blank : ① Cartesian(=rectangular) ② polar

EXAMPLE 1. Plot the point having the given polar coordinates.

① $(4, 225°)$

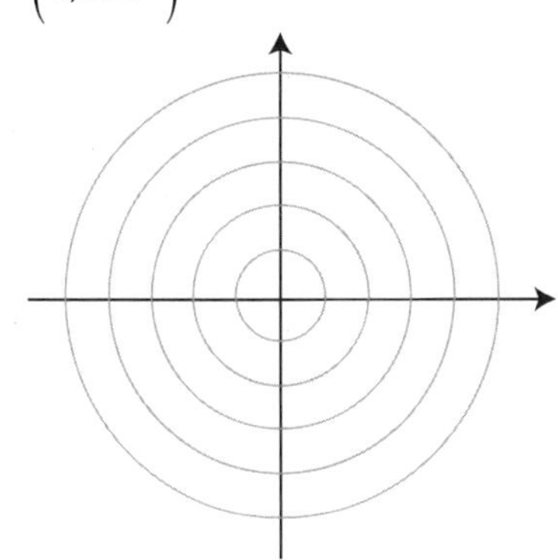

② $(-2, -330°)$

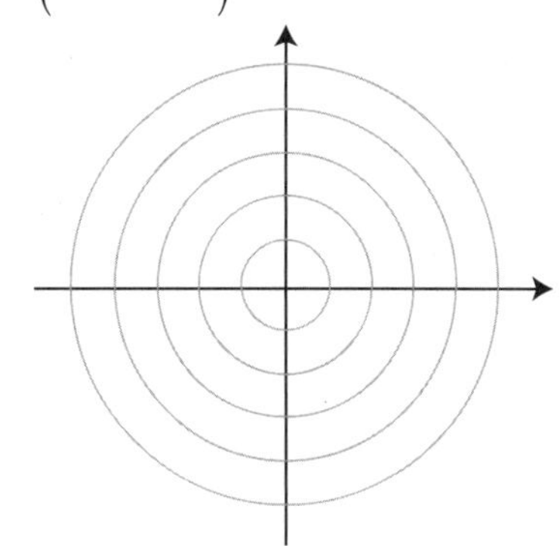

③ $\left(3, \frac{7\pi}{4}\right)$

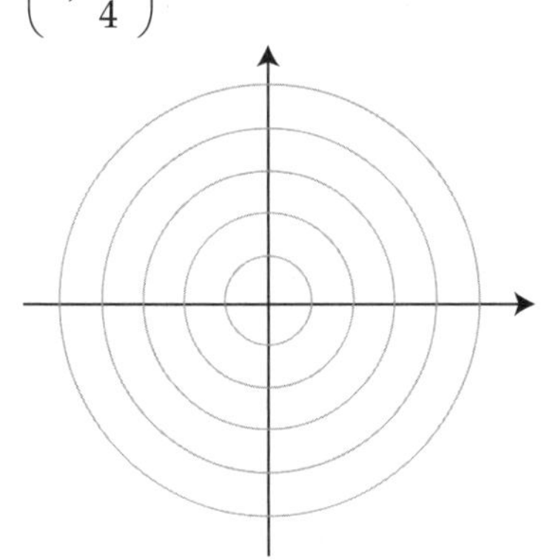

④ $\left(2, -\frac{5\pi}{4}\right)$

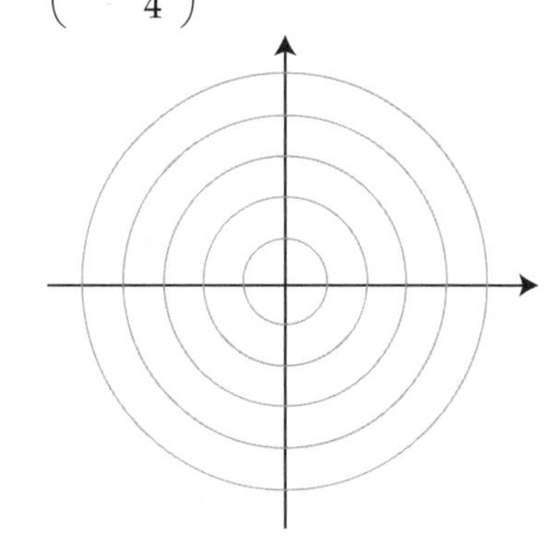

⑤ $\left(2, -\frac{5\pi}{6}\right)$

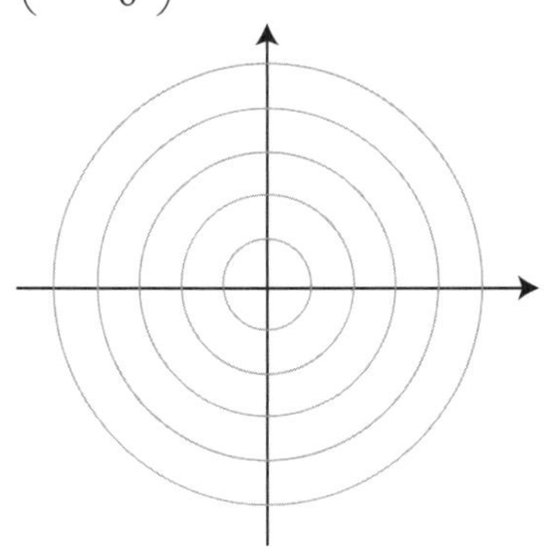

⑥ $\left(1, \frac{13\pi}{3}\right)$

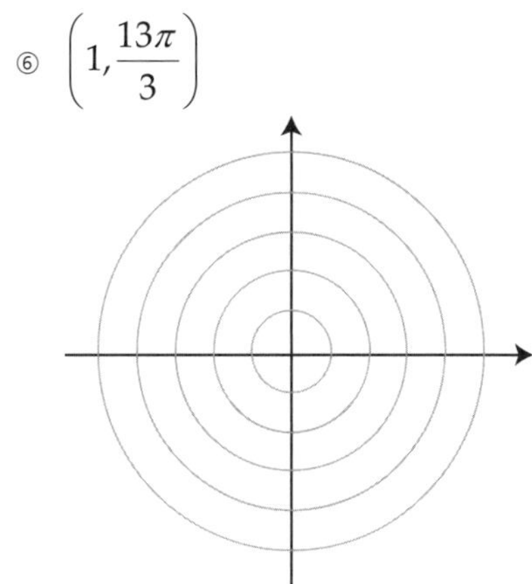

⑦ $\left(-3, \frac{11\pi}{6}\right)$

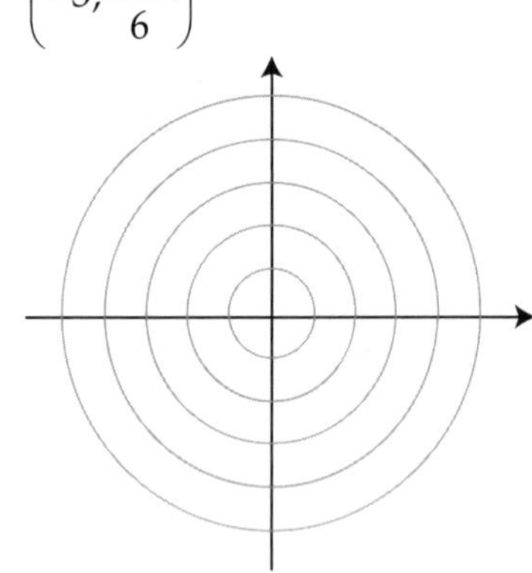

⑧ $\left(-3, 3\pi\right)$

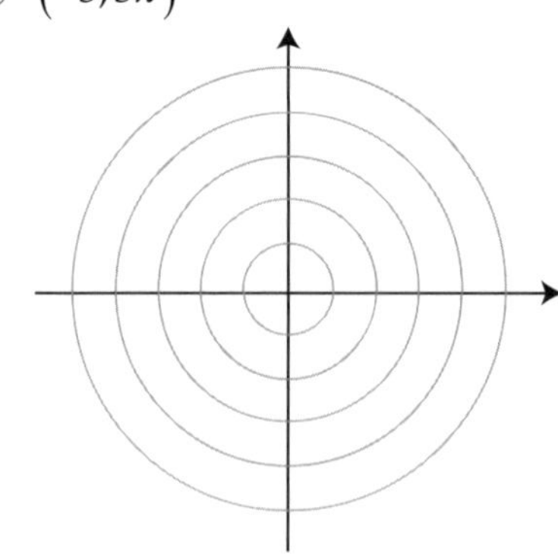

⑨ $\left(-4, -\frac{7\pi}{4}\right)$

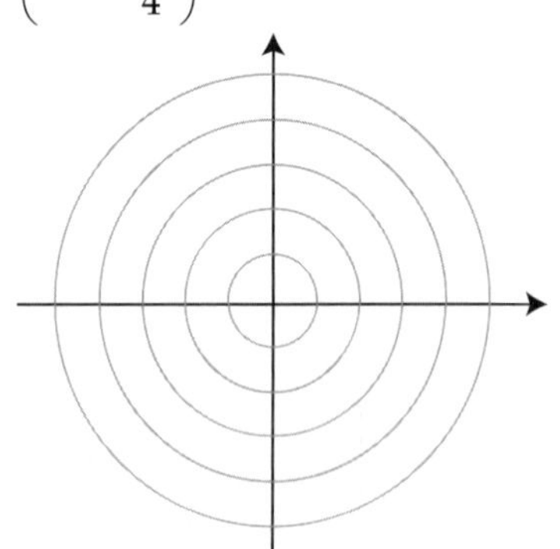

⑩ $\left(-2, -\frac{5\pi}{3}\right)$

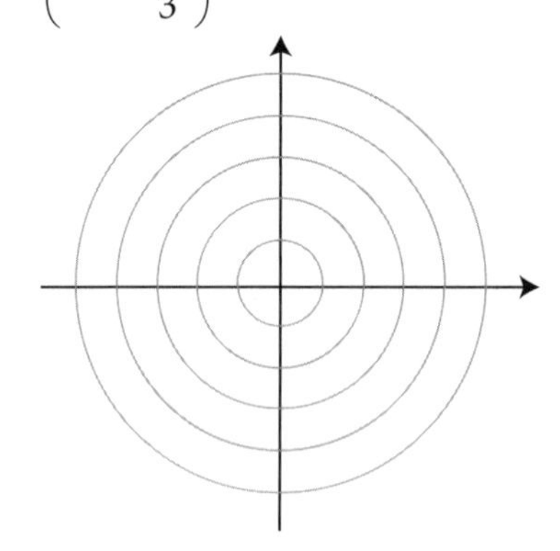

EXAMPLE 2. Locate the given point and write the coordinates in three other ways.

① $\left(3, 60^\circ\right)$

② $\left(4, 90^\circ\right)$

③ $\left(2, \frac{5\pi}{6}\right)$

④ $\left(3, \frac{\pi}{4}\right)$

2. Convert between Polar to Cartesian (r, θ) ⇔ (x, y)

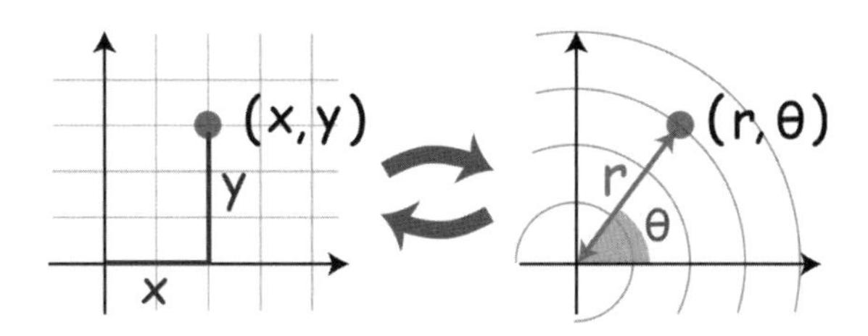

Use the trigonometric Function to find:

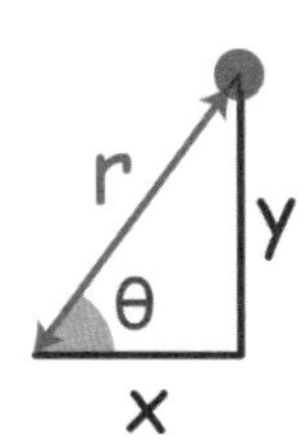

$\sin\theta = \dfrac{①\square}{\square}$ $\cos\theta = \dfrac{②\square}{\square}$

$r = \sqrt{③\square^2 + ④\square^2}$ $\tan\theta = \dfrac{⑤\square}{\square}$

※ **Formulas for Polar Coordinates**

$$x = r \times \cos\theta \qquad y = r \times \sin\theta$$

$$r = \sqrt{x^2 + y^2} \qquad \tan\theta = \frac{y}{x}$$

(remember; $\tan^{-1}$ gives you only the reference angle)

Blank : ① $\frac{y}{r}$ ② $\frac{x}{r}$ ③ x ④ y ⑤ $\frac{y}{x}$

EXAMPLE 3. Convert the given polar coordinates of the point to rectangular coordinates.

① $(6, 225^\circ)$

② $(7, 210^\circ)$

③ $\left(-7, \dfrac{5\pi}{3}\right)$

④ $\left(6, \dfrac{5\pi}{4}\right)$

⑤ $\left(3, -\dfrac{\pi}{6}\right)$

EXAMPLE 4. Convert the rectangular coordinates of the point to polar coordinates (r, θ) with $r > 0$ and $0 \le \theta < 2\pi$.

① $(-1,1)$

② $(6,-6\sqrt{3})$

③ $(5\sqrt{3},5)$

④ $(4,0)$

⑤ $(0,-3)$

⑥ $(-2\sqrt{3},2\sqrt{3})$

⑦ $(-2,4)$

⑧ $(-5,3)$

3. Polar Form → Rectangular Form

We use formulas to convert the equation from polar form to rectangular form.

EXAMPLE 5. Find an equivalent equation in rectangular coordinates.

① $r=1$

② $r\sin\theta=10$

③ $r(1+\cos\theta)=5$

④ $r=\sin\theta$

⑤ $r=10\sin\theta$

⑥ $r\left(\cos\theta-\sin\theta\right)=3$

⑦ $r=\dfrac{2}{4\sin\theta+5\cos\theta}$

⑧ $r=-2\csc\theta$

⑨ $r=3\sec\theta$

EXAMPLE 6. Find an equivalent equation in polar coordinates.

① $y=7$

② $x+y=3$

③ $2x-y=4$

④ $x^2=4-y^2$

⑤ $x^2-y^2=2$

⑥ $x^2+1=y^2$

⑦ $x^2+y^2=1$

AP Style Problem

1. Which of the following is the possible polar coordinate of point A?

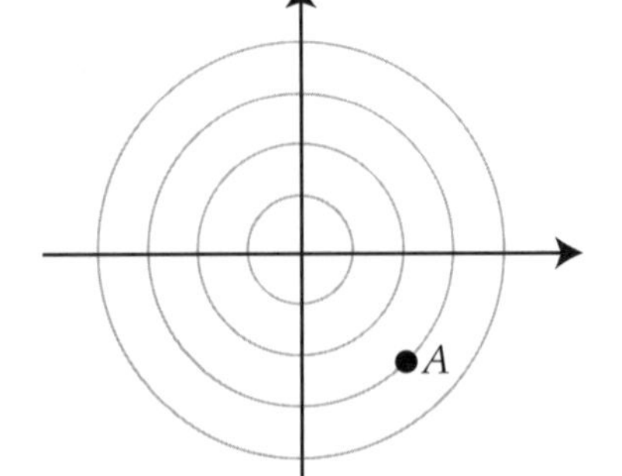

A) $\left(3, \frac{5\pi}{4}\right)$

B) $\left(-3, -\frac{\pi}{4}\right)$

C) $\left(-3, \frac{3\pi}{4}\right)$

D) $\left(-3, \frac{7\pi}{4}\right)$

2. In the polar coordinate system, point P has a polar coordinate $\left(3, \frac{\pi}{3}\right)$. All of the following points give the same location of point P EXCEPT

A) $\left(-3, \frac{4\pi}{3}\right)$

B) $\left(-3, -\frac{2\pi}{3}\right)$

C) $\left(3, -\frac{5\pi}{3}\right)$

D) $\left(-3, -\frac{\pi}{3}\right)$

3. Which of the following expresses the rectangular point $(-\sqrt{2},-\sqrt{6})$ in polar coordinate?

A) $\left(2\sqrt{2},\frac{\pi}{3}\right)$

B) $\left(2\sqrt{2},\frac{4\pi}{3}\right)$

C) $\left(2\sqrt{2},\frac{\pi}{6}\right)$

D) $\left(2\sqrt{2},\frac{7\pi}{6}\right)$

4. Which of the following is equivalent to the polar equation $r=2\csc\theta$ in rectangular form?

A) $x^2+y^2=\frac{4}{x}$

B) $y=\frac{2}{x}$

C) $y=2$

D) $x=2$

7.2 Graphs of Polar Equations

1. Polar Graph

A **polar equation** is a curve with an equation of the form ①____________________. $r = f(\theta)$

Polar curve has many shapes. For example;

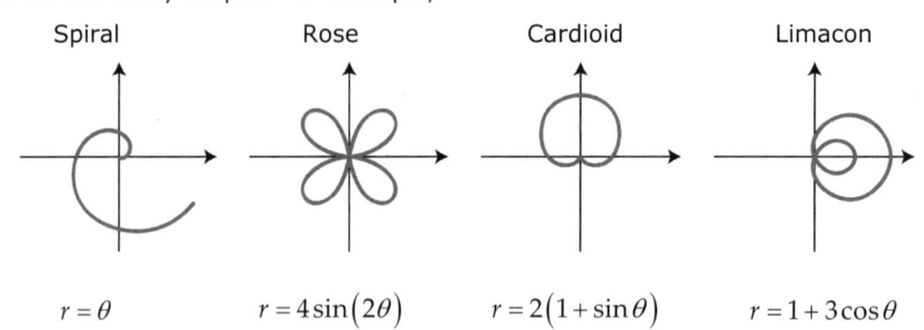

Let's graph some polar equations!

EXAMPLE 1. When $r=1+\cos\theta$, fill in the table and graph the polar.

θ	0	$\frac{\pi}{2}$	π	$\frac{3\pi}{2}$	2π
r	2	1	0	1	2

The shape is a ____________________.

Blank : ① r = f(θ)

EXAMPLE 2. Graph the polar equations.

① $r=1-\sin\theta$

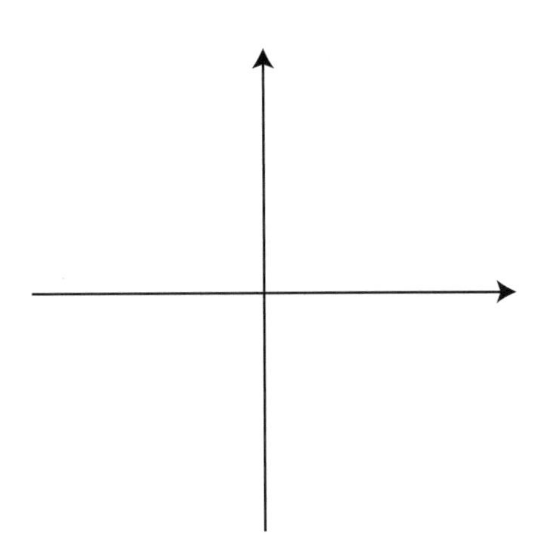

② $r=1-\cos\theta$

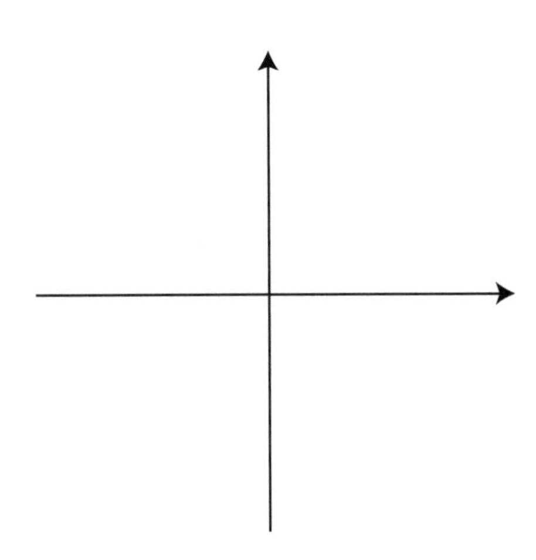

③ $r=2+\sin\theta$

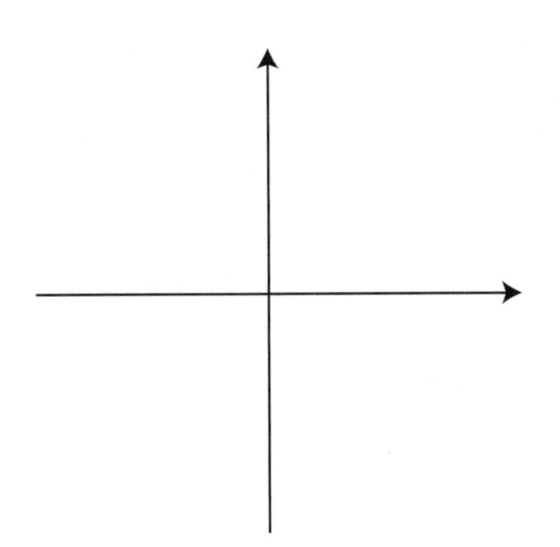

④ $r=3-2\cos\theta$

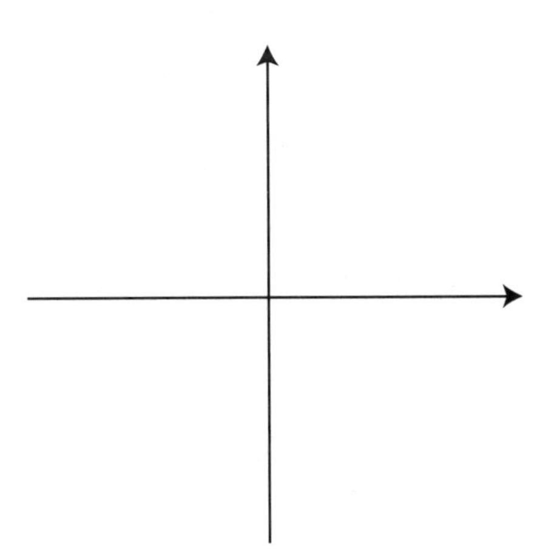

⑤ $r = 2 + 3\cos\theta$

⑥ $r = 2 - 4\sin\theta$

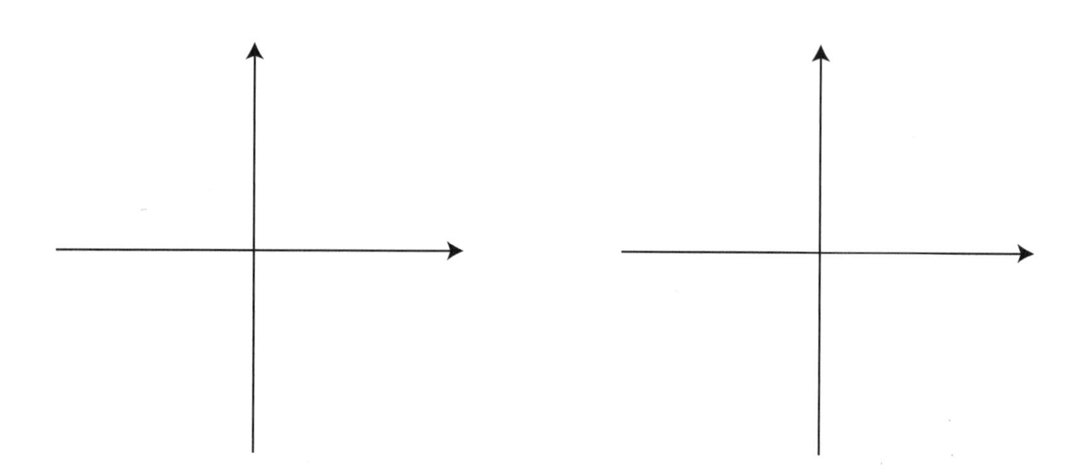

We can notice that

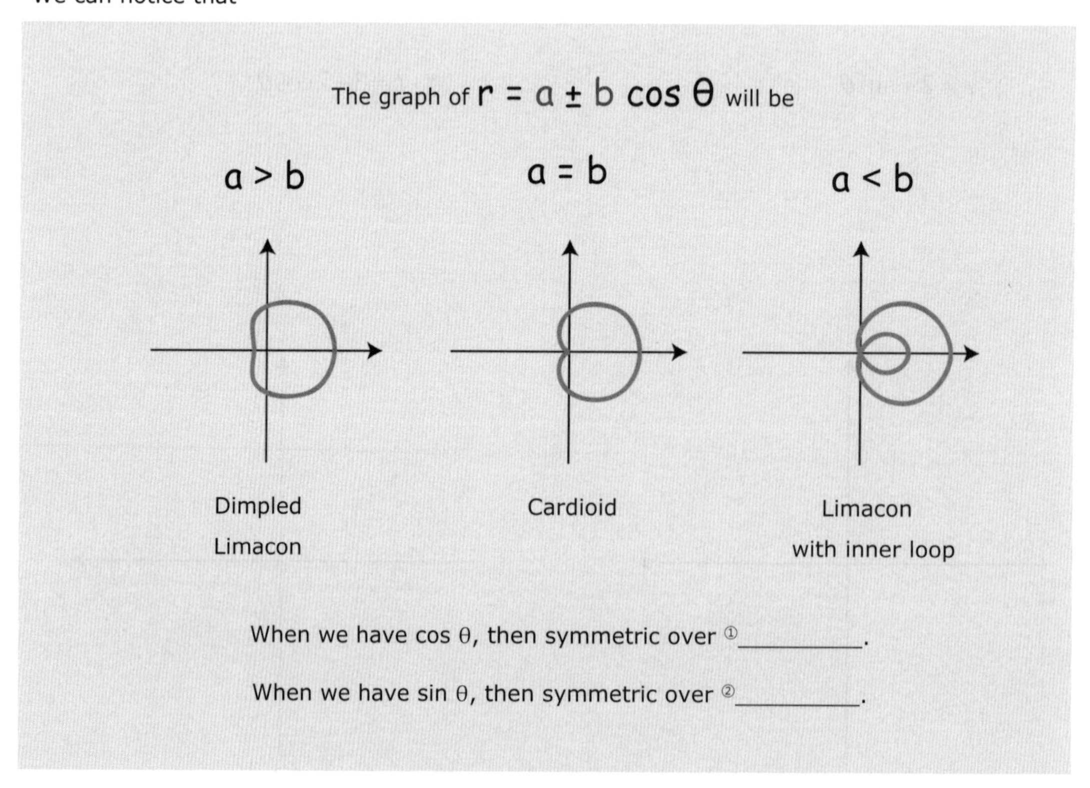

Blank : ① x axis ② y axis

EXAMPLE 3. When $r = 3\cos 2\theta$ fill in the table and graph the polar.

θ	0	$\frac{\pi}{2}$	π	$\frac{3\pi}{2}$	2π
r					

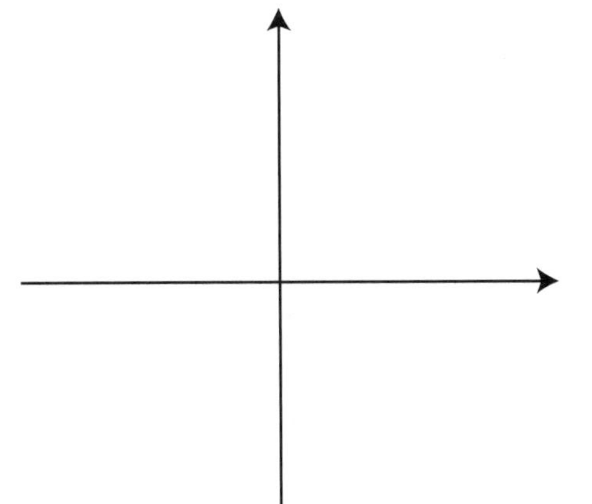

The shape is a _________________.

EXAMPLE 4. Graph the polar equations.

① $r = 4\sin 3\theta$

② $r = 5\sin 2\theta$

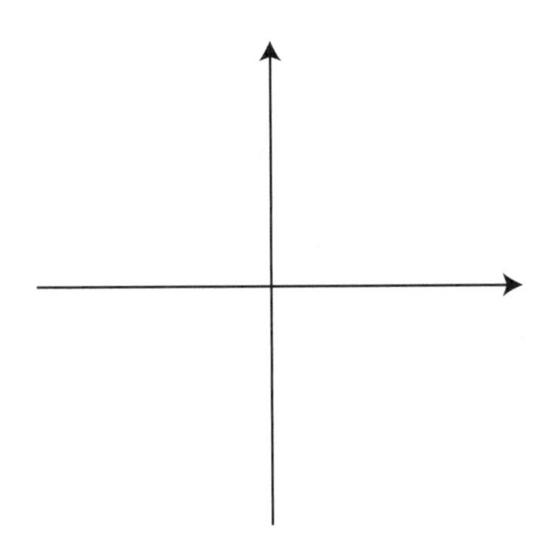

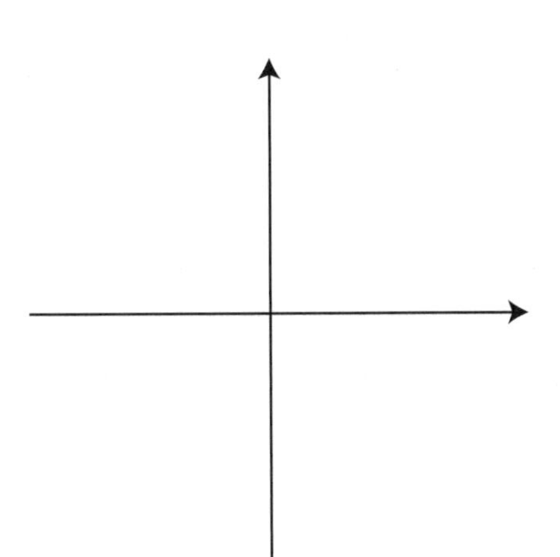

③ $r = 5\sin 4\theta$

④ $r = 2\cos 3\theta$

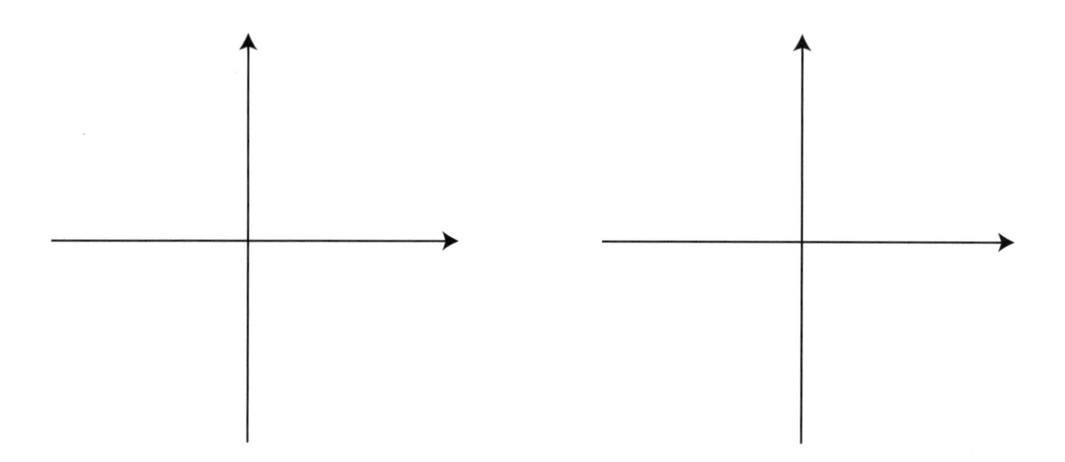

We can notice that

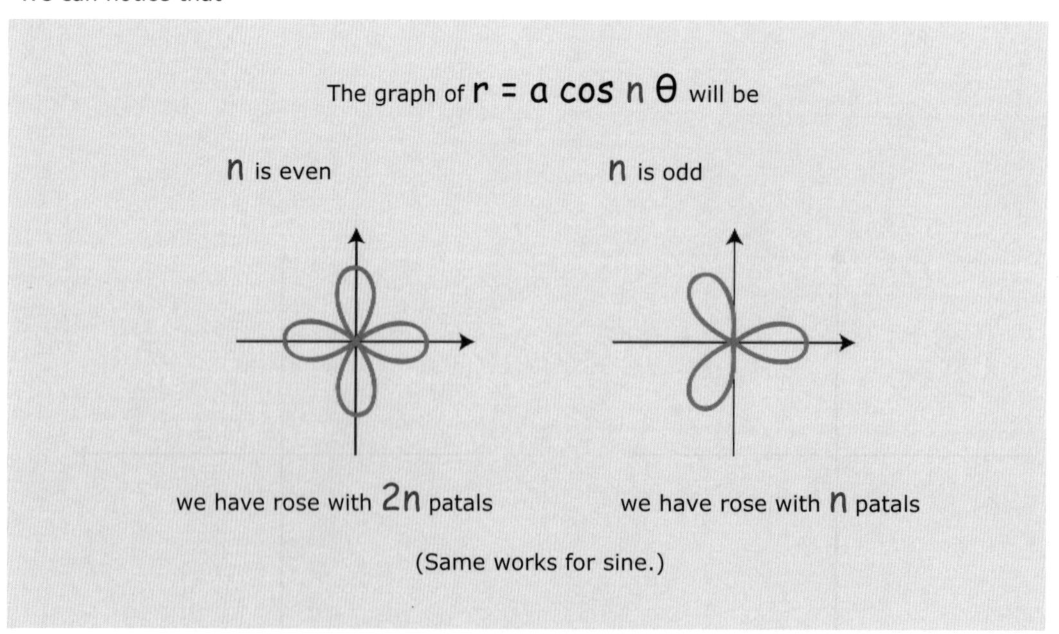

EXAMPLE 5. Graph $r^2 = 9\sin(2\theta)$ in the polar system.

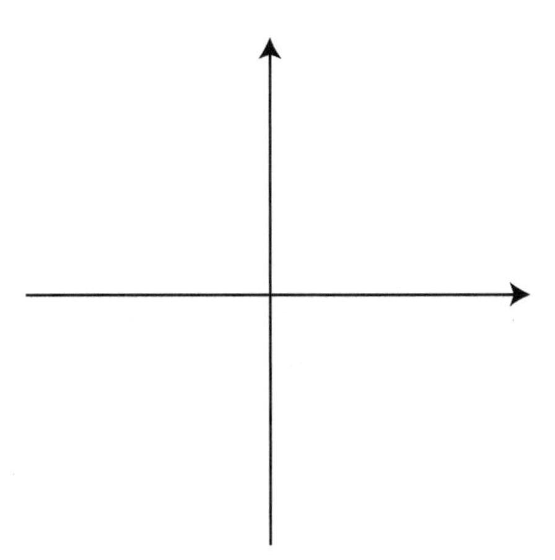

The shape is a ________________.

EXAMPLE 6. Graph $r = 2$ in the polar system.

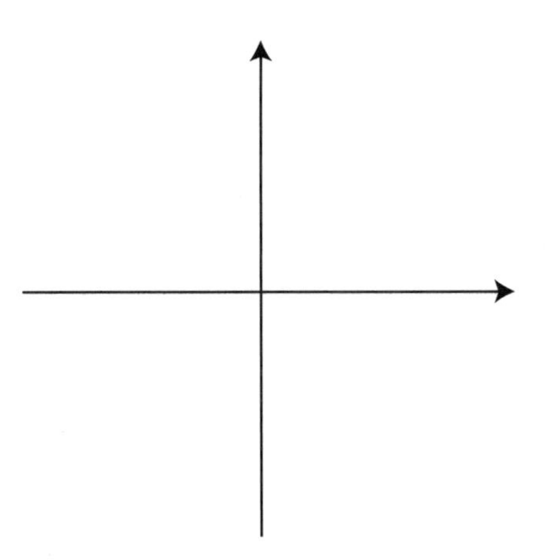

The shape is a ________________.

EXAMPLE 7. Graph $r = 2\cos\theta$ in the polar system.

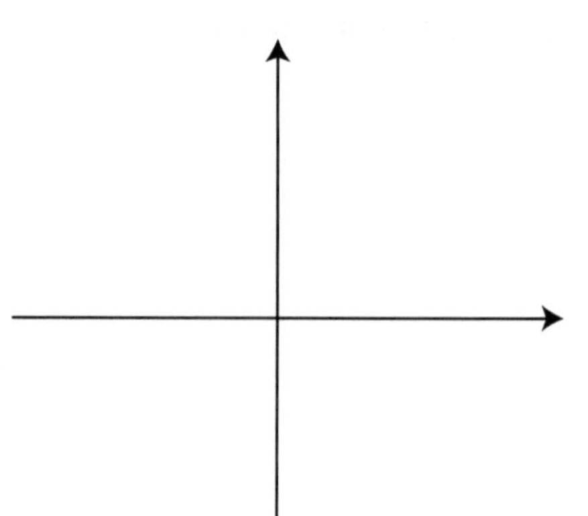

The shape is a ________________.

EXAMPLE 8. Graph $r = \theta$ in the polar system.

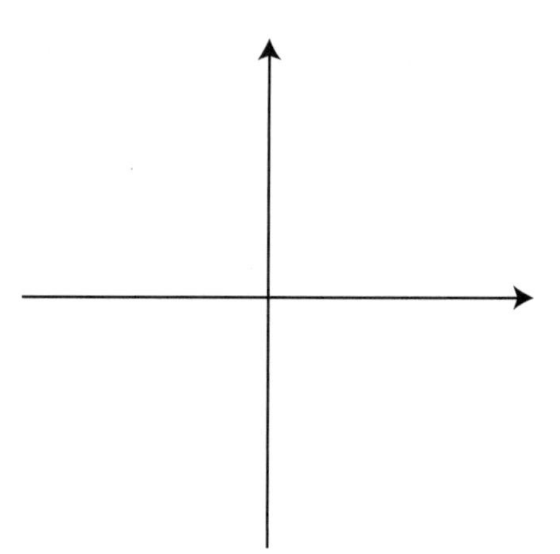

The shape is a ________________.

EXAMPLE 9. Graph $r = \sec\theta$ in the polar system.

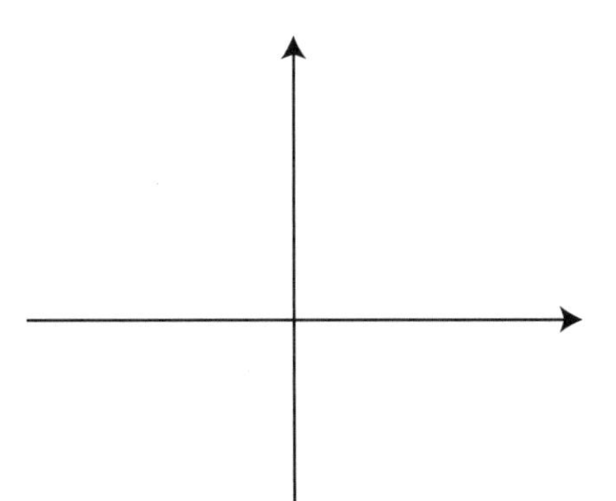

The shape is a ________________.

☺ Polar Graph All in One

Shape	Equation	Notes	
Line	$\theta = c$		
Circle	$r = a$	radius is a	
	$r = a\cos\theta$	diameter is a	

Rose	$r = a \cos n\theta$	If n is even, we have **2n** patals.	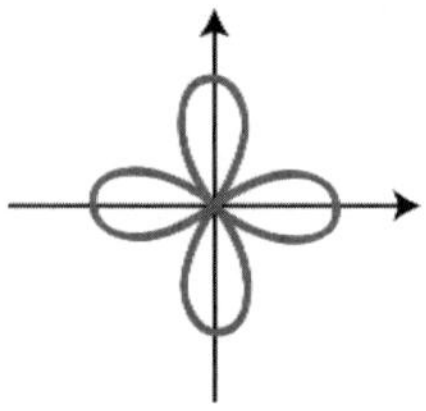
		If n is odd, we have **n** patals.	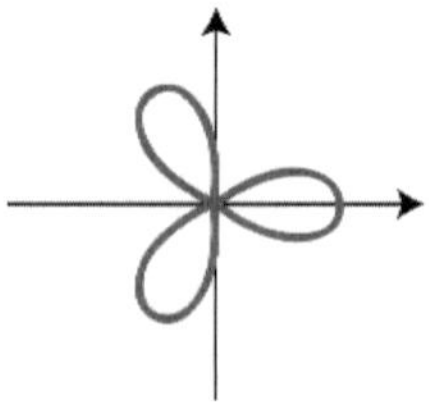
Cardiod or Limacon	$r = a \pm b \cos \theta$	If $a > b$, then Dimpled Limacon	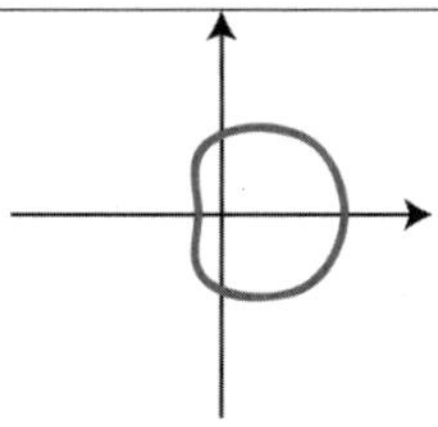
		If $a = b$, then Cardioid	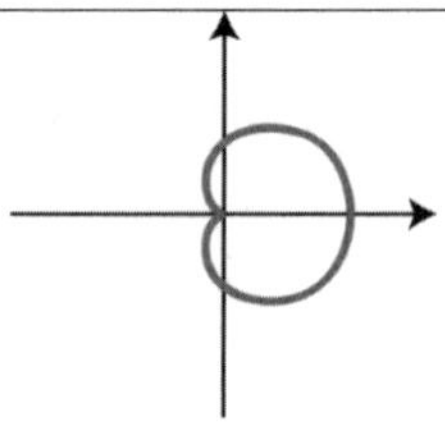
		If $a < b$, then Limacon with inner loop	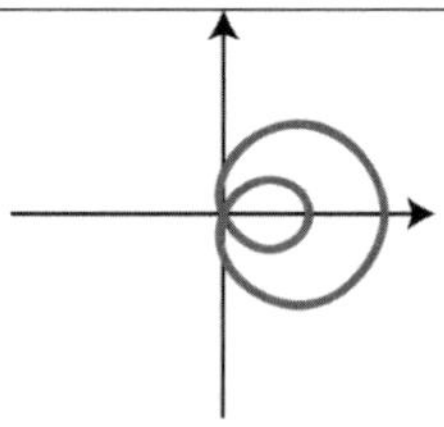
Lemniscate	$r^2 = a^2 \cos 2\theta$		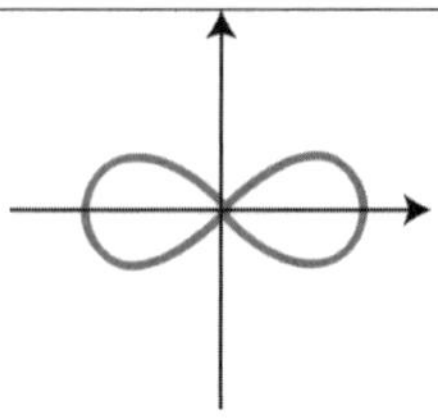
Spiral	$r = \theta$		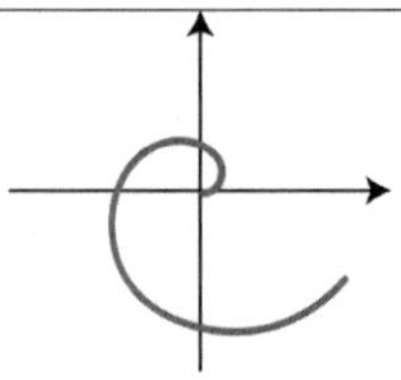

※ Tests for Symmetry in Polar Coordinates

If it is $r = f(\sin\theta)$, then it is symmetric over line $\theta = \frac{\pi}{2}$ (y-axis).

If it is $r = f(\cos\theta)$, then it is symmetric over polar axis(x-axis).

EXAMPLE 10. Match the polar equation with the graphs.

a) $r = 2 - 3\cos\theta$

b) $r = 8\sin\theta$

c) $r = \theta$

d) $r = 3\sin 4\theta$

e) $r = 8\cos 3\theta$

f) $r = 2 + 2\sin\theta$

I.

II.

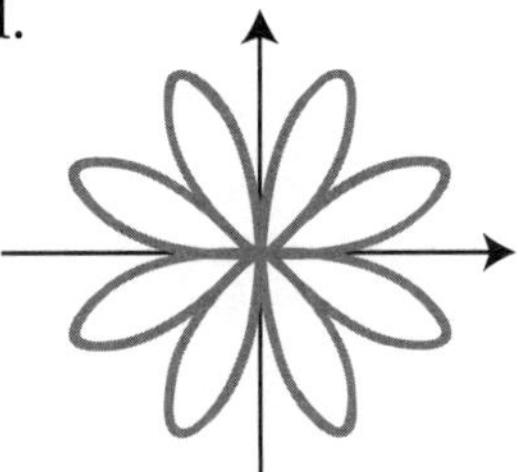

III.

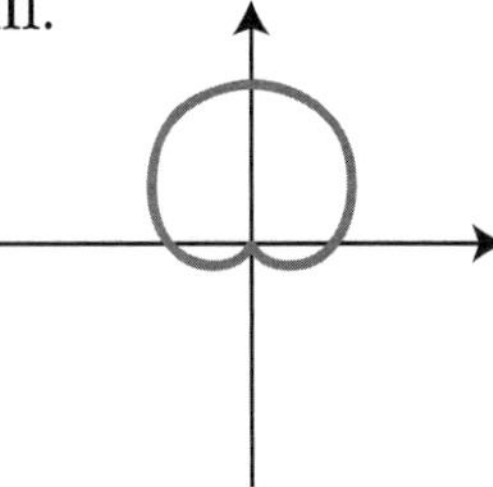

IV.

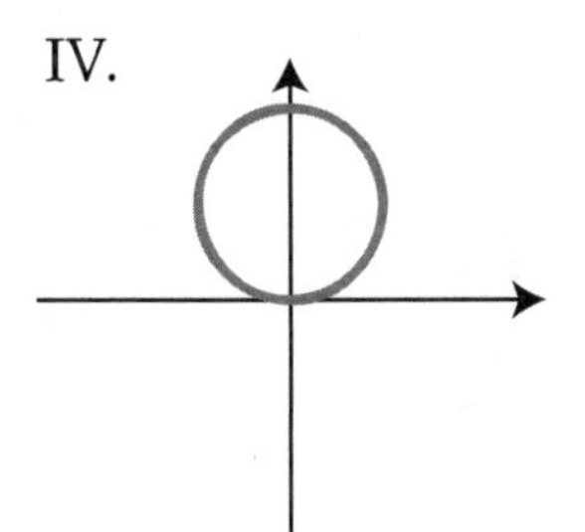

V.

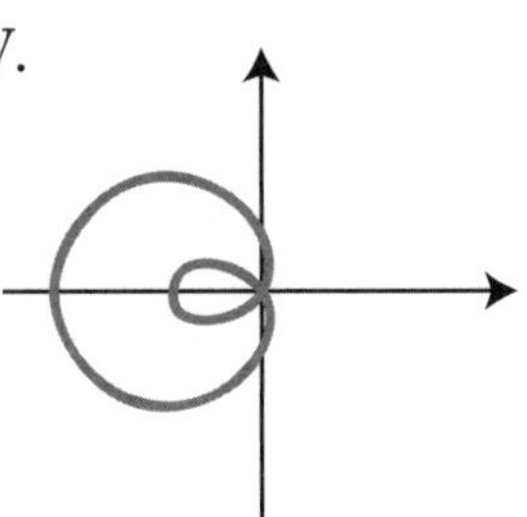

VI.

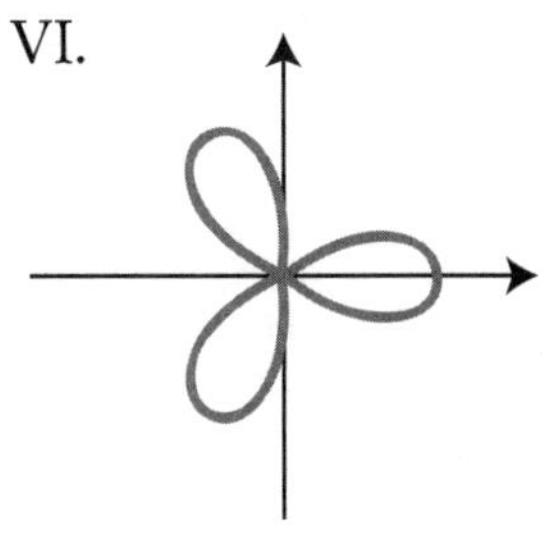

2. Distance from Origin

What can we know from $|r|$ (absolute value of r)?

※r and Distance from Origin

When r is...	Distance from Origin
If $|r|$ is increasing, When r is positive and increasing, When r is negative and decreasing,	then the graph is getting (①closer/farther) to/from the origin.
If $|r|$ is decreasing, When r is positive and decreasing, When r is negative and increasing,	then the graph is getting (②closer/farther) to/from the origin.
If $|r|$ is a minimum,	graph is closest to the origin.
If $|r|$ is a maximum,	graph is farthest from the origin.

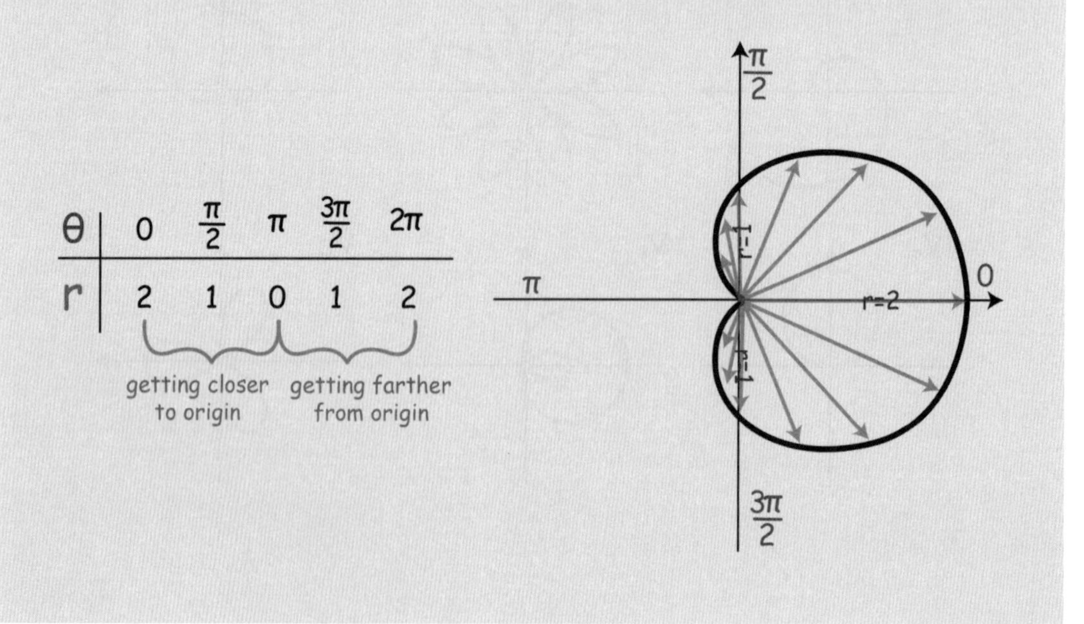

Blank : ① farther ② closer

EXAMPLE 11. Find the angle interval(s) where the polar graph is getting closer to the origin.

① $r=2+2\cos\theta$

θ	0	$\frac{\pi}{2}$	π	$\frac{3\pi}{2}$	2π
r	4	2	0	2	4

② $r=3-3\cos\theta$

θ	0	$\frac{\pi}{2}$	π	$\frac{3\pi}{2}$	2π
r	0	3	6	3	0

③ $r=1+2\sin\theta$

θ	0	$\frac{\pi}{2}$	π	$\frac{3\pi}{2}$	2π
r	1	3	1	-1	1

④ $r=1-2\cos\theta$

θ	0	$\frac{\pi}{2}$	π	$\frac{3\pi}{2}$	2π
r	-1	1	3	1	-1

⑤ $r=4\sin 2\theta,\ 0\le\theta\le\pi$

θ	0		$\frac{\pi}{2}$		π
r	0	4	0	-4	0

AP Style Problem

1. Which of the following is the polar equation for the graph shown below?

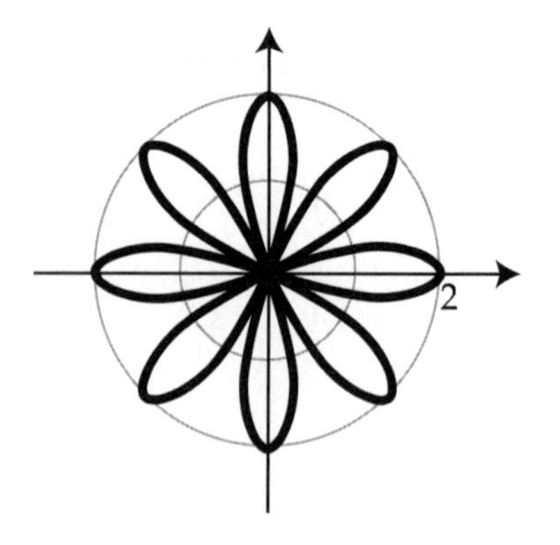

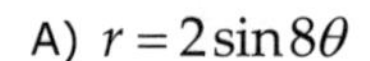
A) $r = 2\sin 8\theta$

B) $r = 2\sin 4\theta$

C) $r = 2\cos 8\theta$

D) $r = 2\cos 4\theta$

2. Which of the following is the graph of $r = 2 - 3\cos\theta$ where $-\dfrac{\pi}{2} \leq \theta \leq \dfrac{\pi}{2}$?

A)

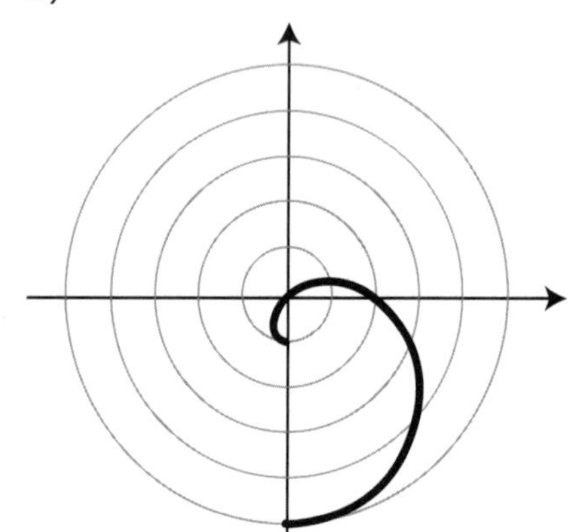

B)

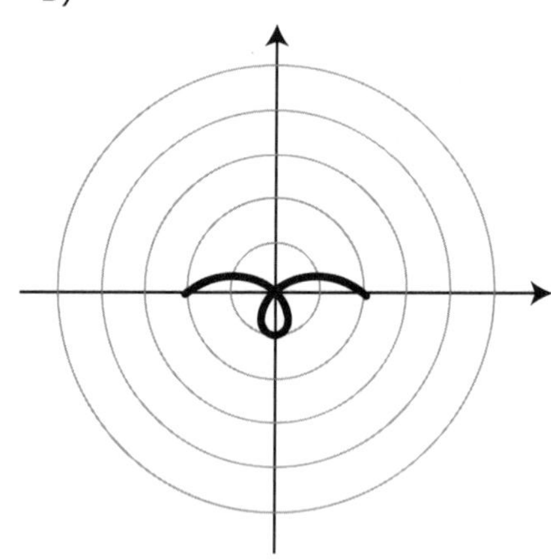

C)

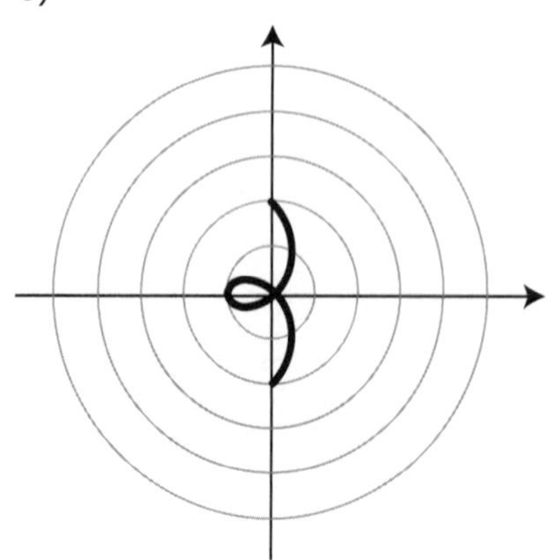

D)

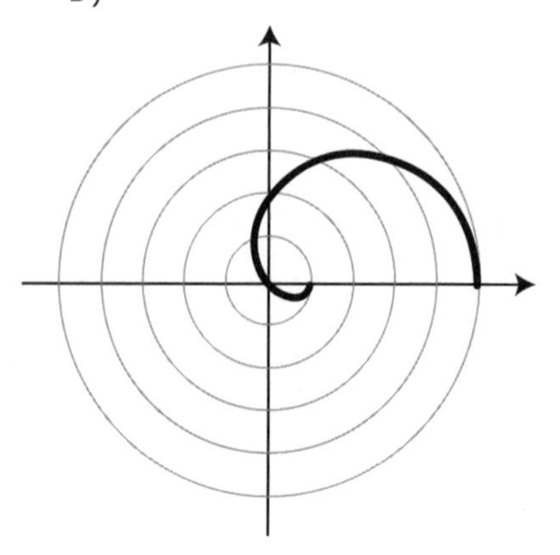

3. The graph of sinusoidal function $g(x)$ is given as shown. Which of the following is the graph of $r = g(\theta)$?

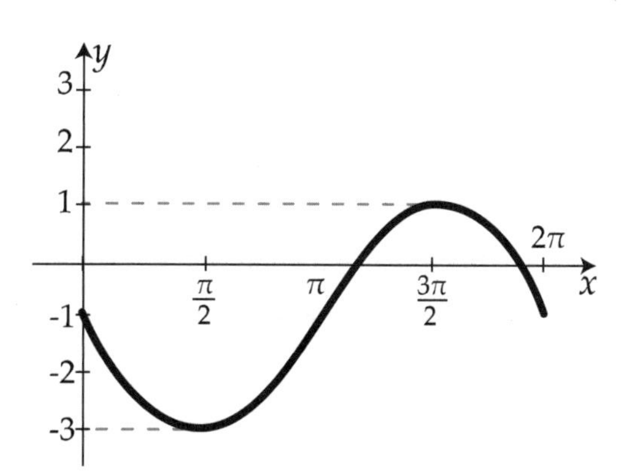

A)

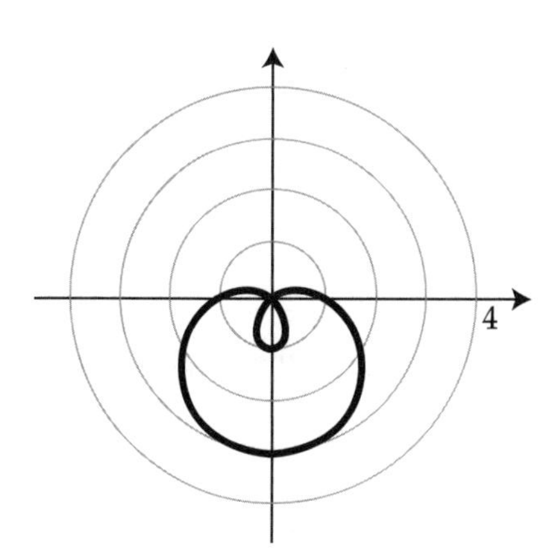

B)

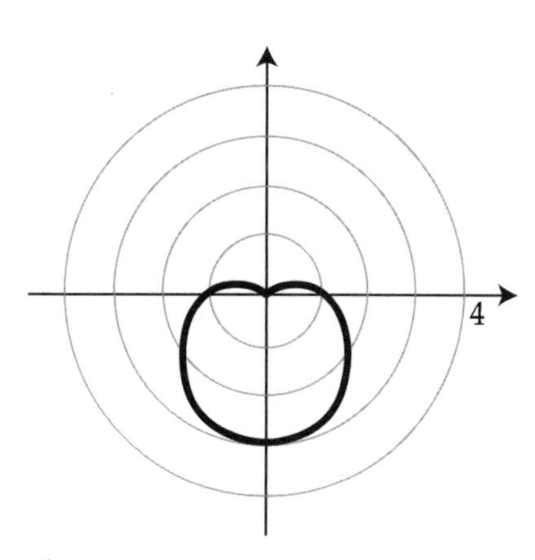

C)

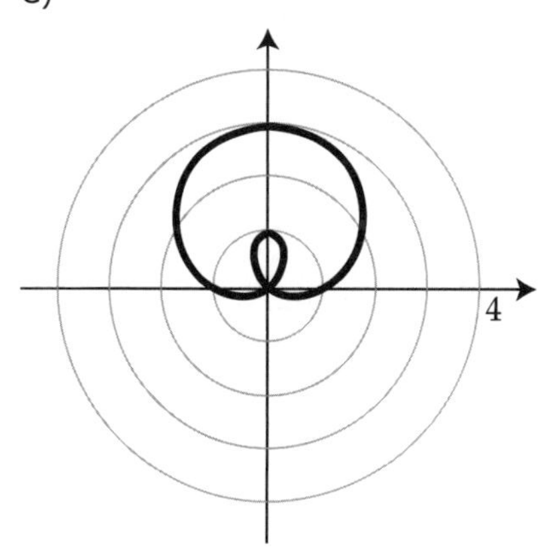

D)

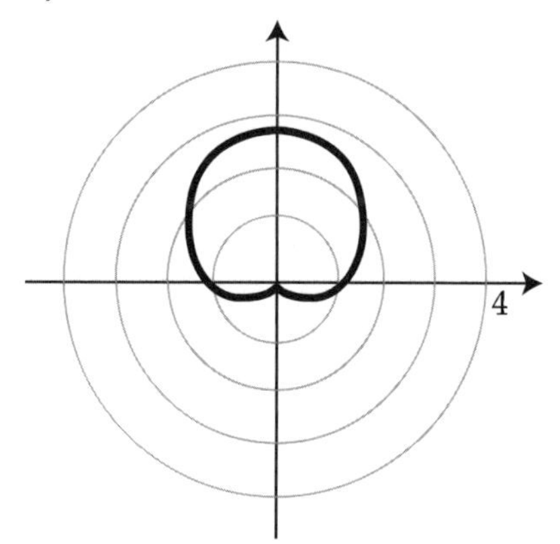

4. Which of the following is true about the graph of $r=2\cos\theta-1$ where $\frac{\pi}{2}\leq\theta\leq\pi$?

A) The graph is getting closer to the origin, because r is positive and increasing.

B) The graph is getting closer to the origin, because r is negative and decreasing.

C) The graph is getting farther from the origin, because r is positive and increasing.

D) The graph is getting farther from the origin, because r is negative and decreasing.

5. The table of $r=3\cos 4\theta$, where $0\leq\theta\leq\pi$ is given. At what angle is the graph farthest from the origin?

θ	0	$\frac{\pi}{8}$	$\frac{\pi}{4}$	$\frac{3\pi}{8}$	$\frac{\pi}{2}$	$\frac{5\pi}{8}$	$\frac{3\pi}{4}$	$\frac{7\pi}{8}$	π
$r=3\cos 4\theta$	3	0	-3	0	3	0	-3	0	3

A) $0, \pi$

B) $0, \frac{\pi}{4}$

C) $0, \frac{\pi}{2}, \pi$

D) $0, \frac{\pi}{4}, \frac{\pi}{2}, \frac{3\pi}{4}, \pi$

6. * The graph of $r=1+2\cos\theta$ is given and the solid line part is called the 'inner loop'. Find the interval of θ that corresponds to the 'inner loop' of r.

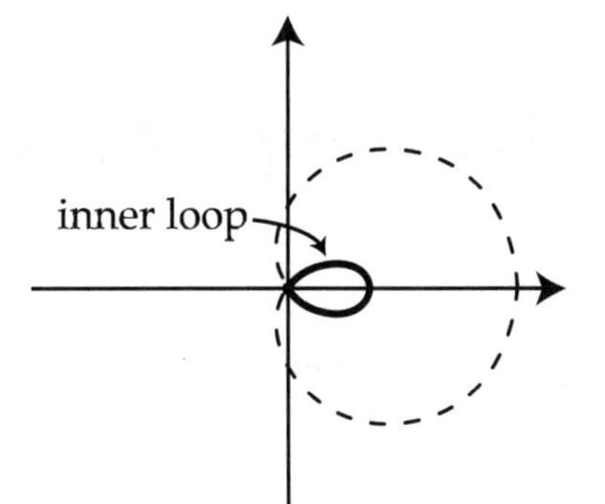

A) $0\le\theta\le\pi$

B) $\dfrac{2\pi}{3}\le\theta\le\dfrac{4\pi}{3}$

C) $\dfrac{4\pi}{3}\le\theta\le\dfrac{5\pi}{3}$

D) $\dfrac{7\pi}{6}\le\theta\le\dfrac{11\pi}{6}$

7.3 Complex Number

1. Complex Number

※ A **complex number**, z, is a number that can be written in the form $z = a + bi$ (Cartesian form), where a and b are real numbers and $i^2 = -1$:

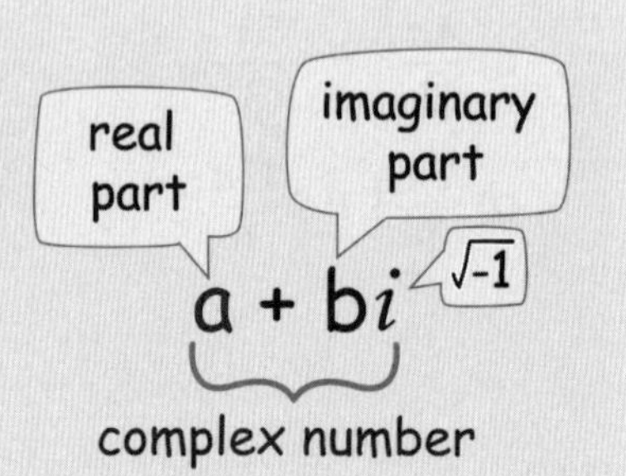

- a is the real part of z.
- bi is the imaginary part of z.
- The complex conjugate of z, written as $z^* =$ ①______

※ If two complex numbers are equal, their real parts are the same and their imaginary parts are the same.

2. Complex Plane

We are familiar with Cartesian plane:
But where do we put a complex number $3+4i$?
We need a new plane!

We call this a ②______ plane or ③______ diagram:
it is a combination of real and imaginary numbers!

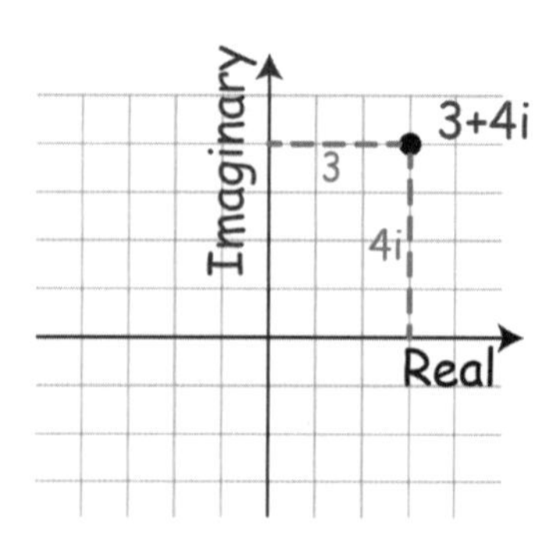

Blank : ① a – bi ② complex ③ Argand

EXAMPLE 1. Draw the complex number in an imaginary plane.

① $z = 1 - 3i$

② $z = -4 - 3i$

③ $z = 6i$

④ $z = -5$

⑤ $z = -2$

⑥ $z = -7i$

3. Polar Form of Complex Numbers

Complex numbers can be represented in ①____________ form.

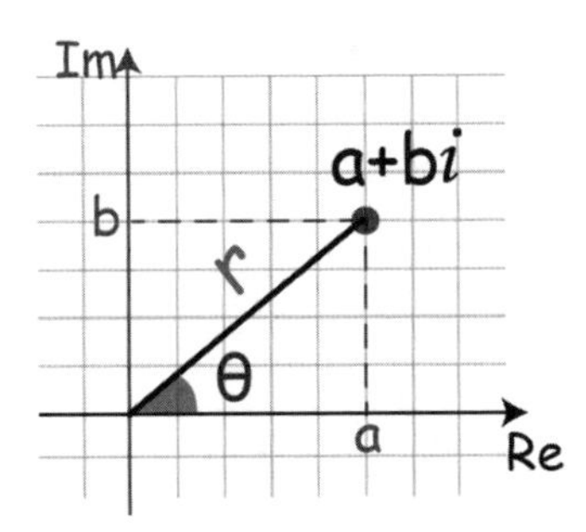

Blank : ① polar

① **modulus** of z = r = |z| = distance of the complex number from the origin

modulus

$|z| = r = |a+bi| =$ ①

② **argument** of z = θ = angle in standard postion

argument

$\tan\theta = \dfrac{b}{a}$

(remember; $\tan^{-1}$ only gives you the reference angle)

③ a complex number can also be described in polar form:

modulus argument

$z =$ ② $= r \text{ cis } \theta$

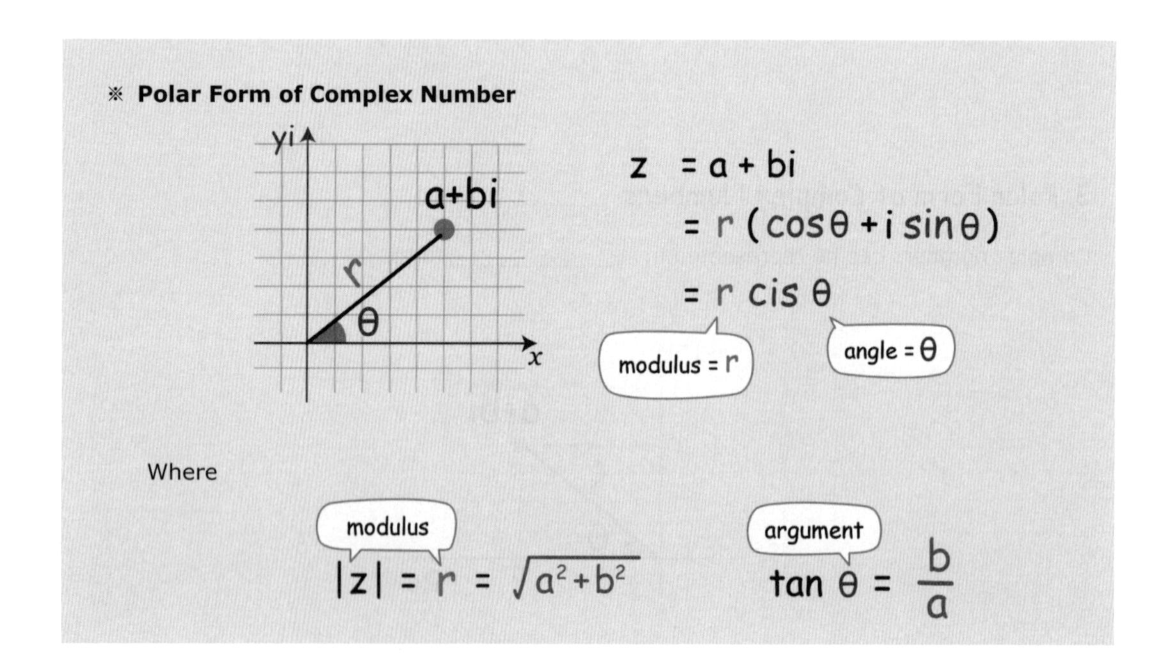

Blank : ① $\sqrt{a^2+b^2}$ ② $r(\cos\theta + i\sin\theta)$

EXAMPLE 2. Use an algebraic method to find the trigonometric form (polar form) with $0 \le \theta < 2\pi$ for

① $z = 1 + i$

② $z = \sqrt{3} - i$

③ $z = -\sqrt{3}i + 1$

④ $z = -2 + 2i$

⑤ $z = -2 + 3i$

⑥ $z = \sqrt{5} - i$

EXAMPLE 3. Convert into Cartesian form.

① $2\left(\cos 120^{\circ} + i\sin 120^{\circ}\right)$

② $3\left(\cos 210^{\circ} + i\sin 210^{\circ}\right)$

③ $2\left(\cos\frac{7\pi}{6} + i\sin\frac{7\pi}{6}\right)$

④ $3\left(\cos\frac{3\pi}{2} + i\sin\frac{3\pi}{2}\right)$

⑤ $4cis23\pi$

⑥ $3cis32\pi$

AP Style Problem

1. Which of the following is the polar form of the complex number $5i-5$?

A) $5\sqrt{2}\left(\cos\frac{3\pi}{4}+i\sin\frac{3\pi}{4}\right)$

B) $5\sqrt{2}\left(\cos\frac{7\pi}{4}+i\sin\frac{7\pi}{4}\right)$

C) $5\sqrt{2}\left(\sin\frac{3\pi}{4}+i\cos\frac{3\pi}{4}\right)$

D) $5\sqrt{2}\left(\sin\frac{7\pi}{4}+i\cos\frac{7\pi}{4}\right)$

Free Response Questions (from ch4-ch7)

1. Trigonometry Graph Modeling

(Calculator) Jay starts a Ferris wheel at the 2 o'clock position. The wheel rotates in counter-clockwise direction, completing on rotation every 4 minutes. The Ferris wheel is 50 feet wide in diameter and the center is 30 feet above the ground.

(a) Find the time t in seconds for Jay to move from the 2 o'clock position to the highest point of the Ferris wheel where $0 \le t \le 240$?

(b) Jay's height above the ground h, in feet, is a function of the number of seconds, t, since Jay boarded the Ferris wheel. The graph of h in one rotation is given as shown with three points A, B and C on the graph. Find the coordinates of the point A, B and C.

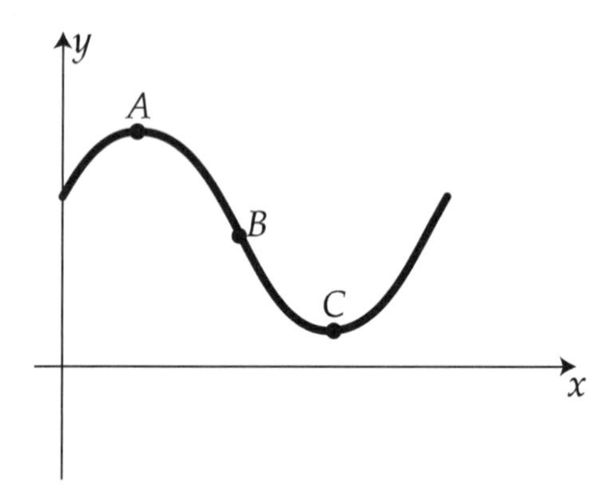

(c) The graph can be modeled by $h = a\cos[b(t-c)]+d$. What are the values a, b, c, and d?

(d) Find the time t in seconds, $0 \le t \le 240$, where the seat is 20 feet above the ground.

(e) In one rotation of the Ferris wheel, find the probability that a randomly selected seat is less than 20 feet above the ground.

2. Polar Graph

(Non-Calculator) The Cartesian and polar graph of $r = 6\cos(2\theta) - 3$ are given as shown.

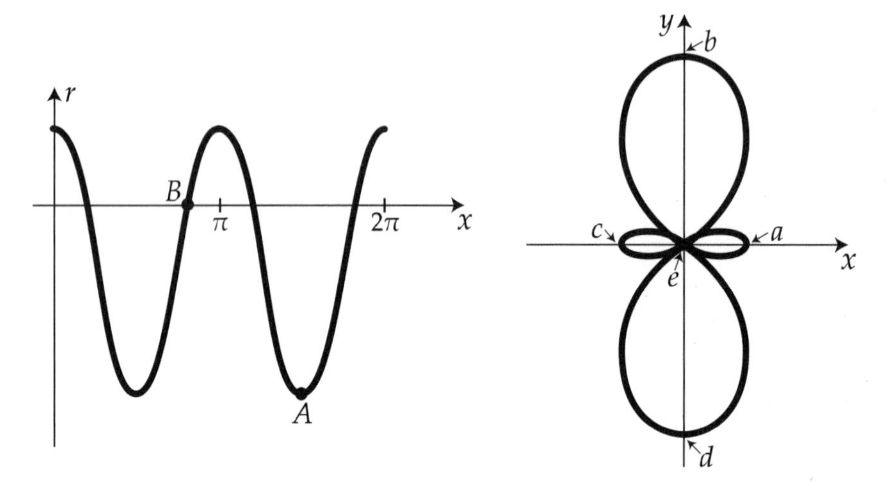

(a) Which point on the polar graph corresponds to the point A on the Cartesian graph?

(b) Find the angle(s) in the interval $0 \le \theta \le 2\pi$ where the graph is farthest from the origin in the polar graph.

(c) Find the angle of point B on the Cartesian graph. Which point on the polar graph corresponds to the point B on the Cartesian graph?

(d) Find the angle(s) in the interval $0 \le \theta \le \pi$ where the graph is getting closer to the origin in the polar graph.

Part 8

Parametric Equation

8.1 Parametric Equations

1. Parametric Equations

A **parametric equation** is where the x and y coordinates are both written in terms of another letter.

$$x = f(t), \qquad y = g(t)$$

That another letter (usually given the letter t or θ) is called a ①___________.

Finding the Cartesian equation from parametric equation
: Try to eliminate the parameter t or θ by using

②______________________ or ③__________________________.

EXAMPLE 1. Find the Cartesian equation for each parametric equation.

① $x=1-2t, \qquad y=2-t$

② $x=\dfrac{t}{3}, \qquad y=\sqrt{t}$

Blank : ① parameter ② substitution ③ trig identity

③ $x = t^2 - 1, \qquad y = -t + 4$

④ $x = 3t, \qquad y = t^2 - 2$

⑤ $x = 2\cos\theta,\ y = 3\sin\theta,\ 0 \le \theta \le 2\pi$

⑥ $x = 5\cos\theta,\ y = 7\sin\theta,\ 0 \le \theta \le 2\pi$

⑦ $x = 2\sec\theta,\ y = 3\tan\theta,\ 0 \le \theta \le 2\pi$

⑧ $x = \sec\theta,\ y = \tan\theta,\ 0 \le \theta \le 2\pi$

2. Graphing Parametric Equations

If we want to graph $x=t^2-2t,\ \ y=t+2;$

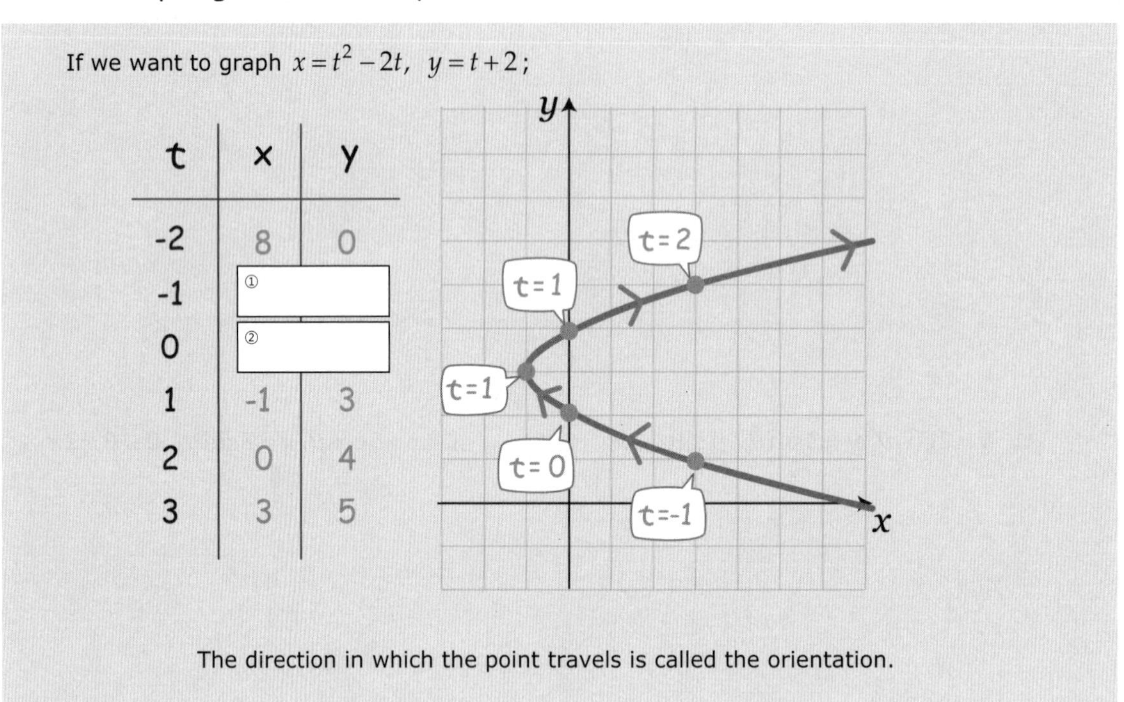

The direction in which the point travels is called the orientation.

EXAMPLE 2. Sketch the parametric function indicating the orientation. State the domain.

① $x=3t+2, y=t+1\ ;\ 0\le t\le 4$

② $x=t-3,\ y=2t^2+4\ ;\ t\ge -2$

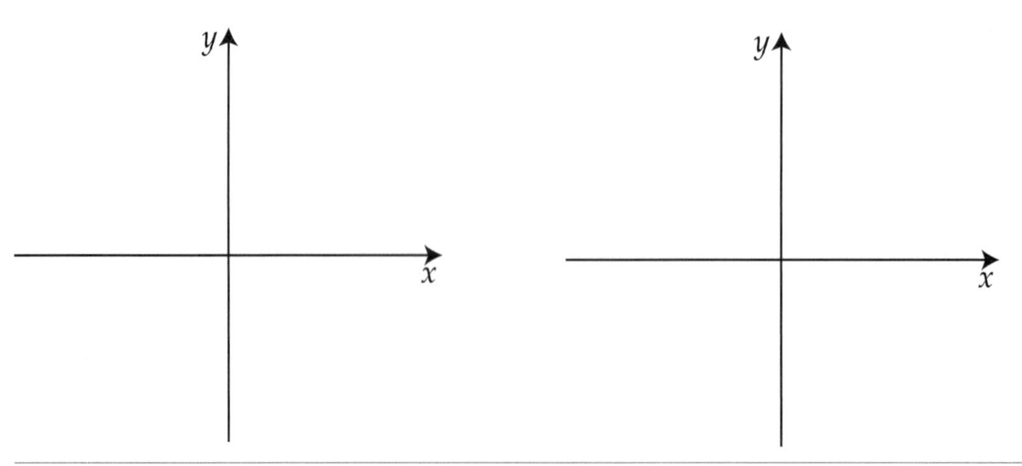

Blank : ① 3 , 1 ② 0 , 2

③ $x=1+t,\ y=t^2-4t\ ;\ -2\le t\le 2$

④ $x=\sqrt{t},\ y=3t+1$

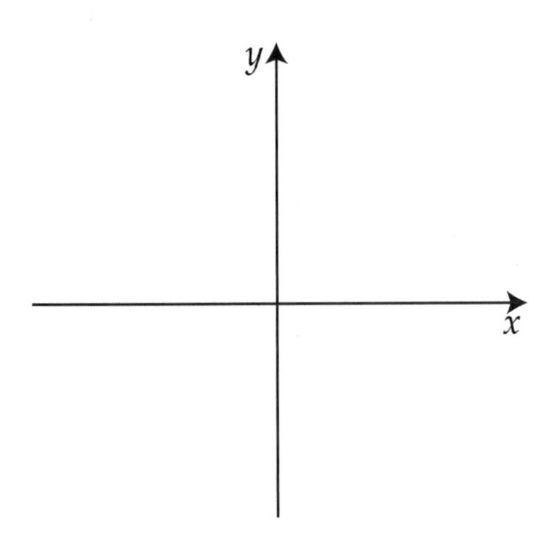

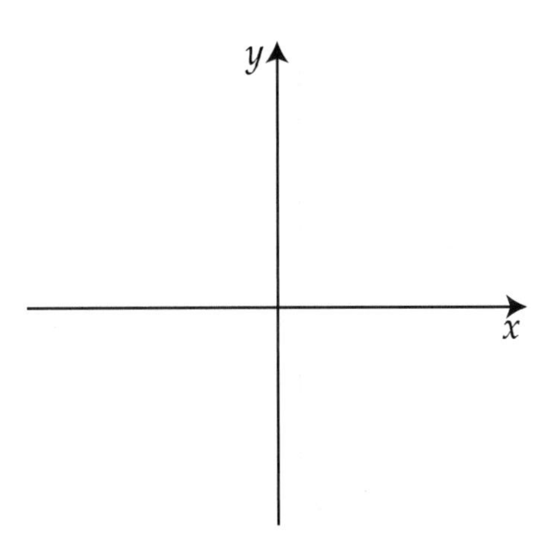

⑤ $x=\sqrt{t-2},\ y=1-t$

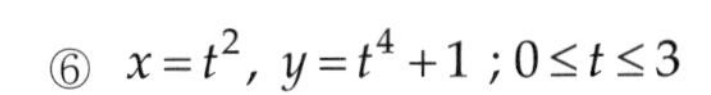
⑥ $x=t^2,\ y=t^4+1\ ;0\le t\le 3$

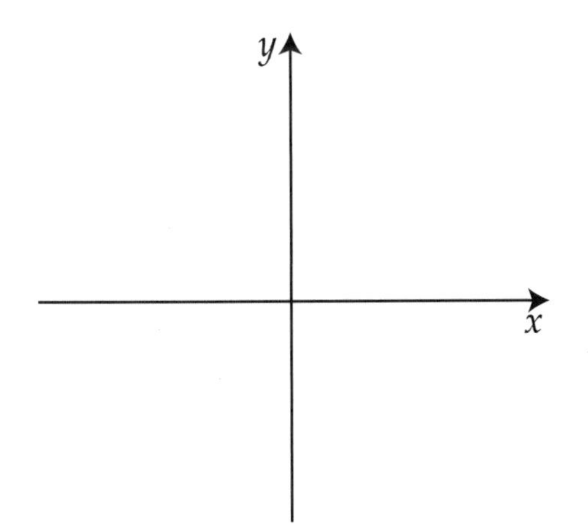

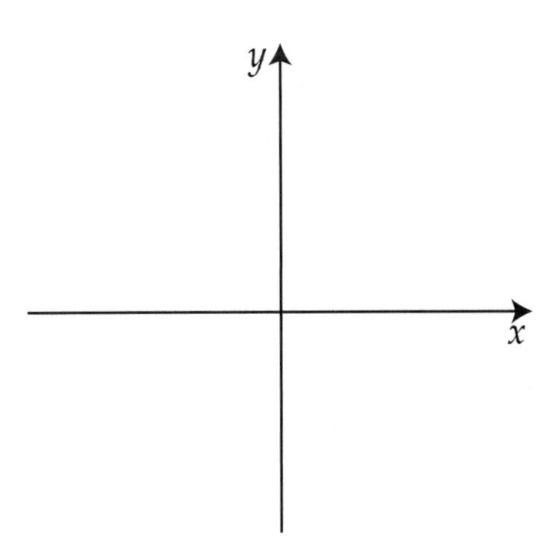

⑦ $x = 2\cos t,\ y = 3\sin t\ ;\ 0 \le t \le 2\pi$

⑧ $x = 2\sin t,\ y = \cos t\ ;\ 0 \le t \le \dfrac{\pi}{2}$

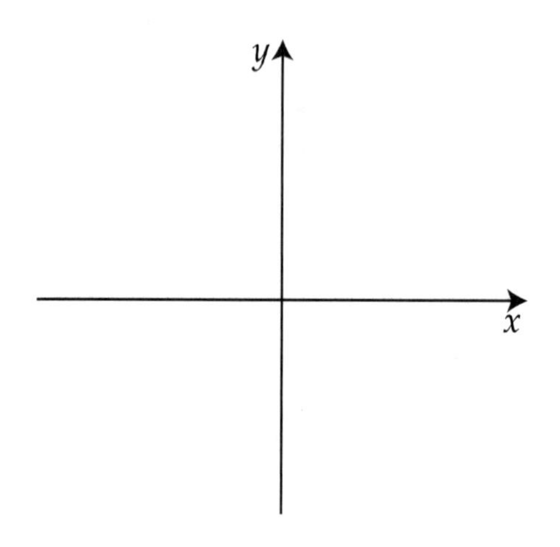

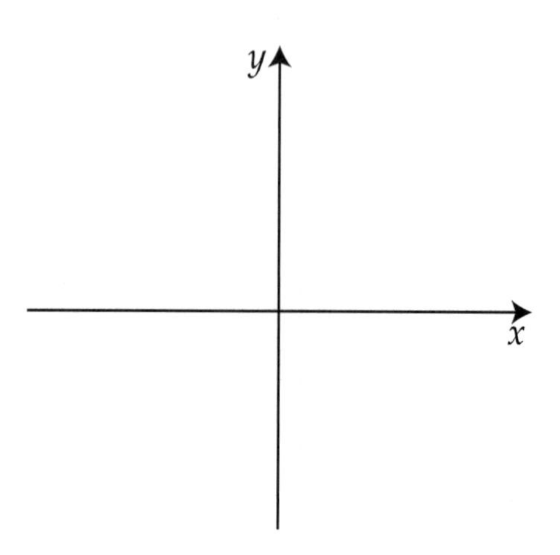

⑨ $x = 2^t,\ y = 2^{-t}\ ;\ t \ge 0$

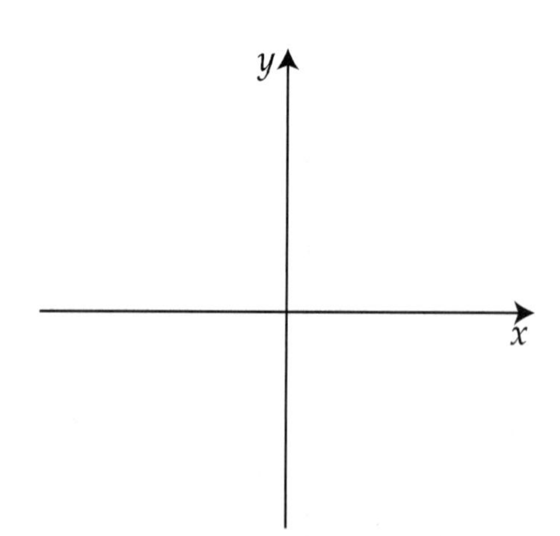

⑩ $x=\sin^2 t,\ y=\sin^4 t\ ;\ 0\le t\le \pi$

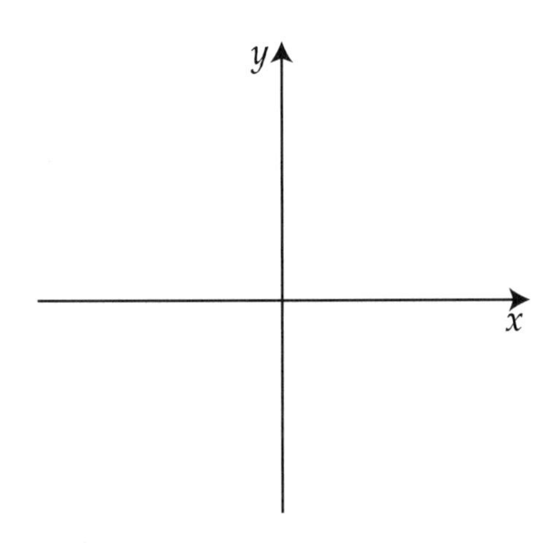

3. Intercepts of Parametric Equations

※ **Intercepts of Parametric Equations**

For parametric equation $x(t)=t^2-2t,\ y(t)=t+2\,;$

To find <u>x intercept</u>, we let ①__________.

$y=0 \Rightarrow$ ② $\boxed{}=0 \Rightarrow t=-2$

then plug in $t=-2$ to $x(t)$;

$x(-2)=$ ③ $\boxed{}$

That gives x intercept at ④______________.

To find <u>y intercept</u>, we let ⑤__________.

$x=0 \Rightarrow$ ⑥ $\boxed{}=0 \Rightarrow t=0,2$

then plug in $t=0,\ 2$ to $y(t)$;

$y(0)=$ ⑦ $\boxed{}$, $y(2)=$ ⑧ $\boxed{}$

That gives y intercept at ⑨______________.

Blank : ① y=0 ② t+2 ③ 8 ④ (8,0) ⑤ x=0 ⑥ t^2-2t ⑦ 2 ⑧ 4 ⑨ (0, 2), (0, 4)

EXAMPLE 3. Find the x intercept and the y intercept.

① $x=3t+2, y=\sqrt{t+1}$

② $x=t-3,\ y=2t^2-2$

③ $x=1+t,\ y=t^2-4t$

8.2 Motions and Parametric Equations

1. Motions and Parametric Equations

Yuna is gliding around on a frozen coordinate plane. At time t (in seconds) Yuna 's position on the coordinate plane is given by $(x(t), y(t))$ where

$$x(t) = t^2 - 2t, \quad y(t) = t + 2.$$

If Yuna travels when $0 \le t \le 3$, she will trace out the parametric curve as shown in the graph.

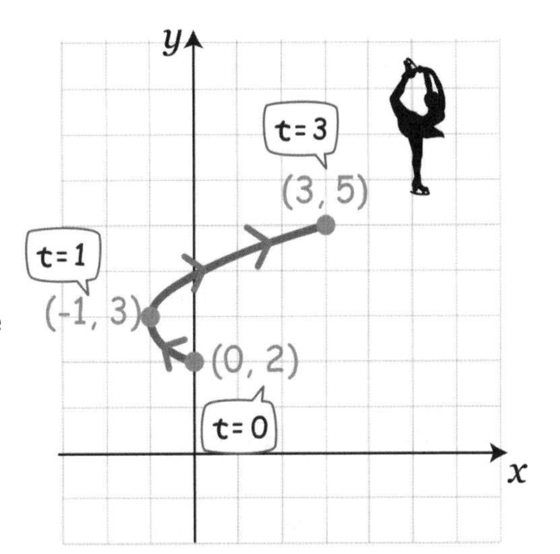

What is Yuna's position at t = 0, t = 1, t = 3?

※ **Position**: the location of the particle in the xy plane.

If position is given by parametric equation $x(t), y(t)$

output of $x(t)$ shows the **horizontal position** of the particle

output of $y(t)$ shows the **vertical position** of the particle.

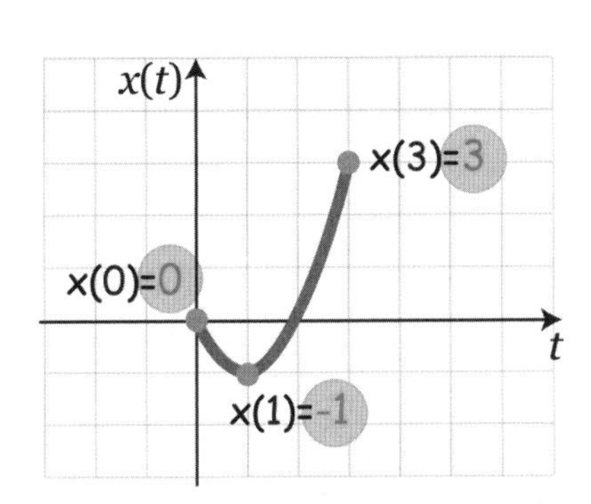

horizontal position $x(t) = t^2 - 2t$

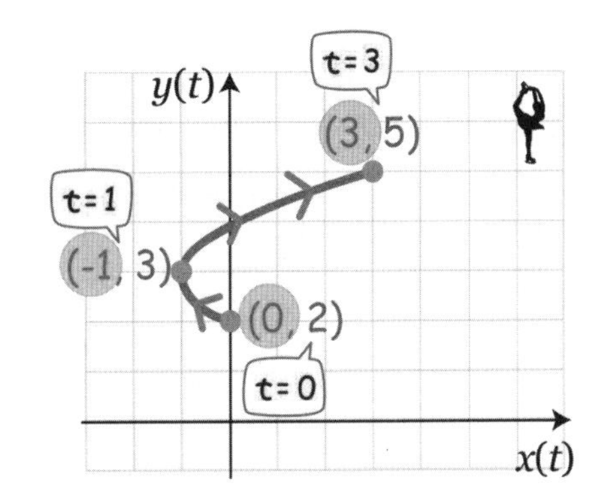

parametric equation

output of $x(t)$ shows the **horizontal position** of the particle

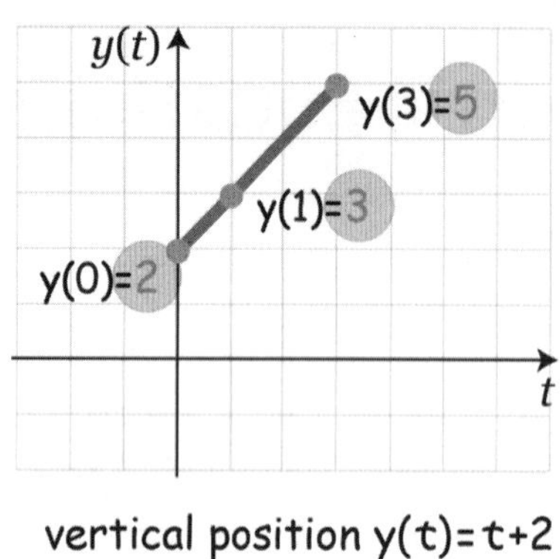

vertical position y(t)=t+2

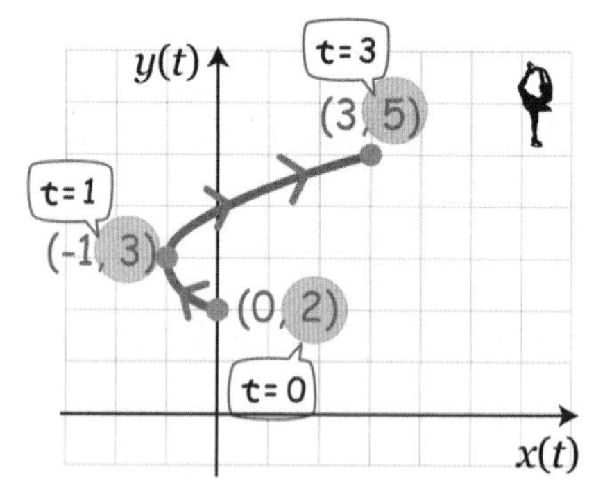

parametric equation

output of $y(t)$ shows the **vertical position** of the particle.

2. Analyzing Motions

※ **Analyzing Motions from parametric equation**

If position is given by parametric equation $x(t), y(t)$

Use output of $x(t)$ to answer all questions about the **horizontal position** of the particle

ex) Farthest right : ①______________

Farthest left : ②______________

moving right : ③____________

moving left : ④____________

Use output of $y(t)$ to answer all questions about the **vertical position** of the particle.

ex) Farthest up : ⑤______________

Farthest down : ⑥______________

moving up: ⑦____________

moving down: ⑧____________

Blank : ① max of x(t) ② min of x(t) ③ x(t) is increasing ④ x(t) is decreasing
⑤ max of y(t) ⑥ min of y(t) ⑦ y(t) is increasing ⑧ y(t) is decreasing

For Yuna's position $x(t)=t^2-2t,\ y(t)=t+2$ where $0\le t\le 3$,

to find the farthest right and farthest left;

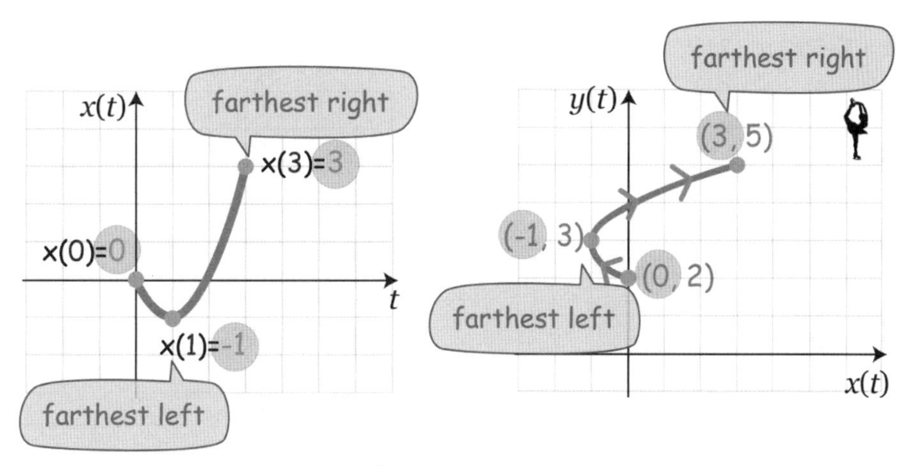

Farthest right : ①____________ Farthest left : ②____________

moving right : ③____________ moving left : ④____________

to find the farthest up and farthest down;

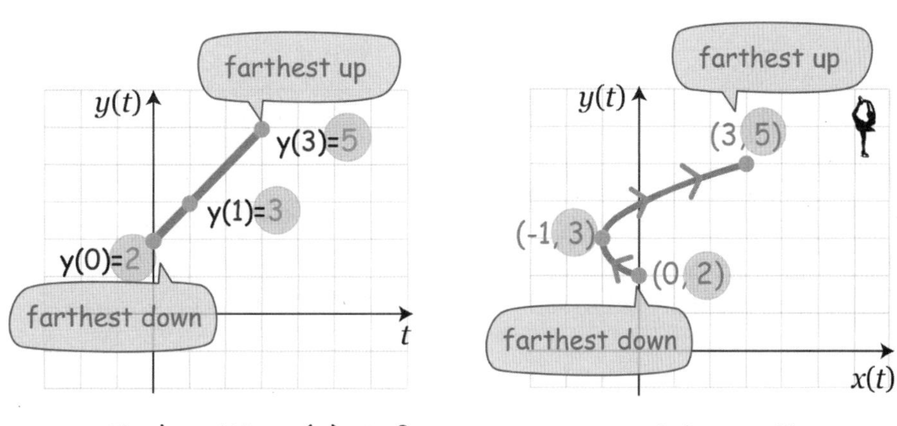

Farthest up : ⑤____________ Farthest down: ⑥____________

moving up: ⑦____________ moving down: ⑧____________

Blank : ① x=3 ② x=-1 ③ 1≤t≤3 ④ 0≤t≤1 ⑤ y=5 ⑥ y=2 ⑦ 0≤t≤3 ⑧ none

EXAMPLE 1. The position is given by parametric equation. Analyze the motion and find the following.

① $x=1-2t,\ y=2-t,\quad 0\le t\le 4$

Farthest right :______________

Farthest down :______________

② $x=-\dfrac{t}{3},\ y=t+3,\ -1\le t\le 2$

Farthest right :______________

Farthest down :______________

③ $x=t^2-1,\quad y=-t+4,\quad -1\le t\le 3$

Farthest left :______________

Farthest up :______________

④ $x=3t,\quad y=t^2-2,\quad -3\le 0\le 2$

Farthest left :______________

Farthest up :______________

⑤ $x = 2\cos\theta,\ y = 3\sin\theta,\ 0 \le \theta \le 2\pi$

Farthest left :_______________

Farthest down :_______________

⑥ $x = 5\cos\theta,\ y = 7\sin\theta,\ 0 \le \theta \le 2\pi$

Farthest left :_______________

Farthest down :_______________

3. Average Rate of Change

If we want to find the slope of the path (=average rate of change) in the interval $t_1 \le t \le t_2$,

we need to find $\dfrac{\text{change of } y}{\text{change of } x} = \boxed{①\quad}$

☺ Δ ('delta') means 'difference of~' or 'change of~'

※ **Average Rate of Change of Parametric equations**

$$\text{Slope of the path} = \frac{\Delta y}{\Delta x} = \frac{\frac{\Delta y}{\Delta t}}{\frac{\Delta x}{\Delta t}} = \frac{\frac{y(t_1)-y(t_2)}{t_1-t_2}}{\frac{x(t_1)-x(t_2)}{t_1-t_2}}$$

For Yuna's position $x(t)=t^2-2t,\ y(t)=t+2$,

to find the slope of the path over $1\le t\le 3$;

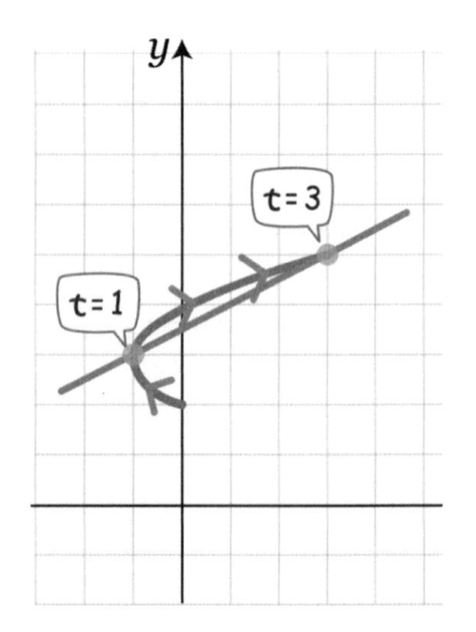

$$\frac{\Delta y}{\Delta t}=\frac{y(3)-y(1)}{\boxed{②}}=\boxed{③}\ \text{and}\ \frac{\Delta x}{\Delta t}=\frac{\boxed{④}}{3-1}=\frac{\boxed{⑤}}{2}=\boxed{⑥}$$

$$\frac{\Delta y}{\Delta x}=\frac{\frac{\Delta y}{\Delta t}}{\frac{\Delta x}{\Delta t}}=\frac{\boxed{⑦}}{\boxed{\ }}=\boxed{⑧}$$

EXAMPLE 2. Find the average rate of change of the path in the given interval.

① $x=3t+2, y=t+1$; $0\le t\le 4$

② $x=2t-3,\ y=4-6t$; $1\le t\le 3$

③ $x=1+t,\ y=t^2-4t$; $-2\le t\le 2$

④ $x=\sqrt{t},\ y=3t+1$; $1\le t\le 4$

Blank : ① $\frac{\Delta y}{\Delta x}$ ② 3 − 1 ③ 1 ④ x(3) − x(1) ⑤ 4 ⑥ 2 ⑦ $\frac{1}{2}$ ⑧ $\frac{1}{2}$

EXAMPLE 3. Which of the following would have the same average rate of change as $f(t)=(2t-3,\ 4-5t)$?

I. $f(t)=(4t-1, 10t+20)$

II. $f(t)=\left(\frac{3}{5}t+8,\ \frac{7}{2}-\frac{3}{2}t\right)$

III. $f(t)=(5t+7, 2t+7)$

8.3 Lines and Circles in Parametric Form

1. Lines in Parametric Form

Parameterization: Converting a Cartesian equation to parametric equation.

※ **Parameterization of line**

If we have a point (a, b) (= initial condition) and slope $m = \dfrac{\Delta y}{\Delta x} = \dfrac{\frac{\Delta y}{\Delta t}}{\frac{\Delta x}{\Delta t}}$

a **line** can be parameterized by

$$x(t) = \frac{\Delta x}{\Delta t}t + a \quad , \quad y(t) = \frac{\Delta y}{\Delta t}t + b$$

(from point: a, b; from slope: $\frac{\Delta x}{\Delta t}$, $\frac{\Delta y}{\Delta t}$)

ex) Find a parameterization of a linear path that starts at $(-4,3)$ and has a slope of 2.

The starting point $(-4,3)$ gives us $x(0) =$ ① ______, $y(0) =$ ② ______

And *slope* $2 = \dfrac{\Delta y}{\Delta x} = \dfrac{\frac{\Delta y}{\Delta t}}{\frac{\Delta x}{\Delta t}} = \dfrac{③\square}{\square}$, so $\dfrac{\Delta x}{\Delta t} =$ ④ $\square$, $\dfrac{\Delta y}{\Delta t} =$ ⑤ $\square$

(pick two numbers which has ratio of 2)

Therefore, a line can be parameterized by ⑥ ________________________________.

Blank : ① -4 ② 3 ③ $\frac{2}{1}$ ④ 1 ⑤ 2 ⑥ x=t–4, y=2t+3

EXAMPLE 1. Find a parameterization of a linear path that;

① starts at $(-4,3)$ and has a slope of $-\frac{1}{3}$

② starts at $(-4,3)$ and has a slope of 2

③ Passes through (2, 5) and (4, -8)

④ Passes through $(5,0)$ and $(-2,-3)$

2. Circles in Parametric Form

What is the Cartesian equation of $x=2\cos t-1, y=2\sin t+3$?

Since $\cos t=$ ①______________ and $\sin t=$ ②______________,

If we use the trig identity $\sin^2 t+\cos^2 t=1$,

then we will have ③__________________________.

Blank : ① $\frac{x+1}{2}$ ② $\frac{y-3}{2}$ ③ $\frac{(x+1)^2}{2^2}+\frac{(y-3)^2}{2^2}=1 \Rightarrow (x+1)^2+(y-3)^2=2^2$

※ **Parameterization of Circle**

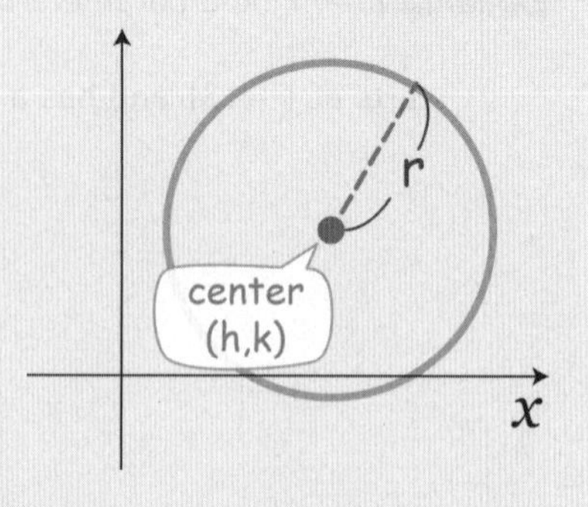

If a circle has a center at point (h, k) and radius r

a **circle** can be parameterized by

center (h,k)

$$x(t) = r\cos t + h, \quad y(t) = r\sin t + k$$

radius r

EXAMPLE 2. Find a parameterization of a circular path that;

① has a center at $(-4, 3)$ and radius 3

② has a center at $(11, -5)$ and radius 7

③ has a center at $(0, 3)$ and radius 7

④ has a center at $(-5, 0)$ and radius 4

⑤ $(x+1)^2+(y-2)^2=25$

⑥ $(x+3)^2+y^2=9$

⑦ unit circle

EXAMPLE 3. Graph the following

① $x=\cos t-1,\ y=\sin t+3$

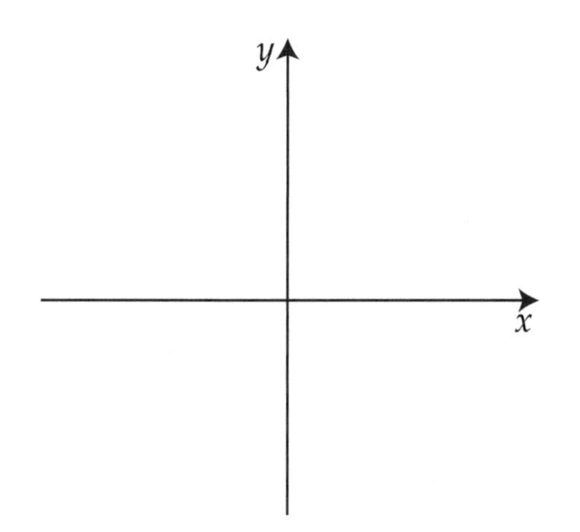

② $x=\cos t+1,\ y=\sin t-1$

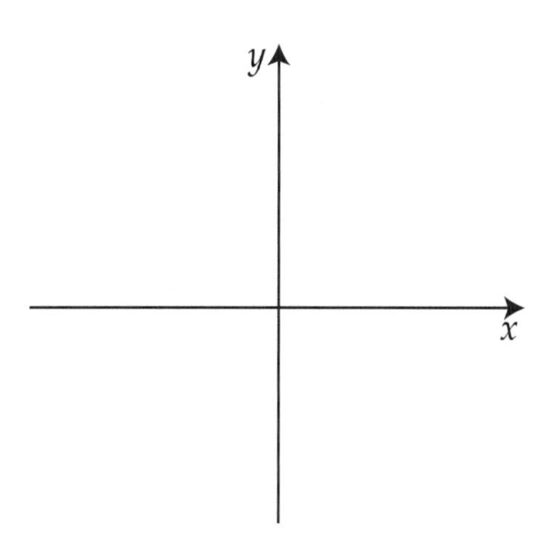

③ $x = 2\cos t,\ y = 2\sin t$

④ $x = 5\cos t - 2,\ y = 5\sin t + 3$

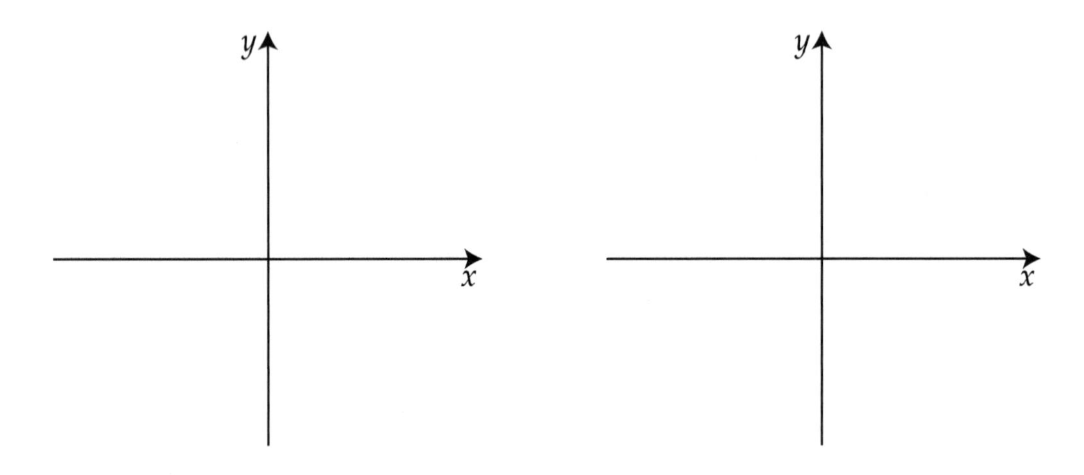

Part 9

Conic Section

9.1 Conic Sections and Parabolas

1. Conic Section

What shape do we have if we cut a cone ..

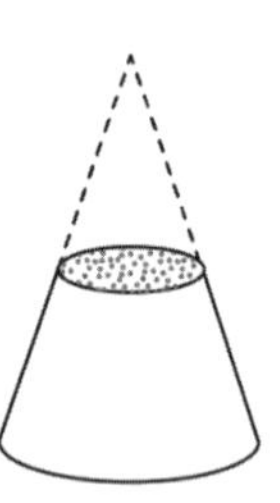
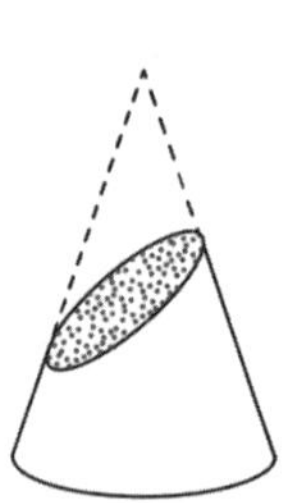
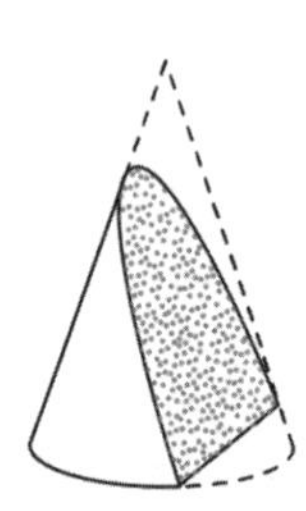
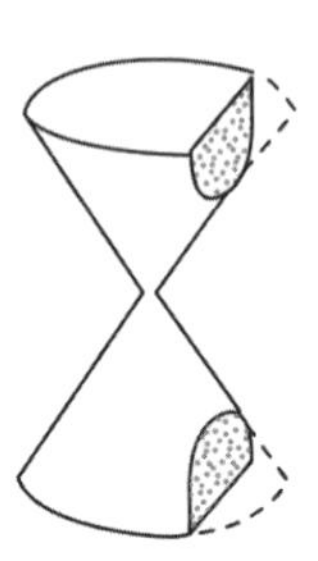

straight through	slight angle	parallel to edge of cone	steep angle
:① ____________	:② ______________	:③ ______________	:④ ______________

These curves are related!

2. Parabola

A ⑤ ________________ is a curve where any point is at an **equal distance** from *a straight line*(⑥ ___________________) and *a fixed point*(⑦ _________________).

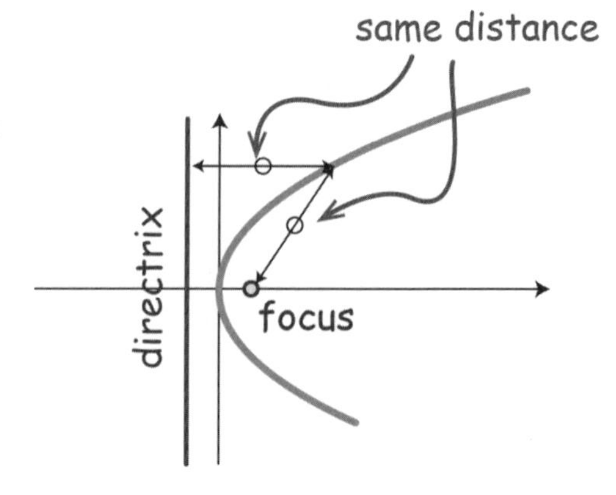

Blank : ① circle ② ellipse ③ parabola ④ hyperbola ⑤ parabola ⑥ directrix ⑦ focus

☺Vocabulary

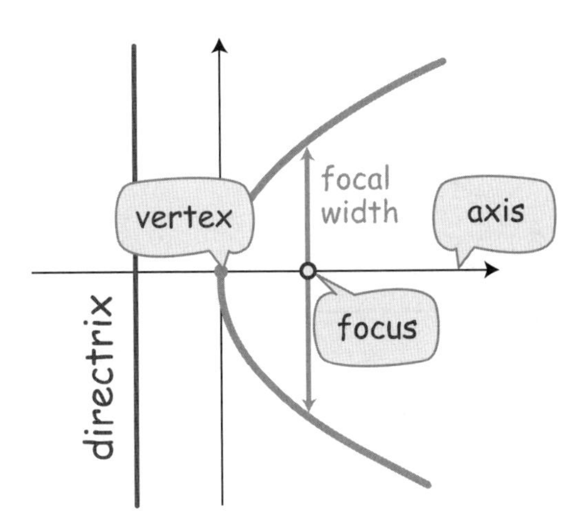

•① _______________ :a fixed point

•② _______________ :a fixed straight line

•③ _______________ : turning point

•④ _______________ of Symmetry

•⑤ ____________________ :the distance from the vertex to the focus of the parabola.

•⑥ ____________________ (= focal diameter)

:a line that passes through the focus and has endpoints on the parabola.

Blank : ① focus ② directrix ③ vertex ④ axis ⑤ focal length ⑥ focal width

3. Equation of Parabola

A **parabola** is the set of all points in a plane that are equidistant from a fixed line, the directrix , and a fixed point, the focus.

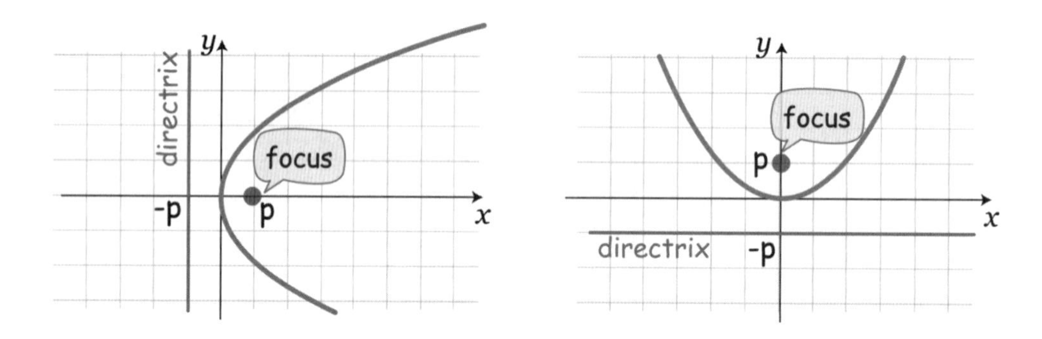

$4px = y^2$	Formula	$4py = x^2$
①	**..it opens..**	up
②	**Focus**	(0, p)
③	**Directrix**	y = -p
④	**Vertex**	(0, 0)
⑤	**Axis of Symmetry**	y-axis
⑥	**Focal length**	$\lvert p \rvert$
⑦	**Focal width**	$\lvert 4p \rvert$

Blank : ① to the right ② (p, 0) ③ x = -p ④ (0, 0) ⑤ x axis ⑥ |p| ⑦ |4p|

EXAMPLE 1. Graph the parabola. Label the focus, directrix and focal width.

① $y = \frac{1}{4}x^2$

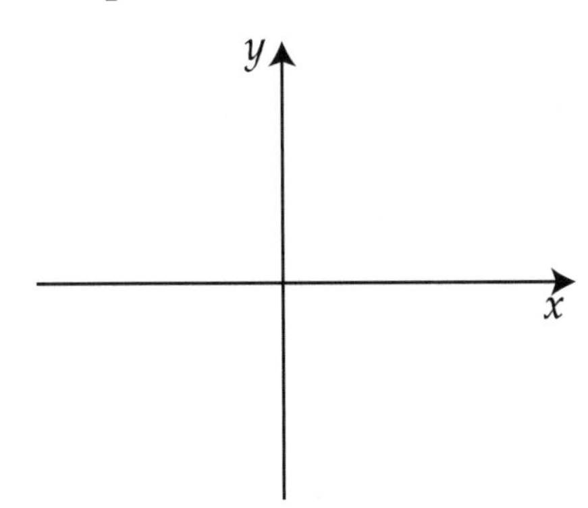

Focus:

directrix:

focal width:

② $16y = x^2$

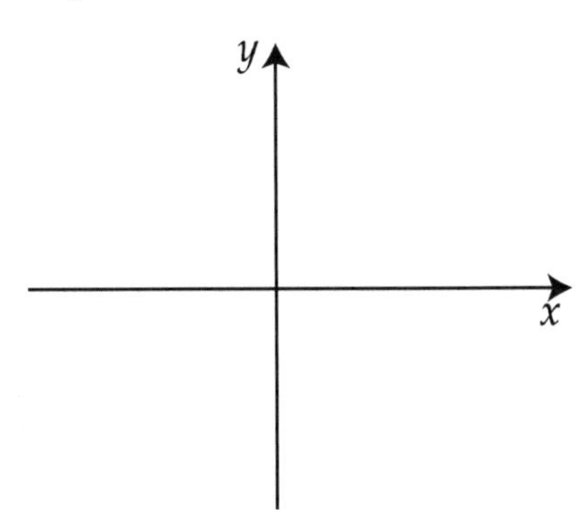

Focus:

directrix:

focal width:

③ $x = 8y^2$

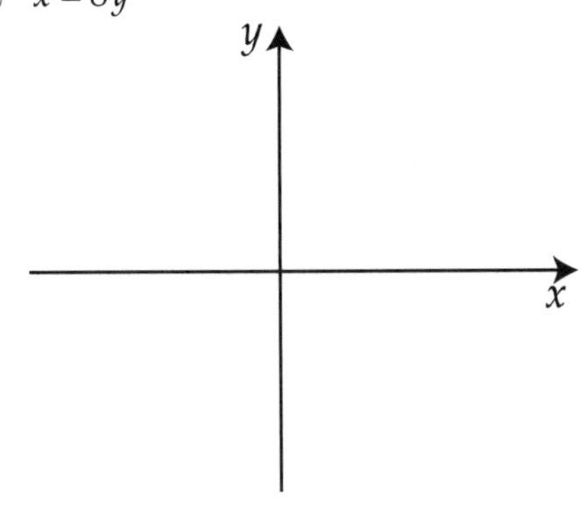

Focus:

directrix:

focal width:

④ $8x = y^2$

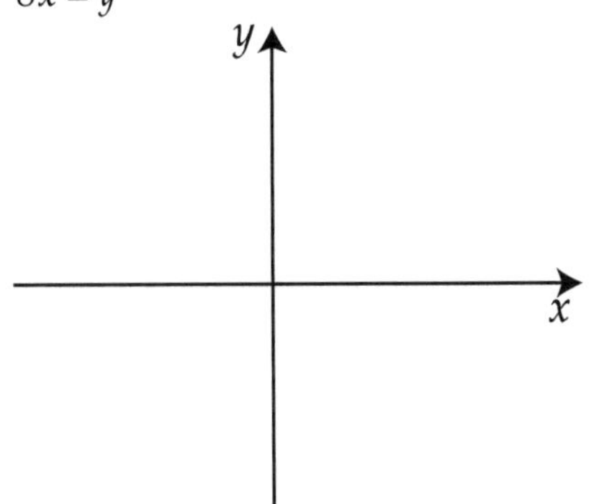

Focus:

directrix:

focal width:

⑤ $y^2+6x=0$

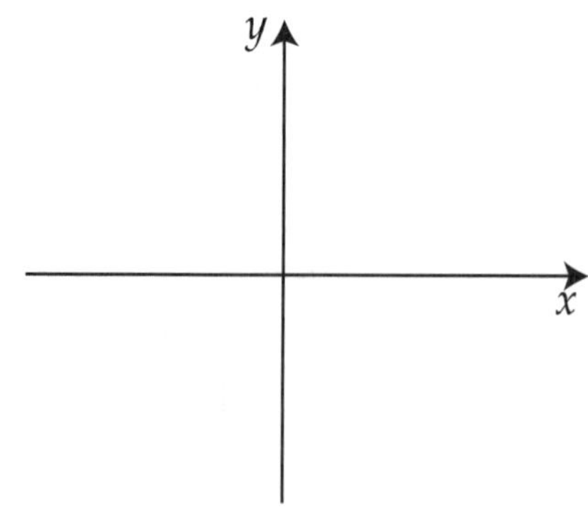

Focus:

directrix:

focal width:

⑥ $0=x^2+16y$

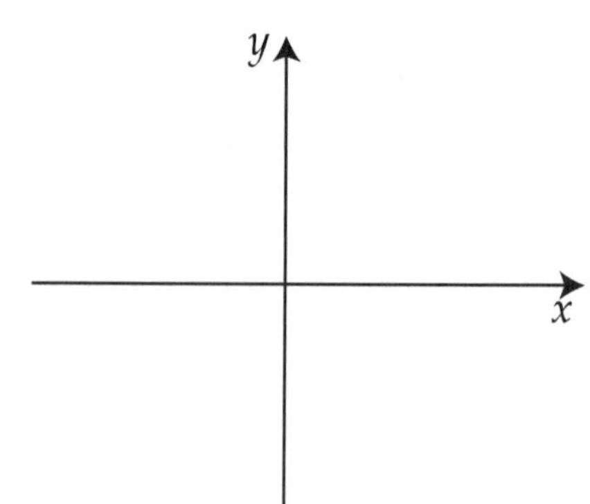

Focus:

directrix:

focal width :

EXAMPLE 2. Find the standard form of the equation for the parabola.

① Vertex at the origin, focus at (0, -2)

② Focus at (0, 3), directrix y = -3

③ Focus at (7, 0), directrix x = -7

④ Vertex at the origin, focus at (-8, 0)

⑤ Vertex at the origin, opens to the right, focal width = 12

⑥ Vertex at the origin, opens downward, focal width = 16

⑦

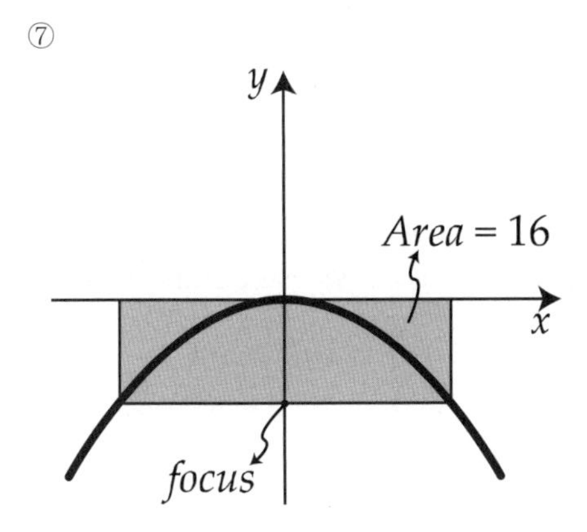

9.2 Ellipses

1. Ellipses

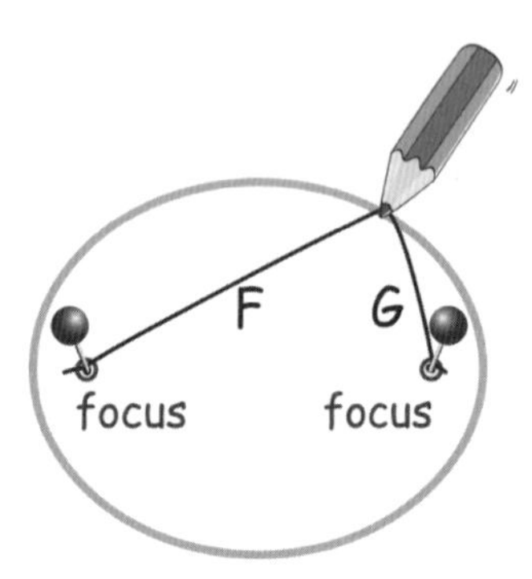

An ①__________ usually looks like a squashed circle.

We can draw an ellipse using a string whose ends are attached to two nails. The pencil is moved all the way around while always keeping the string tight.

(The length of the sting F + G always be the ②______)

※ Geometric Definition of Ellipse

An **ellipse** is the set of all points (x, y) in a plane, the sum of whose distances from two distinct fixed point (foci) is constant.

☺Vocabulary

•③______________ :The points at which an ellipse makes its sharpest turns.

•④______________ :two fixed points, together they are called ⑤_________.(pronounced "fo-sigh")

•⑥_________________ :the longest diameter (at the widest part of the ellipse).

•⑦_________ ________ :the shortest diameter(at the narrowest part of the ellipse).

Blank : ① ellipse ② same ③ vertex ④ focus ⑤ foci ⑥ major axis ⑦ minor axis

2. Equation of Ellipses

By placing an ellipse on an x-y plane, the equation of the curve is:

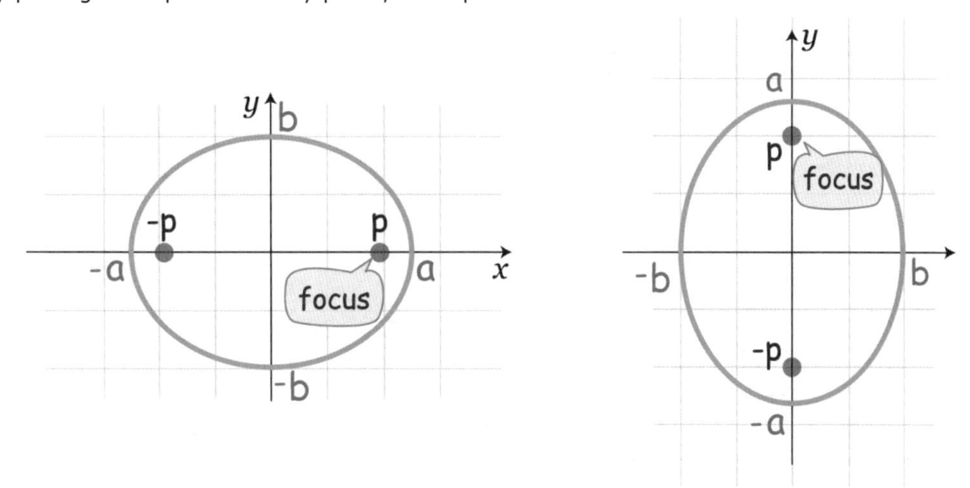

$\frac{x^2}{a^2}+\frac{y^2}{b^2}=1$ vertex	**Standard Equation**	$\frac{y^2}{a^2}+\frac{x^2}{b^2}=1$ vertex
①	**center**	(0, 0)
②	**Foci**	(0, ±p)
③	**Vertices**	(0, ±a)
④	**major Axis**	|2a|
⑤	**minor Axis**	|2b|
⑥	**Pythagorean Relation**	$p=\sqrt{a^2-b^2}$

Blank : ① (0, 0) ② (±p, 0) ③ (±a, 0) ④ |2a| ⑤ |2b| ⑥ $p=\sqrt{a^2-b^2}$

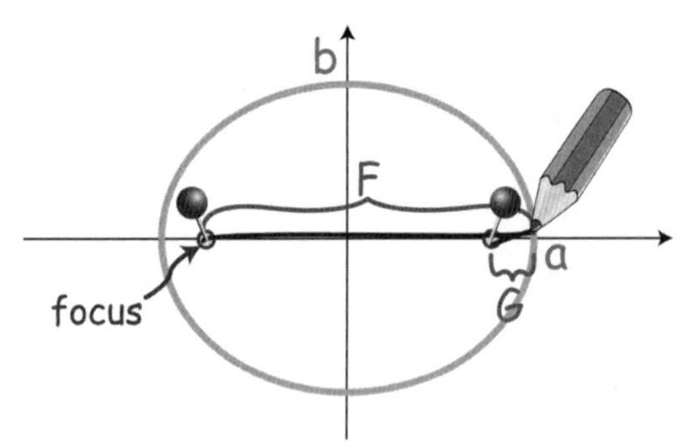

While we are drawing an ellipse if you fixed a pencil on a (=the vertex) *then we can say*

F + G = ①_______

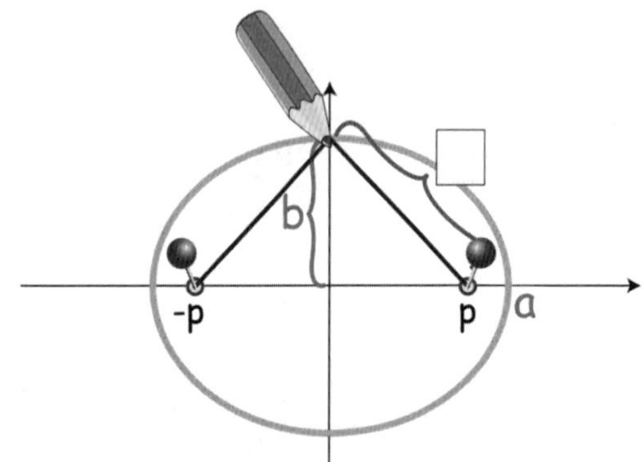

If *you fix a pencil on* b then we can say

$p^2 =$ ② $-$ ③

$p =$ ④

※ **Pythagorean Relation in Ellipse** (helps to find foci of ellipse)

$$\text{focus } p = \sqrt{a^2 - b^2}$$

3. How to graph Ellipses

EXAMPLE 1. Graph the ellipse. Label the Foci, vertices.

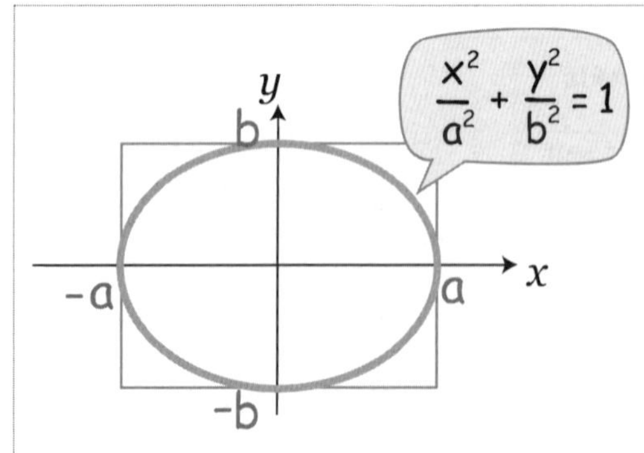

i) Mark points a units along x directions from the center and points b units along y directions from the center.

ii) Draw an ellipse through these points

Blank : ① 2a ② a^2 ③ b^2 ④ $\sqrt{a^2 - b^2}$

① $\frac{x^2}{16}+\frac{y^2}{25}=1$

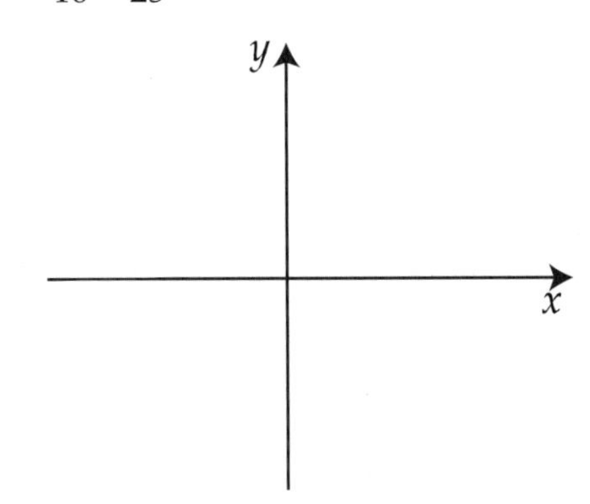

Foci:

Vertices:

② $\frac{x^2}{49}+\frac{y^2}{3}=1$

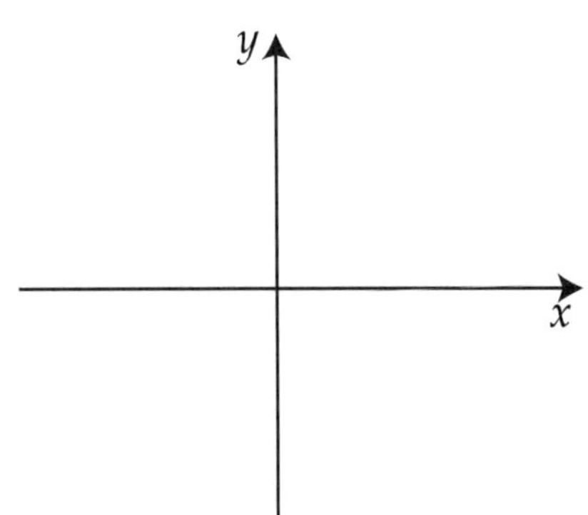

Foci:

Vertices:

③ $4x^2+16y^2=64$

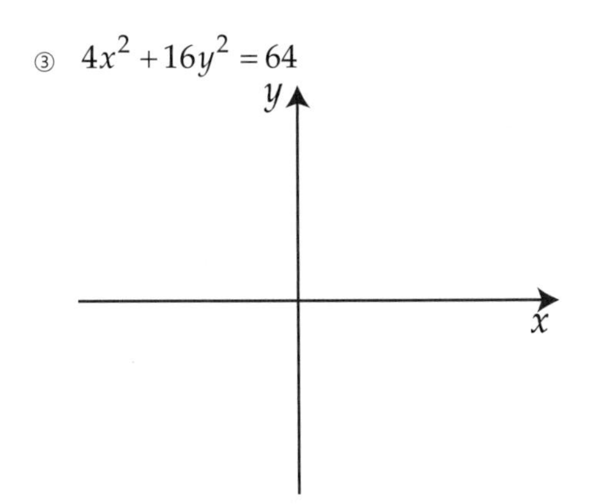

Foci:

Vertices:

④ $16x^2+9y^2=144$

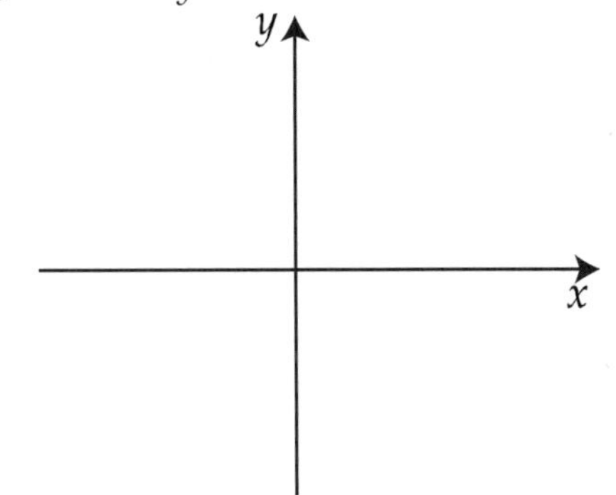

Foci:

Vertices:

⑤ $16x^2 + 4y^2 = 1$

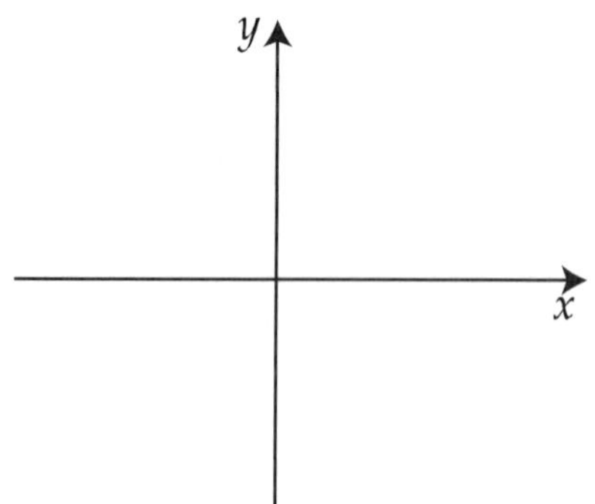

Foci:

Vertices:

⑥ $x^2 + 9y^2 = 1$

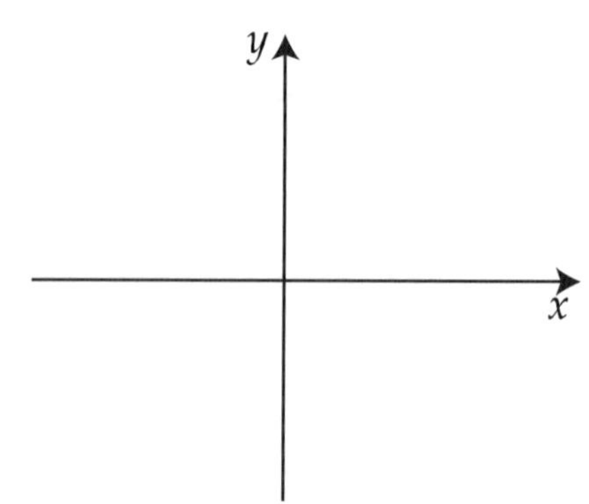

Foci:

Vertices:

⑦ $20y^2 = 5 - 4x^2$

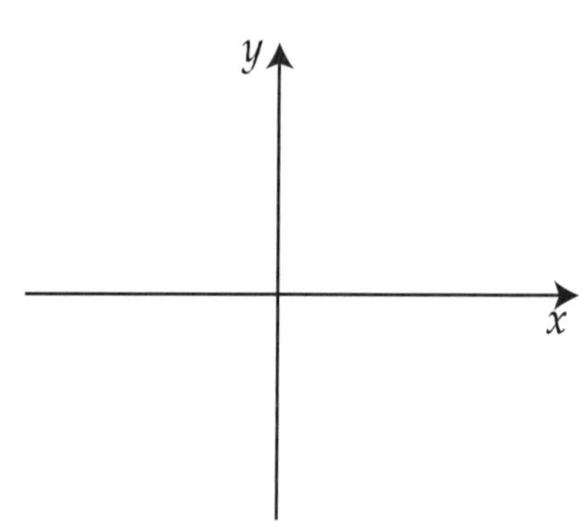

Foci:

Vertices:

EXAMPLE 2. Find the standard form of the equation for the ellipse.

① Vertices at $(\pm10, 0)$ and foci at $(\pm5, 0)$

② Vertices at $(\pm4, 0)$ and foci at $(\pm3, 0)$

③ Major axis is of length 16 and foci at $(0, \pm2)$

④ Major axis is of length 6 and foci at $(0, \pm2)$

⑤ Minor axis is of length 8 and foci at $(\pm3, 0)$

⑥ Minor axis is of length 10 and foci at $(0, \pm7)$

⑦ Minor axis endpoints $(\pm 2, 0)$, major axis length 22

⑧ An ellipse with intercepts $(\pm 3, 0)$ and $(0, \pm 8)$, center at origin

9.3 Hyperbolas

1. Hyperbolas

※ **Geometric Definition of Hyperbola**

An **Hyperbola** is the set of all points (x, y) in a plane, the difference of whose distances from two distinct fixed point (foci) is constant.

☺Vocabulary

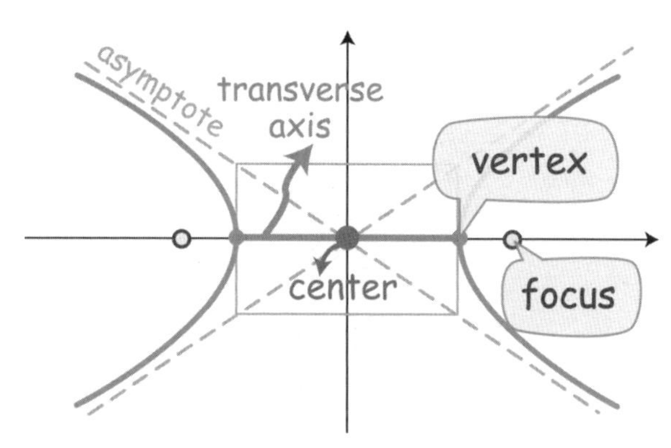

•① ______________ : The points at which an hyperbola makes a turns.

•② ______________ : two fixed points, together they are called ③__________.

•④ ______________ axis : line goes through vertex to another vertex

Blank : ① vertex ② focus ③ foci ④ transverse axis

2. Equation of Hyperbolas

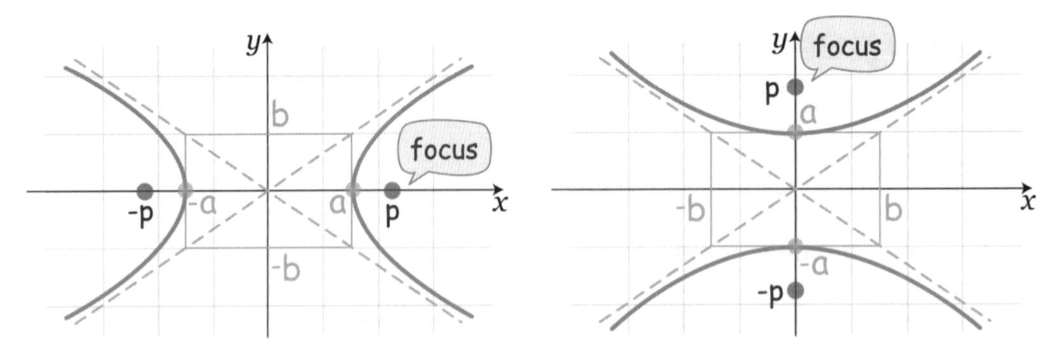

$\frac{x^2}{a^2} - \frac{y^2}{b^2} = 1$ (x) vertex	**Standard Equation**	$\frac{y^2}{a^2} - \frac{x^2}{b^2} = 1$ (y) vertex
①	**center**	(0, 0)
②	**Foci**	(0, ±p)
③	**Vertices**	(0, ±a)
④	**Transverse axis**	|2a|
⑤	**Pythagorean Relation**	$p=\sqrt{a^2+b^2}$
⑥	**Asymptotes**	$y=\pm\frac{a}{b}x$

Blank : ① (0, 0) ② (±p, 0) ③ (±a, 0) ④ |2a| ⑤ $p=\sqrt{a^2+b^2}$ ⑥ $y=\pm\frac{b}{a}x$

☺ Other way of finding Asymptotes

Asymptotes can be found by replacing the 1 by ①________:

$$\frac{x^2}{a^2}-\frac{y^2}{b^2}=1 \rightarrow \frac{x^2}{a^2}-\frac{y^2}{b^2}=0 \rightarrow y= \boxed{②}$$

$$\frac{y^2}{a^2}-\frac{x^2}{b^2}=1 \rightarrow \frac{y^2}{a^2}-\frac{x^2}{b^2}=0 \rightarrow y= \boxed{③}$$

$$\frac{x^2}{a^2}-\frac{y^2}{b^2}=0$$

switch to 0

3. How to graph Hyperbolas

EXAMPLE 1. Graph the hyperbola. Label the Foci, vertices, and asymptotes.

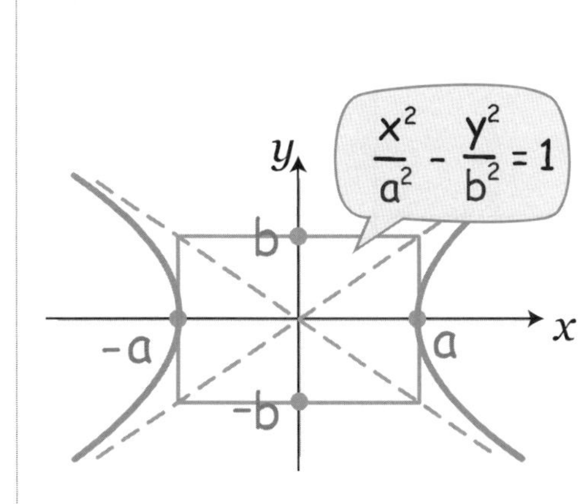

i) Mark points a units along x directions from the center and points b units along y directions from the center.

ii) draw a rectangle and asymptotes as shown

ii) Draw a hyperbola. (shape depends on the first term)

Blank : ① 0 ② $y=\pm\frac{b}{a}x$ ③ $y=\pm\frac{a}{b}x$

① $\frac{x^2}{25}-\frac{y^2}{36}=1$

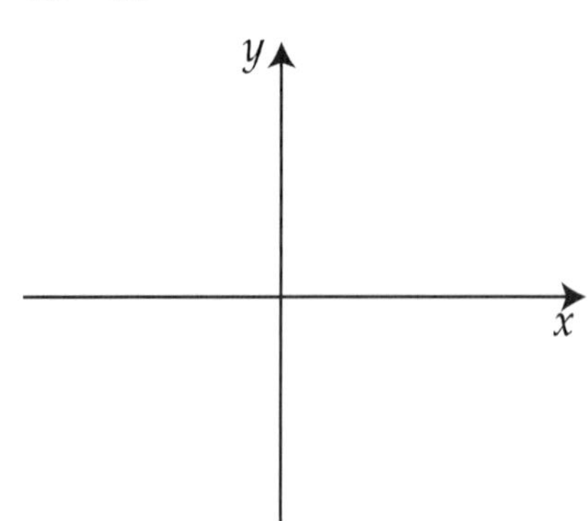

Foci:

Vertices:

Asymptotes:

② $\frac{x^2}{16}-\frac{y^2}{4}=1$

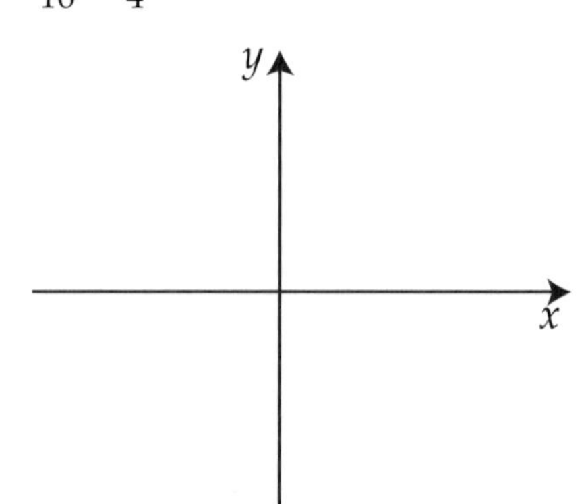

Foci:

Vertices:

Asymptotes:

③ $16y^2-x^2=1$

Foci:

Vertices:

Asymptotes:

④ $y^2-9x^2=1$

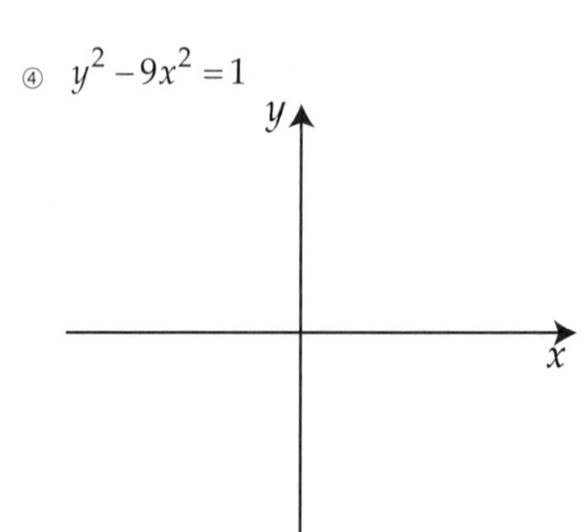

Foci:

Vertices:

Asymptotes:

⑤ $x^2 - 4y^2 = 5$

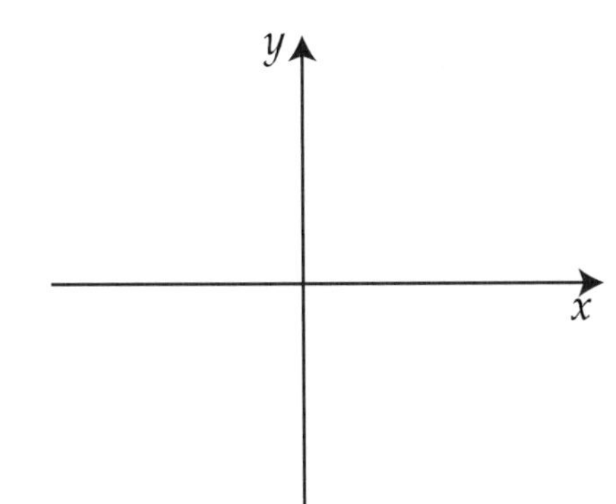

Foci:

Vertices:

Asymptotes:

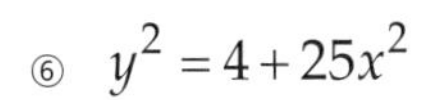

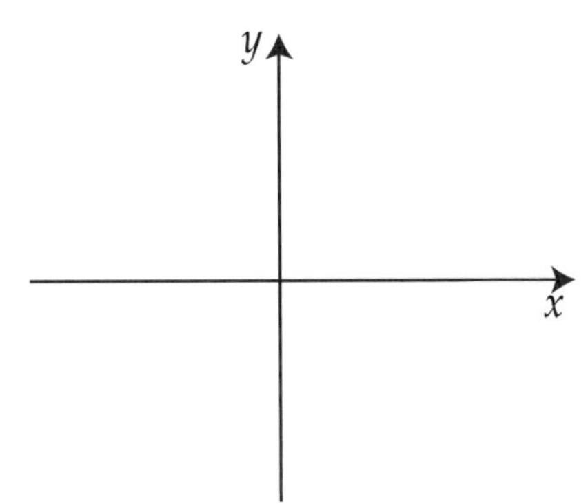

Foci:

Vertices:

Asymptotes:

EXAMPLE 2. Find the standard form of the equation for the hyperbola.

① Vertices at (±2, 0), foci at (±4, 0)

② Vertices at (0, ±4), foci at (0, ±10)

③ Foci at $(0, \pm 9)$, transverse axis with length 10

④ Foci at $(\pm 6, 0)$, transverse axis with length 4

⑤ Vertices at $(0, \pm 4)$, asymptotes at $y = \pm\frac{2}{5}x$

⑥ Vertices at $(\pm 6, 0)$, asymptotes at $y = \pm\frac{2}{3}x$

9.4 Transformation of Conics

1. Transformation of Conic Sections

☺ Reminder

Find the center of $x^2+y^2-10x+8y+16=0$. ①____________________________

※ **Transformation of Conic Sections**

x replaced by x - h

y replaced by y - k

then the center (0, 0) is shifted to (h, k)

2. Transformation of Parabola

※ **Transformation of Parabola**

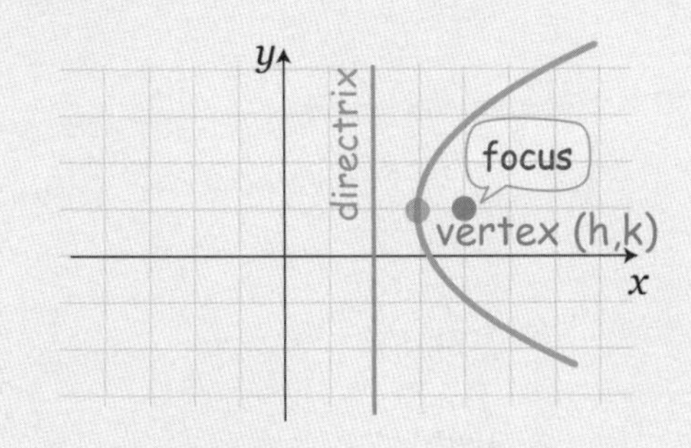

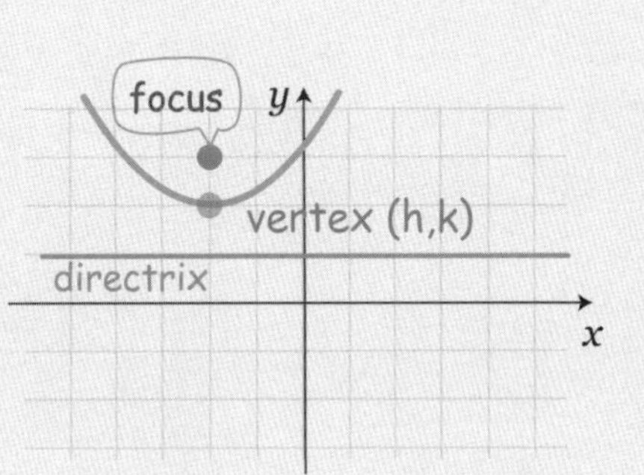

focus

$4p(x-h)=(y-k)^2$

x

focus

$4p(y-k)=(x-h)^2$

y

Blank : ① $(x-5)^2+(x+4)^2=5^2$, center : (5, -4)

EXAMPLE 1. Find the vertex, foci and directrix of the parabola.

① $(x-2)^2=8(y-3)$

Vertex:

Focus:

directrix:

② $(y-5)^2=16(x+2)$

Vertex:

Focus:

directrix:

③ $x=-\frac{1}{2}(y+7)^2-2$

Vertex:

Focus:

directrix:

④ $y=-(x-3)^2-1$

Vertex:

Focus:

directrix:

EXAMPLE 2. Find the standard form of the equation for the parabola.

① Focus at (3, -9), directrix $x = -1$

② Focus at (4, -9), directrix $y = -15$

③ Vertex at (8, 3), opens upward, focal width = 10

④ Vertex at (1, -4), opens to the left, focal width = 16

3. Transformation of Ellipses

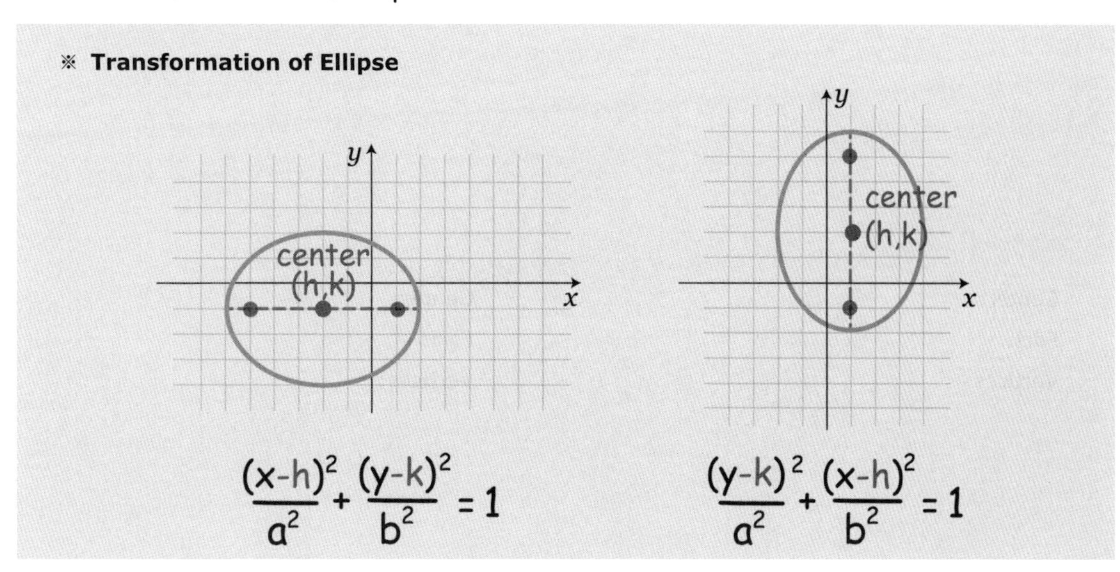

EXAMPLE 3. Find the center, foci, and vertices of the ellipse.

① $\dfrac{(x+4)^2}{16}+\dfrac{(y-2)^2}{25}=1$

Center:

Foci:

Vertices:

② $\dfrac{(x-2)^2}{36}+\dfrac{(y+1)^2}{9}=1$

Center:

Foci:

Vertices:

③ $4(x-2)^2+8(y+1)^2=2$

Center:

Foci:

Vertices:

④ $4(x+3)^2+9(y-2)^2=1$

Center:

Foci:

Vertices:

EXAMPLE 4. Find the standard form of the equation for the ellipse.

① An ellipse with foci at (2, 5) and (2, -1); major axis length of 10

② An ellipse with foci at (-1, 1) and (5, 1); major axis length of 10

③ An ellipse with major axis from (-1, -4) to (7, -4); minor axis from (3, -7) to (3, -1).

4. Transformation of Hyperbolas

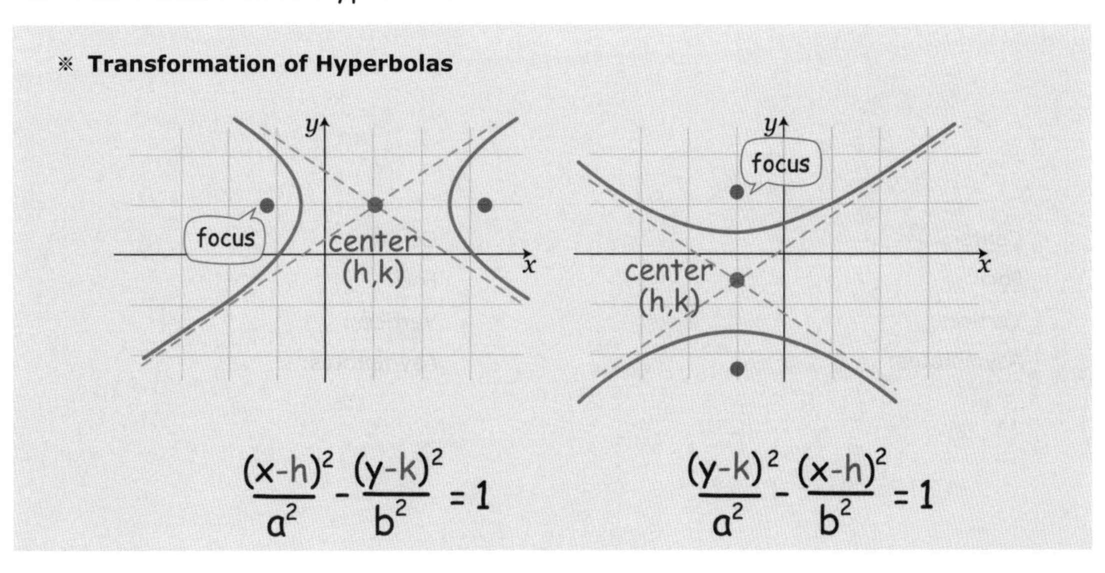

EXAMPLE 5. Find the center, foci, vertices, and asymptotes of the hyperbola.

① $\dfrac{(x+5)^2}{16}-\dfrac{(y-1)^2}{9}=1$

Center:
Foci:
Vertices:
Asymptotes:

② $\dfrac{(x+4)^2}{4}-\dfrac{(y+4)^2}{25}=1$

Center:
Foci:
Vertices:
Asymptotes:

③ $4(y-5)^2-(x-1)^2=16$

Center:
Foci:
Vertices:
Asymptotes:

④ $4(y-3)^2-(x-4)^2=1$

Center:
Foci:
Vertices:
Asymptotes:

EXAMPLE 6. Find the standard form of the equation for the hyperbola.

① Center (4, -3), focus (11, -3), vertex (6, -3)

② Center (-2, 5), focus (-2, 12), vertex (-2, 8)

③ Vertices at (7, -2), (5, -2), asymptotes at $y = 4x - 26, y = -4x + 22$

④ Vertices at (-1, 9),(-1, 3), asymptotes at $y - 6 = \pm\frac{3}{2}(x+1)$

5. Identifying Conics

※ **Identifying Conics**

$$Ax^2 + Cy^2 + Dx + Ey + F = 0$$

If $AC = 0$, then it is parabola. (If A or C is 0)

If $AC > 0$, then it is Ellipse. (If A,C have same sign.)

If $AC < 0$, then it is hyperbola. (If A,C have different sign.)

EXAMPLE 7. Determine whether the equation represents an ellipse, a parabola, a hyperbola. Then find the vertex (or vertices) and focus(or foci).

① $x^2+6x-4y+1=0$ ② $y^2+2y-x=0$

③ $2x^2+3y^2-8x+6y+5=0$

④ $4x^2+3y^2+8x-6y=5$

⑤ $x^2-y^2-2x-2y-1=0$

⑥ $y^2-4x^2-4y-8x-4=0$

9.5 Conic Sections in Parametric Form

1. Parabola in Parametric Form

Parameterization: Converting a Cartesian equation to parametric equation.

※ **Parameterization of parabola**

Parabolas can be parameterized by solving for x or y;

If you solve for x, replace y with t and write $(x(t), y(t)) = (f(t), t)$

If you solve for y, replace x with t and write $(x(t), y(t)) = (t, f(t))$

ex) Find a parameterization of a parabola $y-4=3(x+2)^2$

If we solve for y; ①____________________.

Replace x with t and write $(x(t), y(t)) = (t, f(t))$

Therefore, it can be parameterized by ②____________________.

EXAMPLE 1. Parameterize the given parabola.

① $x+(y+2)^2=0$

② $y-(x-1)^2=0$

③ $(x+2)^2=\frac{1}{2}(y+1)$

④ $x-3=2(y-2)^2$

Blank : ① $y = 4+3(x+2)^2$ ② $x=t,\ y=4+3(t+2)^2$

2. Ellipse in Parametric Form

☺ Reminder: Identify the parametric equation $x=2\cos t-1, y=2\sin t+3$.

What is the Cartesian equation of $x=2\cos t-1, y=4\sin t+3$?

Since $\cos t=$ ① ____________ and $\sin t=$ ② ____________,

If we use the trig identity $\sin^2 t+\cos^2 t=1$,

then we will have ③ ____________________.

※ **Parameterization of Ellipse**

If an ellipse has a center at point (h, k), horizontal radius a, and vertical radius b, an **ellipse** can be parameterized by

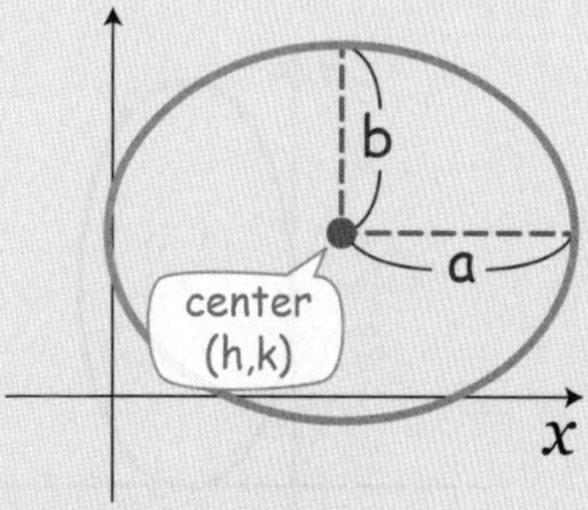

$$x(t)=a\cos t+h,\quad y(t)=b\sin t+k$$

EXAMPLE 2. Parameterize the given ellipse.

① has a center at $(-4,3)$, horizontal radius 3 and vertical radius 5

② has a center at $(0,-5)$, horizontal radius 2 and vertical radius 7

Blank : ① $\frac{x+1}{2}$ ② $\frac{y-3}{4}$ ③ $\frac{(x+1)^2}{2^2}+\frac{(y-3)^2}{4^2}=1$

③ $(x+2)^2+9y^2=9$

④ $4(y-2)^2=100-25(x-5)^2$

⑤

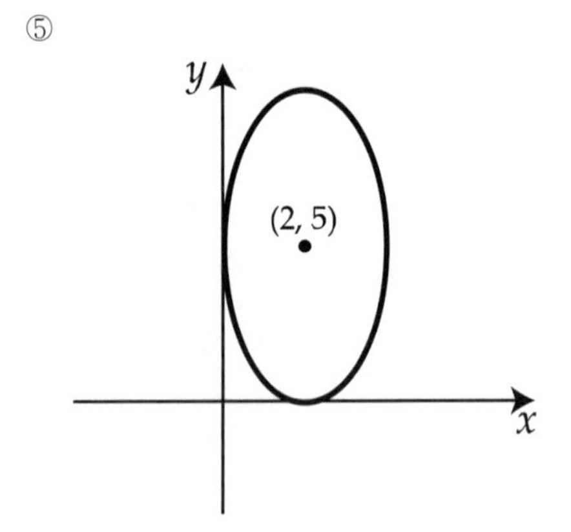

⑥

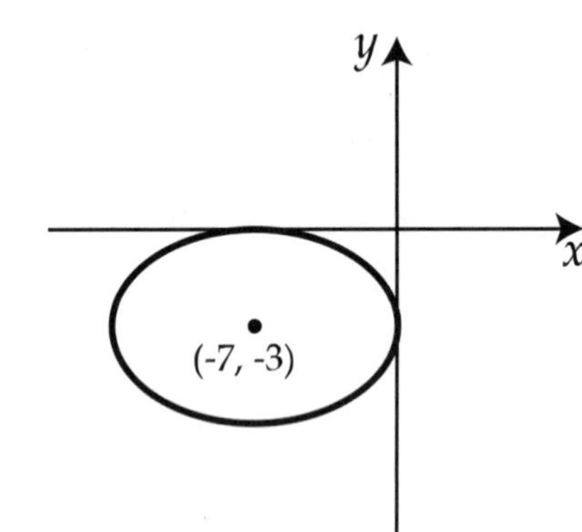

EXAMPLE 3. Graph the following.

① $x=2\cos t-2,\ \ y=3\sin t$

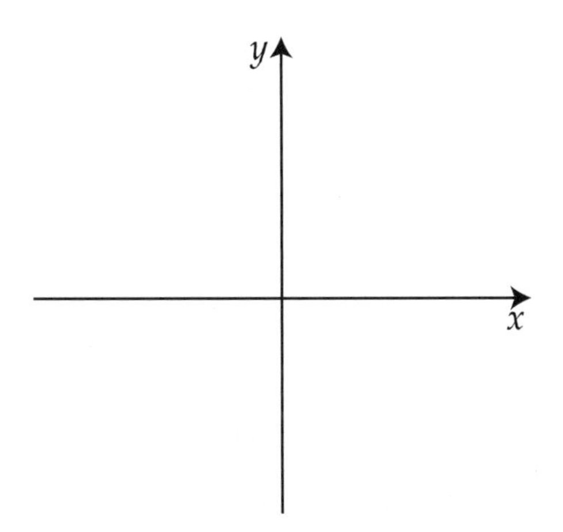

② $x=3\cos t+1,\ \ y=\sin t-3$

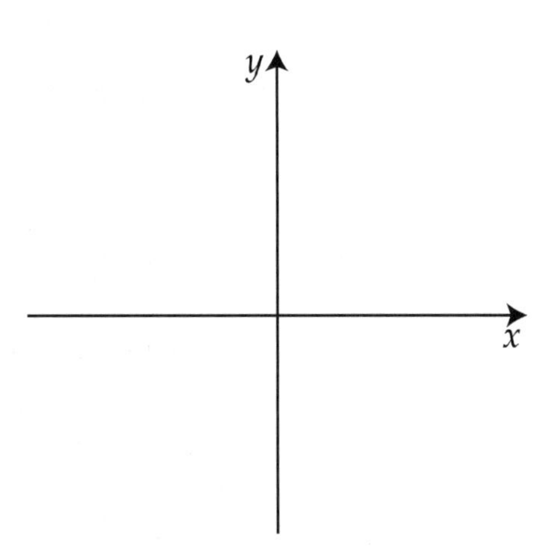

③ $x=4\cos t-1,\ \ y=3\sin t-3$

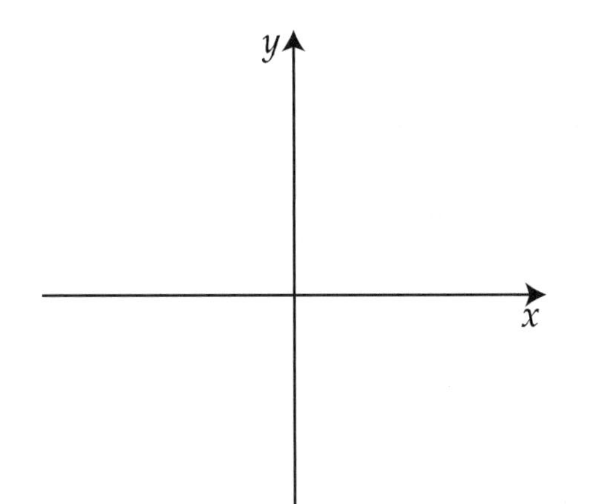

④ $x=\cos t-1,\ \ y=4\sin t$

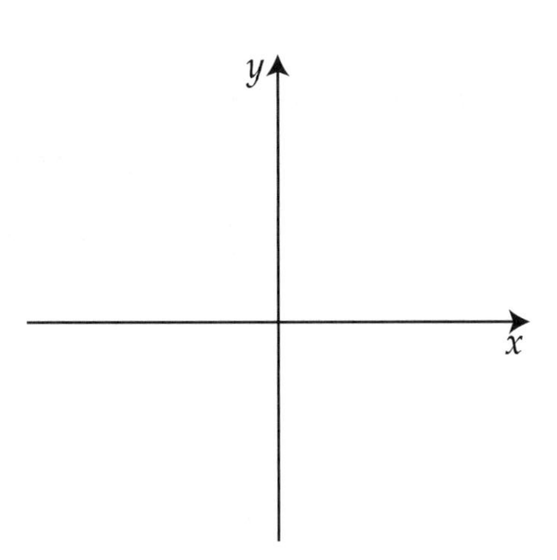

3. Hyperbola in Parametric Form

What is the Cartesian equation of $x = 2\sec t - 1, y = 4\tan t + 3$?

Since $\sec t =$ ① __________ and $\tan t =$ ② __________,

If we use the trig identity $1 + \tan^2 t = \sec^2 t$,

then we will have ③ ____________________.

※ **Parameterization of Hyperbola**

A **hyperbola** can be parameterized by ;

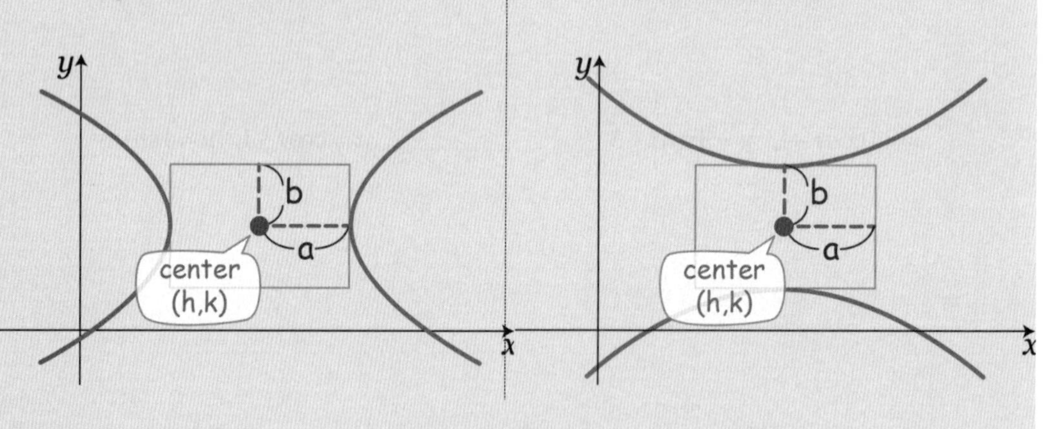

☺ Easy way to remember:

If $\sec t$ is at x, then the hyperbola is in shape of 

If $\sec t$ is at y, then the hyperbola is in shape of 

Blank : ① $\frac{x+1}{2}$ ② $\frac{y-3}{4}$ ③ $1 + \frac{(y-3)^2}{4^2} = \frac{(x+1)^2}{2^2} \Rightarrow \frac{(x+1)^2}{2^2} - \frac{(y-3)^2}{4^2} = 1$

EXAMPLE 4. Parameterize the given hyperbola.

① $\frac{(y-2)^2}{4}-\frac{(x+1)^2}{9}=1$

② $\frac{y^2}{9}-\frac{(x-5)^2}{36}=1$

③ $4x^2=36+9(y+3)^2$

④ $(x+3)^2-4(y-2)^2=16$

EXAMPLE 5. Graph the following.

① $x=2\sec t+1,\ y=3\tan t-3$

② $x=2\tan t+2,\ y=3\sec t-5$

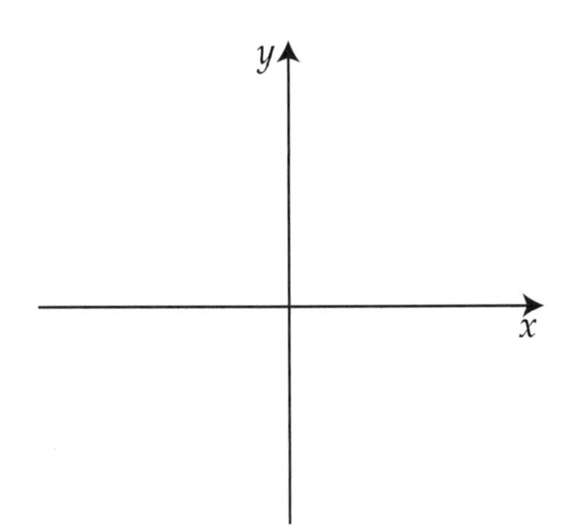

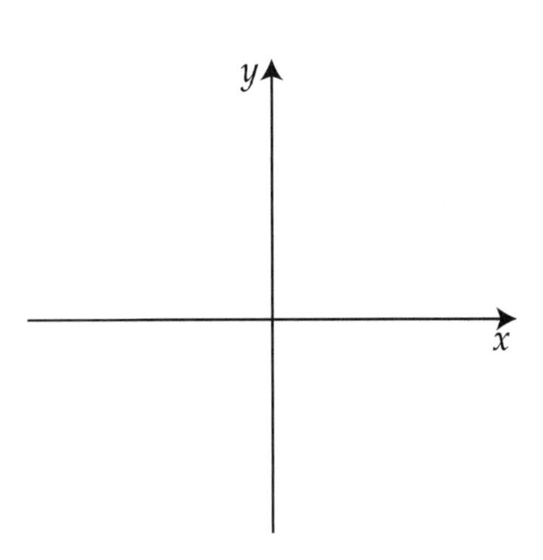

③ $x=\tan t-1,\ \ y=4\sec t-2$

④ $x=3\sec t,\ \ y=4\tan t+4$

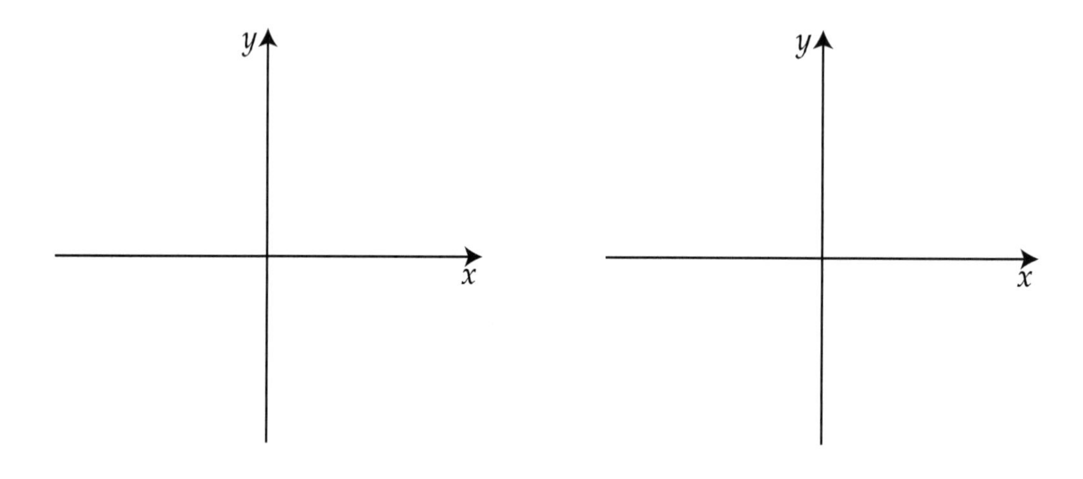

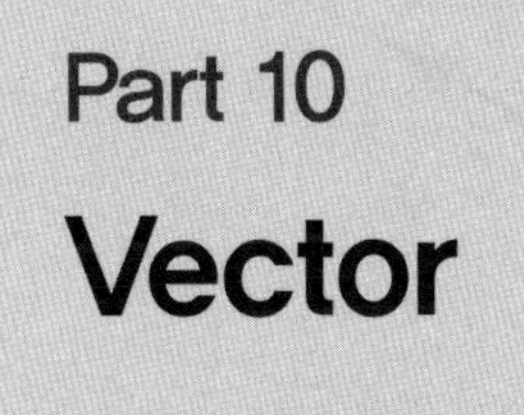

Part 10

Vector

10.1 Vectors in Two Dimensions

1. Vector

A **vector** has ①______________ (how long it is) and ②________________:

A **scalar** has only ③______________ (no directions).

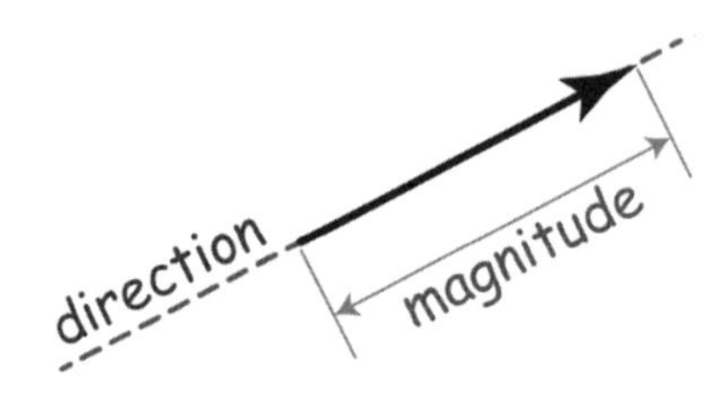

The length of the line shows its *magnitude*
and the arrowhead points in the *direction*.

- Notation :④ ____________________

- Magnitude notation : ⑤__________

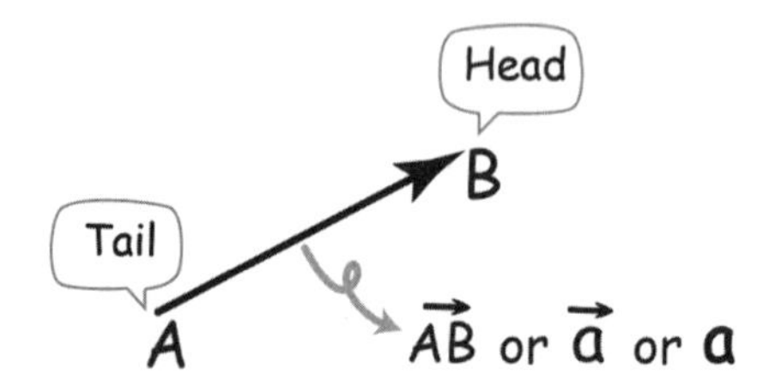

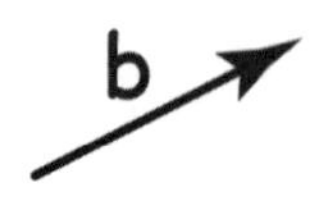

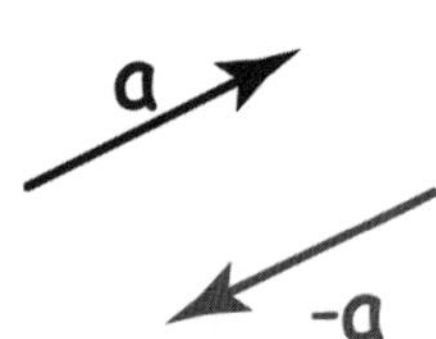

- Opposite vector : $-\vec{a}$

two vectors that have **equal length** and the **opposite direction.**

- Equivalent vectors : $\vec{a}=\vec{b}$

two vectors that have **equal length** and the **same direction.**

Blank : ① magnitude ② direction ③ magnitude ④ $\overrightarrow{AB}=\vec{a}=a$ ⑤ $|\overrightarrow{AB}|=|\vec{a}|=|a|$

2. Adding two vectors (Resultant of vectors)

We can **add** two vectors in two different ways;

1) "Head to tail" method	2) "Tail to tail" method (Parallelogram Method)
: by simply joining them head-to-tail	: by making a Parallelogram

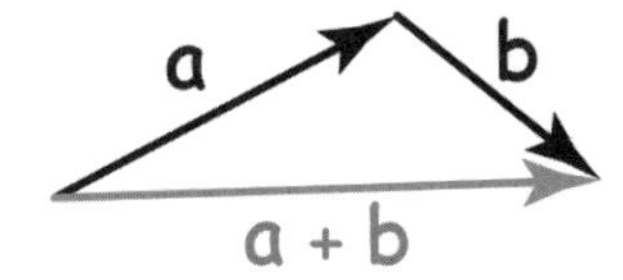

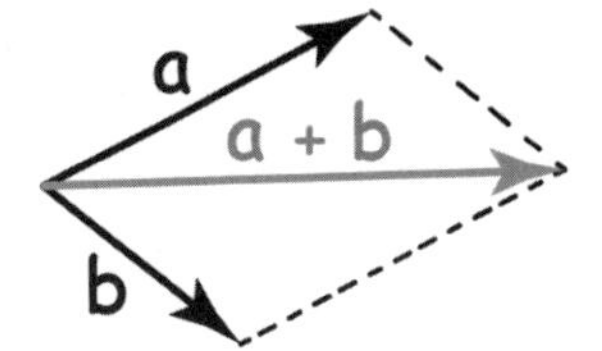

The sum of the vectors is called the ①______________.

3. Subtracting two vectors

We can also **subtract** one vector from another:
first we reverse the direction of the vector we want to subtract,
then add them as usual:

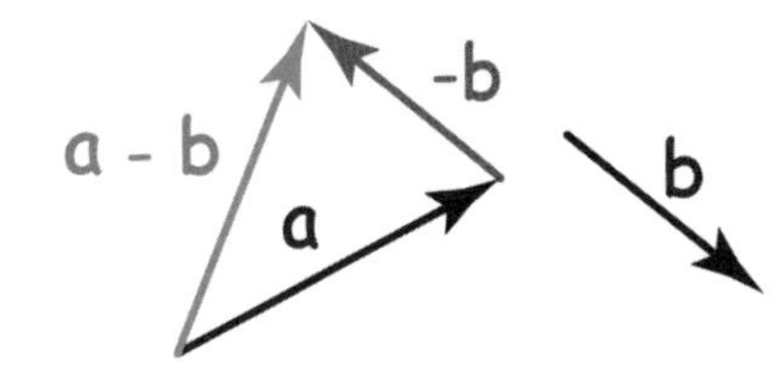

4. Scale Multiplication

When we multiply a vector by a scalar it is called "**scaling**" a vector, because we change how big or small the vector is.
It changes only the magnitude NOT the direction.

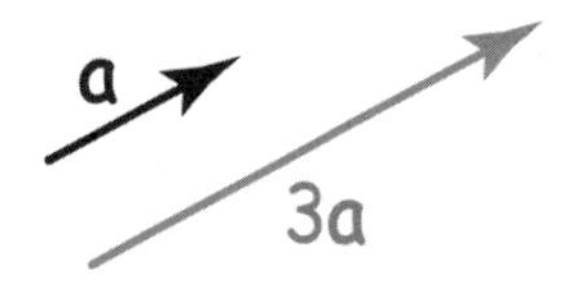

Vectors are ②_______________ if and only if they are *scalar multiples* of each other.

Blank : ① resultant ② parallel

EXAMPLE 1. Use the vectors **a**, **b**, **c** and **d** in the accompanying figure to graph the vector.

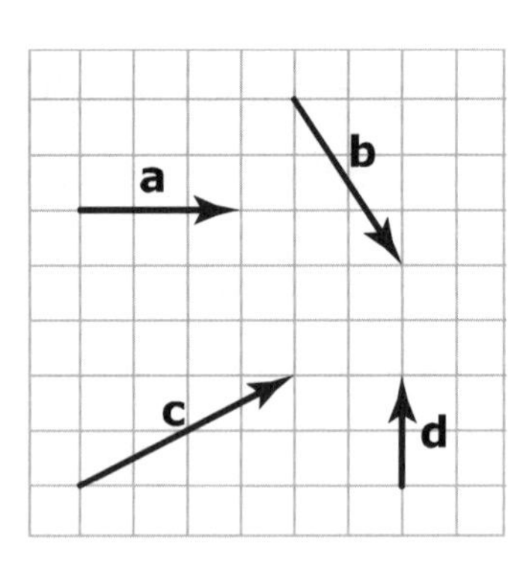

① **b** + **c**

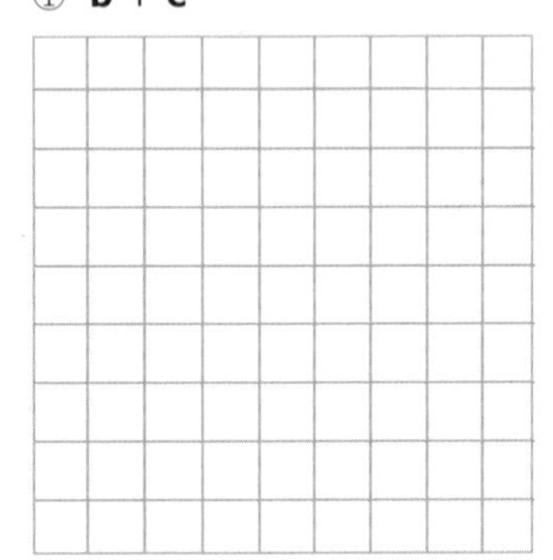

② 3**d**

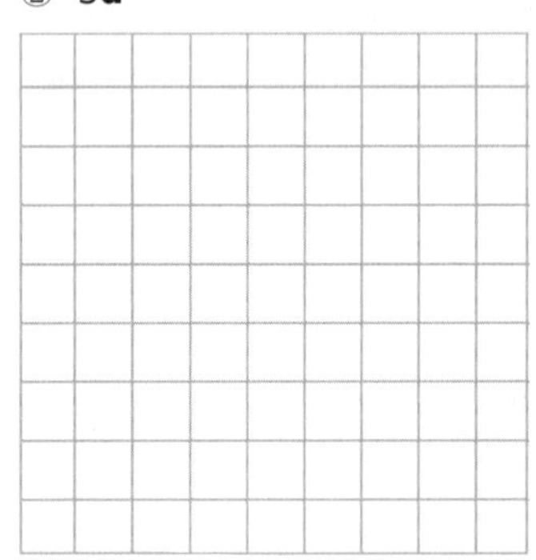

③ 2**b**

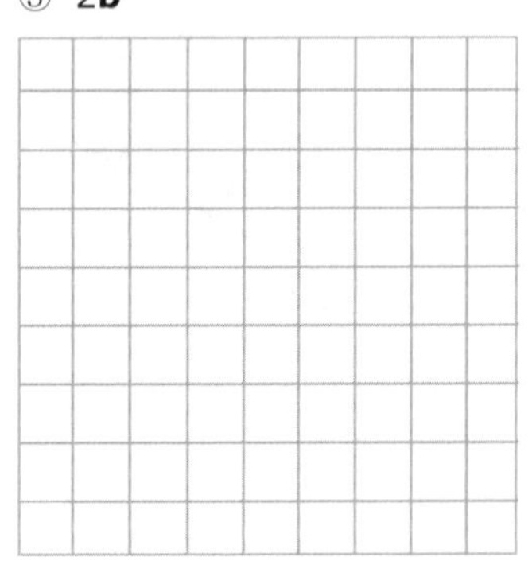

④ **a** + **b**

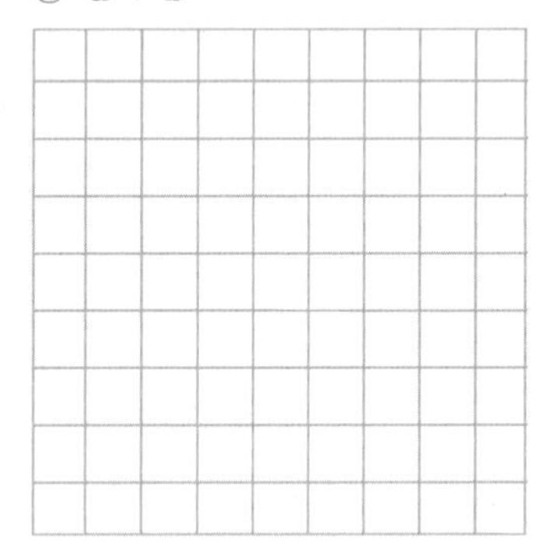

⑤ **c** - **d**

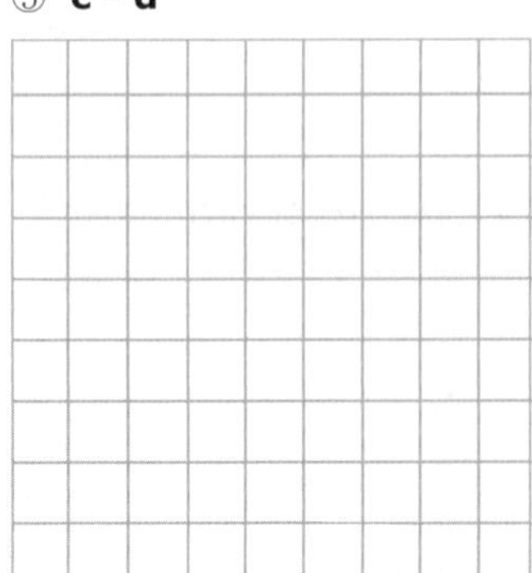

⑥ **b** - **c**

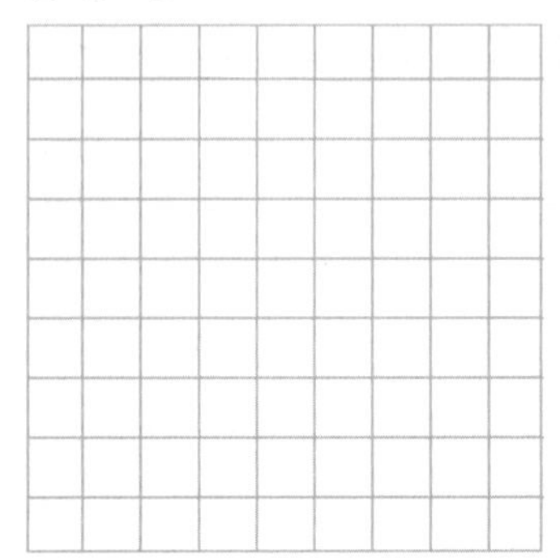

⑦ $2\mathbf{a} - 3\mathbf{d}$

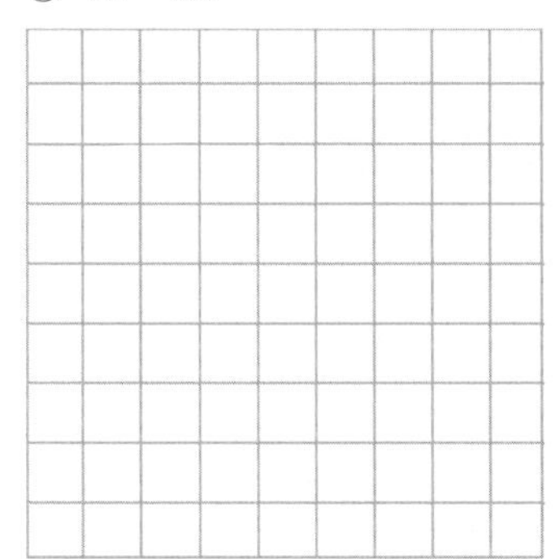

⑧ $2\mathbf{b} - \mathbf{c}$

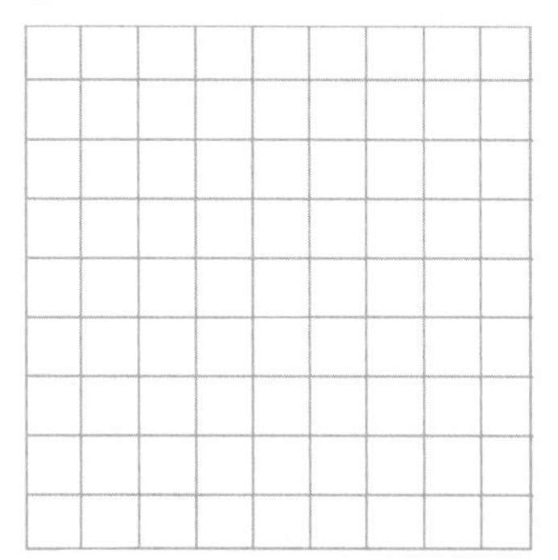

EXAMPLE 2. True or false?

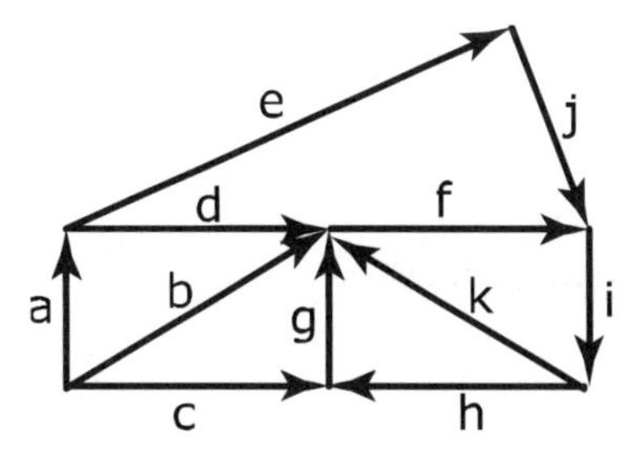

① $\mathbf{a} + \mathbf{d} = \mathbf{b}$

② $\mathbf{c} + \mathbf{g} = \mathbf{b}$

③ $\mathbf{b} + \mathbf{g} = \mathbf{c}$

④ $\mathbf{k} + \mathbf{g} = \mathbf{h}$

⑤ $\mathbf{a} + \mathbf{c} = \mathbf{b}$

⑥ $\mathbf{g} + \mathbf{f} + \mathbf{i} + \mathbf{h} = 0$

⑦ $\mathbf{a} + \mathbf{e} + \mathbf{j} + \mathbf{i} + \mathbf{h} = \mathbf{c}$

5. Vectors in the Plane

When we put a vector in a Cartesian coordinate, we can represent a vector using numbers.

A ①_____________ vector is a vector that starts at the origin (0, 0).

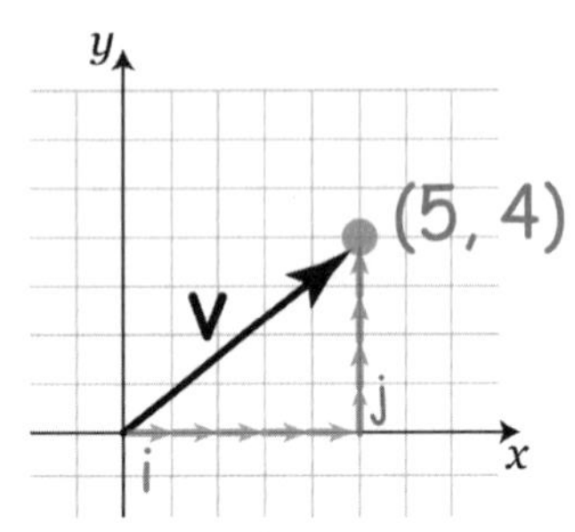

V = ② [ ] = ③ [ ]

The vectors **i** is called horizontal unit vector with magnitude of 1.

The vectors **j** is called vertical unit vector with magnitude of 1.

※ **Position Vectors**

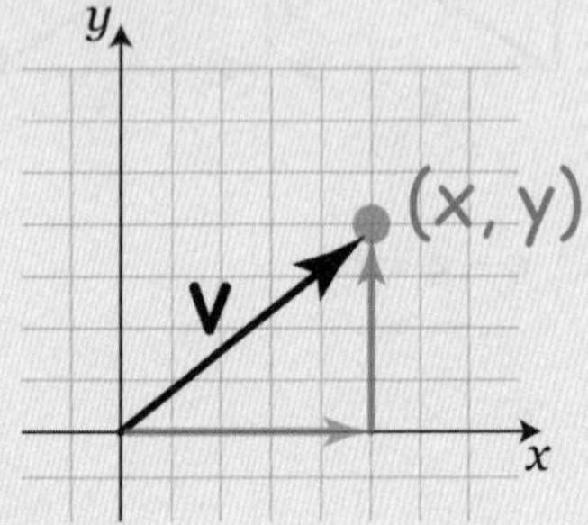

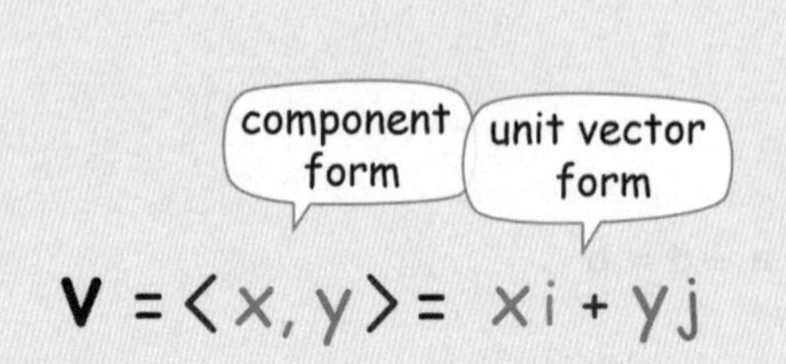

EXAMPLE 3. Write the illustrated vectors in component form and in unit vector form:

①

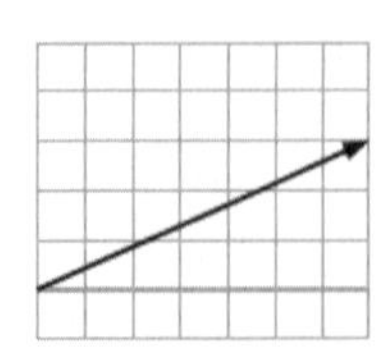

②

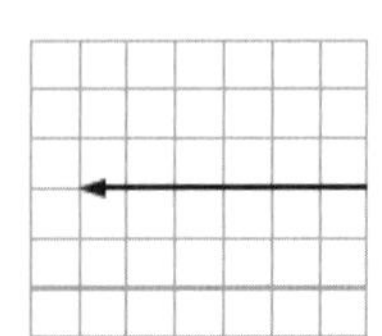

Blank : ① position ②<5, 4> ③ 5i+4j

③

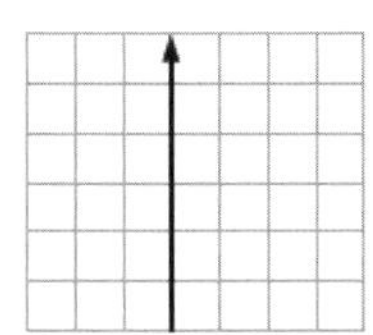

④

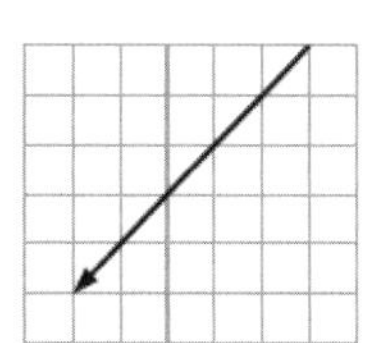

6. Displacement Vector

A displacement vector is a vector between two points.

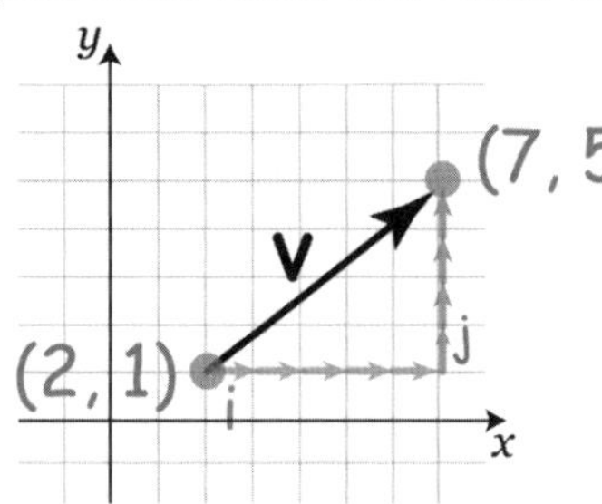

V = ①

= ②

※ **Displacement Vector** (Head minus Tail Rule)

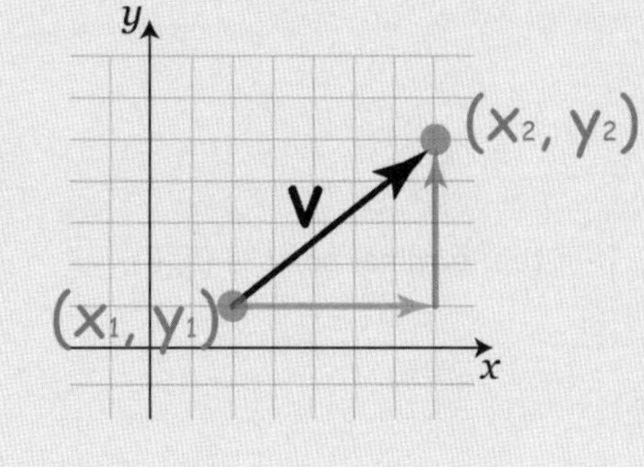

HMT

$$\mathbf{V} = \langle x_2 - x_1,\ y_2 - y_1 \rangle$$

$$= (x_2 - x_1)\,i + (y_2 - y_1)\,j$$

Blank : ① <7–2, 5–1> = <5, 4> ② 5i+4j

EXAMPLE 4. Given points $A(-1, 2)$, $B(3, 4)$, and $C(4, -5)$, find the vector of:

① vector **b**

② vector **c**

③ C from O

④ B from O

⑤ $\overrightarrow{AO}$

⑥ $\overrightarrow{CO}$

⑦ $\overrightarrow{BA}$

⑧ $\overrightarrow{AC}$

⑨ $\overrightarrow{BC}$

⑩ B from A

⑪ A from C

EXAMPLE 5. Find the terminal point of $v = 3i - 2j$ if the initial point is $(-2, 1)$.

EXAMPLE 6. Find the initial point of $v = \langle -3, 1 \rangle$ if the terminal point is $(5, 0)$.

7. Vector Algebra

※ Sum Vector of V and U

If vector $\mathbf{v} = \langle v_1, v_2 \rangle$, $\mathbf{u} = \langle u_1, u_2 \rangle$

$$v + u = \langle v_1 + u_1,\ v_2 + u_2 \rangle$$

※ Scaling Vector

$$k\mathbf{v} = \langle kv_1, kv_2 \rangle$$

※ Parallel Vectors

Parallel vectors have same direction but different lengths.
Numerically, vectors are parallel if and only if they are *scalar multiples* of each other

If **a** is parallel to **b**, then **a** = k**b**.

EXAMPLE 7. Consider the vector $\mathbf{u} = \langle 2, -1 \rangle$, $\mathbf{v} = \langle 3, 0 \rangle$. Find;

① $\mathbf{u} + \mathbf{v}$

② $\mathbf{u} - \mathbf{v}$

③ $2\mathbf{u} + 3\mathbf{v}$

④ $-\mathbf{u} + 4\mathbf{v}$

EXAMPLE 8. Consider the vector $\mathbf{u} = 3\mathbf{i} - \mathbf{j}$, $\mathbf{v} = -\mathbf{i} + 3\mathbf{j}$. Find;

① $\mathbf{u} + \mathbf{v}$

② $\mathbf{u} - \mathbf{v}$

③ $-\mathbf{u} - 4\mathbf{v}$

④ $\mathbf{u} - 3\mathbf{v}$

※ Magnitude

Magnitude is the length of the vector (= distance from head to tail)

Magnitude of v =<x, y> is

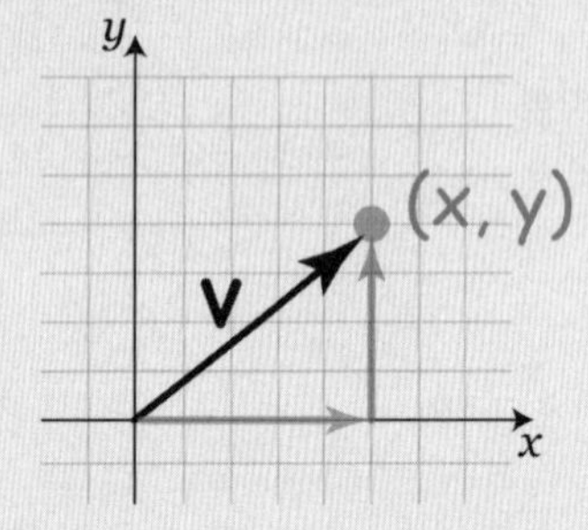

Magnitude of **V**

$$|\mathbf{v}| = \sqrt{x^2 + y^2}$$

※ Unit Vector in the same direction

You can find a unit vector(vector with magnitude of 1) **in the same direction** by dividing out the ①____________ of the vector.

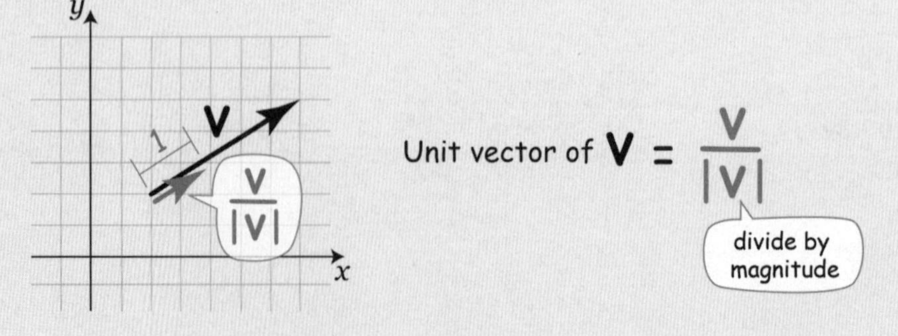

Blank : ① magnitude

EXAMPLE 9. Find the magnitude of the given vector and a unit vector in the same direction of the given vector.

① **a** = <2, 0>

② $-3i + j$

③ **v** = $5i - 12j$

④ **b** = <-1, 5>

⑤ vector from P = (5, 8) to Q = (6, 9)

⑥ vector from P = (-1, 3) to Q = (2, 7)

8. Horizontal and Vertical Component of Vector

※ **Horizontal and Vertical Components of Vector** (Vector using Direction Angle θ)

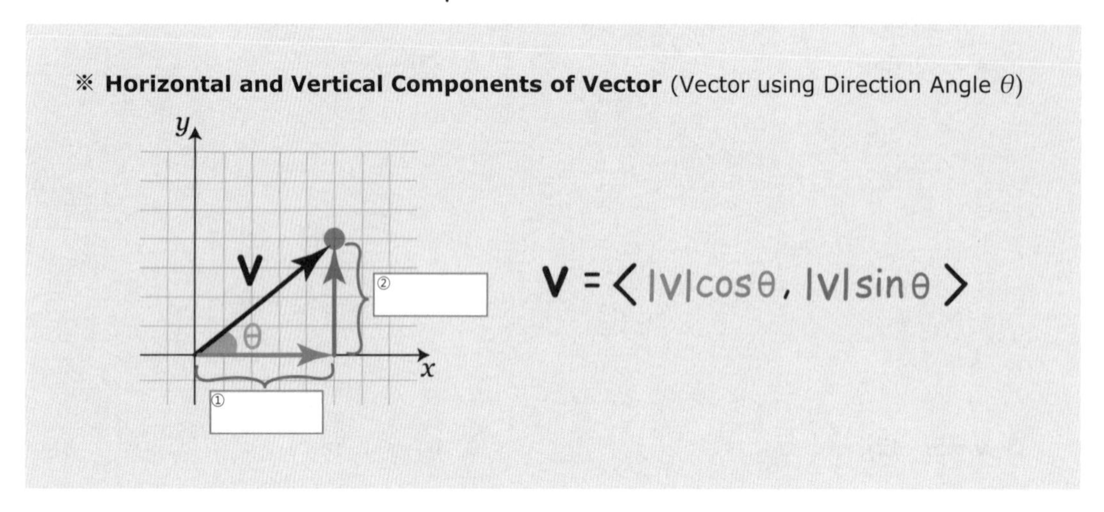

EXAMPLE 10. Write the vector **v** in the form ai+bj, given its magnitude and the direction.

① $|\mathbf{v}| = 5,\ \theta = 60°$

② $|\mathbf{v}| = 3,\ \theta = 150°$

③ $|\mathbf{v}| = 3,\ \theta = 240°$

④ $|\mathbf{v}| = 24,\ \theta = 225°$

Blank : ① $|v| \cos \theta$ ② $|v| \sin \theta$

※ When v = <x, y> is given,

Magnitude of **v**

$$|v| = \sqrt{x^2 + y^2}$$

direction angle

$$\tan\theta = \frac{y}{x}$$

EXAMPLE 11. Find the magnitude and direction (in degrees) of each vector, and rewrite the vector with horizontal and vertical components.

① **a** = $\langle 3,3 \rangle$

② **b** = $\langle 1,\sqrt{3} \rangle$

③ **c** = $-3\sqrt{3}i + 3j$

④ **d** = $-5i - 5j$

⑤ **e** = $-i - 5j$

⑥ **f** = $i - 3j$

10.2 Finding Resultant Vector using Trig

1. Law of Sine and Law of Cosine

※ **Law of Sines**

Law of Sines shows that in any triangle the lengths of the sides are ①__________________ to the sines of the corresponding opposite angles,

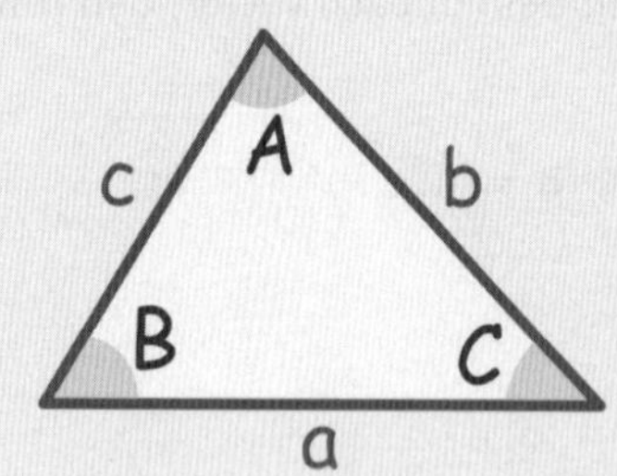

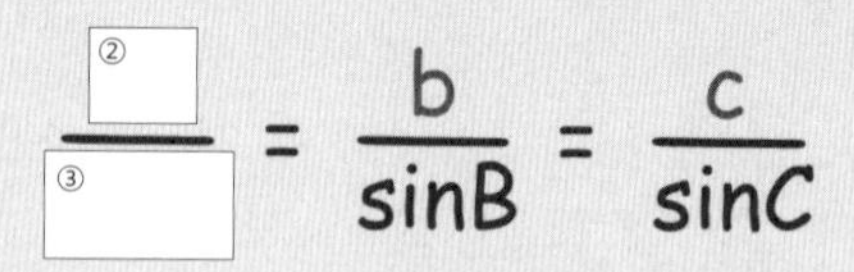

$$\frac{②\square}{③\square} = \frac{b}{\sin B} = \frac{c}{\sin C}$$

EXAMPLE 1. Use the Law of Sines to find the measurement indicated.

① Find AB.

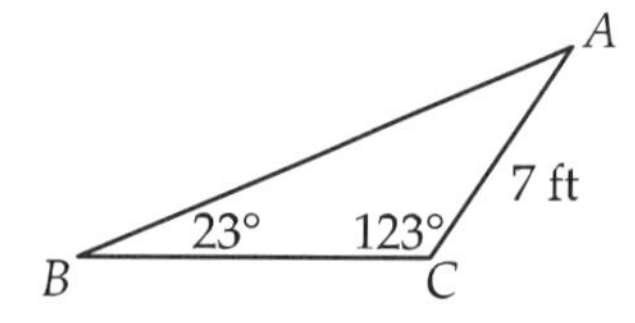

② Find AC

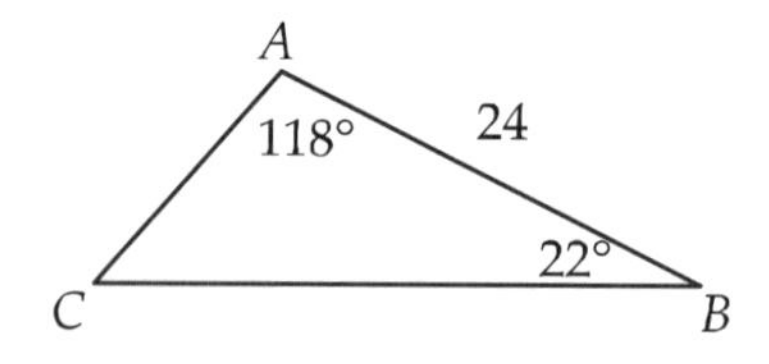

Blank : ① proportional ② a ③ sinA

③ Find m∠C

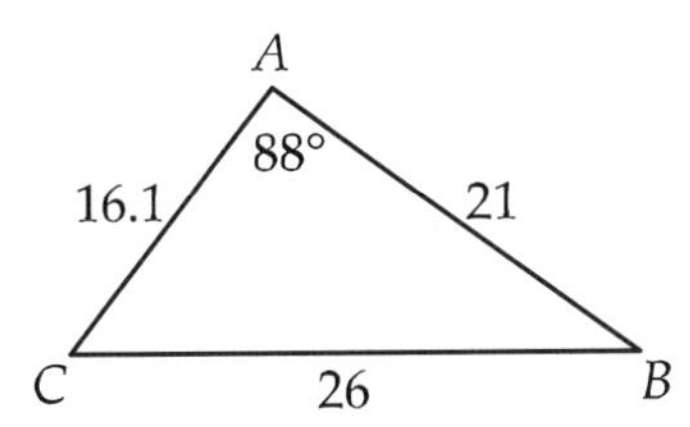

④ Find m∠C

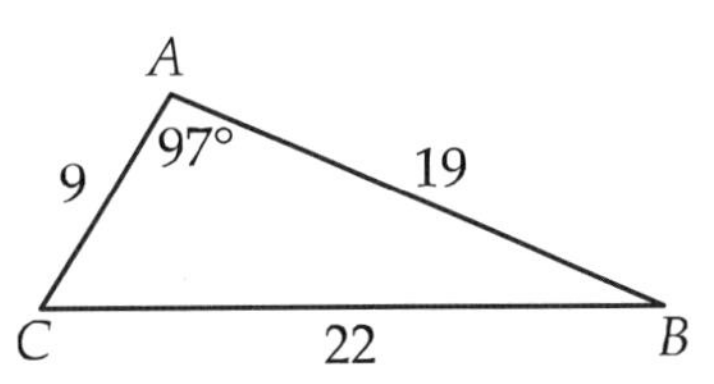

※ **Law of Cosines**

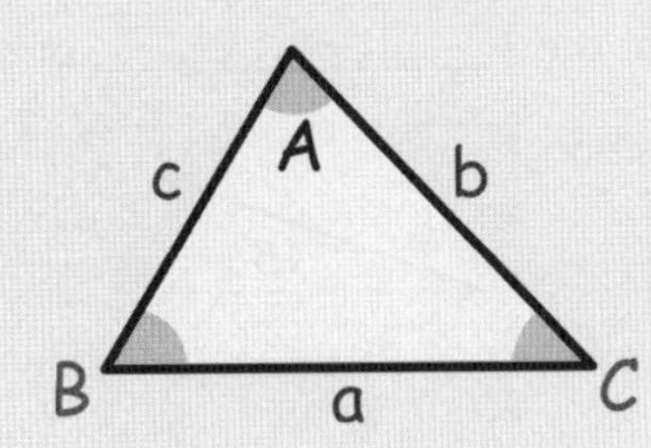

$$a^2 = b^2 + c^2 - 2bc\cos A$$

$$b^2 = a^2 + c^2 - 2ac\cos B$$

$$c^2 = a^2 + b^2 - 2ab\cos C$$

EXAMPLE 2. Use the Law of Cosines to find the measurement indicated.

① Find BA.

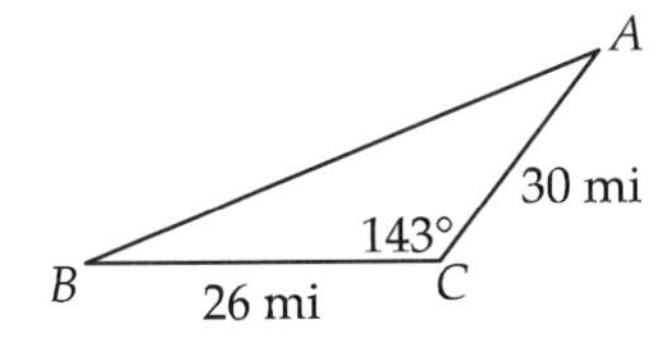

② Find AC

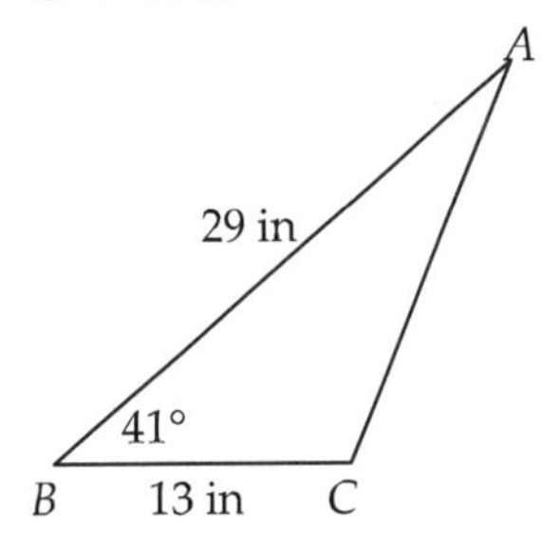

③ Find m∠C

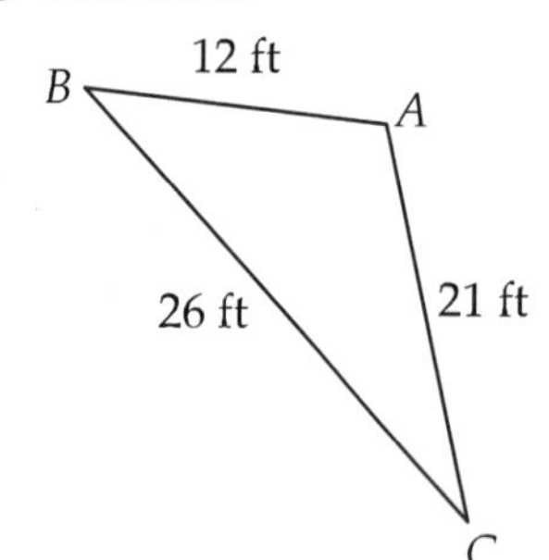

④ Find m∠A

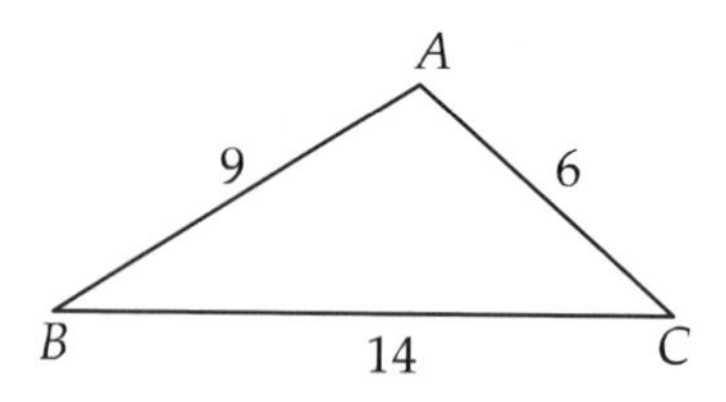

☺ Tip:

1) Speed, tension, weight, and force represent the ① ________________ of the vector.

2) 15° to the horizontal = 15° in standard position

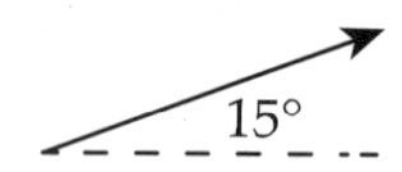

EXAMPLE 3. Victor runs with the football at a speed of 2 m/sec due east and throws the ball at a speed of 8 m/sec in the direction 43° to the horizontal as shown in the figure. What is the resultant speed and direction of the ball?

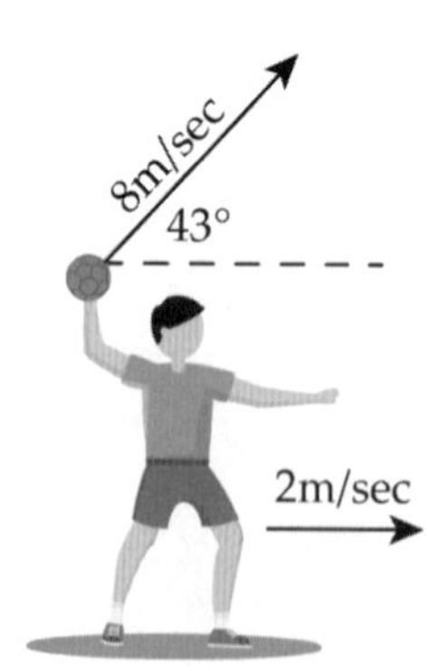

Blank : ① magnitude

EXAMPLE 4. A jet is flying through a wind that is blowing 10 km/h in the direction E12°N as shown. The jet's speed with no wind is 580 km/h in the direction E38°S. What is the true speed and direction of the jet?

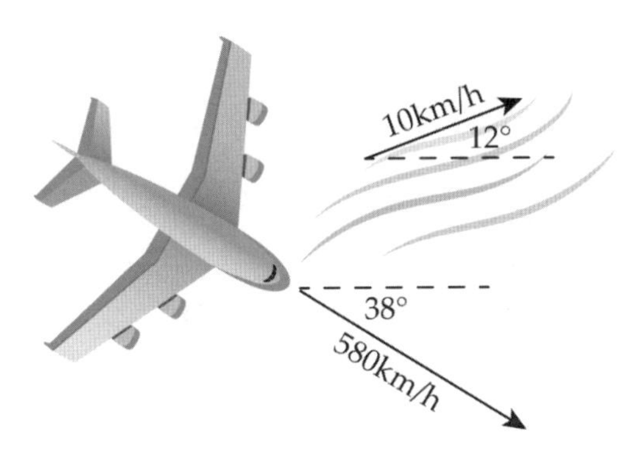

EXAMPLE 5. A river flows due east at 3 mph. A boat attempts to travel with a speed of 25 mph at the angle 27° to the horizontal as shown in the figure. Find the true speed and the direction of the boat.

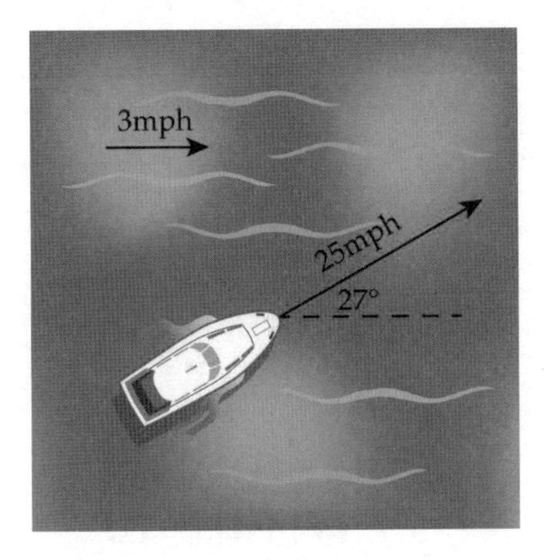

10.3 The Dot Product

1. Dot Products and Angles

You can calculate the dot Product of two vectors in two ways:

1) Dot Product using magnitudes and angle

Where:

|**u**| is the magnitude (length) of vector **u**

|**v**| is the magnitude (length) of vector **v**

θ is the angle between **u** and **v**

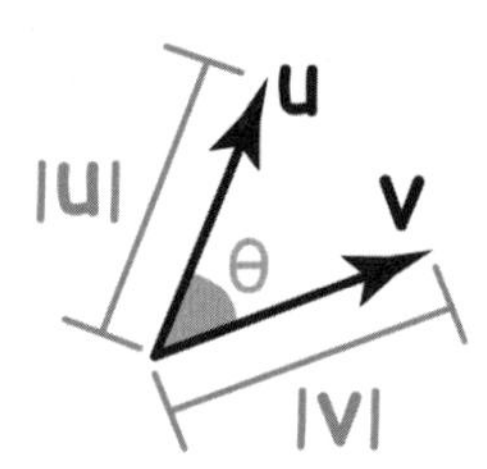

So we multiply the length of **u** times the length of **v**, then multiply by the cosine of the angle between **u** and **v**.

v · u = ①

2) Dot Product using coordinates

We multiply the x's, multiply the y's, then add.

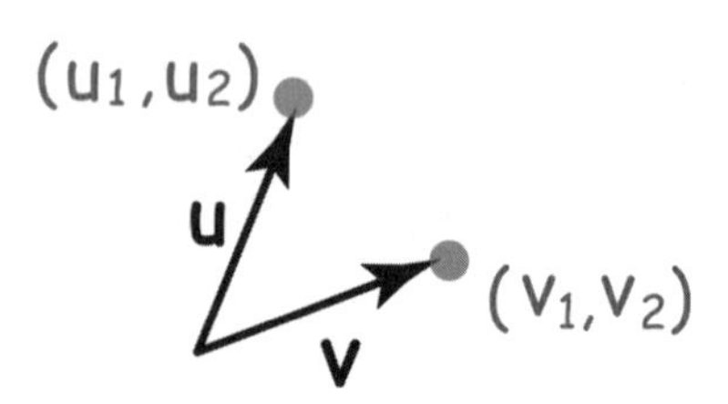

v · u = ②

※ **Dot (Scalar) Product of Vectors**

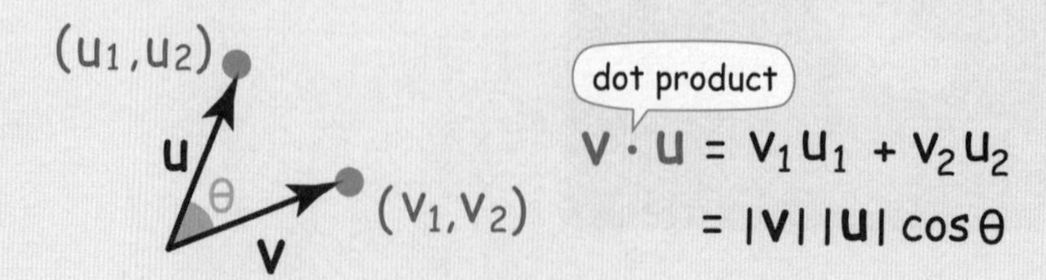

Blank : ① $|v|\ |u| \cos\theta$ ② $v_1u_1 + v_2u_2$

Notice that the scalar product of two vectors is a ①__________ number

NOT a vector. ($\mathbf{v} \cdot \mathbf{u} \neq \mathbf{vu}$)

EXAMPLE 1. Find $\mathbf{a} \cdot \mathbf{b}$.

① $\mathbf{a} = \langle 2,4 \rangle$, $\mathbf{b} = \langle 2,5 \rangle$

② $\mathbf{a} = \langle 5,-10 \rangle$, $\mathbf{b} = \langle 6,5 \rangle$

③ $\mathbf{a} = 6i + 2j$, $\mathbf{b} = 2i + 4j$

④ $\mathbf{a} = 4i + 8j$, $\mathbf{b} = -2i + 3j$

⑤ $\mathbf{a} = 6j$, $\mathbf{b} = i + 4j$

⑥ $\mathbf{a} = -i + 2j$, $\mathbf{b} = -j$

⑦ $|\mathbf{a}| = 2$, $|\mathbf{b}| = 3$,

Angle between **a** and **b** is $\frac{2\pi}{3}$

⑧ $|\mathbf{a}| = 3$, $|\mathbf{b}| = 2$,

Angle between **a** and **b** is $\frac{5\pi}{6}$

Blank : ① scalar

2. Angle between two vectors

If v · u = |v||u| cos θ , then cos θ =① ____________________.

※ **Angle between two vectors**

The angle between two Vector **v** = $\langle v_1, v_2 \rangle$ and **u** = $\langle u_1, u_2 \rangle$ is

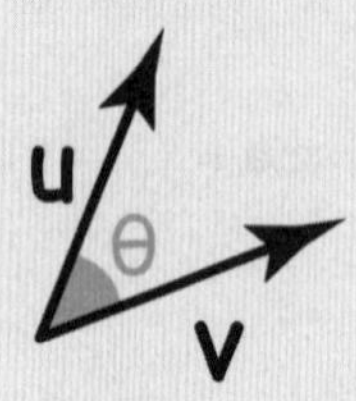

$$\cos\theta = \frac{\mathbf{v} \cdot \mathbf{u}}{|\mathbf{v}|\,|\mathbf{u}|}$$

EXAMPLE 2. Find the angle between the two vectors to the nearest degree.

① **a** = $\langle 2,4 \rangle$, **b** = $\langle 2,1 \rangle$

② **a** = $\langle 5,-1 \rangle$, **b** = $\langle 0,5 \rangle$

③ **a** = $4i+8j$, **b** = $-2i+j$

④ **a** = $6i-2j$, **b** = $2i+4j$

⑤ **a** = $6j$, **b** = $i+4j$

⑥ **a** = $-i+2j$, **b** = $-i$

Blank : ① $\frac{\mathbf{v} \cdot \mathbf{u}}{|\mathbf{v}||\mathbf{u}|}$

3. Properties of Dot Product

※ **Facts about Dot product**

① If vectors **V** and **U** are perpendicular , ①______________

② If vectors **V** and **U** are parallel ,② ______________

③ $\mathbf{v} \cdot \mathbf{v}$ = ③______________

④ $(k\mathbf{v}) \cdot \mathbf{u}$ = ④______________ (k is scalar)

⑤ $\mathbf{v} \cdot (\mathbf{a} + \mathbf{b})$ = ⑤______________ (distributive)

⑥ $\mathbf{v} \cdot \mathbf{u}$ = ⑥______________ (commutative)

⑦ $\mathbf{v} \cdot \mathbf{u} \cdot \mathbf{w}$ (⑦possible/impossible)

EXAMPLE 3. Determine whether the vectors u and v are parallel, orthogonal, or neither.

Orthogonal means perpendicular.

① $\mathbf{u} = \langle 10,0 \rangle$, $\mathbf{v} = \langle 0,-9 \rangle$

② $\mathbf{u} = \langle 7,2 \rangle$, $\mathbf{v} = \langle 21,6 \rangle$

Blank : ① $v \cdot u = 0$ ② $v \cdot u = |v||u|$ ③ $|v|^2$ ④ $k(v \cdot u)$ ⑤ $v \cdot a + v \cdot b$ ⑥ $u \cdot v$ ⑦impossible

③ $\mathbf{u} = i + \sqrt{3}j$, $\mathbf{v} = i - 2j$

④ $\mathbf{u} = 2i + 4j$, $\mathbf{v} = 4i - 2j$

⑤ $\mathbf{u} = 3i + 2j$, $\mathbf{v} = 6i + 4j$

⑥ $\mathbf{u} = \langle 8, 4 \rangle$, $\mathbf{v} = \langle 10, 7 \rangle$

⑦ $\mathbf{u} = \langle 1, -2 \rangle$, $\mathbf{v} = \langle -4, 8 \rangle$

10.4 Motions in Vectors

1. Motions and Parametric Equations

Yuna is gliding around on a frozen coordinate plane. At time t (in seconds) Yuna 's position on the coordinate plane is given by $s(t) = \left\langle t^2 - 2t, t+2 \right\rangle$ and Yuna 's velocity vector is given by $v(t) = \left\langle 2t-2, 1 \right\rangle$. If Yuna travels when $0 \le t \le 3$, she will trace out the parametric curve like the one sketched below.

What is Yuna's position vector at t = 0, t = 1, t = 3?

What is Yuna's velocity vector at t = 0, t = 1, t = 3? What can we find out from the velocity vector?

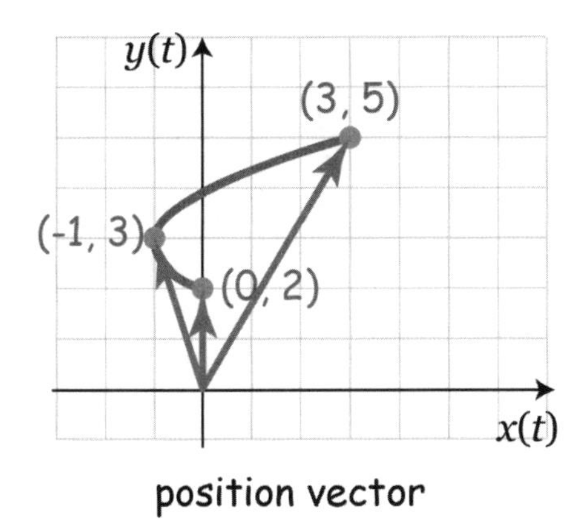

position vector

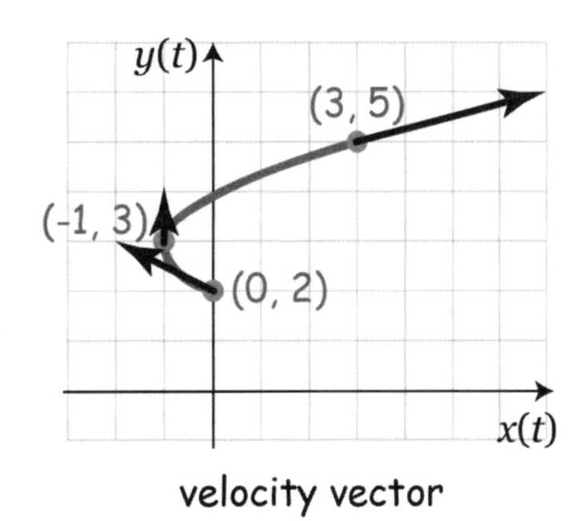

velocity vector

※ **Velocity Vector** $v(t) = \langle x(t), y(t) \rangle$

Velocity vector shows the velocity of a particle moving in a plane in each time t.

	Velocity	**Speed**
Concept	shows you (①how fast / direction) so it is (②vector/scalar)	shows you (③how fast / direction) so it is (④vector/scalar)
Sign	Velocity can be positive or negative. Sign shows the direction.	Speed is always positive; **speed =** $\lvert V(t) \rvert$
What can we find from Velocity Vector $v(t) = \langle x(t), y(t) \rangle$	If horizontal velocity x(t) is positive = moving ⑤________ negative = moving ⑥________ If vertical velocity y(t) is positive = moving ⑦________ negative = moving ⑧________	Speed is the ⑨______________ of the velocity vector.

Blank : ① both 'how fast' and 'direction' ② vector ③ how fast ④ scalar
⑤ right ⑥ left ⑦ up ⑧ down ⑨ magnitude

EXAMPLE 1. A particle moves in the xy-plane with velocity vector $x(t)$, $y(t)$ such that $x(t) = 2t - 4$ and $y(t) = -t^2 + 2t$ in the time interval $0 \le t \le 5$.

① Find the velocity vector of the particle at $t = 1$.

② Find the velocity vector of the particle at $t = 3$.

③ Is the particle moving to the left or to the right when $t = 1$?

④ Is the particle moving to the left or to the right when $t = 3$?

⑤ Is the particle moving up or down when $t = 1$?

⑥ Is the particle moving up or down when $t = 3$?

⑦ How fast is the particle moving when $t = 1$?

⑧ How fast is the particle moving when $t = 3$?

Part 11

Matrices

11.1 Algebra of Matrices

1. Matrix

※ A **Matrix** is an array of numbers (plural: Matrices)

※ **Order of Matrix** (=Dimension of Matrix = ①________ of Matrix)

To show how many rows and columns a matrix has we often write

numbers of ②____________× numbers of ③____________.

3 columns

2 rows $\begin{bmatrix} 6 & 4 & 24 \\ 1 & -9 & 7 \end{bmatrix}$ 2 X 3

$\begin{bmatrix} 2 & 3 \\ 0 & 4 \\ 3 & -1 \end{bmatrix}$ ④

$\begin{bmatrix} -1 & -3 \\ 2 & 4 \end{bmatrix}$ ⑤

※ Each **element** is shown by a lower case letter with a "subscript" of row, column:

$$A = \begin{bmatrix} a_{11} & a_{12} & a_{13} \\ a_{21} & a_{22} & a_{23} \end{bmatrix}$$

a_{rc} row column

2. Operation of Matrices

※ To **add** or **subtract** two matrices: add the numbers in the matching positions.

2-1=1

$$\begin{bmatrix} 2 & -5 \\ 1 & 9 \end{bmatrix} - \begin{bmatrix} 1 & 3 \\ -2 & -4 \end{bmatrix} = \begin{bmatrix} 1 & -8 \\ 3 & 13 \end{bmatrix}$$

Blank : ① size ② row ③ column ④ 3 x 2 ⑤ 2 x 2

※ We can **multiply** a matrix **by some value**:

$3 \times 1 = 3$

$$3 \times \begin{bmatrix} 1 & 3 \\ -2 & -4 \end{bmatrix} = \begin{bmatrix} 3 & 9 \\ -6 & -12 \end{bmatrix}$$

EXAMPLE 1. Find the sum or difference of the matrix, if possible.

① $\begin{bmatrix} -1 & 5 \\ 2 & 4 \end{bmatrix} + 2\begin{bmatrix} 2 & 8 \\ -4 & 5 \end{bmatrix}$

② $\begin{bmatrix} 1 & 5 & 7 \\ -4 & 0 & 5 \end{bmatrix} - 2\begin{bmatrix} 7 & -1 & 4 \\ -2 & 4 & 4 \end{bmatrix}$

③ $-\begin{bmatrix} 2 & -3 \\ 4 & -1 \\ 8 & 7 \end{bmatrix} - 3\begin{bmatrix} 0 & 8 \\ 4 & -2 \\ 3 & 4 \end{bmatrix}$

④ $\begin{bmatrix} 1 & 4 & 0 \end{bmatrix} + 2\begin{bmatrix} 4 \\ -3 \\ 1 \end{bmatrix}$

⑤ $2\begin{bmatrix} 2 & 3 \\ -8 & 2 \end{bmatrix} + 3\begin{bmatrix} 2 & 4 \\ 4 & -2 \\ 5 & 7 \end{bmatrix}$

EXAMPLE 2. Solve the equation.

① $\begin{bmatrix} 2 & 6 \\ 3 & 1 \end{bmatrix} + X = \begin{bmatrix} 4 & 7 \\ 0 & 9 \end{bmatrix}$

② $\begin{bmatrix} 0 & 3 & 9 \\ -1 & 5 & -3 \end{bmatrix} - X = \begin{bmatrix} 2 & -11 & 6 \\ 3 & 4 & 8 \end{bmatrix}$

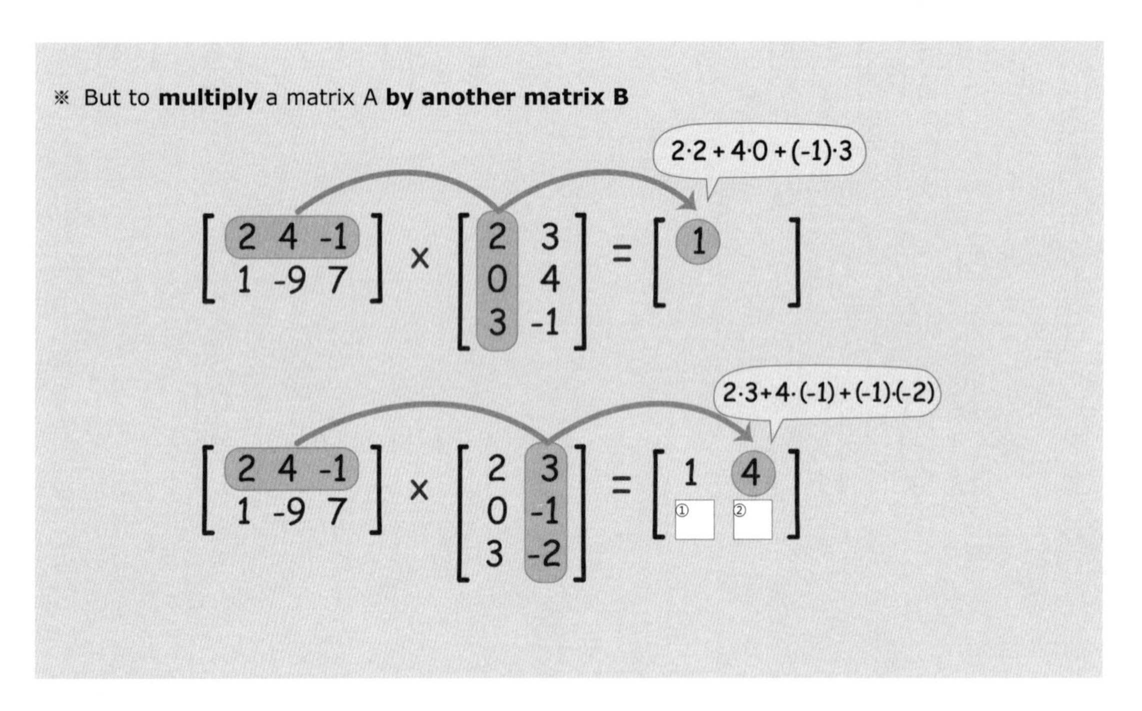

EXAMPLE 3. Find the product, if possible.

① $\begin{bmatrix} -1 & 0 \\ 2 & 4 \end{bmatrix} \times \begin{bmatrix} 2 & 8 \\ -4 & 1 \end{bmatrix}$

② $\begin{bmatrix} -1 & 3 \\ 3 & 2 \end{bmatrix} \times \begin{bmatrix} -2 & 0 \\ -1 & 3 \end{bmatrix}$

Blank : ① 23 ② -2

③ $\begin{bmatrix} -2 & 3 & 5 \end{bmatrix} \times \begin{bmatrix} 0 & 1 \\ -1 & 2 \\ 2 & 3 \end{bmatrix}$

④ $\begin{bmatrix} 0 & -3 & 1 \\ 5 & -1 & 0 \end{bmatrix} \times \begin{bmatrix} 1 & 2 \\ 0 & 1 \\ 1 & -1 \end{bmatrix}$

⑤ $\begin{bmatrix} 0 & 7 \\ 2 & 1 \end{bmatrix} \times \begin{bmatrix} -4 & 3 & 0 \\ 4 & 2 & -2 \end{bmatrix}$

⑥ $\begin{bmatrix} -6 & 2 & 9 \end{bmatrix} \times \begin{bmatrix} 4 \\ 0 \\ -3 \end{bmatrix}$

⑦ $\begin{bmatrix} -4 & 1 & -5 \\ 0 & 0 & 4 \end{bmatrix} \times \begin{bmatrix} 0 & 2 & -5 \\ 1 & 3 & 7 \end{bmatrix}$

⑧ $\begin{bmatrix} 2 & -8 \\ -1 & 3 \\ 2 & -6 \end{bmatrix} \times \begin{bmatrix} 5 & 0 \\ -1 & 7 \\ 2 & 1 \end{bmatrix}$

※ **Product of Matrices**

Matrix product: $A \times B = AB$

size: $m \times n$ $n \times p$ $m \times p$

(same: n)

Product of Matrices AB is defined
only when the number of columns in A is equal to the number of rows in B.

EXAMPLE 4. If matrices A, B, C, D and E have dimensions of 2×3, 2×3, 3×2, 2×2, and 3×3, respectively, what are the dimensions of following operations?

① BA

② AEC

③ BCD

④ $D(A-3B)$

⑤ $BC+2D$

⑥ $(AE-B)C$

⑦ $(BC-D)A$

EXAMPLE 5. Which of the following matrices can be multiplied by themselves?

$$A=\begin{bmatrix} 1 & 1 \\ 3 & 1 \\ 1 & 2 \end{bmatrix} \quad B=\begin{bmatrix} 3 & -1 & 1 \\ 1 & -1 & 2 \\ 2 & 1 & -3 \end{bmatrix} \quad C=\begin{bmatrix} 1 & 1 & 4 \end{bmatrix} \quad D=\begin{bmatrix} 1 & 0 \\ 0 & 1 \end{bmatrix}$$

3. Properties of Matrices

※ **Properties of Matrix**

$(AB)C = A(BC)$ — Association Property

$C(A + B) = CA + CB$

$(A + B)C = AC + BC$ — Distributive Property

BUT Matrix **Multiplication** is ***NOT Commutative***!

$$AB \neq BA$$

EXAMPLE 6. If A and B are two matrices, find $(A+B)^2$.

EXAMPLE 7. If A and B are two matrices, find $(A+B)(A-B)$.

4. Identity Matrix

The "Identity Matrix" is the matrix equivalent of the number "1":

$$I_3 = \begin{bmatrix} 1 & 0 & 0 \\ 0 & 1 & 0 \\ 0 & 0 & 1 \end{bmatrix}$$

3 x 3 Identity Matrix

It is "square" matrix (has same number of rows as columns),

It has 1s on the diagonal and 0s everywhere else.

Its symbol is the capital letter I.

It is a special matrix, because when we multiply by it, the original is unchanged:

$$AI = IA = A$$

EXAMPLE 8. Find the product.

① $\begin{bmatrix} -1 & 0 \\ 2 & 4 \end{bmatrix} \times \begin{bmatrix} 1 & 0 \\ 0 & 1 \end{bmatrix}$

② $\begin{bmatrix} 1 & 0 & 2 \\ 0 & 2 & -4 \\ 3 & -5 & 4 \end{bmatrix} \times \begin{bmatrix} 1 & 0 & 0 \\ 0 & 1 & 0 \\ 0 & 0 & 1 \end{bmatrix}$

11.2 Determinant and Inverse Matrix

1. Determinant of Matrices

The ①______________________of a matrix is a special number that can be calculated from a **square matrix**.

※ **Determinant of 2x2 Matrix:**

$$|A| = \begin{vmatrix} a & b \\ c & d \end{vmatrix} = ad - bc$$

Determinant of Matrix A

⊖ ⊕

EXAMPLE 1. Find the determinant.

① $\begin{vmatrix} 7 & 3 \\ 5 & 2 \end{vmatrix}$

② $\begin{vmatrix} 2 & 3 \\ -1 & 2 \end{vmatrix}$

③ $\begin{vmatrix} 4 & 5 \\ -1 & 1 \end{vmatrix}$

④ $\begin{vmatrix} 12 & -7 \\ -4 & 3 \end{vmatrix}$

⑤ $\begin{vmatrix} 2 & 2 \\ 5 & 5 \end{vmatrix}$

⑥ $\begin{vmatrix} 2 & -4 \\ 3 & -6 \end{vmatrix}$

Blank : ① determinant

2. Inverse of Matrices

When you multiply a Matrix by its ①______________ matrix, you get the Identity Matrix.

$$A \times A^{-1} = A^{-1} \times A = I$$

※ **Inverse of a Matrix of 2 x 2**

$$A^{-1} = \begin{bmatrix} a & b \\ c & d \end{bmatrix}^{-1} = \frac{1}{|A|}\begin{bmatrix} d & -b \\ -c & a \end{bmatrix}$$

switch! negative~

Determinant

$$= \frac{1}{ad-bc}\begin{bmatrix} d & -b \\ -c & a \end{bmatrix}$$

If det $\boldsymbol{A} \neq 0$, then $\boldsymbol{A}^{-1}$ exists

If det $\boldsymbol{A}$ = ②________, then $\boldsymbol{A}^{-1}$ does not exist.

EXAMPLE 2. Find the inverse, if it exists, for the matrix.

① $\begin{bmatrix} 5 & 3 \\ 3 & 2 \end{bmatrix}$

② $\begin{bmatrix} 10 & 1 \\ -1 & 0 \end{bmatrix}$

③ $\begin{bmatrix} -5 & -1 \\ 6 & 0 \end{bmatrix}$

④ $\begin{bmatrix} -2 & 4 \\ 4 & -4 \end{bmatrix}$

Blank : ① inverse ② 0

⑤ $\begin{bmatrix} 4 & -8 \\ 3 & -6 \end{bmatrix}$

⑥ $\begin{bmatrix} -2 & 4 \\ 1 & -2 \end{bmatrix}$

3. Determinant and Parallelogram Area

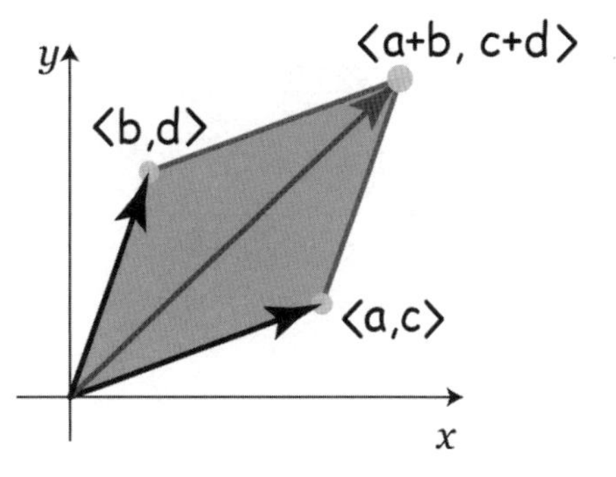

If we have vector <a, c> and vector <b, d>, by find the resultant vector <a+b, b+d>, we can construct a parallelogram as shown.

Let's try to find the area of the parallelogram.

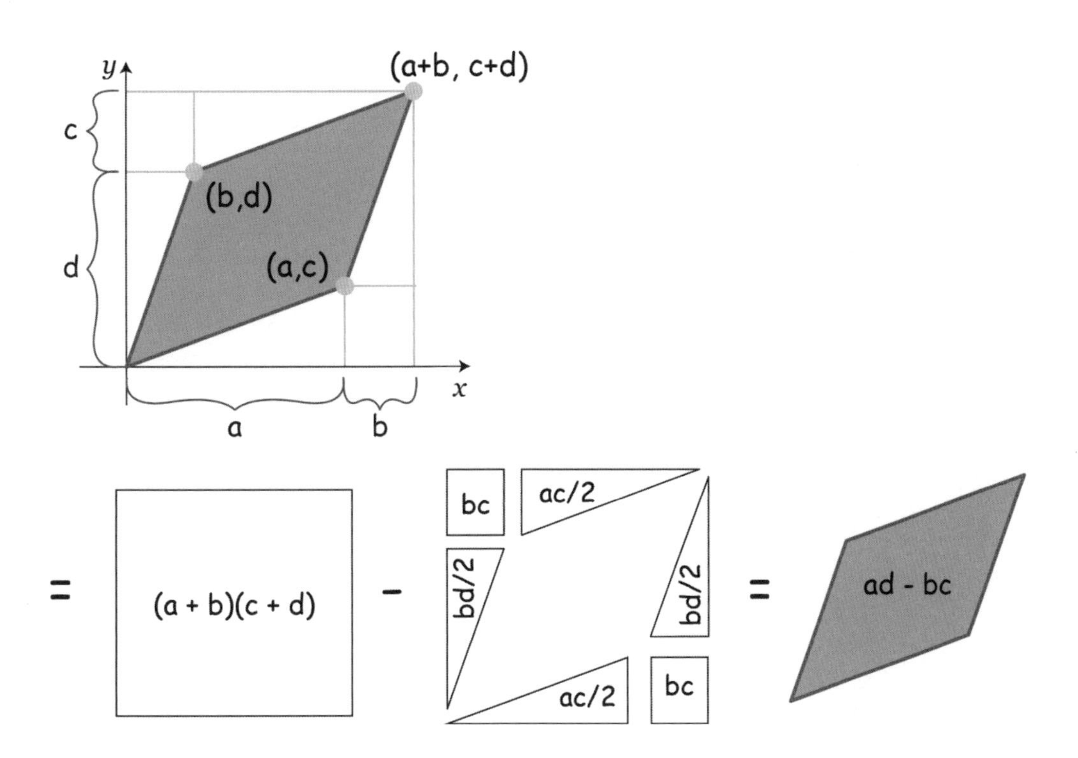

$ad - bc$ is the ① ____________ of the matrix $\begin{bmatrix} a & b \\ c & d \end{bmatrix}$.

※ Determinant and Parallelogram Area

The area of the parallelogram constructed by two vectors <a, c>,<b, d> is;

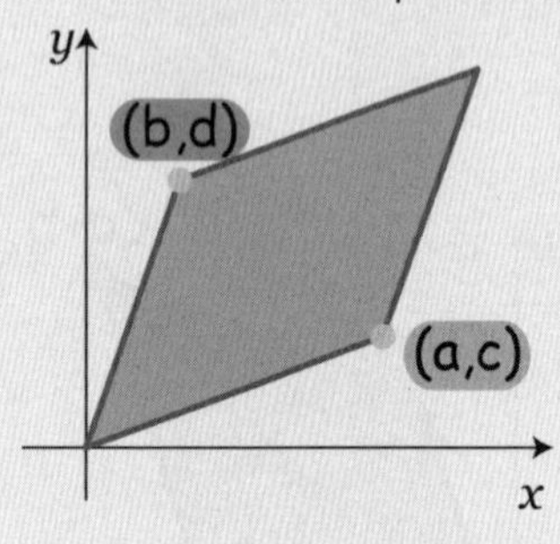

Area of ▱ $= \begin{vmatrix} a & b \\ c & d \end{vmatrix} = ad - bc$

EXAMPLE 3. Find the area of the parallelogram constructed by two vectors;

① $\langle 3,2 \rangle$ and $\langle -3,7 \rangle$

② $\langle 1,-2 \rangle$ and $\langle -2,5 \rangle$

③ $i + 2j$ and $3i - 4j$

④ $2i$ and $3i - 5j$

☺ Tip: If you have negative determinant, take the absolute value and make it positive. (Area is always positive!)

Blank : ① determinant

11.3 Linear Transformation

1. Vectors as a Matrix

Vector can also be written as matrices.

A vector can be written as either a row matrix or a column matrix.

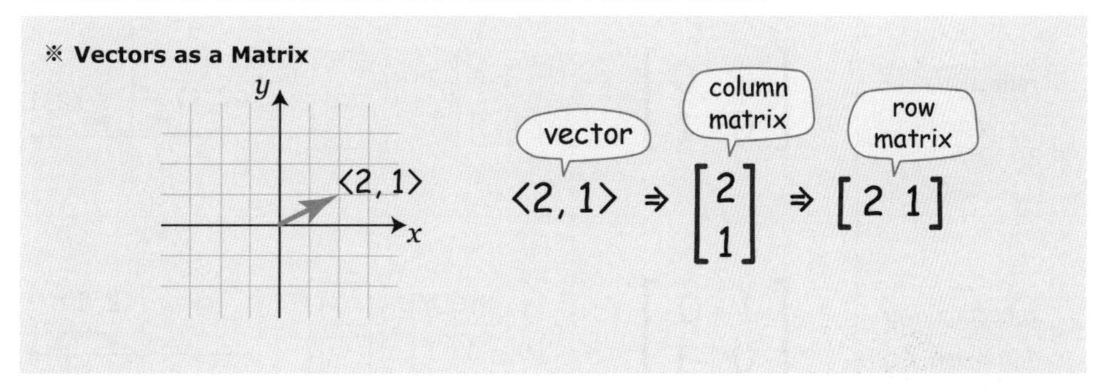

Writing vectors as matrices is useful when working with transformations.

2. Linear Transformation of Matrix

☺ Reminder: If $f(x) = 2x$, then $f(2) =$ ① ________

A 2 × 2 matrix can be used to apply a linear transformation to points or vectors on a Cartesian grid.

※ Linear Transformation

Let $\vec{a}$: input vector, $\vec{b}$: output vector, A is a 2x2 matrix for linear transformation L, then we can say;

input vector

2x2 Matrix

$$L(\vec{a}) = A \cdot \vec{a} = \vec{b}$$

linear Transformation

multiply

Blank : ① 4

Linear Transformation L	A (=2x2 Matrix)	Transformation of vector $\langle 2,1 \rangle$	
identity matrix	$\begin{bmatrix} 1 & 0 \\ 0 & 1 \end{bmatrix}$	$\begin{bmatrix} 1 & 0 \\ 0 & 1 \end{bmatrix}\begin{bmatrix} 2 \\ 1 \end{bmatrix}=\begin{bmatrix} ① \\ \square \end{bmatrix}$	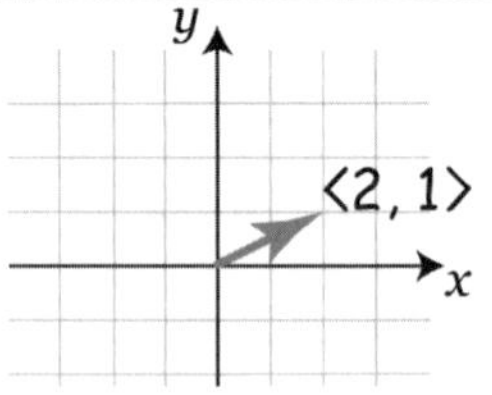 y, x, <2, 1>
reflection over y axis	$\begin{bmatrix} -1 & 0 \\ 0 & 1 \end{bmatrix}$	$\begin{bmatrix} -1 & 0 \\ 0 & 1 \end{bmatrix}\begin{bmatrix} 2 \\ 1 \end{bmatrix}=\begin{bmatrix} ② \\ \square \end{bmatrix}$	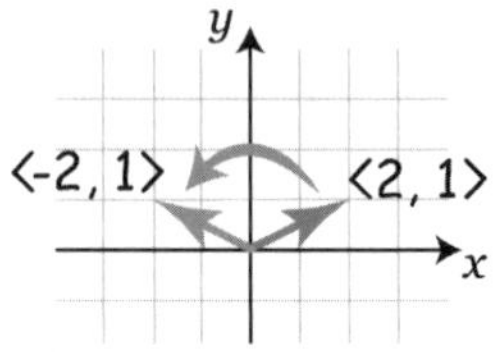 y, x, <-2, 1>, <2, 1>
reflection over x axis	$\begin{bmatrix} 1 & 0 \\ 0 & -1 \end{bmatrix}$	$\begin{bmatrix} 1 & 0 \\ 0 & -1 \end{bmatrix}\begin{bmatrix} 2 \\ 1 \end{bmatrix}=\begin{bmatrix} ③ \\ \square \end{bmatrix}$	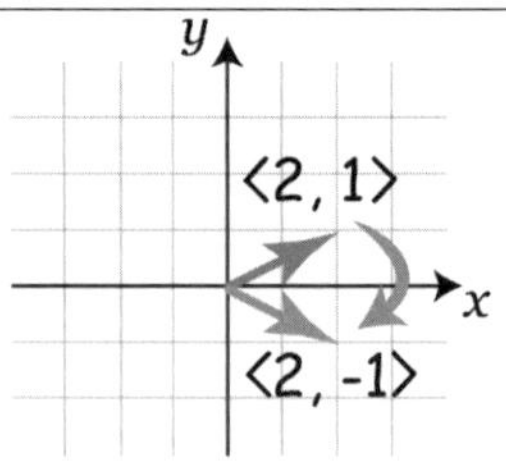 y, x, <2, 1>, <2, -1>
reflection over origin	$\begin{bmatrix} -1 & 0 \\ 0 & -1 \end{bmatrix}$	$\begin{bmatrix} -1 & 0 \\ 0 & -1 \end{bmatrix}\begin{bmatrix} 2 \\ 1 \end{bmatrix}=\begin{bmatrix} ④ \\ \square \end{bmatrix}$	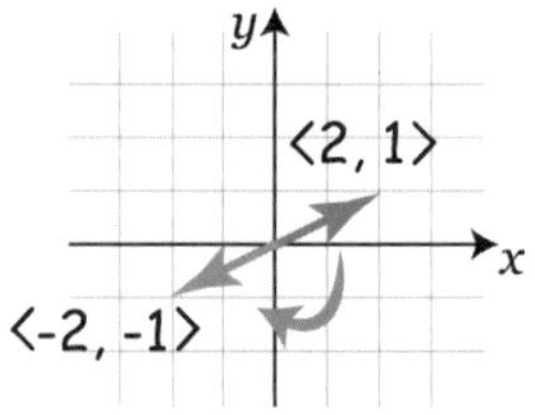 y, x, <2, 1>, <-2, -1>
reflection over y = x	$\begin{bmatrix} 0 & 1 \\ 1 & 0 \end{bmatrix}$	$\begin{bmatrix} 0 & 1 \\ 1 & 0 \end{bmatrix}\begin{bmatrix} 2 \\ 1 \end{bmatrix}=\begin{bmatrix} ⑤ \\ \square \end{bmatrix}$	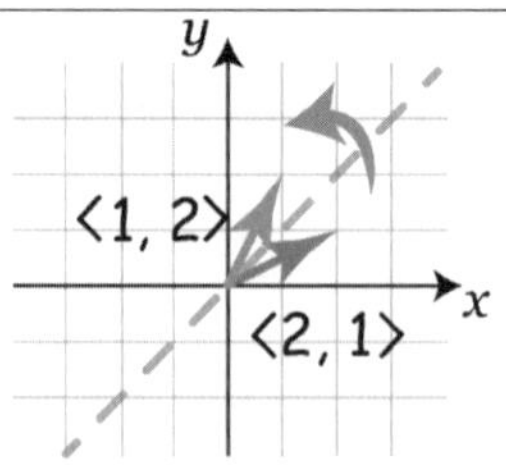 y, x, <1, 2>, <2, 1>
Enlargement by scale factor k	$\begin{bmatrix} k & 0 \\ 0 & k \end{bmatrix}$	$\begin{bmatrix} 2 & 0 \\ 0 & 2 \end{bmatrix}\begin{bmatrix} 2 \\ 1 \end{bmatrix}=\begin{bmatrix} ⑥ \\ \square \end{bmatrix}$	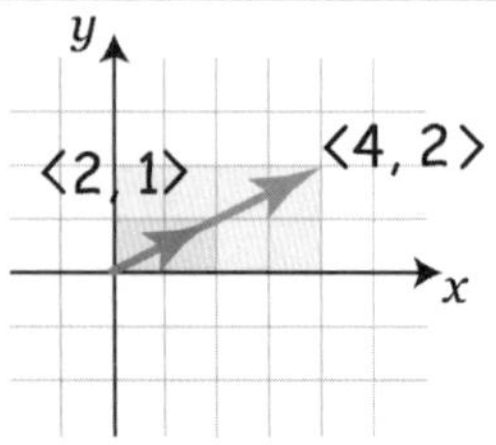 y, x, <2, 1>, <4, 2>

Rotation $\theta°$ counterclockwise about (0,0)	$\begin{bmatrix} \cos\theta & -\sin\theta \\ \sin\theta & \cos\theta \end{bmatrix}$	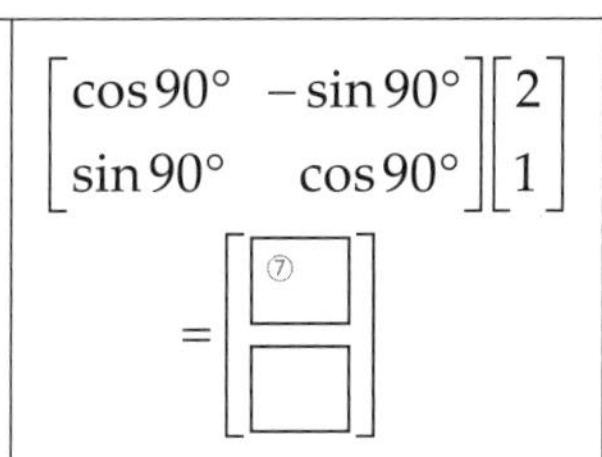	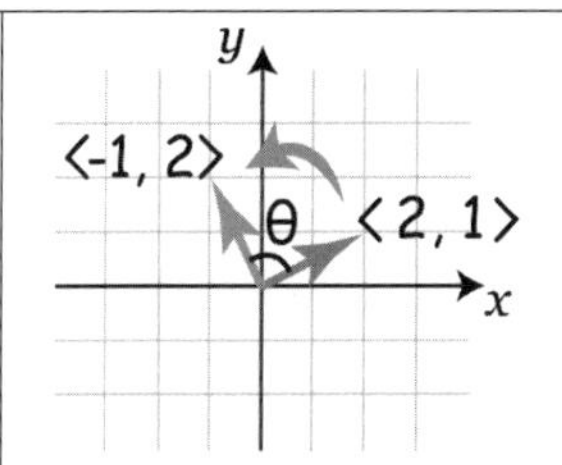

EXAMPLE 1. If input vector $\vec{a} = \langle 4, -2 \rangle$ is given, find the output vector $\vec{b}$ after the linear transformation L.

① L = Reflection over x axis

② L = Reflection over y axis

③ L = Reflection over origin

④ L = Reflection over y = x

Blank : ① $\begin{bmatrix} 2 \\ 1 \end{bmatrix}$ ② $\begin{bmatrix} -2 \\ 1 \end{bmatrix}$ ③ $\begin{bmatrix} 2 \\ -1 \end{bmatrix}$ ④ $\begin{bmatrix} -2 \\ -1 \end{bmatrix}$ ⑤ $\begin{bmatrix} 1 \\ 2 \end{bmatrix}$ ⑥ $\begin{bmatrix} 4 \\ 2 \end{bmatrix}$ ⑦ $\begin{bmatrix} -1 \\ 2 \end{bmatrix}$

⑤ L = Rotation $60°$ counterclockwise about (0,0)

⑥ L = Rotation $30°$ counterclockwise about (0,0)

⑦ $L = \begin{bmatrix} 1 & 3 \\ 4 & -1 \end{bmatrix}$

⑧ $L = \begin{bmatrix} -2 & 0 \\ 0 & 3 \end{bmatrix}$

3. Composite Transformation of Matrix

☺ Reminder: If $f(x) = 2x$ and $g(x) = x + 2$, then $f(g(2)) =$ ① ________

※ **Composite Transformation**: Linear transformation A followed by linear transformation B

$$B(A\vec{a}) = (B \cdot A)\vec{a} = \vec{b}$$

Blank : ① 8

ex) If input vector $\vec{a} = \langle 2,1 \rangle$ is given, find the output vector $\vec{b}$ after reflecting over origin(A) followed by enlarging by scale factor 2 (B).

Transformation $A = \begin{bmatrix} -1 & 0 \\ 0 & -1 \end{bmatrix}$, transformation $B = \begin{bmatrix} 2 & 0 \\ 0 & 2 \end{bmatrix}$, we are finding $\vec{b} = B(A\vec{a})$.

There are two ways to solve;

$A\vec{a} = \begin{bmatrix} -1 & 0 \\ 0 & -1 \end{bmatrix} \begin{bmatrix} ① \\ \square \end{bmatrix} = \begin{bmatrix} ② \\ \square \end{bmatrix}$

Then $\vec{b} = B(A\vec{a}) = \begin{bmatrix} 2 & 0 \\ 0 & 2 \end{bmatrix} \begin{bmatrix} -2 \\ ③ \end{bmatrix} = \begin{bmatrix} ④ \\ \square \end{bmatrix}$

$BA = \begin{bmatrix} 2 & 0 \\ 0 & 2 \end{bmatrix} \begin{bmatrix} -1 & 0 \\ 0 & -1 \end{bmatrix} = \begin{bmatrix} ⑤ & \square \\ \square & \square \end{bmatrix}$

Then $\vec{b} = B(A\vec{a}) = (BA)\vec{a}$

$= \begin{bmatrix} -2 & 0 \\ ⑥ & \square \end{bmatrix} \begin{bmatrix} ⑦ \\ \square \end{bmatrix} = \begin{bmatrix} ⑧ \\ \square \end{bmatrix}$

EXAMPLE 2. If input vector $\vec{a} = \langle 4,-2 \rangle$ is given, find the output vector $\vec{b}$ after the linear transformation A followed by linear transformation B.

① A = Reflection over x axis

$B = \begin{bmatrix} 1 & 3 \\ 3 & 1 \end{bmatrix}$

② $A = \begin{bmatrix} 2 & -1 \\ -1 & 2 \end{bmatrix}$

B = Reflection over y axis

Blank : ① $\begin{bmatrix} 2 \\ 1 \end{bmatrix}$ ② $\begin{bmatrix} -2 \\ -1 \end{bmatrix}$ ③ -1 ④ $\begin{bmatrix} -4 \\ -2 \end{bmatrix}$ ⑤ $\begin{bmatrix} -2 & 0 \\ 0 & -2 \end{bmatrix}$ ⑥ $0 \quad -2$ ⑦ $\begin{bmatrix} 2 \\ 1 \end{bmatrix}$ ⑧ $\begin{bmatrix} -4 \\ -2 \end{bmatrix}$

③ $A = \begin{bmatrix} 2 & -1 \\ -1 & 2 \end{bmatrix}$

B = Reflection over y = x

④ A = Reflection over origin

$B = \begin{bmatrix} 0 & 1 \\ 1 & -1 \end{bmatrix}$

⑤ A = Rotation $30°$ counterclockwise about (0,0)

B = Reflection over y axis

⑥ A = Reflection over y = x

B = Rotation $60°$ counterclockwise about (0,0)

4. Finding Inverse Transformation

☺ Reminder: If $f(3)=2$, then $f^{-1}(2)=$ ① ________

ex) If output vector $\vec{b}=\langle 2,1\rangle$ is given, find the input vector $\vec{a}$ where the linear transformation L = reflection over origin.

Transformation $L=\begin{bmatrix} -1 & 0 \\ 0 & -1 \end{bmatrix}$ and since $L(\vec{a})=\vec{b}$ then ② ____________.

$$\vec{a}=\begin{bmatrix} -1 & 0 \\ 0 & -1 \end{bmatrix}^{-1}\begin{bmatrix} 2 \\ 1 \end{bmatrix}=\begin{bmatrix} \text{③}\square & \square \\ \square & \square \end{bmatrix}\begin{bmatrix} 2 \\ 1 \end{bmatrix}=\begin{bmatrix} \text{④}\square \\ \square \end{bmatrix}$$

※ **Inverse Transformation**

Let $\vec{a}$: input vector, $\vec{b}$: output vector, A is a 2x2 matrix for linear transformation L, then we can say;

If $L(\vec{a})=\vec{b}$, then $\vec{a}=L^{-1}(\vec{b})=A^{-1}\cdot\vec{b}$

Blank : ① 3 ② $\vec{a}=L^{-1}(\vec{b})$ ③ $\begin{bmatrix} -1 & 0 \\ 0 & -1 \end{bmatrix}$ ④ $\begin{bmatrix} -2 \\ -1 \end{bmatrix}$

EXAMPLE 3. If output vector $\vec{a} = \langle 4, -2 \rangle$ is given, find the input vector $\vec{a}$ where the linear transformation is L.

① L = Reflection over x axis

② L = Reflection over y axis

③ $L = \begin{bmatrix} 0 & 3 \\ 4 & -1 \end{bmatrix}$

④ $L = \begin{bmatrix} -2 & 0 \\ 0 & 3 \end{bmatrix}$

11.4 Matrices Modeling Context

1. Markov Process

☺ Reminder:

In a certain city, the weather could be either sunny or cloudy.

Tomorrow will be sunny 30% of the time, then what is the probability it will be cloudy tomorrow?

In another certain city, the weather could be either sunny or cloudy. If today is sunny, tomorrow will be sunny 80% of the time. If today is cloudy, tomorrow will be cloudy 70% of the time.

We can organize the info as;

Table:

today / tomorrow	sunny	cloudy
sunny	0.8	①
cloudy	②	③

diagram:

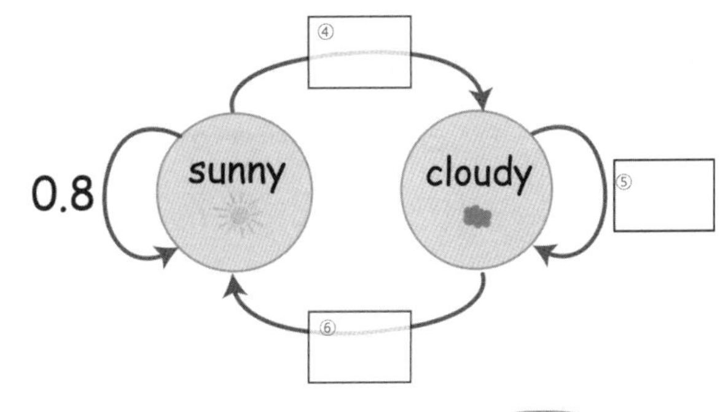

Transition Matrix :

$$T = \begin{bmatrix} 0.8 & 0.3 \\ 0.2 & 0.7 \end{bmatrix}$$

Transition Matrix

Blank : ① 0.3 ② 0.2 ③ 0.7 ④ 0.2 ⑤ 0.7 ⑥ 0.3

Suppose that the probability that it will be sunny on a certain Monday is 0.2 and the probability that it will be cloudy is 0.8.

A) Determine the probability that it will be cloudy on the next day, Tuesday.

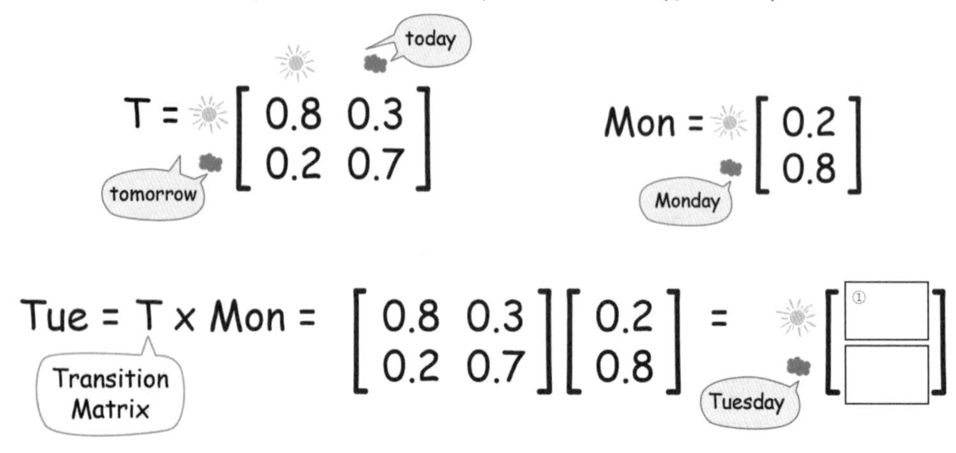

Therefore, the probability that Tuesday will be cloudy is ②_______.

B) Determine the probability that it will be cloudy on Wednesday.

$$\text{Wed} = T \times \text{Tue} = \begin{bmatrix} 0.8 & 0.3 \\ 0.2 & 0.7 \end{bmatrix}\begin{bmatrix} 0.8 & 0.3 \\ 0.2 & 0.7 \end{bmatrix}\begin{bmatrix} 0.2 \\ 0.8 \end{bmatrix}$$

Transition Matrix

$$= \begin{bmatrix} 0.8 & 0.3 \\ 0.2 & 0.7 \end{bmatrix}^2 \begin{bmatrix} 0.2 \\ 0.8 \end{bmatrix} = \begin{bmatrix} ③\square \\ \square \end{bmatrix}$$

Wednesday

Therefore, the probability that Wednesday will be cloudy is ④_______.

Blank : ① $\begin{bmatrix} 0.4 \\ 0.6 \end{bmatrix}$ ② 60% ③ $\begin{bmatrix} 0.5 \\ 0.5 \end{bmatrix}$ ④ 50%

※ **Markov Property**

Markov property says that the future stage or outcome of each event depends only on the current stage.

We can use transition probability matrix to model how the stage change.
(The column of the transition probability matrix must sum to 1)

Next Stage = Transition Probability Matrix x Initial Stage

m Stage from now = (Transition Probability Matrix)^m x Initial Stage

EXAMPLE 1. Commuters to a city commute either by car or public transportation. Suppose each year 20% of the car commuters change to public transportation and 10% of those using public transportation change to cars. If 60% of the commuters now use public transportation and 40% use cars this year,

(a) find the percentage of the population who use cars next year.

(b) find the percentage of the population who use cars two years from now.

EXAMPLE 2. Over the years, it has been observed that each year, 3% of those living in a certain city's urban area move to the suburbs and 5% of those in the suburbs move into the urban area. Suppose that there are 6000 people lives in the suburb and 4000 people lives in the urban area this year.

(a) Find the population living in the urban next year.

(b) Find the population living in the urban three years from now.

Calculator Skills (for Ti 84, Ti nspire CAS)

1. Calculator Skills for Ti-84

1) Basics

Clear All data	**2nd** – + – 4 – **ENTER** **2nd** – + – 7 – 1 – 2
Insert a symbol	**2nd** – **DEL**(insert) – **ENTER**
Recall the expression	**2nd** – **ENTER**(entry) or use ▼ or ▲ button to highlight the expression, then **ENTER**
BASIC SETUP	**MODE** Normal Sci Eng Float 0123456789 Radian Degree Func Par Pol Seq Connected Dot Sequential Simul Real a+bi re^θi Full Horiz G-T
Degree Radian mode	**MODE** - ▼ - ▼
To **return to the 'home' screen** at any time	**2nd** – **MODE**(QUIT)
To enter a **rational expression**	(numerator)/(denominator)
Negative vs **subtract** sign	negative button:(-) subtract button:-
To change a **decimal →fraction**	**MATH** – **ENTER** – **ENTER**

To find or enter the **absolute value**	**MATH** – ▶ (NUM) – 1:abs
sin cos tan	**sin**, **cos**, **tan**
csc sec cot	1 / **sin**, 1 / **cos**, 1 / **tan**
arcsin, arccos, arctan	**2nd** – **sin**, **2nd** – **cos**, **2nd** – **tan**

2) Graphing functions

To **graph** an equation	**Y=** – enter the equation – **GRAPH**

3) Window adjusting

To **change the viewing window** for a graph	**WINDOW** enter values and desired scales WINDOW Xmin=-10 Xmax=10 Xscl=1 Ymin=-10 Ymax=10 Yscl=1 Xres=1 (xscl,yscl means distance between tick marks)
Going back to the **standard size** of the window setting	**ZOOM**–6:ZStandard

EXAMPLE 1. Graph the function $y = x^3 - 12x^2 + 30x + 30$, $-10 \le x \le 10$, $-20 \le y \le 60$.

4) Finding Zeros

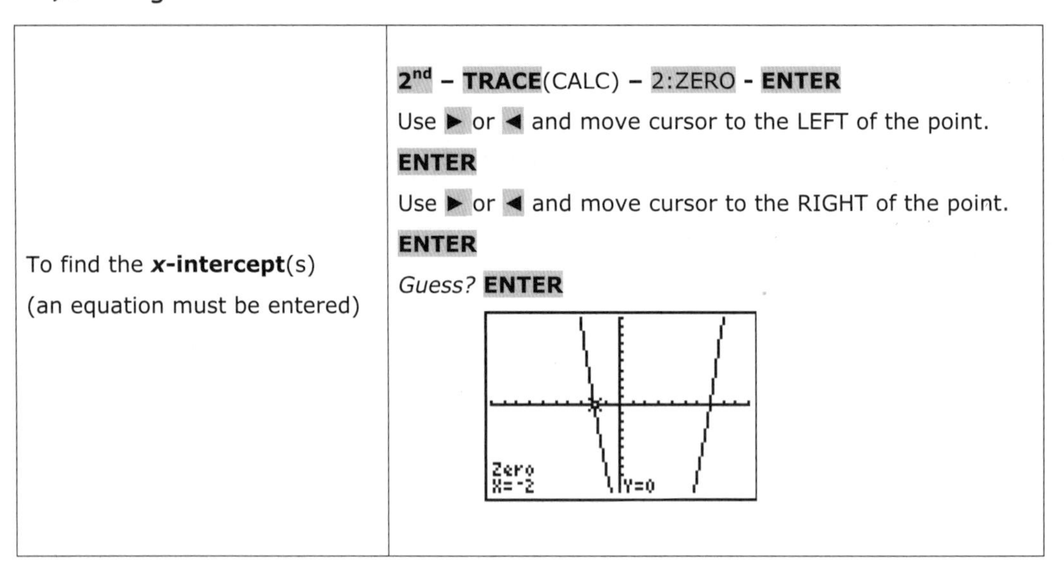

To find the ***x*-intercept**(s) (an equation must be entered)	**2nd** – **TRACE**(CALC) – 2:ZERO - **ENTER** Use ▶ or ◀ and move cursor to the LEFT of the point. **ENTER** Use ▶ or ◀ and move cursor to the RIGHT of the point. **ENTER** *Guess?* **ENTER** Zero X=-2 Y=0

EXAMPLE 2. Graph the function $y = x^3 - 12x^2 + 30x + 30$. Find the Zeros.

5) Finding Max or Min

To find the **maximum (minimum) point** (an equation must be entered.)	**2nd** – **TRACE**(CALC) – 3:MAXIMUM – **ENTER** Use ▶ or ◀ and move cursor to the LEFT of the point. **ENTER** Use ▶ or ◀ and move cursor to the RIGHT of the point. **ENTER** **ENTER**

EXAMPLE 3. Graph the function $y = x^3 - 12x^2 + 30x + 30$. Find the max and min.

6) Finding Intersections

To find the **intersection of 2 graphs** (2 equations must be entered)	**2nd** – **TRACE**(CALC) – 5:INTERSECT – **ENTER** Use ▶ or ◀ to move cursor close to the point. **ENTER** – **ENTER** – **ENTER** (press enter 3 times) NOTE: Your *x* value must be within your viewing window. 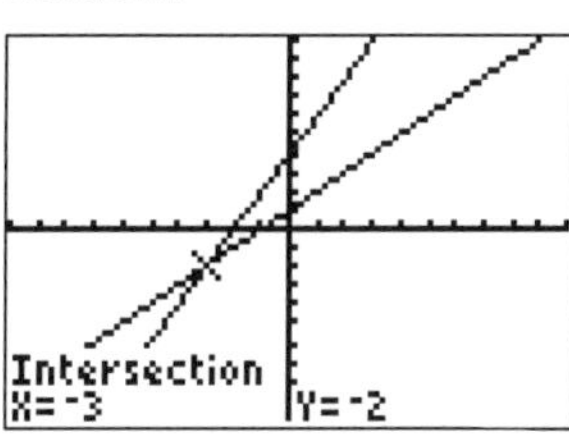

To **solve an equation by graphing** (2 equations must be entered)	**Y=** Enter left-hand side of equation in $y1$ = ; right-hand side in $y2$ = Graph and find the point(s) of intersection.
Window setting [-2π, 2π] when you have a trig functions	**Zoom – trig**

EXAMPLE 4. Find the least positive x that satisfy $\frac{1}{4}+\sin(\pi x)=4^{-x}$.

EXAMPLE 5. Find the x value of the intersection of $y=2$ and $y=\frac{20}{1+x^2}$.

7) Regressions

Enter Data	**STAT** and then select 1:Edit. Enter all of your "x data" into L1 and all of your "y data" into L2. L1 | L2 | L3 2 5 | 1 | ------ 7 | 5 8 | 4 9 | 5 1 | 1 5 | 4 L2(7) =
Calculating Regressions	EDIT CALC TESTS 1:1-Var Stats 2:2-Var Stats 3:Med-Med 4:LinReg(ax+b) 5:QuadReg 6:CubicReg 7↓QuartReg Once you have your data in, go to **STAT** and then the ►CALC menu up top. Finally, select 4:LinReg (or other regression model you want) and press enter. LinReg(ax+b) Xlist:L1 Ylist:L2 FreqList: Store RegEQ: Calculate Put L1 in X list, L2 is Y list. Make sure you 'store regression equation' in your y1(**ALPHA-trace – y1)** LinReg y=ax+b a=.5224489796 b=.2857142857 r²=.6430141287 r=.8018816177 If you can't see r^2 and r, then go to **mode** go up to STAT Diagnostic-On

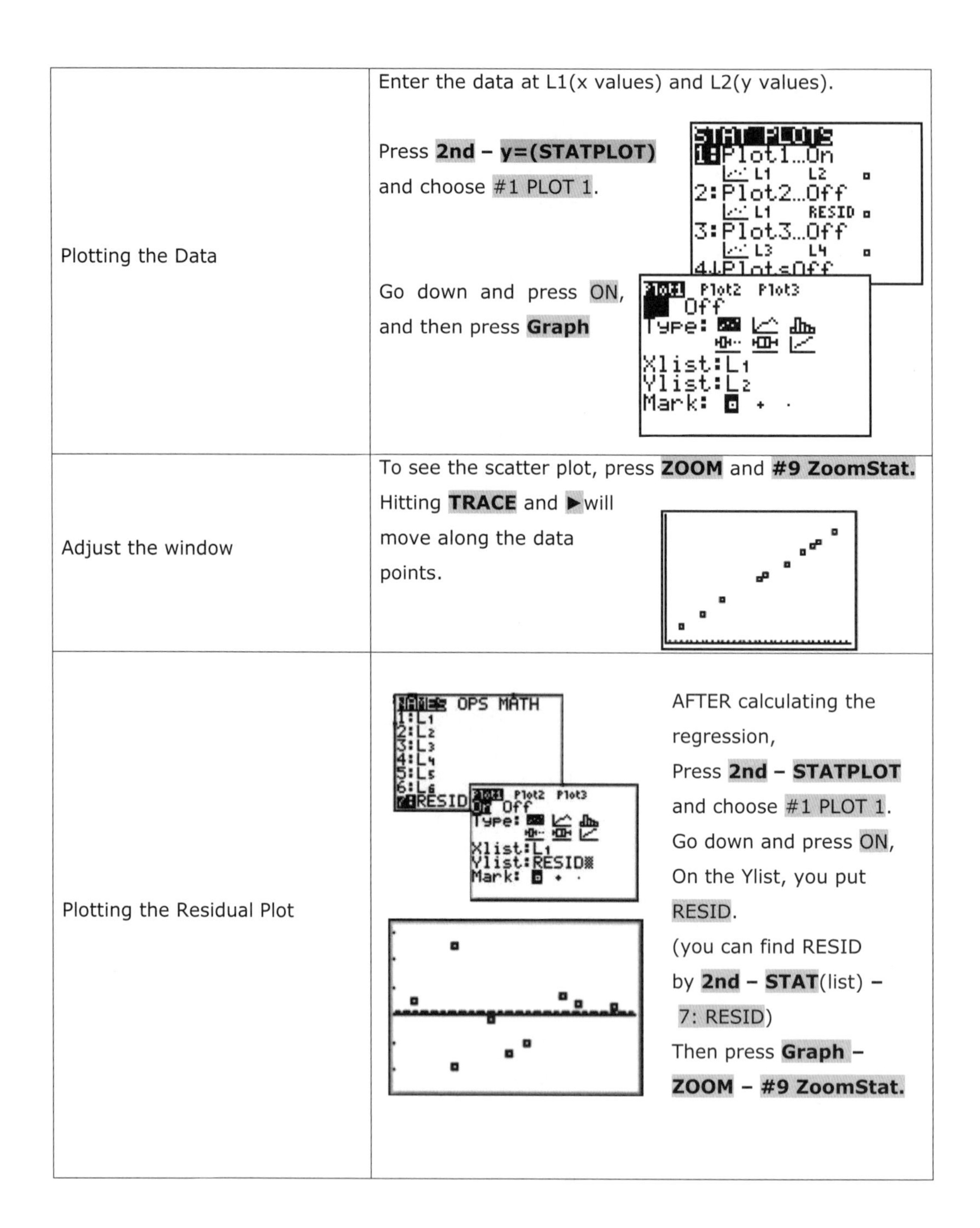

Plotting the Data	Enter the data at L1(x values) and L2(y values). Press **2nd** – **y=(STATPLOT)** and choose #1 PLOT 1. Go down and press ON, and then press **Graph**
Adjust the window	To see the scatter plot, press **ZOOM** and **#9 ZoomStat.** Hitting **TRACE** and ►will move along the data points.
Plotting the Residual Plot	AFTER calculating the regression, Press **2nd** – **STATPLOT** and choose #1 PLOT 1. Go down and press ON, On the Ylist, you put RESID. (you can find RESID by **2nd** – **STAT**(list) – 7: RESID) Then press **Graph** – **ZOOM** – **#9 ZoomStat.**

8) Graphing Polar or Parametric Curves

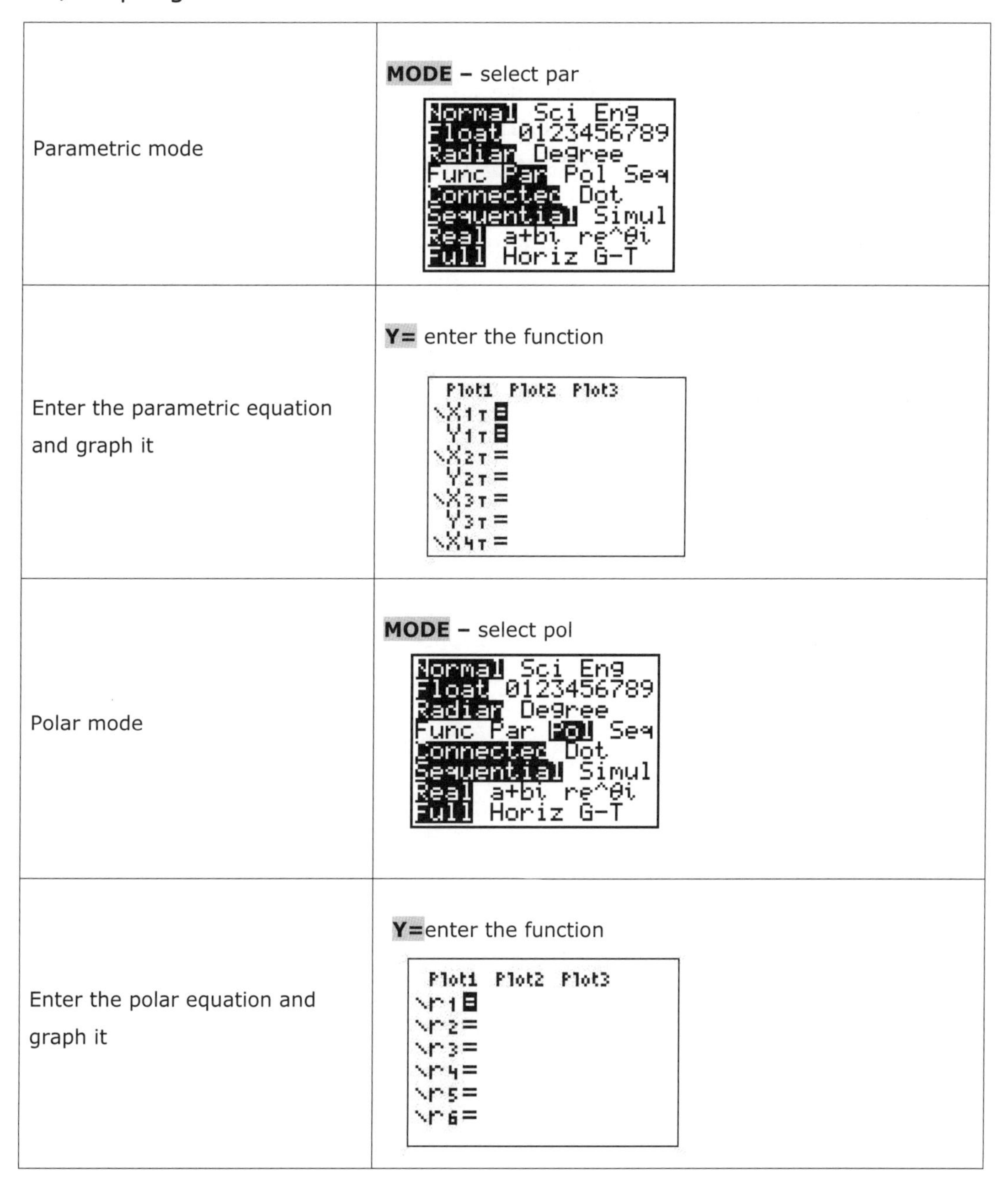

Parametric mode	**MODE** – select par
Enter the parametric equation and graph it	**Y=** enter the function
Polar mode	**MODE** – select pol
Enter the polar equation and graph it	**Y=**enter the function

EXAMPLE 6. Graph polar curve $r = 2 - 2\sin\theta$.

9) Matrix

Setting up a Matrix A	**2nd** – **x⁻¹**(MATRIX) ▶ – ▶ – EDIT –1:[A] – **ENTER** Define the size of the matrix and enter the elements. Use **ENTER** to move to the next spaces. **2nd** – **MODE**(QUIT)
Paste the matrix into the home screen	**2nd** – **x⁻¹**(MATRIX) choose a matrix you want to paste **ENTER**
Finding **determinant**	**2nd** – **x⁻¹**(MATRIX) ▶ –MATH –1:[A] det – **ENTER**

EXAMPLE 7. Calculate the followings .

① $\begin{bmatrix} 2 & 4 \\ 7 & -8 \end{bmatrix}\begin{bmatrix} 4 & -10 & 7 \\ 8 & -2 & -4 \end{bmatrix}$

② $\begin{vmatrix} 7 & 0 & -1 \\ 5 & -3 & 3 \\ -7 & 8 & 1 \end{vmatrix}$

2. Calculator Skills for Ti-nspire CAS

1) Basics

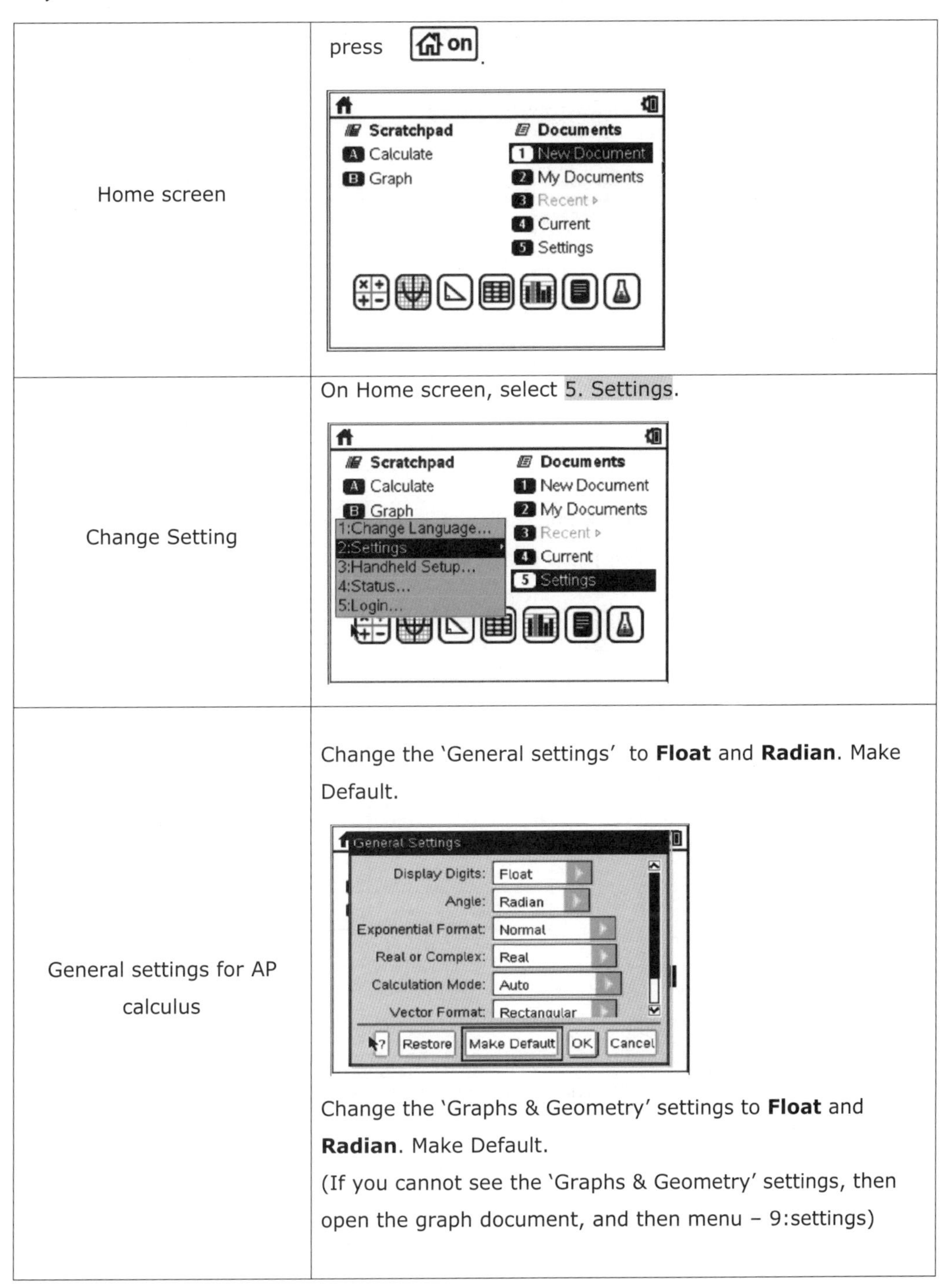

Home screen	press [on].
Change Setting	On Home screen, select 5. Settings.
General settings for AP calculus	Change the 'General settings' to **Float** and **Radian**. Make Default. Change the 'Graphs & Geometry' settings to **Float** and **Radian**. Make Default. (If you cannot see the 'Graphs & Geometry' settings, then open the graph document, and then menu – 9:settings)

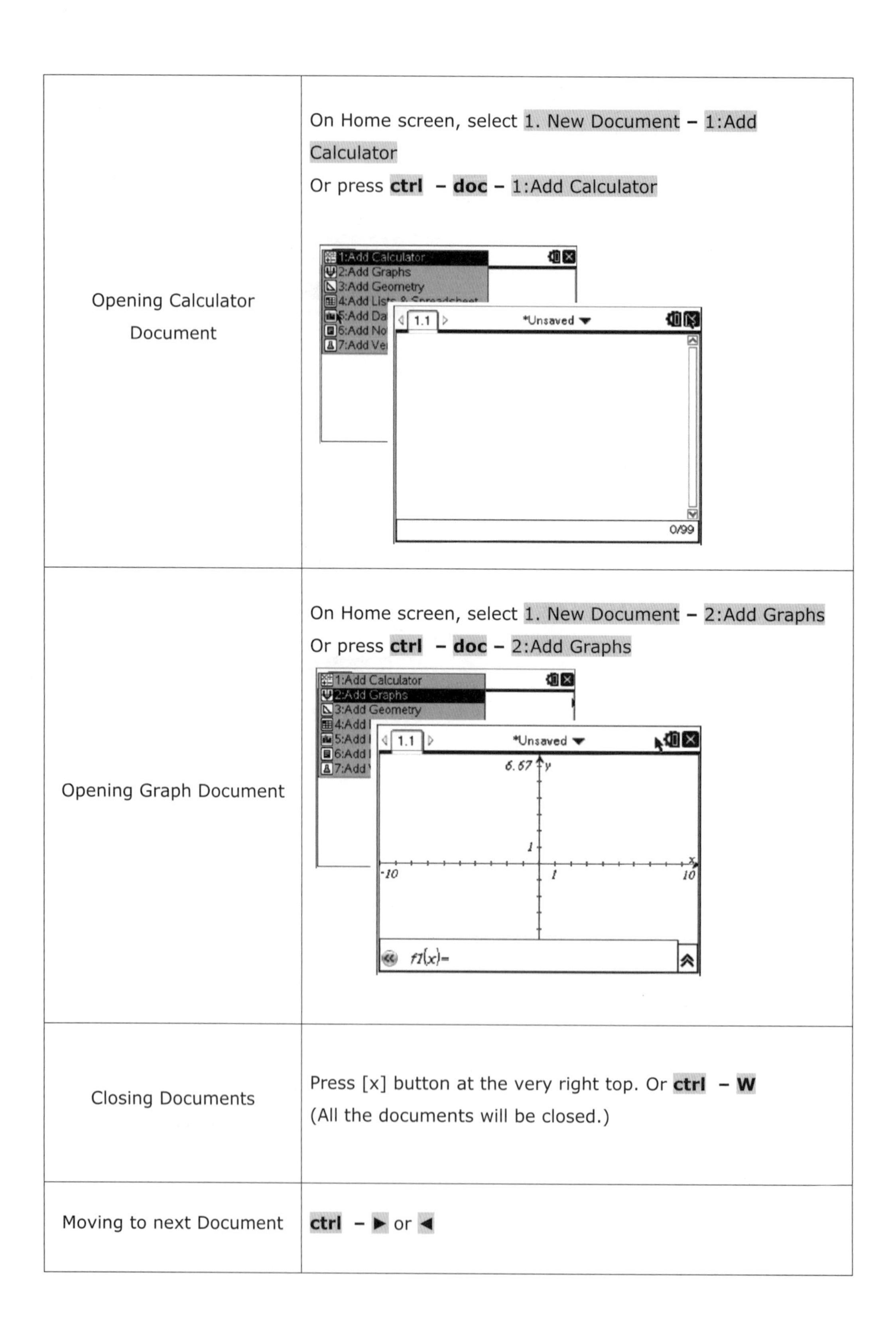

Opening Calculator Document	On Home screen, select 1. New Document – 1:Add Calculator Or press **ctrl** – **doc** – 1:Add Calculator
Opening Graph Document	On Home screen, select 1. New Document – 2:Add Graphs Or press **ctrl** – **doc** – 2:Add Graphs
Closing Documents	Press [x] button at the very right top. Or **ctrl** – **W** (All the documents will be closed.)
Moving to next Document	**ctrl** – ► or ◄

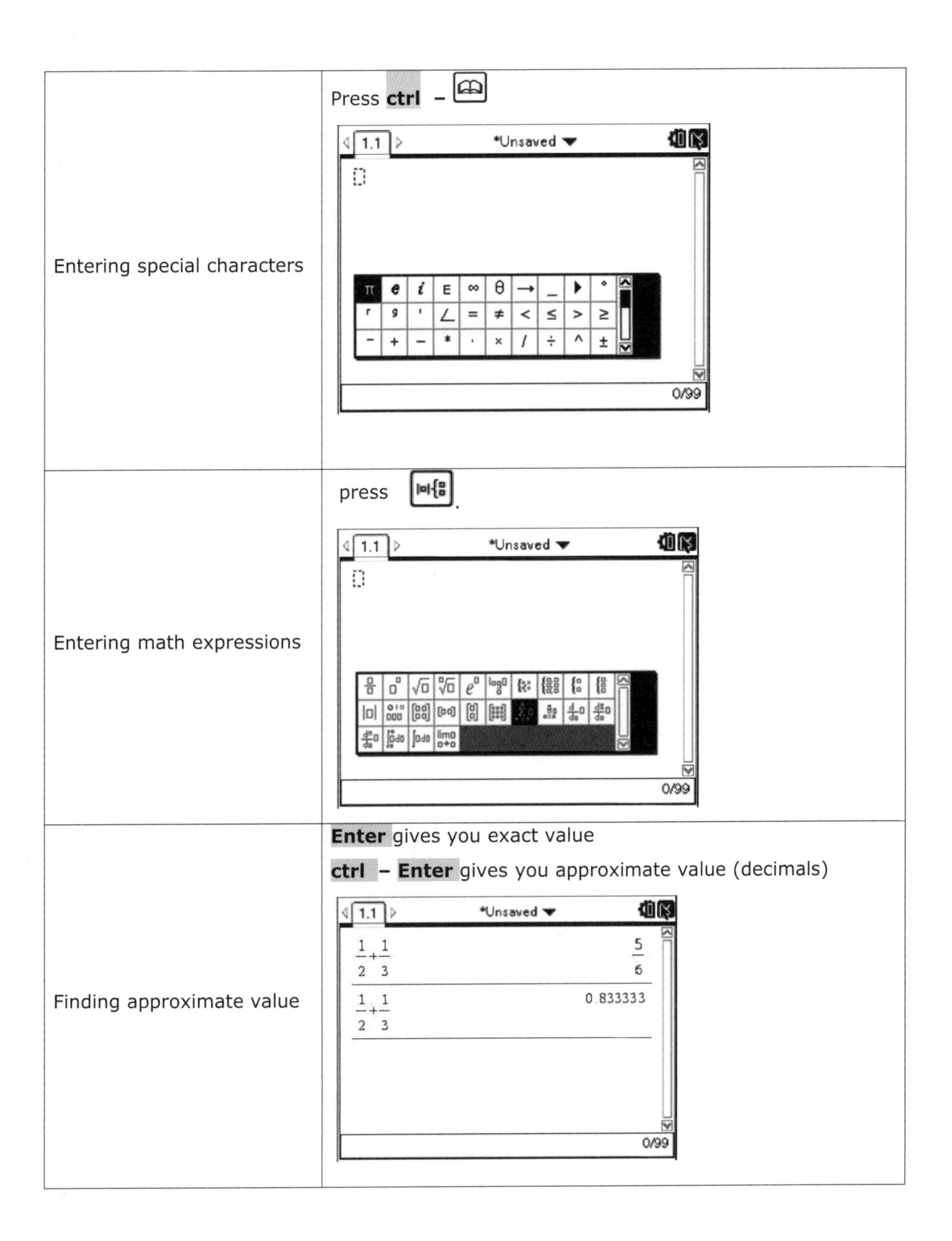

Entering special characters	Press **ctrl** – [catalog key]
Entering math expressions	press [templates key].
Finding approximate value	**Enter** gives you exact value **ctrl** – **Enter** gives you approximate value (decimals)

2) Graphing functions

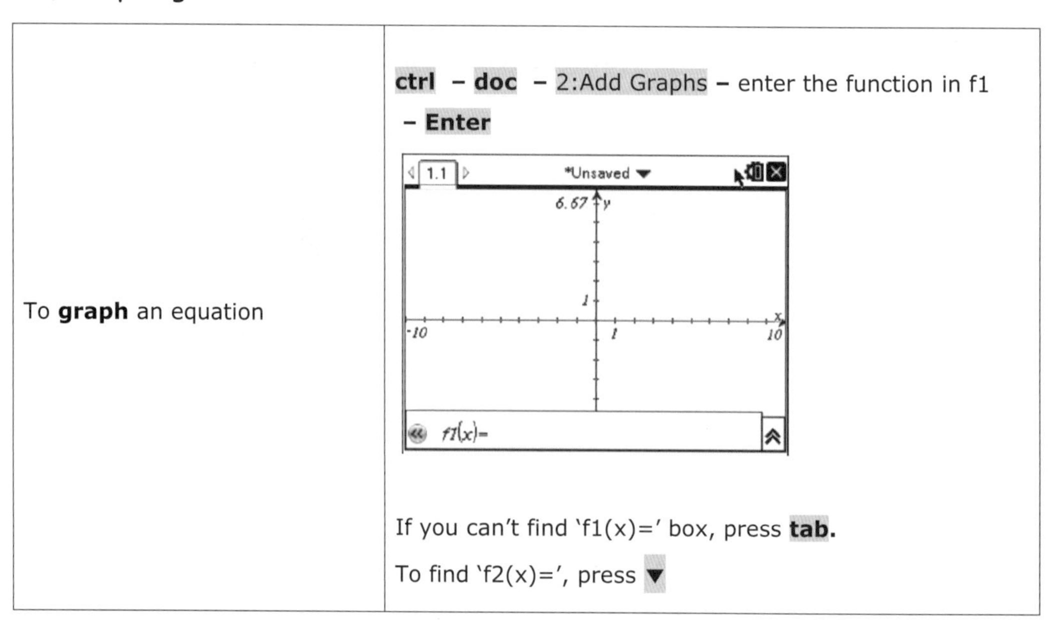

To **graph** an equation	**ctrl** – **doc** – 2:Add Graphs – enter the function in f1 – **Enter** If you can't find 'f1(x)=' box, press **tab.** To find 'f2(x)=', press ▼

3) Window adjusting

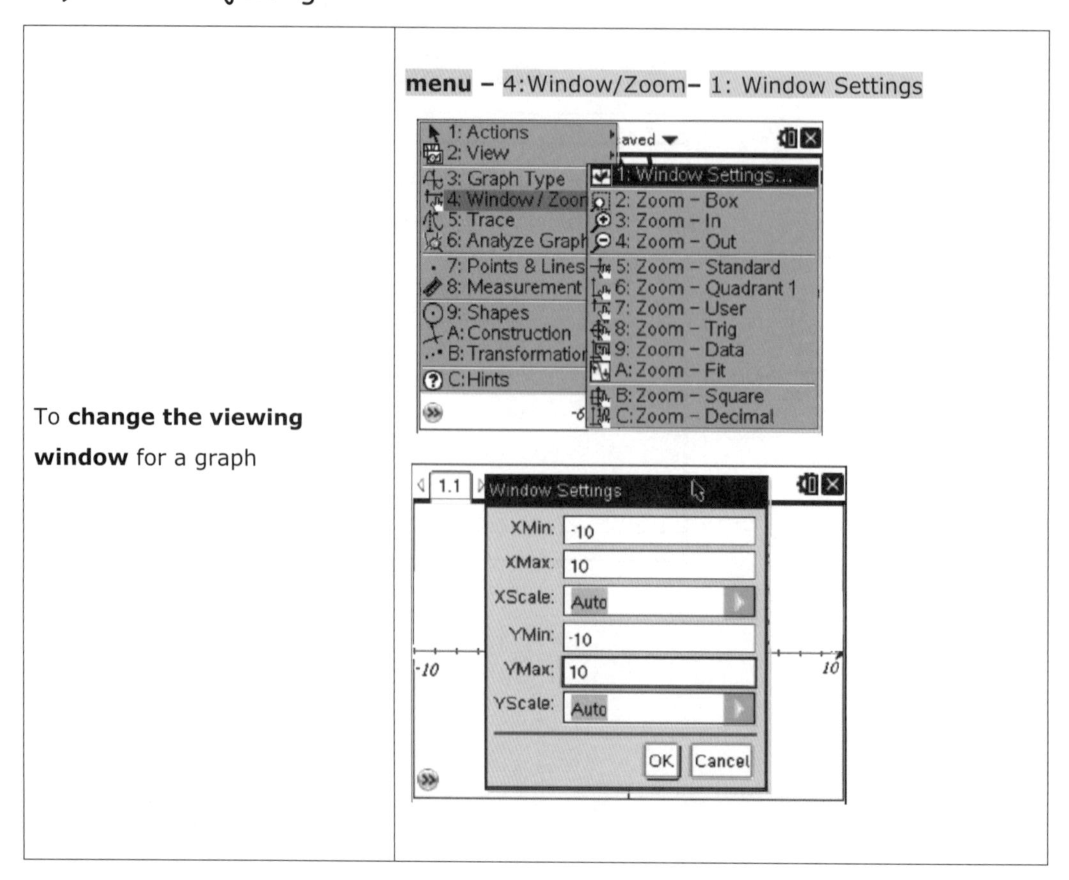

To **change the viewing window** for a graph	**menu** – 4:Window/Zoom– 1: Window Settings

Going back to the **standard size** of the window setting	**menu** – 4:Window/Zoom– 5: Zoom Standard

EXAMPLE 1. Graph the function $y = x^3 - 12x^2 + 30x + 30$, $-10 \le x \le 10$, $-20 \le y \le 60$.

4) Finding Zeros

To find the ***x*-intercept**(s) (an equation must be entered)	**menu** – 6: Analyze Graph– 1: Zero 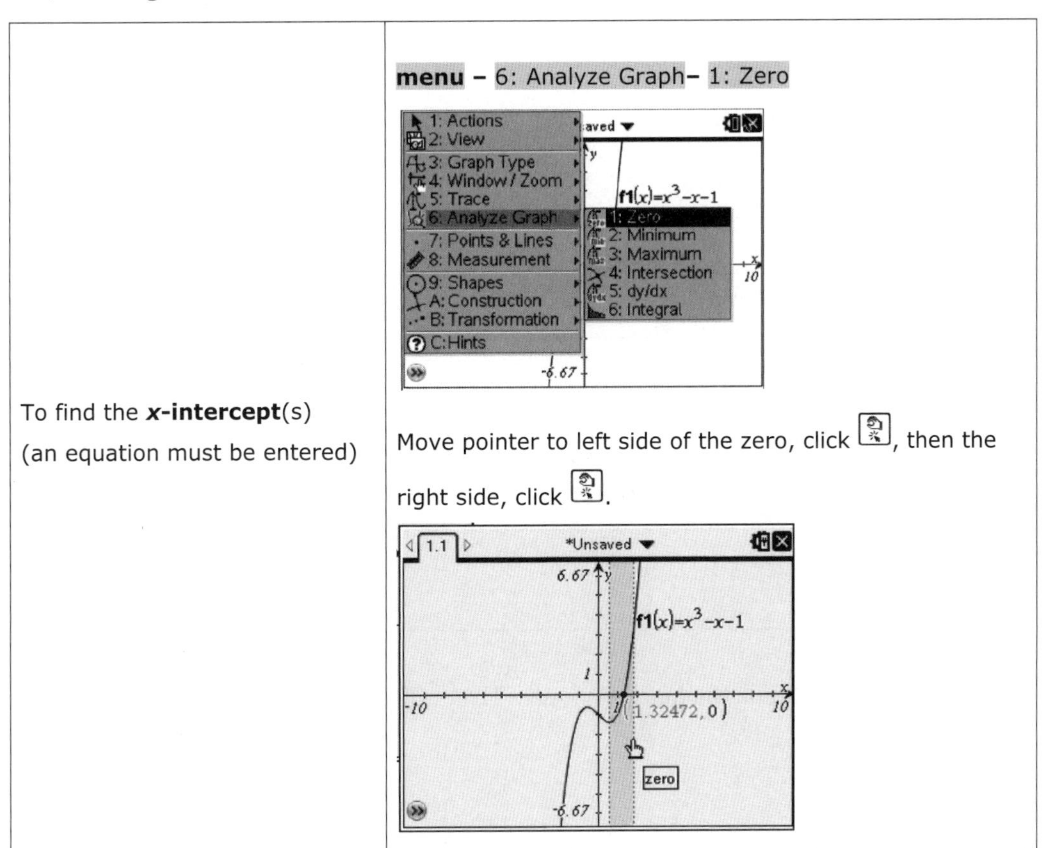 Move pointer to left side of the zero, click , then the right side, click .

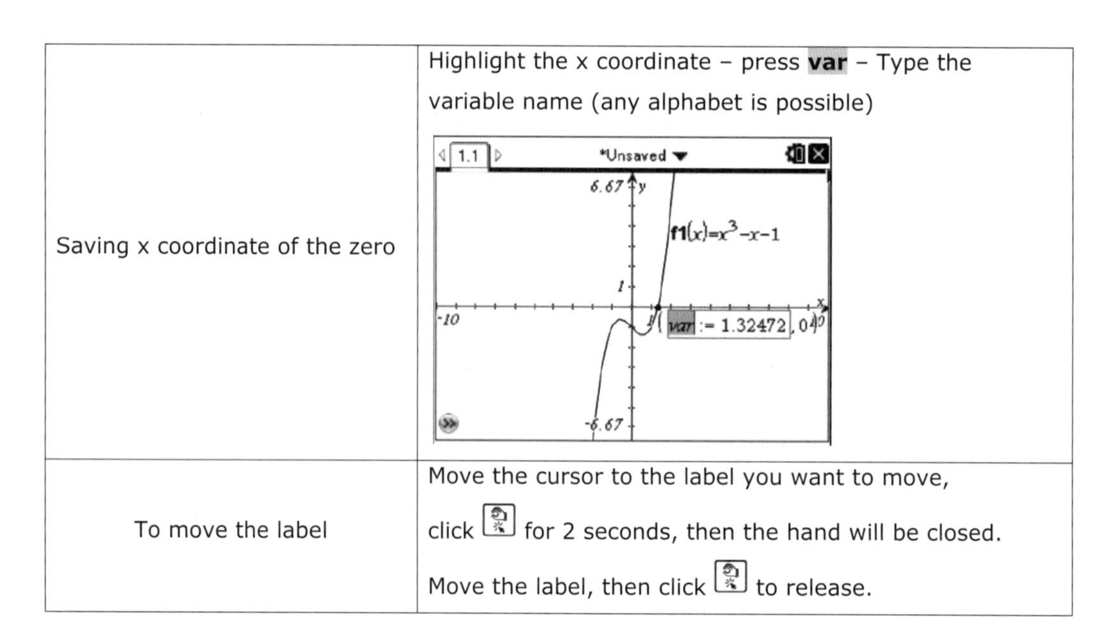

Saving x coordinate of the zero	Highlight the x coordinate – press **var** – Type the variable name (any alphabet is possible)
To move the label	Move the cursor to the label you want to move, click for 2 seconds, then the hand will be closed. Move the label, then click to release.

EXAMPLE 2. Graph the function $y = x^3 - 12x^2 + 30x + 30$. Find the Zeros.

5) Finding Max or Min

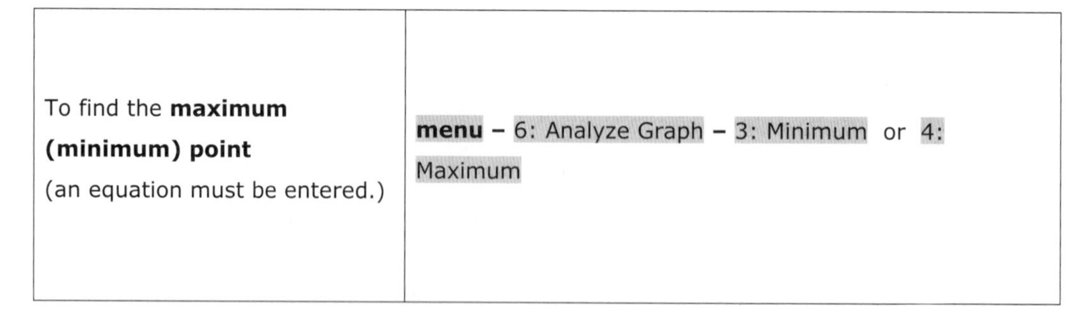

To find the **maximum (minimum) point** (an equation must be entered.)	**menu** – 6: Analyze Graph – 3: Minimum or 4: Maximum

EXAMPLE 3. Graph the function $y = x^3 - 12x^2 + 30x + 30$. Find the max and min.

6) Finding Intersections

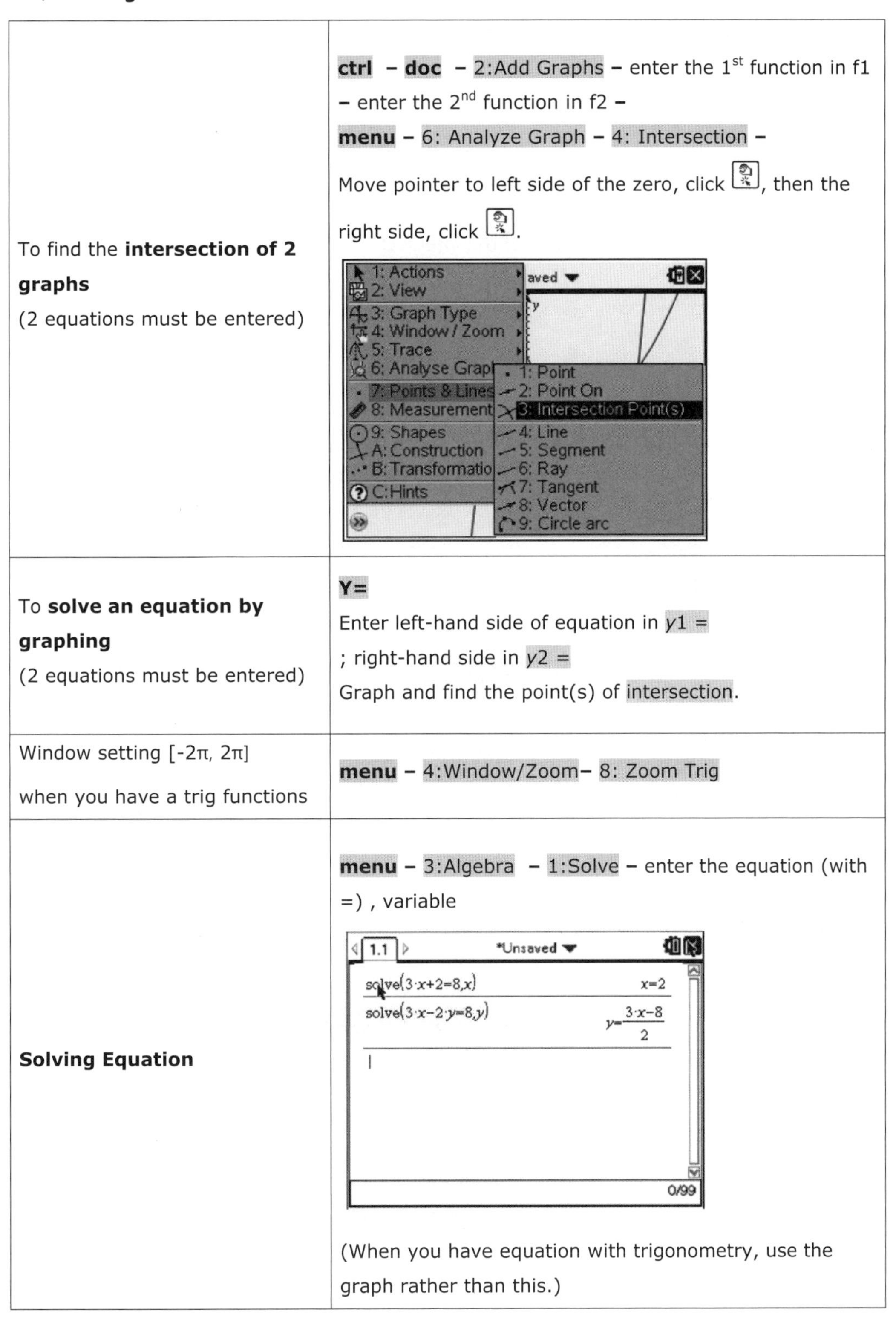

To find the **intersection of 2 graphs** (2 equations must be entered)	**ctrl** – **doc** – 2:Add Graphs – enter the 1st function in f1 – enter the 2nd function in f2 – **menu** – 6: Analyze Graph – 4: Intersection – Move pointer to left side of the zero, click, then the right side, click.
To **solve an equation by graphing** (2 equations must be entered)	**Y=** Enter left-hand side of equation in $y1 =$; right-hand side in $y2 =$ Graph and find the point(s) of intersection.
Window setting [-2π, 2π] when you have a trig functions	**menu** – 4:Window/Zoom– 8: Zoom Trig
Solving Equation	**menu** – 3:Algebra – 1:Solve – enter the equation (with =) , variable (When you have equation with trigonometry, use the graph rather than this.)

EXAMPLE 4. Find the least positive x that satisfy $\frac{1}{4}+\sin(\pi x)=4^{-x}$.

EXAMPLE 5. Find the x value of the intersection of $y=2$ and $y=\frac{20}{1+x^2}$.

7) Regressions

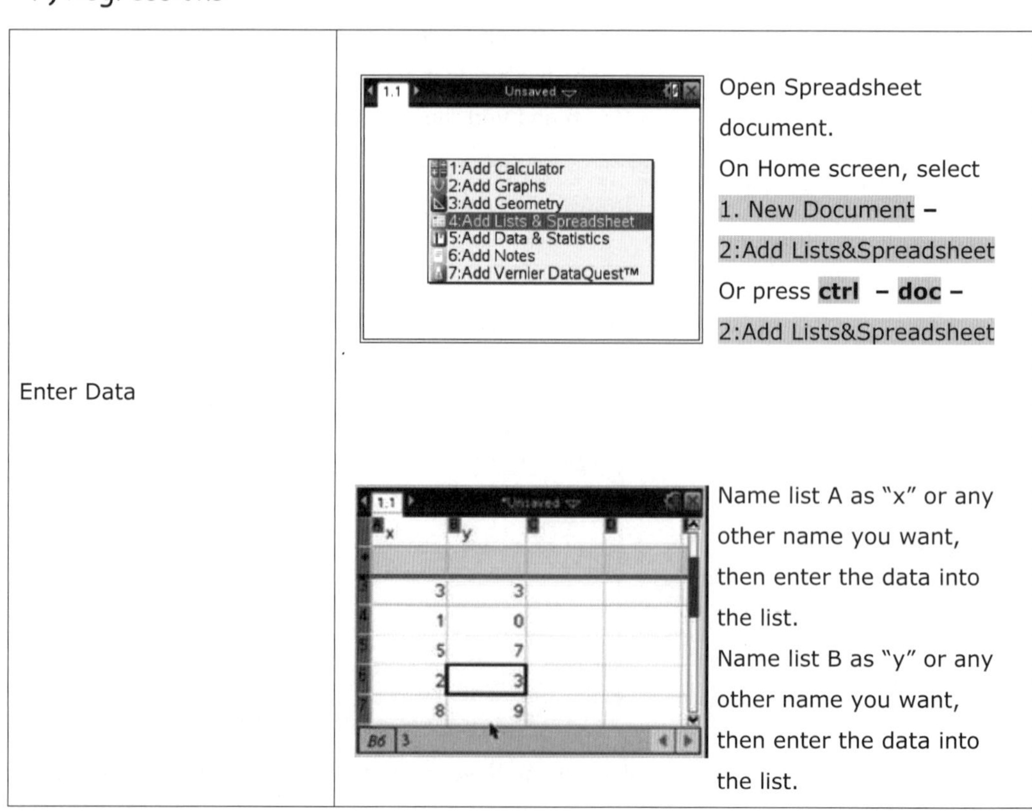

Enter Data	Open Spreadsheet document. On Home screen, select 1. New Document – 2:Add Lists&Spreadsheet Or press **ctrl** – **doc** – 2:Add Lists&Spreadsheet Name list A as "x" or any other name you want, then enter the data into the list. Name list B as "y" or any other name you want, then enter the data into the list.

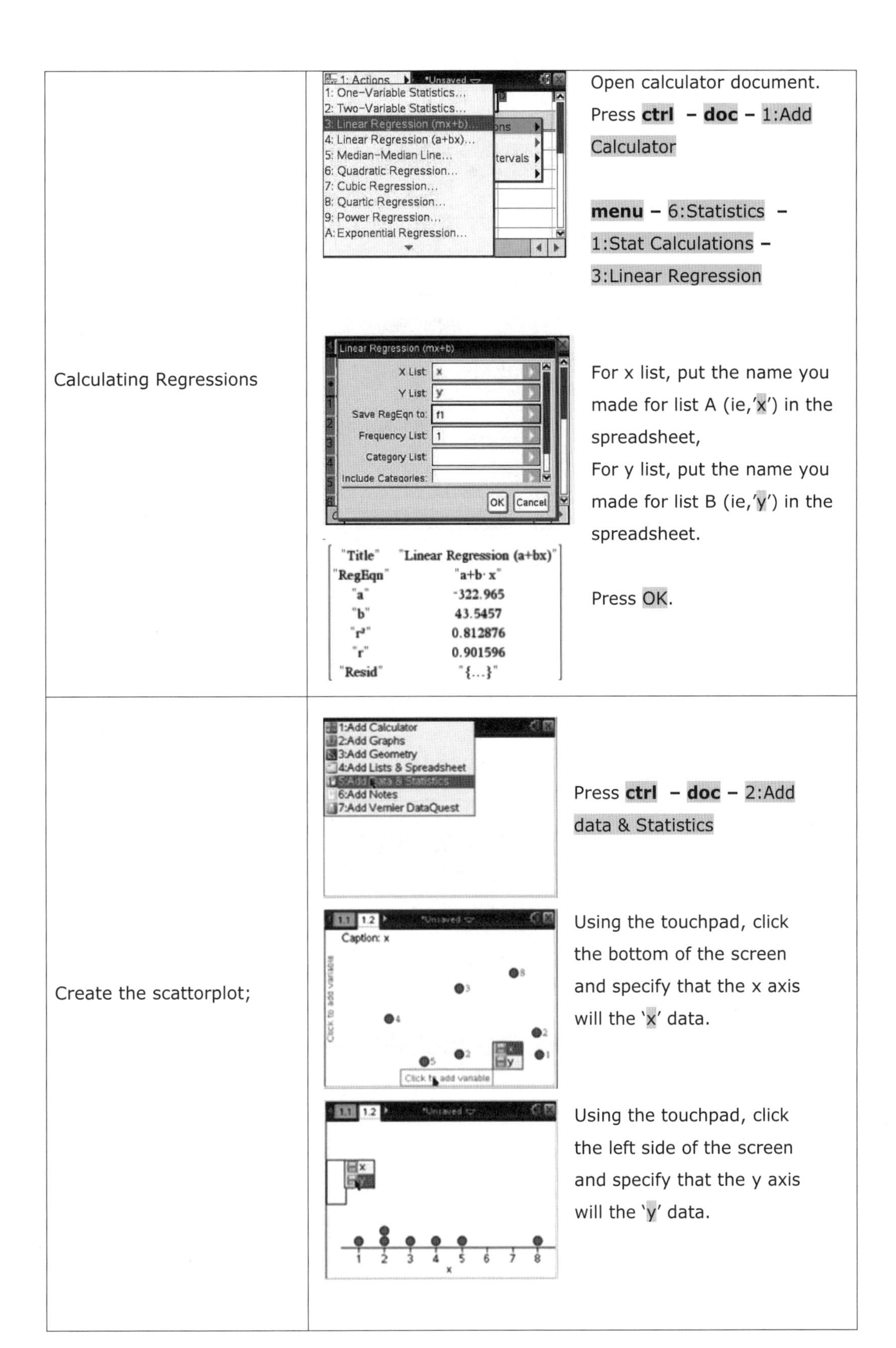

Calculating Regressions	Open calculator document. Press **ctrl** – **doc** – 1:Add Calculator **menu** – 6:Statistics – 1:Stat Calculations – 3:Linear Regression For x list, put the name you made for list A (ie,'x') in the spreadsheet, For y list, put the name you made for list B (ie,'y') in the spreadsheet. Press OK.
Create the scattorplot;	Press **ctrl** – **doc** – 2:Add data & Statistics Using the touchpad, click the bottom of the screen and specify that the x axis will the 'x' data. Using the touchpad, click the left side of the screen and specify that the y axis will the 'y' data.

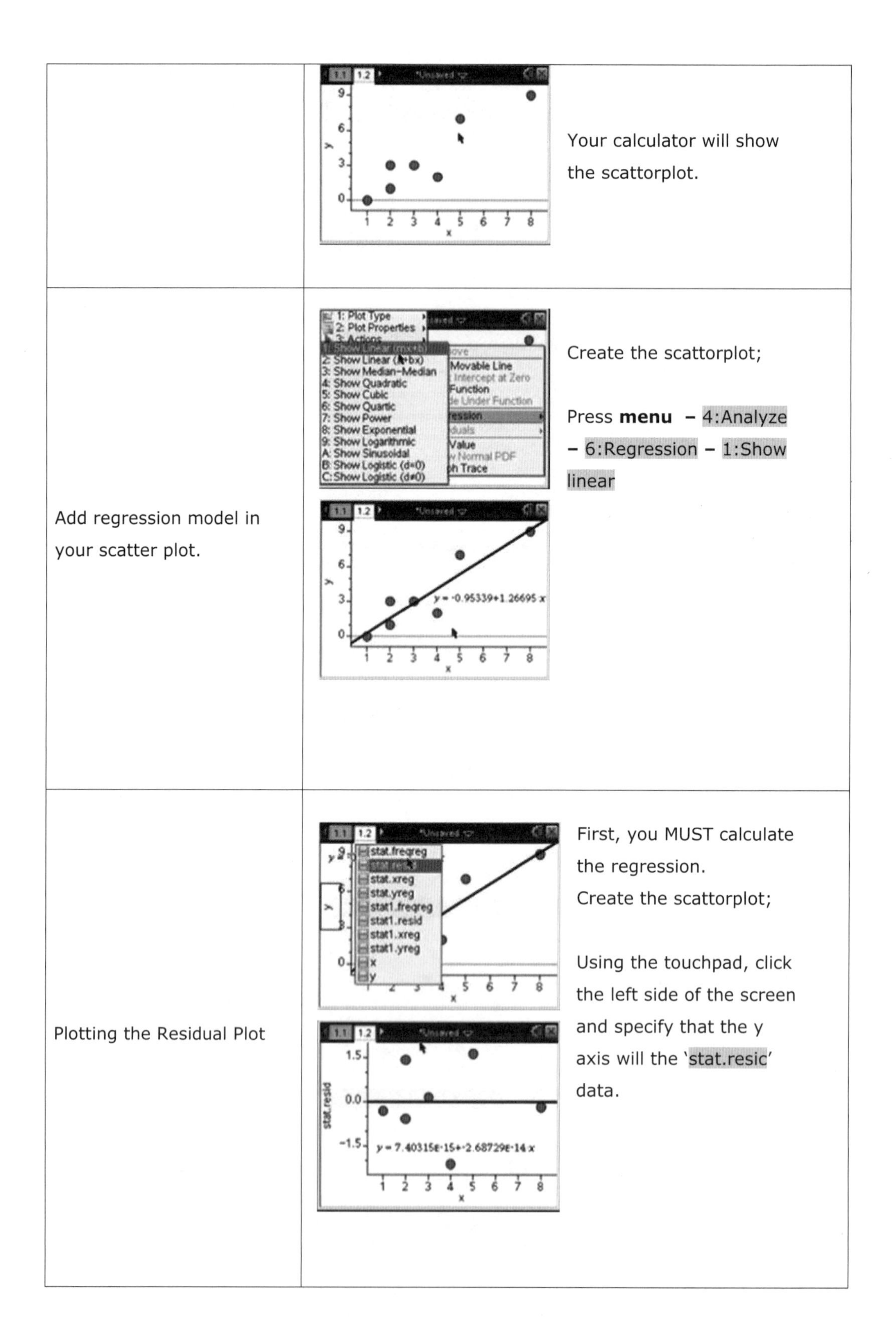

		Your calculator will show the scattorplot.
Add regression model in your scatter plot.		Create the scattorplot; Press **menu** – 4:Analyze – 6:Regression – 1:Show linear
Plotting the Residual Plot		First, you MUST calculate the regression. Create the scattorplot; Using the touchpad, click the left side of the screen and specify that the y axis will the 'stat.resic' data.

8) Graphing Polar or Parametric Curves

Enter the parametric equation and graph it	**ctrl** – **doc** – 2:Add Graphs – **menu** – 3: Graph Type – 2: Parametric – Enter the parametric equation Note: Do not put x as your variable! Put t !
Enter the polar equation and graph it	**ctrl** – **doc** – 2:Add Graphs – **menu** – 3: Graph Type – 3: Polar – Enter the polar equation Note: Do not put x as your variable! Put θ 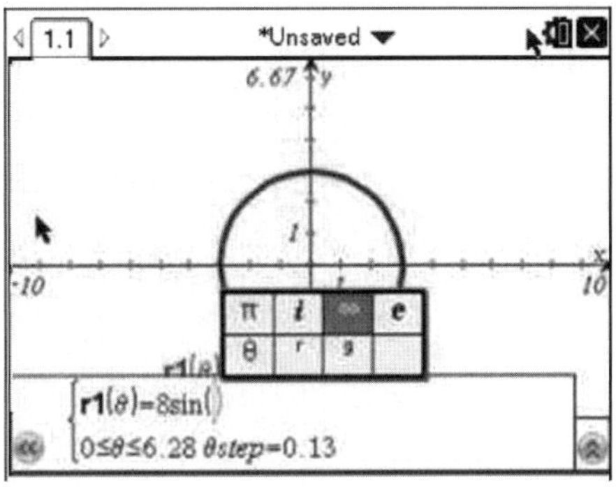 !

EXAMPLE 6. Graph polar curve $r = 2 - 2\sin\theta$.

10) Matrix

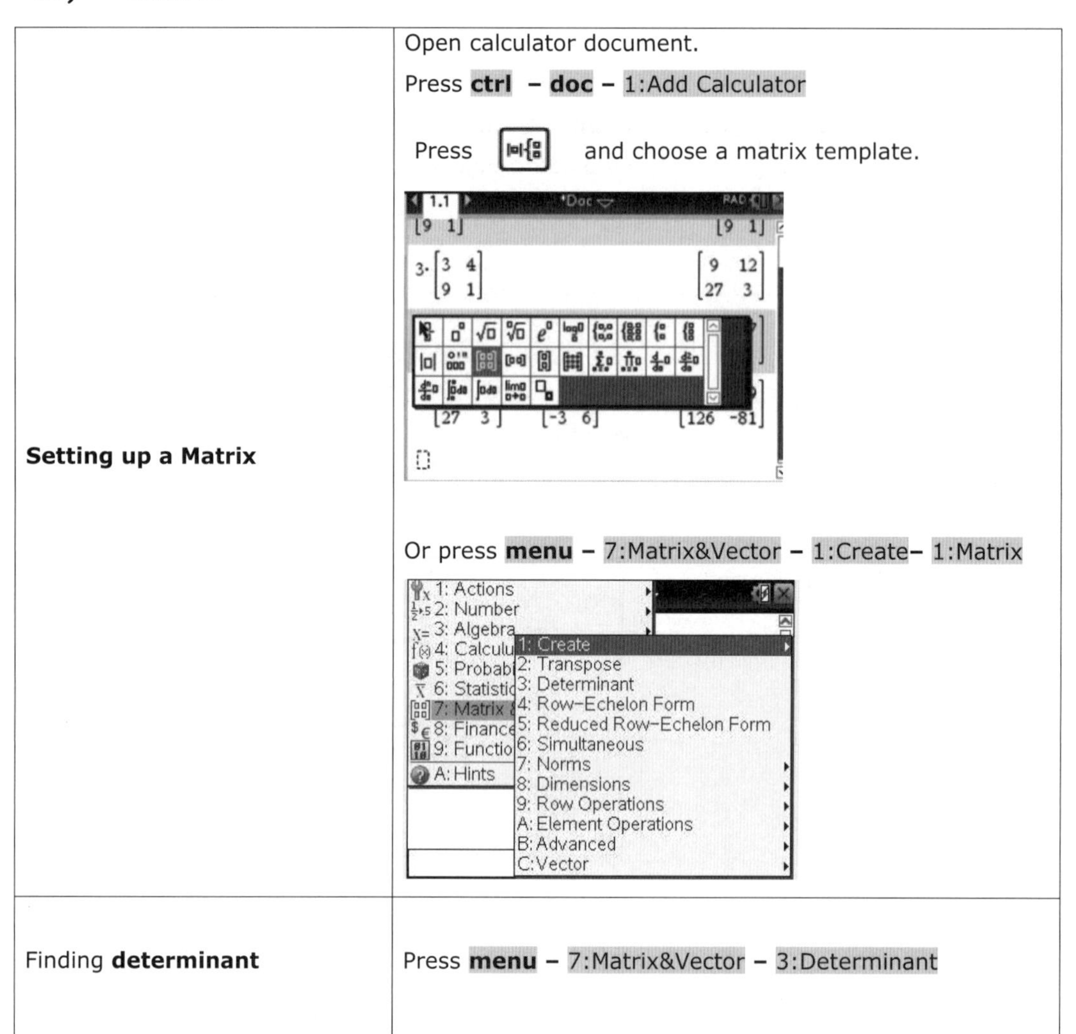

Setting up a Matrix	Open calculator document. Press **ctrl** – **doc** – 1:Add Calculator Press [template key] and choose a matrix template. Or press **menu** – 7:Matrix&Vector – 1:Create– 1:Matrix
Finding **determinant**	Press **menu** – 7:Matrix&Vector – 3:Determinant

EXAMPLE 7. Calculate the followings .

③ $\begin{bmatrix} 2 & 4 \\ 7 & -8 \end{bmatrix}\begin{bmatrix} 4 & -10 & 7 \\ 8 & -2 & -4 \end{bmatrix}$

④ $\begin{vmatrix} 7 & 0 & -1 \\ 5 & -3 & 3 \\ -7 & 8 & 1 \end{vmatrix}$

Mia's AP Precalculus

Multiple Choice Practice Test

Section A. Do not us calculator

1. Which of the following is in the domain of the function $f(x)=\dfrac{3}{\sqrt{x^2-2x-8}}$?

A) -3

B) 1

C) -2

D) 4

2. The polynomial has zeros $1-i, -3i,$ and 0. What is the least possible degree of the polynomial?

A) 4

B) 5

C) 6

D) 8

3. The values of a function g are given as shown. What can we conclude about the function g?

x	-5	-2	4	5
$g(x)$	-7	-1	11	13

A) g is linear since average rate of change is 0.

B) g is quadratic since average rate of change is constant.

C) g is linear since rate of change of the average rates of change is 0.

D) g is quadratic since rate of change of the average rates of change is 2.

4. The graph of $f(x)$ has domain [-2, 0] and the range [0, 4]. What is the domain and range of $g(x) = f(1-x)+1$?

A) domain [-1, 1] and the range [-1, 3].

B) domain [1, 3] and the range [-1, 3].

C) domain [-1, 1] and the range [1, 5].

D) domain [1, 3] and the range [1, 5].

5. The values of an exponential function $f(x)$ are given in the table.

x	0	2	4
$f(x)$	1000	?	2000

What is the value of $f(2)$?

A) 1200

B) $1000\sqrt{2}$

C) $1000\sqrt{5}$

D) 1500

6. Find the quadrant in which θ lies given that $\sin\theta\tan\theta<0$ and $\cos\theta\csc\theta>0$.

A) quadrant I

B) quadrant II

C) quadrant III

D) quadrant IV

7. Which of the following table could be rational function $f(x)=\dfrac{x^2}{x(x-1)}$?

A)

x	0	0.9	0.999	1	1.001	1.1
$f(x)$	und	9	999	und	-1001	-11

B)

x	0	0.9	0.999	1	1.001	1.1
$f(x)$	und	9	999	und	-1001	-11

C)

x	0	0.9	0.999	1	1.001	1.1
$f(x)$	0	-9	-999	und	1001	11

D)

x	0	0.9	0.999	1	1.001	1.1
$f(x)$	und	-9	-999	und	1001	11

8. Find the value of $2^{3\log_2 3-\log_2 9}$.

A) 3

B) 12

C) 27

D) 81

9. For $g(x)$, y is proportional over the equal length of x. Which of the following could be $g(x)$?

A)

x	y
1	0
1/2	1
1/4	2
1/8	3

B)

x	$\ln y$
1	2
2	5
3	8
4	11

C)

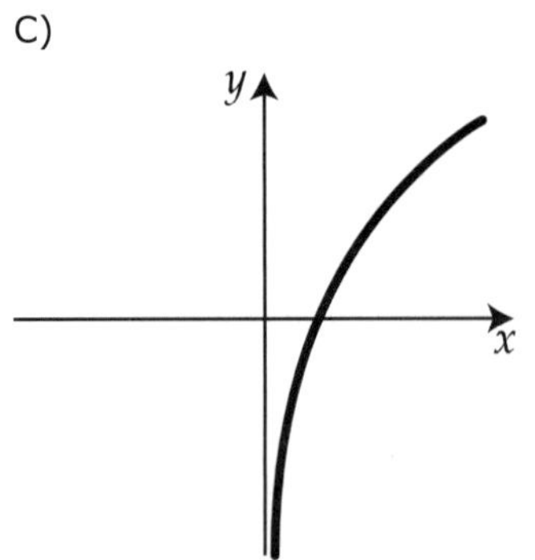

D)

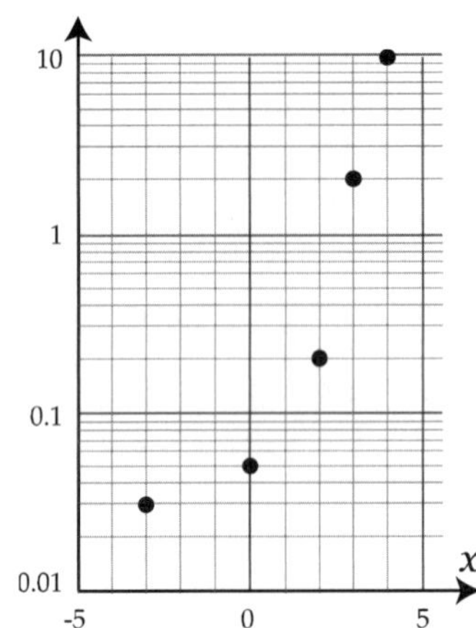

10. $\sec\left(\frac{\pi}{3}-\frac{\pi}{4}\right)=$

A) $\sqrt{6}-\sqrt{2}$

B) $\sqrt{6}+\sqrt{2}$

C) $\frac{1}{\sqrt{6}+\sqrt{2}}$

D) $\frac{1}{\sqrt{6}-\sqrt{2}}$

11.A rational function h is defined by $h(x)=\dfrac{(x-3)^2}{(x+1)(x-5)^3}$. What are all intervals on which $h(x)\le 0$?

A) $(-\infty,-1)\cup(-1,3)\cup(3,5)\cup(5,\infty)$

B) $(-\infty,-1)\cup(-1,5)\cup(5,\infty)$

C) $(-1,3)\cup(3,5)$

D) $(-1,5)$

12.For $f(x)=-2(1.2)^x$, which of the following is true?

I. $f(x)$ has a same graph with -2.4^x

II. $\lim_{x\to-\infty} f(x)=0$

III. $\lim_{x\to\infty} f(x)=\infty$

A) II

B) I, II

C) II, III

D) I, II, III

13.Which of the following is a function that decreases at an increasing rate?

A) $y=-e^x$

B) $y=-e^{-x}$

C) $y=\ln(-x)$

D) $y=-\ln x$

14.What are all values of θ that satisfy $2\cos\theta - 1 \geq 0$ where $0 \leq \theta \leq 2\pi$?

A) $0 \leq \theta \leq \frac{\pi}{3}, \frac{5\pi}{3} \leq \theta \leq 2\pi$

B) $\frac{\pi}{3} \leq \theta \leq \frac{5\pi}{3}$

C) $0 \leq \theta \leq \frac{\pi}{6}, \frac{11\pi}{6} \leq \theta \leq 2\pi$

D) $\frac{\pi}{6} \leq \theta \leq \frac{11\pi}{6}$

15. Angel A and B are in standard position is given in a unit circle as shown, where $B > A$. Which of the following is true?

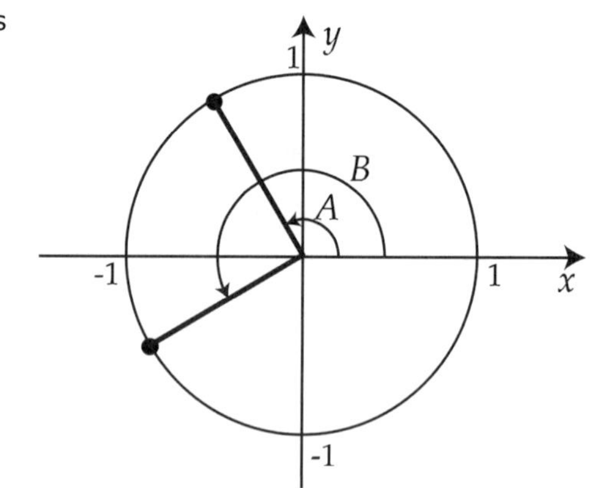

A) $\sin A < \sin B$

B) $\cos A < \cos B$

C) $\tan A < \tan B$

D) none of these

16.Describe the transformation which transform the graph of $y = x^2 + 5$ to the graph of $y = 3x^2 + 14$.

A) Vertical dilation by the factor of 3, then vertical translation of 9

B) Vertical dilation by the factor of 3, then vertical translation of -1

C) Vertical dilation by the factor of $\frac{1}{3}$, then vertical translation of 9

D) Vertical dilation by the factor of $\frac{1}{3}$, then vertical translation of -1

17.The graph shown has equation $y = a\sin\left(bx - \frac{\pi}{6}\right) + c$. Find the values of a, b and c.

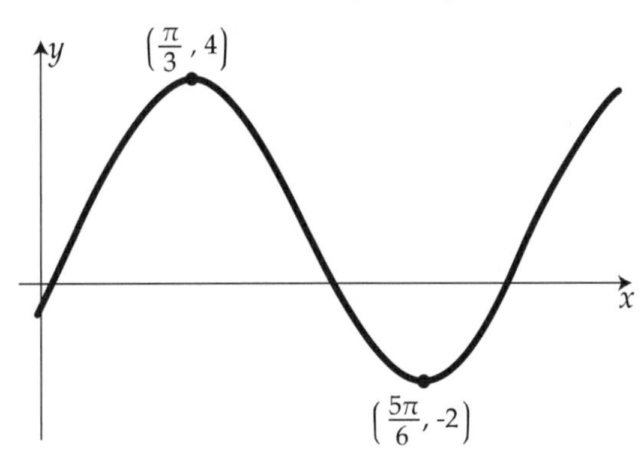

A) $a = 1, b = 2, c = 3$

B) $a = 3, b = 2, c = 1$

C) $a = 1, b = 4, c = 3$

D) $a = 3, b = 4, c = 1$

18.The temperature outside a house during a 14-hour period is given by $T(x)$. Of the following, on which interval is the rate of the temperature negative and decreasing?

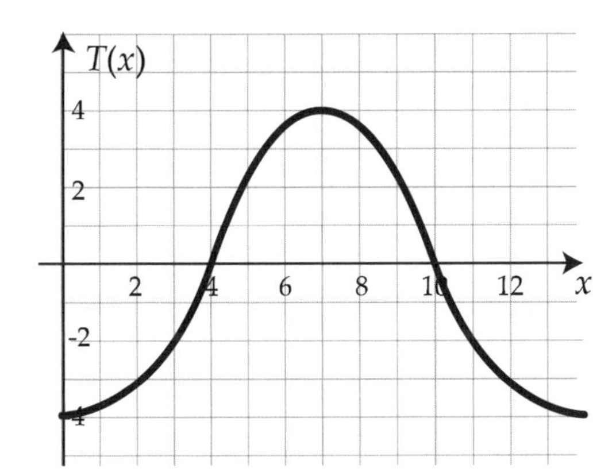

A) $0 \le x \le 4$

B) $4 \le x \le 7$

C) $7 \le x \le 10$

D) $10 \le x \le 13$

19.The polynomial $x^2 + (k+1)x - 3$ has a factor $x + k$. Find k.

A) -2

B) -3

C) 3

D) 3/2

20.What is $\lim_{x \to \frac{3\pi}{2}^-} \tan x$?

A) -∞

B) ∞

C) 1

D) 0

21.What is the range of $y = 2\cos^{-1} 2x$?

A) $0 \le y \le \pi$

B) $0 \le y \le 2\pi$

C) $0 \le y \le \frac{\pi}{2}$

D) $0 \le y \le \frac{\pi}{4}$

22.Find the x coordinates of the point of intersection of $f(x) = \log(2x-1) + \log x$ and $g(x) = 1$.

A) $x = \frac{5}{2}$

B) $x = -2$

C) $x = -2, \frac{5}{2}$

D) $x = -2, 5$

23. The way a paddlewheel rotates models a sinusoidal function. The paddle blade makes one revolution clockwise in 3 seconds and reaches its maximum heicht of 22 feet from the water level. The center of the paddlewheel is 5 feet above the water level. Find an equation $h(t)$ that describing height of the paddlewheel when it starts at the very top and t is in seconds

A) $h(t) = -17\cos\left(\frac{3\pi t}{2}\right) + 5$

B) $h(t) = 17\cos\left(\frac{3\pi t}{2}\right) + 5$

C) $h(t) = -17\cos\left(\frac{2\pi t}{3}\right) + 5$

D) $h(t) = 17\cos\left(\frac{2\pi t}{3}\right) + 5$

24. Which of the rational function satisfy $\lim_{x\to\infty} f(x) = 0$?

A) $y = \frac{x^5 + 1}{x^3 - 1}$

B) $y = \frac{2x + 1}{x - 1}$

C) $y = \frac{x + 1}{x^3 - 1}$

D) $y = \frac{x^2 + 1}{x - 1}$

25. Which of the following is the polar equation for the graph shown below?

A) $r = 3\sin 5\theta$

B) $r = 3\sin 10\theta$

C) $r = 3\cos 5\theta$

D) $r = 3\cos 10\theta$

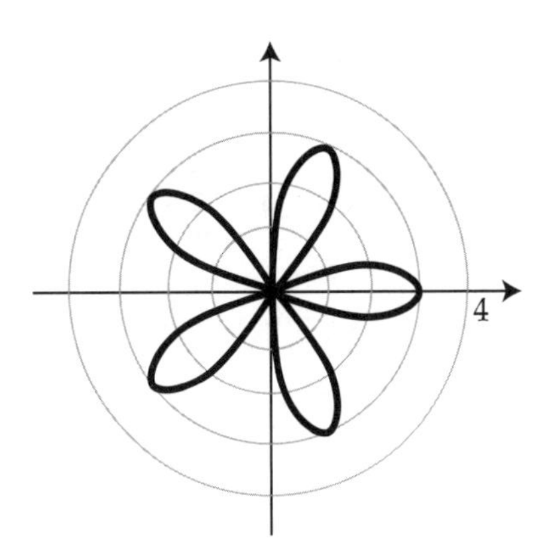

26.What are all values of x, for $-\frac{\pi}{2} \le x \le \frac{\pi}{2}$, where $3\cot^2 2x - 1 = 0$?

A) $\frac{\pi}{3}, \frac{\pi}{6}$

B) $-\frac{\pi}{6}, \frac{\pi}{6}$

C) $-\frac{\pi}{3}, \frac{\pi}{3}$

D) $-\frac{\pi}{3}, -\frac{\pi}{6}, \frac{\pi}{6}, \frac{\pi}{3}$

27.For a geometric sequence b_n, $\frac{b_8}{b_2} = 64$. What is $\frac{b_{33}}{b_{30}}$?

A) 2

B) 4

C) 8

D) 16

28.Which of the following is equivalent to $\left(x + \frac{2}{x}\right)^4$?

A) $x^4 + 8x^2 + 24 + \frac{32}{x^2} + \frac{16}{x^4}$

B) $x^4 + 8x^2 + 32 + \frac{24}{x^2} + \frac{16}{x^4}$

C) $x^4 + 8x + 24 + \frac{32}{x} + \frac{16}{x^4}$

D) $x^4 + 8x + 32 + \frac{24}{x} + \frac{16}{x^4}$

Section B. Use calculator

29.How many point of inflection are there for $y = x^4 + x^3 - 4x^2 + 7$?

A) one

B) two

C) three

D) four

30.Polonium-220 has a half-life of 72days. Calculate the time required for the Polonium-210 to decay to 20% of its initial value.

A) 167 days

B) 156 days

C) 143 days

D) 112 days

31.Which of the following is true about the hole of $R(x) = \dfrac{x^2 - 2x - 8}{x^3 - 5x^2 - 14x}$?

A) $R(x)$ has no holes

B) $R(x)$ has a hole at $(-2, 0)$

C) $R(x)$ has a hole at $(-2, -\frac{1}{3})$

D) $R(x)$ has a hole at $(7, \frac{4}{11})$

32.Which of the following is NOT equivalent to 1?

A) $\sin x \cos x(\tan x + \cot x)$

B) $2\cos^2 x - \cos 2x$

C) $(\sin x + \cos x)^2 - \sin 2x$

D) $(\sin^2 x - 1)(1 + \tan^2 x)$

33.An exponential model and a quadratic model were calculated from the data. These are the residual plot for these models.

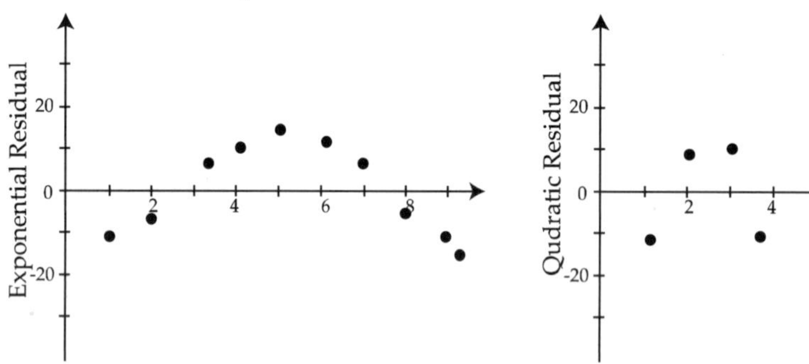

A) The exponential model is more appropriate than the quadratic model.

B) The quadratic model is more appropriate than the exponential model.

C) Both exponential and quadratic model are appropriate.

D) There is not enough information.

34. Find the absolute maximum for the graph $y = x^3 - 3x^2 - 5.2x + 7$ in the interval $2 \le x \le 3$.

A) -7.4

B) -9.2

C) 8.8

D) 9.2

35.The population of deer is modeled by the function $D(t) = 22\cos\left(\frac{\pi}{11}t\right) + 42$, where t is the number of months after 1 January 2024. Which of the following describes the deer population on 1 January 2026?

A) The deer population is decreasing at increasing rate

B) The deer population is decreasing at decreasing rate

C) The deer population is increasing at increasing rate

D) The deer population is increasing at decreasing rate

36.The polynomial $P(x) = ax^5 + bx^3 + c$ where $a < 0, b \neq 0$ and $c \neq 0$. Which of the following is true about the graph of P?

I. $\lim_{x \to \infty} P(x) = \infty$

II. $\lim_{x \to -\infty} P(x) = \infty$

III. $P(x)$ has absolute maximum.

A) I

B) II

C) I, II

D) II, III

37.What is the period of $y = |\sin 3x - \cos x|$?

A) 0.786

B) 1.571

C) 3.142

D) 6.283

38.For the polar equation $r = 3\cos 5\theta - 2$ where $0 \le \theta \le 1$, find the interval where the polar graph is getting farther from the origin.

A) $0 \le \theta < 0.168$

B) $0.168 < \theta < 0.628$

C) $0.628 < \theta \le 1$

D) $0 \le \theta < 0.168$, $0.628 < \theta \le 1$

39.In 2000, there were approximately 200 million people living in Dallas, Texas. The population was increasing by 0.3 percent per decade. If the population continues to increase at the same rate, how many million people will be living in Houston in t *years*?

A) $200(1.003)^{\frac{t}{10}}$

B) $200(1.003)^{10t}$

C) $200(1.3)^{\frac{t}{10}}$

D) $200(1.3)^{10t}$

40.Which of the following expresses the rectangular point (-1, 2) in polar coordinate?

A) $(\sqrt{5}, \pi + \arctan(-2))$

B) $(\sqrt{5}, \pi + \arctan 2)$

C) $(\sqrt{5}, \pi - \arctan(-2))$

D) $(\sqrt{5}, \pi - \arctan 2)$

Answers

1. Functions

※ Decimal answers are rounded to the nearest thousandth.

1.1 Function

Ex1.	① Function ② Not a function ③ Function ④ Not a function ⑤ Not a function ⑥ Function ⑦ Not a function ⑧ Not a function ⑨ Not a function ⑩ Function ⑪ Function ⑫ Not a function
Ex2.	① $5,\ 5+4h+h^2,\ 4+h$ ② $3,\ 3+2h,\ 2$ ③ $\frac{2}{3}, \frac{h+2}{h+3}, \frac{1}{3(h+3)}$ ④ $1, \frac{1}{h+1}, -\frac{1}{h+1}$ ⑤ $\sqrt{3}, \sqrt{3+h}, \frac{\sqrt{3+h}-\sqrt{3}}{h}$
Ex3.	① $\mathbb{R}$ ② $\mathbb{R}, x \le \frac{4}{3}$ ③ $\mathbb{R}, x \ne \pm 2$ ④ $\mathbb{R}, x \ne 6\ or\ -1$ ⑤ $\mathbb{R}$ ⑥ $\mathbb{R}$ ⑦ $\mathbb{R}, x \ge -4, x \ne 5$ ⑧ $\mathbb{R}, x \ge 2, x \ne 7$ ⑨ $\mathbb{R}, x > 5$ ⑩ $\mathbb{R}, x > 1$ ⑪ $\mathbb{R}, x \le -1\ or\ x \ge 4$ ⑫ $\mathbb{R}$ ⑬ $\mathbb{R}, -2 < x < 2$ ⑭ $\mathbb{R}, x \ne \pm\sqrt{5}$ ⑮ $\mathbb{R}, x \ge -3, x \ne 2$ ⑯ $\mathbb{R}, x > 3$
Ex4.	① $Domain: (2,5) \cup (5,8]$, $Range: [-1,4)$ ② $Domain:\ \mathbb{R}$, $Range: (-\infty,4]$ ③ $Domain: (-\infty,-4) \cup (-4,4) \cup (4,\infty)$, $Range:\ (-\infty,-2] \cup (0,\infty)$ ④ $Domain:\ (-\infty,0) \cup (0,\infty)$, $Range:\ (-\infty,2) \cup (2,\infty)$
Ex5.	① $y = \frac{4-x^2}{2}$, Function ② $y = \frac{1}{x^2-1}$, Function ③ $y = -2 \pm \sqrt{4-x^2}$, Not a function ④ $y = \pm 3\sqrt{\frac{x^2}{4}-1}$, Not a function ⑤ $y = \sqrt[3]{x}$, Function ⑥ $y = \pm\sqrt[4]{x}$, Not a function
Ex6.	① $f(x) = 4(x-1)^2 + (x-1)$ ② $f(x) = \left(\frac{x}{2}\right)^2 - 1$ ③ $f(x) = \log 2(x+1)$
AP	1. C 2. D 3. A 4. D

1.2 Rate of Change

Ex1.	① 11 ② 2 ③ $-\frac{3}{7}$ ④ 1 ⑤ $\frac{2\sqrt{3}-3}{3}$ ⑥ $\frac{\sqrt{5}-1}{4}$
Ex2.	① –2 ② $\frac{15}{8}$ ③ $-\frac{3}{7}$ ④ –1 ⑤ 0 ⑥ $\frac{1}{4}$
Ex3.	① linear, increasing ② linear, decreasing ③ quadratic, open down ④ quadratic, open up ⑤ quadratic, open up ⑥ linear, decreasing ⑦ quadratic, open down
AP	1. C 2. C 3. D 4. C 5. B 6. C

1.3 Analyzing Functions

Ex1.	① even ② odd ③ odd ④ neither ⑤ even, odd ⑥ even ⑦ neither
Ex2.	① even ② odd ③ odd ④ even ⑤ neither ⑥ neither ⑦ even ⑧ neither ⑨ odd ⑩ even ⑪ odd ⑫ odd ⑬ odd ⑭ neither
Ex3.	(see table below)
Ex4.	① relative max : 4(at x = -4) relative min : -3(at x = 1) ② relative max : 4(at x = 0), 2(at x = 6) relative min : 0(at x = 4) ③ absolute max : 4(at x = -4) absolute min : none ④ absolute max : 4(at x = 0) absolute min : none ⑤ absolute max : 1(at x = -2) absolute min : -3(at x = 1)

Ex3.

interval	A	B	C	D	E	F	G	H
rate of change	+	–	–	+	–	+	+	–

⑥ absolute max : 4(at x = 0)

absolute min : 0(at x = -2)

⑦ $[-4,1]\cup(4,\infty)$

⑧ $(-4,0]\cup[4,6]$

⑨ $(-\infty,-4]\cup[1,4)$

⑩ $(-7,-4)\cup[0,4]$

Ex5.

interval	A	B	C	D	E	F	G	H
rate of change	dec	dec	inc	inc	inc	inc	dec	dec

Ex6.

① 3

② 2

③ 4

④ 3

Ex7.

① $[0,3]\cup[4,5]\cup[12,14]$

② $[7,10]\cup[14,16]$

③ $[5,7]\cup[10,12]$

④ $[3,4]$

⑤ $[3,4]$

⑥ $[7,10]\cup[14,16]$

※ Answers with open brackets () are also correct.

AP

1. C	2. 1)B	2.2) D
2.3) B	3. B	4. C
5. D	6. B	

1.4 Piecewise Functions

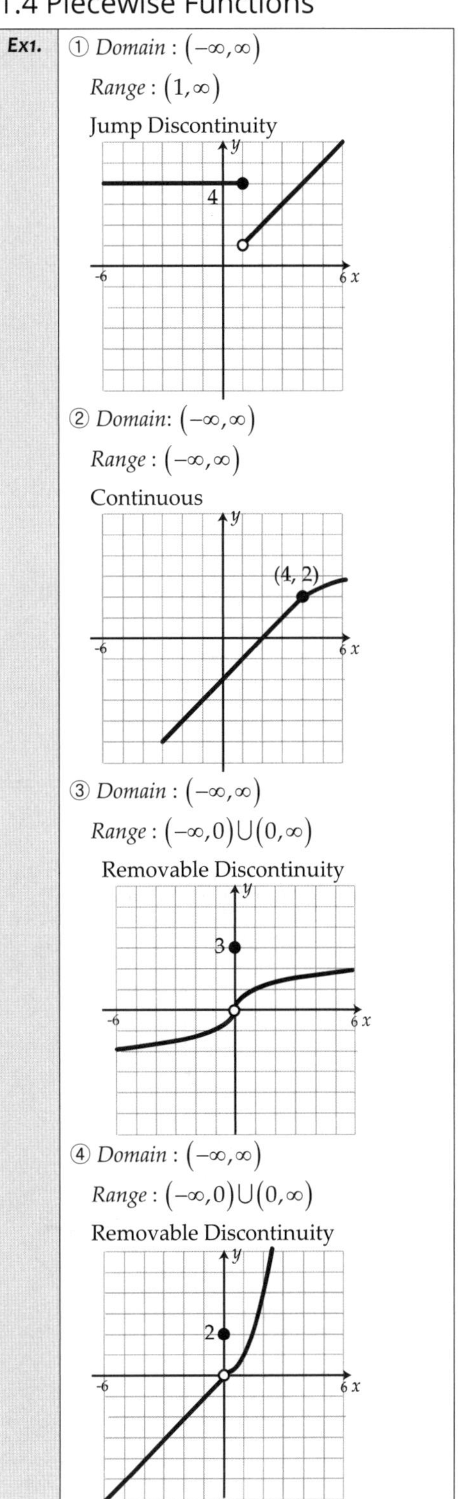

Ex2.	① $f(x)=\begin{cases}-2x-3, & x\le 0\\ 2, & x>0\end{cases}$ ② $f(x)=\begin{cases}1, & x<-2\\ -x+2, & x\ge -2\end{cases}$ ③ $f(x)=\begin{cases}1, & x=1\\ \frac{1}{2}x-2, & x\ne 1\end{cases}$ ④ $f(x)=\begin{cases}-3, & x=0\\ x^2, & x\ne 0\end{cases}$ ⑤ $f(x)=\begin{cases}-2, & x\le -2\\ x, & -2<x\le 2\\ 2, & x>2\end{cases}$
Ex3.	① $f(0)=2$, $f(1)=2$, $f(3)=3$ ② $g(27)=3$, $g(-2)=-\frac{1}{2}$, $g(0)=0$ ③ $f(-4)=9$, $f(1)=2$, $f(2)=4$
Ex4.	$b=2$
AP	1. B

1.5 Transforming Function

Ex1.

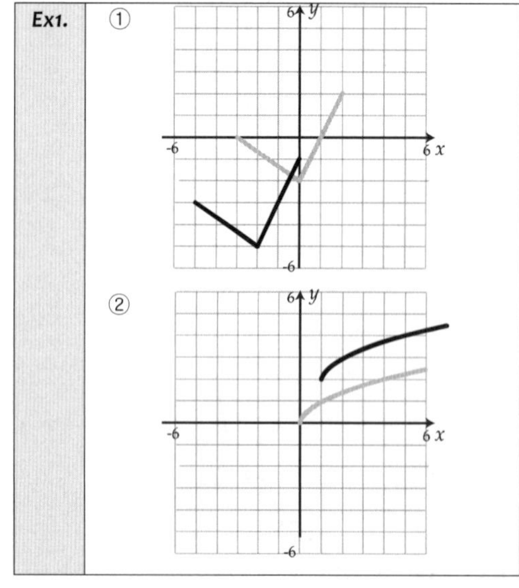

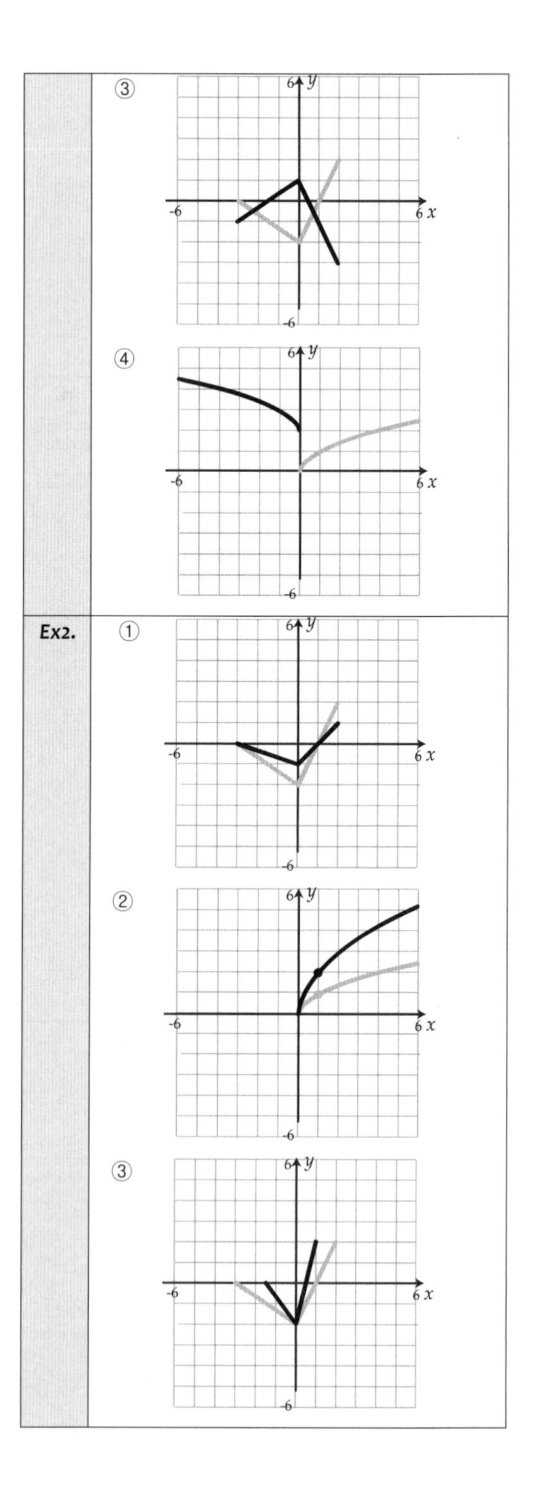

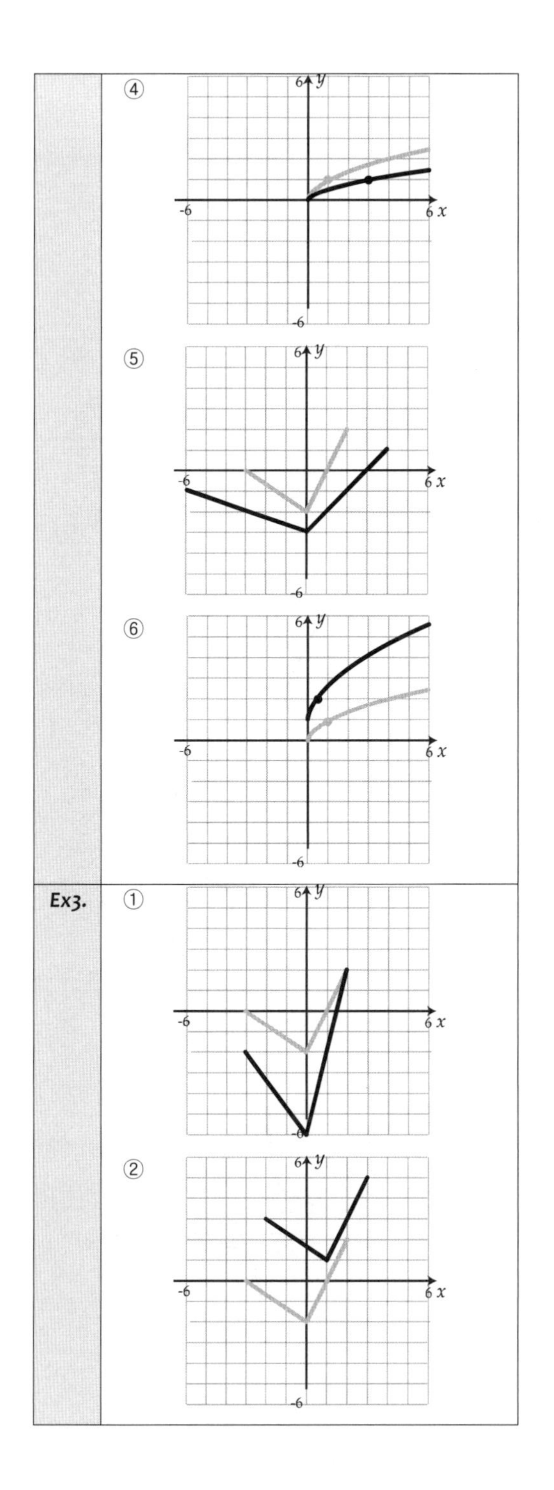
④
⑤
⑥
Ex3.
①
②
6
y
-6
6
x
-6

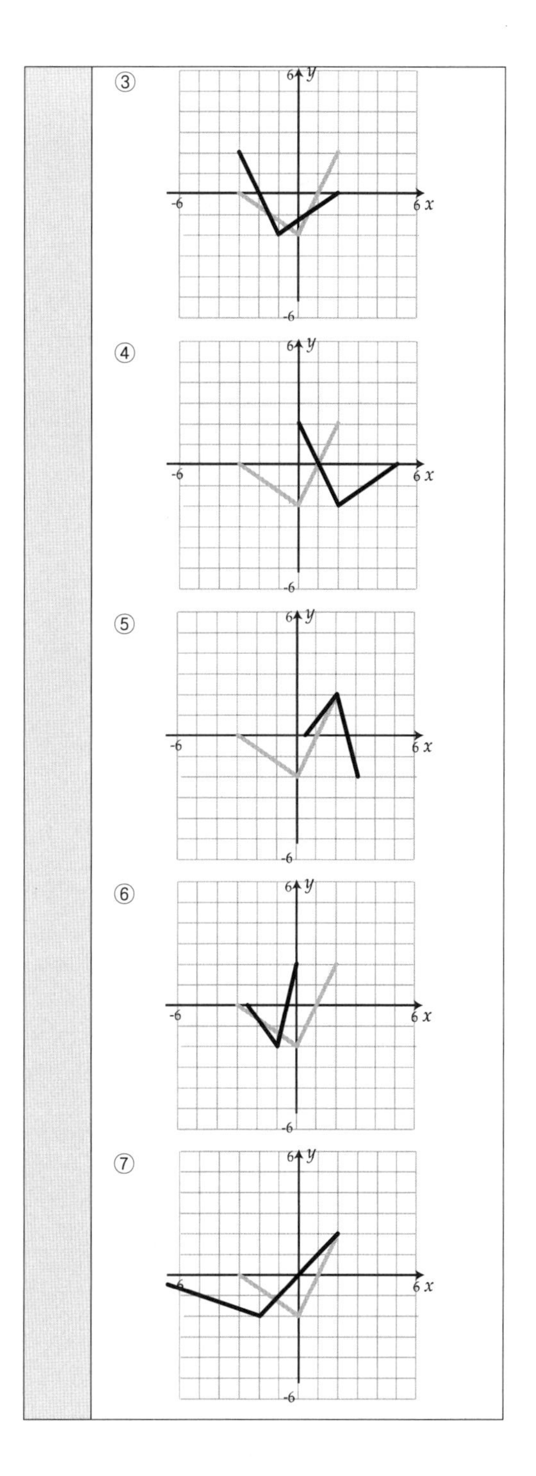
③
④
⑤
⑥
⑦
6
y
-6
6
x
-6

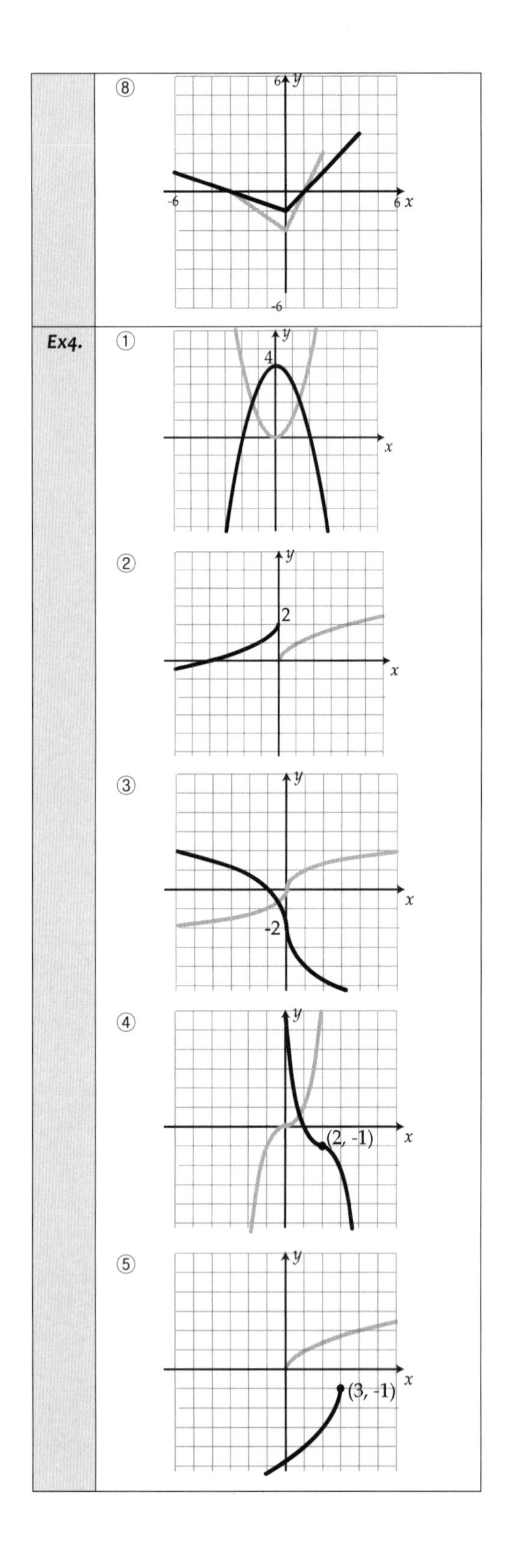

⑧
Ex4.
①
②
③
④
(2, -1)
⑤
(3, -1)

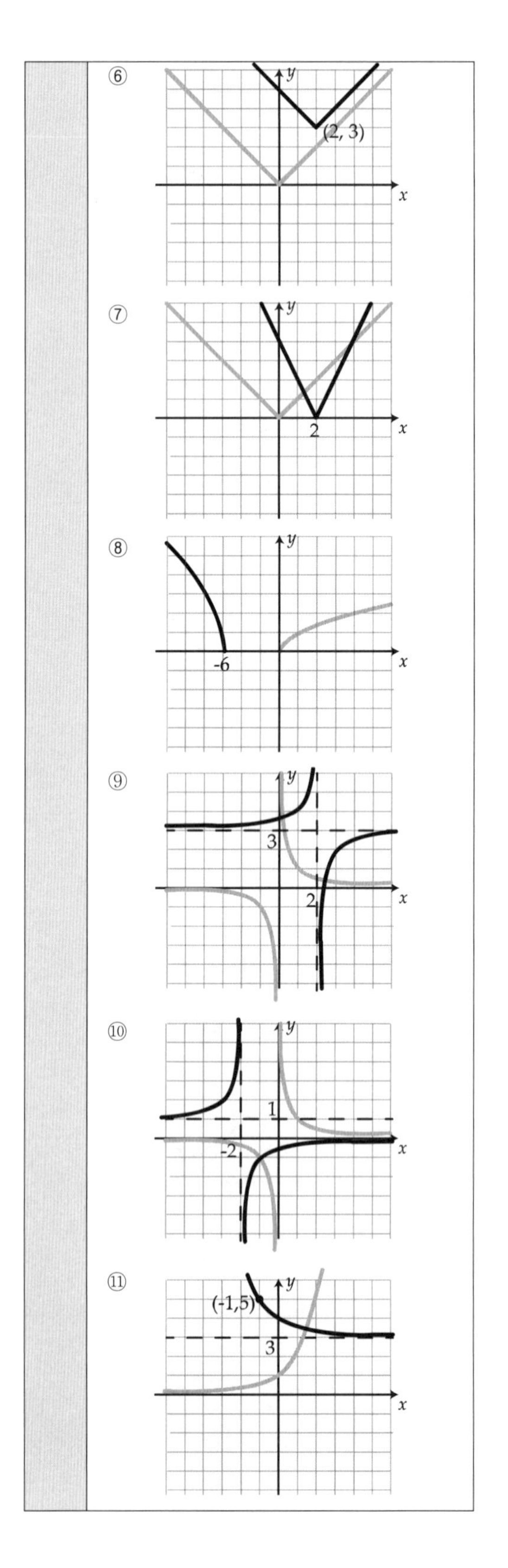

⑥
(2, 3)
⑦
⑧
-6
⑨
⑩
⑪
(-1,5)

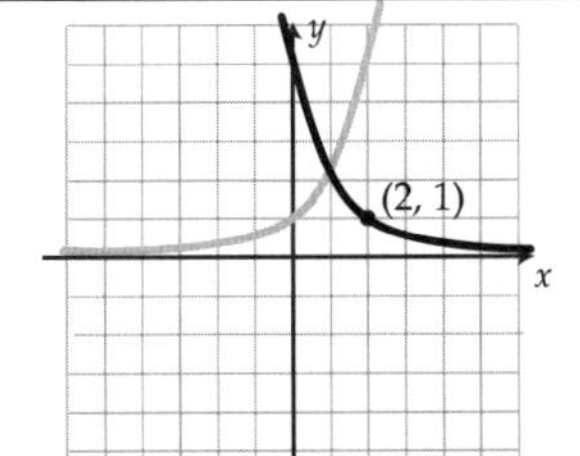

⑬

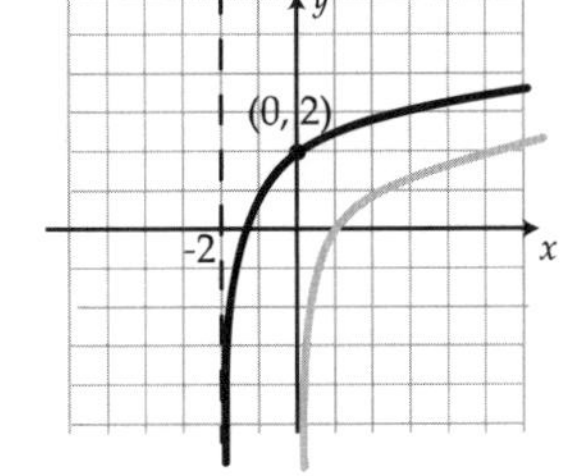

⑭

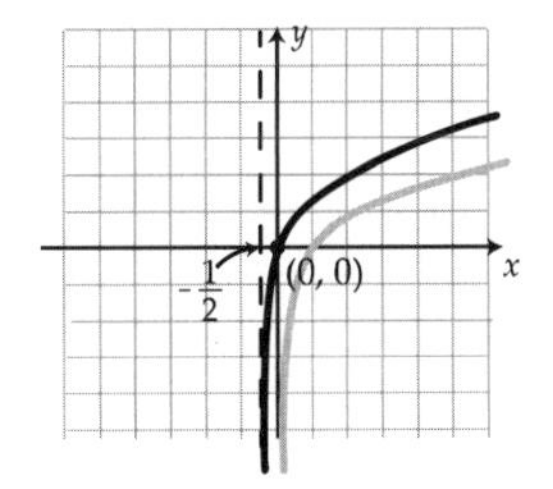

Ex5.

① $(4,-4)$

② $(4,-1)$

③ $(6,-1)$

④ $(3,-4)$

⑤ $(3,-2)$

⑥ $(2,-5)$

⑦ $(-5,-2)$

⑧ $(-2,-2)$

Ex6.

① $y=\sqrt{x+3}-2$

② $y=\sqrt{x-2}+7$

③ $y=\frac{1}{2}\sqrt{x}+3$

④ $y=2\sqrt{x+1}+4$

⑤ $y=\sqrt{2x-1}$

⑥ $y=\sqrt{\frac{1}{2}(x+2)}-1$

⑦ $y=3\sqrt{-x+2}+6$

⑧ $y=-2\sqrt{x+2}-4$

⑨ $y=-\sqrt{3(x-1)}-3$

Ex7.

① $f(x)=\begin{cases} x-2, & x\ge 2 \\ 2-x & x<2 \end{cases}$

② $f(x)=\begin{cases} 2-x, & x\le 2 \\ x-2, & x>2 \end{cases}$

③ $f(x)=\begin{cases} 1, & x\ge 1 \\ -1, & x<1 \end{cases}$

④ $f(x)=\begin{cases} 1, & x\ge 0 \\ -1, & x<0 \end{cases}$

⑤ $f(x)=\begin{cases} x^2-3x-4, & x\le -1 \text{ or } x\ge 4 \\ -\left(x^2-3x-4\right), & -1<x<4 \end{cases}$

⑥ $f(x)=\begin{cases} x^2-x, & x\le 0 \text{ or } x\ge 1 \\ x-x^2, & 0<x<1 \end{cases}$

⑦ $f(x)=\begin{cases} 5, & x\le 3 \\ 2x-1, & x>3 \end{cases}$

Ex8.

①

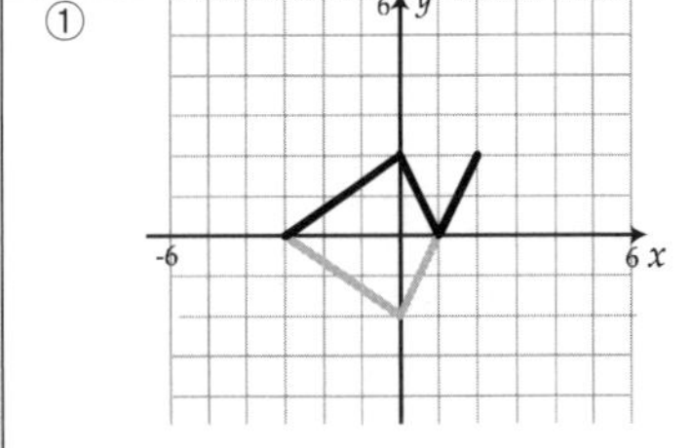

②

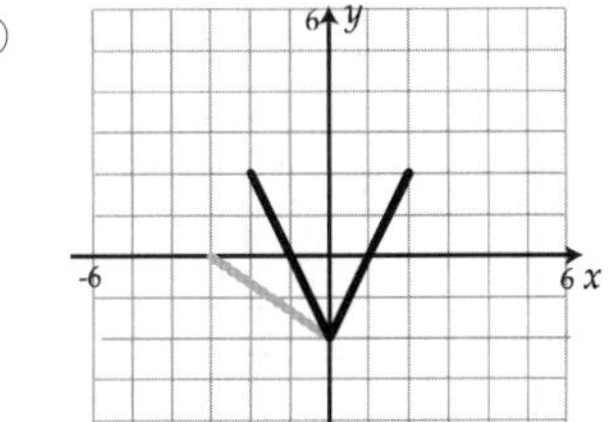

③

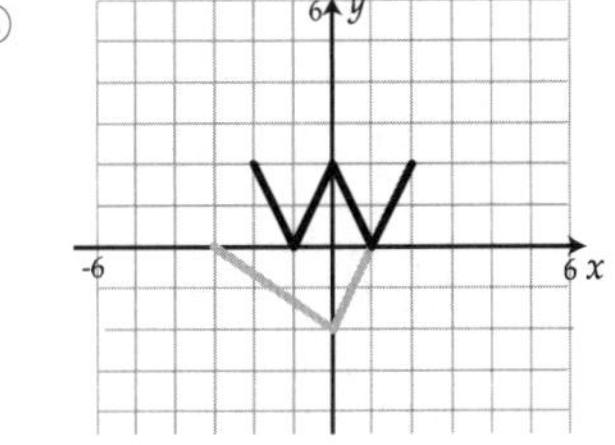

AP	1. D	2. C	3. A
	4. A	5. C	6. D

1.6 Composing Function

Ex1.	① $2\sqrt{x}+3$ ② $\sqrt{2x+3}$ ③ $4x+9$ ④ $\sqrt[4]{x}$ ⑤ 3 ⑥ 5 ⑦ $4\sqrt{x}+9$
Ex2.	① 11 ② 10 ③ 13 ④ 6
Ex3.	① 1 ② 0 ③ 0 ④ 1
Ex4.	① $f(g(x))=\frac{2x}{x(2+3x)}$ $D:x\neq 0,-\frac{2}{3}$ $g(f(x))=\frac{2(x+3)}{x}$ $D:x\neq 0,-3$ ② $f(g(x))=\frac{x+4}{2}$ $D:x\neq -4$ $g(f(x))=\frac{2x}{1+2x}$ $D:x\neq -\frac{1}{2},0$ ③ $f(g(x))=\sqrt[4]{1-x}$ $D:x\leq 1$ $g(f(x))=\sqrt{1-\sqrt{x}}$ $D:0\leq x\leq 1$ ④ $f(g(x))=(x-2)^{\frac{1}{4}}$ $D:x\geq 2$ $g(f(x))=\sqrt{\sqrt{x}-2}$ $D:x\geq 4$ ⑤ $f(g(x))=x$ $D:x\geq 0$ $g(f(x))=\|x\|$ $D:\mathbb{R}$ ⑥ $f(g(x))=x-3$ $D:x\geq 3$ $g(f(x))=\sqrt{x^2-3}$ $D:x\leq -\sqrt{3}\ or\ x\geq \sqrt{3}$
Ex5.	※ Answers may vary ① $g(x)=x^2+1,\ f(x)=\sqrt{x}$ ② $g(x)=x^3+x+1,\ f(x)=\frac{1}{x^2}$ ③ $g(x)=\sqrt{x+1},\ f(x)=x^2+2x+3$ ④ $g(x)=x-3,\ f(x)=x^2+2x-5$
AP	1. D 2. C 3. B 4. C 5. A

1.7 Inverse Function

Ex1.	① One to one function ② Not a function ③ just function ④ just function ⑤ Not a function ⑥ One to one function ⑦ One to one function ⑧ Not a function ⑨ just function ⑩ One to one function ⑪ Not a function ⑫ One to one function

Ex2.	① 6 ② 10 ③ 10 ④ 11 ⑤ 3 ⑥ 10 ⑦ 11 ⑧ 4
Ex3.	① -1 ② 1 ③ 0 ④ -2 ⑤ 1 ⑥ $\frac{1}{2}$ ⑦ -4 ⑧ 2 ⑨ -1 ⑩ -1
Ex4.	① $f^{-1}=\frac{x-7}{6}$ ② $f^{-1}=\frac{4}{x}$ ③ $f^{-1}=\frac{\sqrt[3]{2(x-2)}+3}{2}$ ④ $f^{-1}=\left(\frac{x}{2}\right)^3-1$ ⑤ $f^{-1}=\sqrt{x-1}$ ⑥ $f^{-1}=-\sqrt{x-1}$ ⑦ $f^{-1}=2-\sqrt{x-3}$ ⑧ $f^{-1}=3-\sqrt{\frac{x}{2}}$ ⑨ $f^{-1}=\frac{-3x-6}{7x+2}$ ⑩ $f^{-1}=\frac{2x+4}{x+3}$ ⑪ $f^{-1}=\frac{2-3x}{x}$ ⑫ $f^{-1}=\frac{2x}{-x-3}$
Ex5.	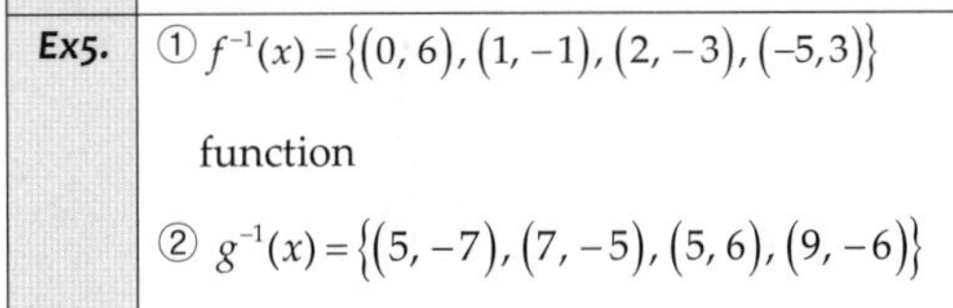 ① $f^{-1}(x)=\{(0,6),(1,-1),(2,-3),(-5,3)\}$ function ② $g^{-1}(x)=\{(5,-7),(7,-5),(5,6),(9,-6)\}$ Not a function
Ex6.	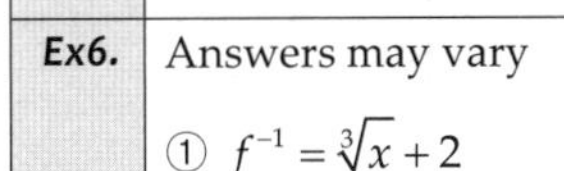 Answers may vary ① $f^{-1}=\sqrt[3]{x}+2$ ② $f^{-1}=x^2+2$ ③ D: $x\geq-2$, $f^{-1}=\sqrt{x+1}-2$ ④ D: $x\geq1$, $f^{-1}=\sqrt{x-2}+1$ ⑤ D: $x\geq-1$, $f^{-1}=x-1$ ⑥ D: $x\geq-1$, $f^{-1}=\sqrt{x-1}-1$
Ex7.	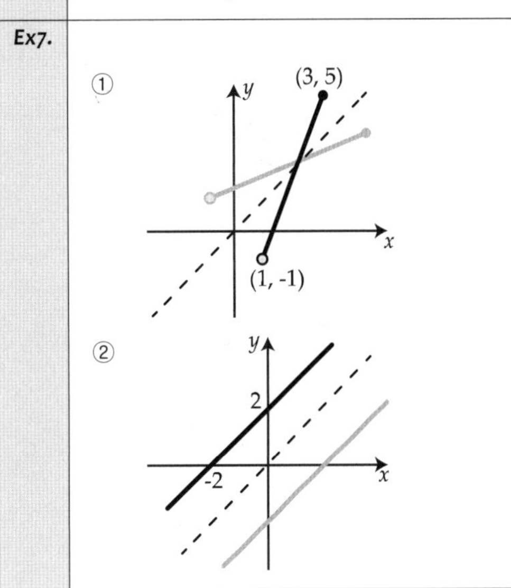

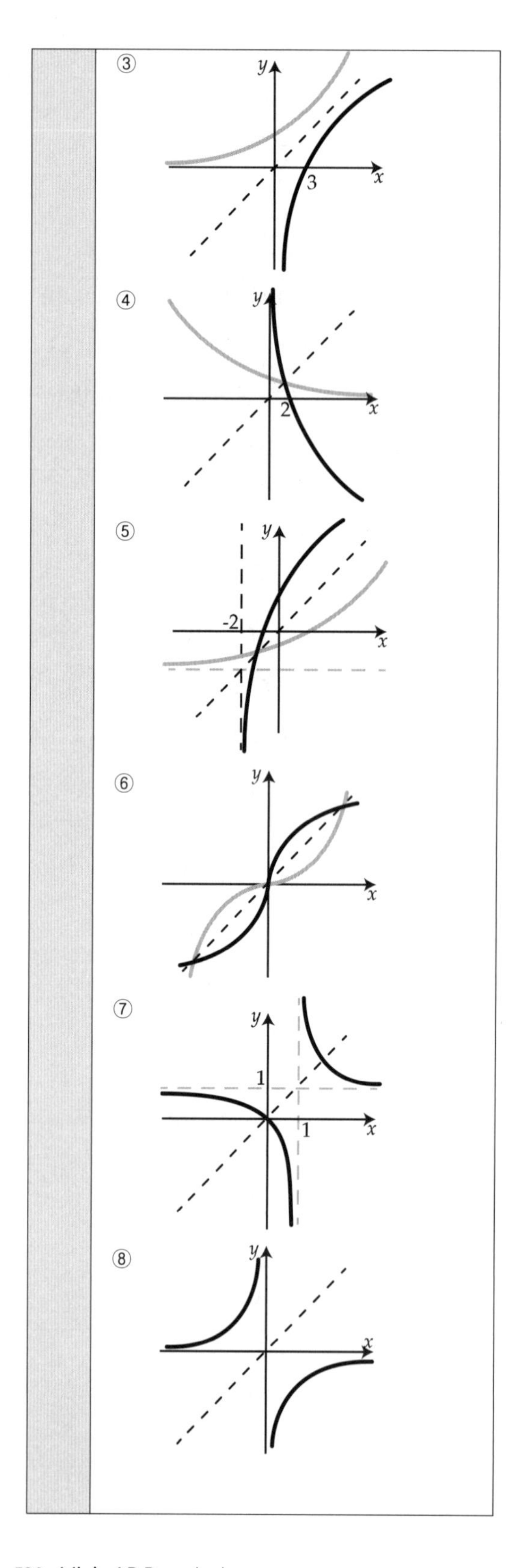

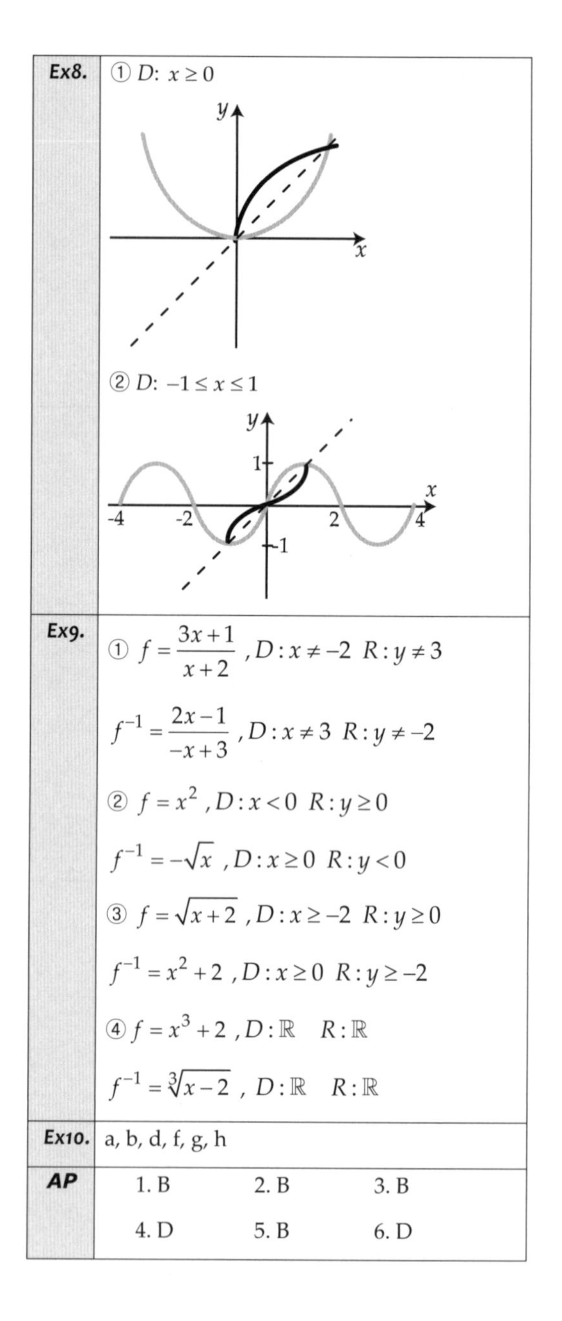

Ex8.	① $D: x \geq 0$ ② $D: -1 \leq x \leq 1$
Ex9.	① $f = \dfrac{3x+1}{x+2}$, $D: x \neq -2$ $R: y \neq 3$ $f^{-1} = \dfrac{2x-1}{-x+3}$, $D: x \neq 3$ $R: y \neq -2$ ② $f = x^2$, $D: x < 0$ $R: y \geq 0$ $f^{-1} = -\sqrt{x}$, $D: x \geq 0$ $R: y < 0$ ③ $f = \sqrt{x+2}$, $D: x \geq -2$ $R: y \geq 0$ $f^{-1} = x^2 + 2$, $D: x \geq 0$ $R: y \geq -2$ ④ $f = x^3 + 2$, $D: \mathbb{R}$ $R: \mathbb{R}$ $f^{-1} = \sqrt[3]{x-2}$, $D: \mathbb{R}$ $R: \mathbb{R}$
Ex10.	a, b, d, f, g, h
AP	1. B 2. B 3. B 4. D 5. B 6. D

1.8 Limit

Ex1.	① i) 3 ii) 1 iii) 1 iv) 1 ② i) 8 ii) 5 iii) 8 iv) DNE ③ i) und ii) 5 iii) 3 iv) DNE ④ i) und ii) 5 iii) 5 iv) 5
Ex2.	① 1 ② ∞ ③ $-\infty$ ④ 0 ⑤ $-\infty$ ⑥ $-\infty$ ⑦ ∞ ⑧ ∞ ⑨ ∞ ⑩ $-\infty$
Ex3.	① -5 ② 2 ③ 8 ④ -4 ⑤ $-\infty$, ∞ ⑥ ∞, $-\infty$
Ex4.	① 19 ② -2 ③ $\frac{1}{4}$ ④ -5 ⑤ $\frac{\sqrt{3}}{6}$ ⑥ $-\frac{1}{6}$
Ex5.	① 0 ② $\frac{2}{3}$ ③ ∞ ④ ∞ ⑤ 3 ⑥ 0
AP	1. D 2. D

Answers

2. Polynomial and Rational Functions

2.1 Polynomial Functions

Ex1.	① Polynomial, 4 ② Not a Polynomial ③ Not a Polynomial ④ Polynomial, 0 ⑤ Polynomial, 5 ⑥ Polynomial, 1 ⑦ Not a Polynomial ⑧ Polynomial, 0
Ex2.	① $\lim_{x\to-\infty} f(x)=\infty$, $\lim_{x\to\infty} f(x)=\infty$ ② $\lim_{x\to-\infty} f(x)=\infty$, $\lim_{x\to\infty} f(x)=-\infty$ ③ $\lim_{x\to-\infty} f(x)=\infty$, $\lim_{x\to\infty} f(x)=-\infty$ ④ $\lim_{x\to-\infty} f(x)=\infty$, $\lim_{x\to\infty} f(x)=\infty$ ⑤ $\lim_{x\to-\infty} f(x)=-\infty$, $\lim_{x\to\infty} f(x)=-\infty$ ⑥ $\lim_{x\to-\infty} f(x)=-\infty$, $\lim_{x\to\infty} f(x)=\infty$
Ex3.	① 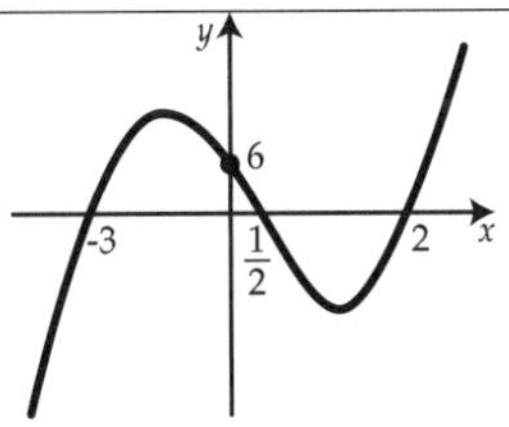 ② 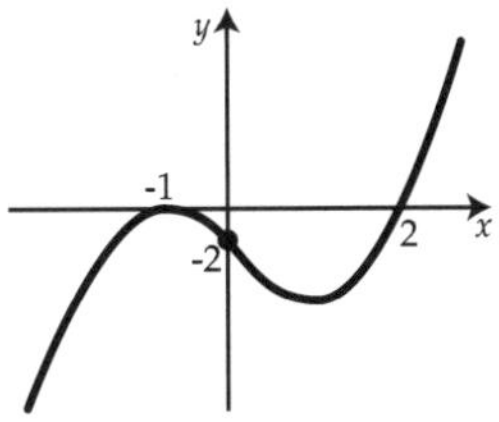

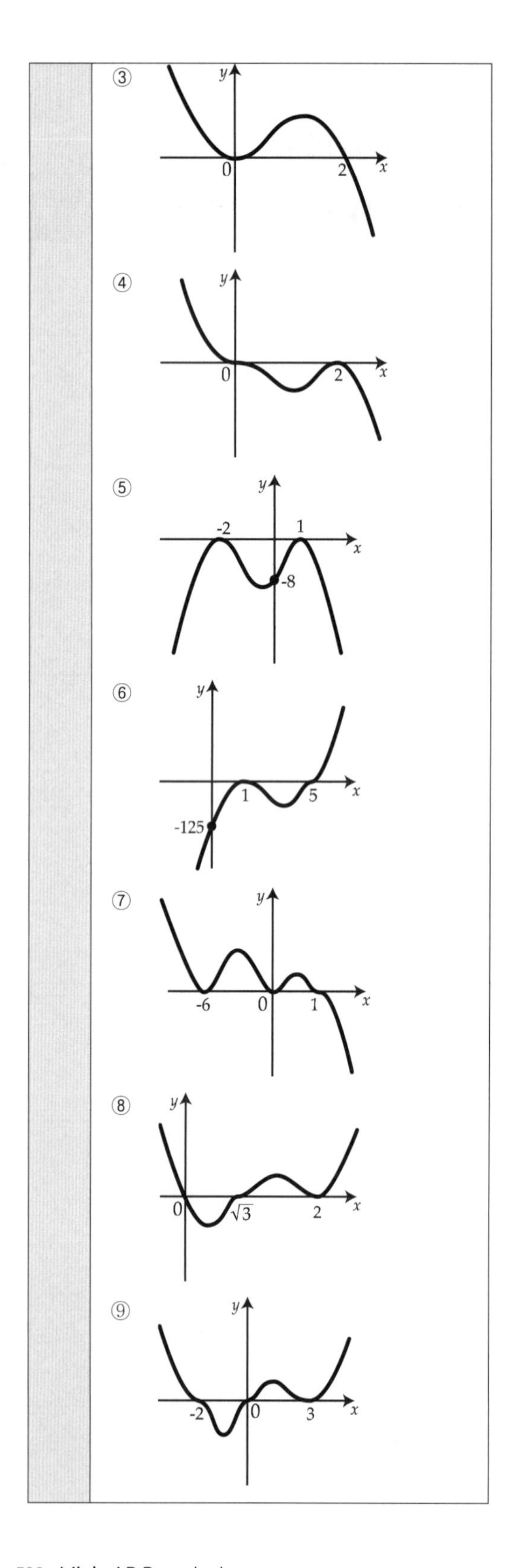
③
y
0
2
x
④
y
0
2
x
⑤
y
-2
1
-8
x
⑥
y
1
5
-125
x
⑦
y
-6
0
1
x
⑧
y
0
√3
2
x
⑨
y
-2
0
3
x

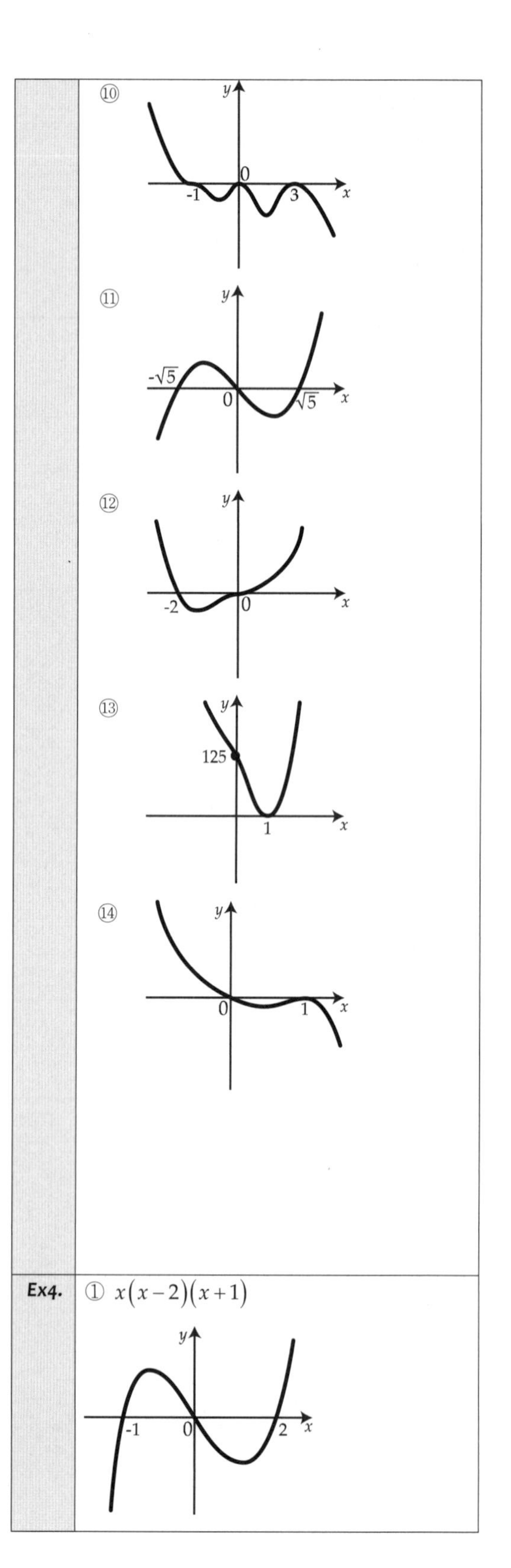
⑩
y
0
-1
3
x
⑪
y
-√5
0
√5
x
⑫
y
-2
0
x
⑬
y
125
1
x
⑭
y
0
1
x

Ex4. ① $x(x-2)(x+1)$

y
-1
0
2
x

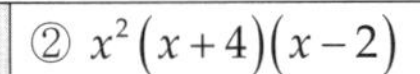

② $x^2(x+4)(x-2)$

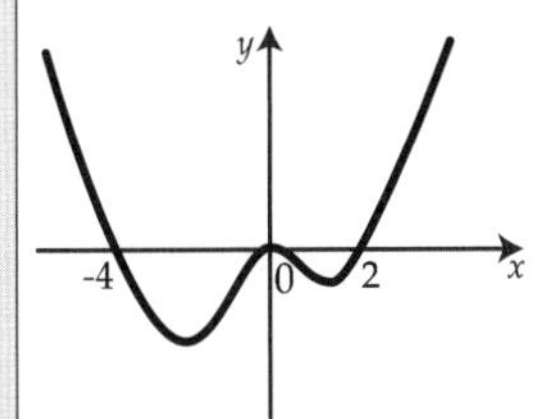

③ $(x+1)^2(x-1)$

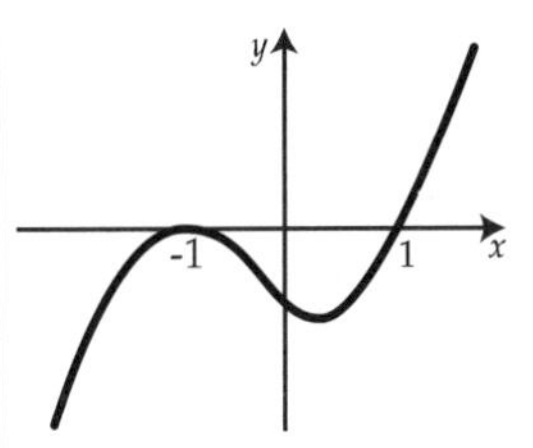

④ $(x-2)^2(x+2)$

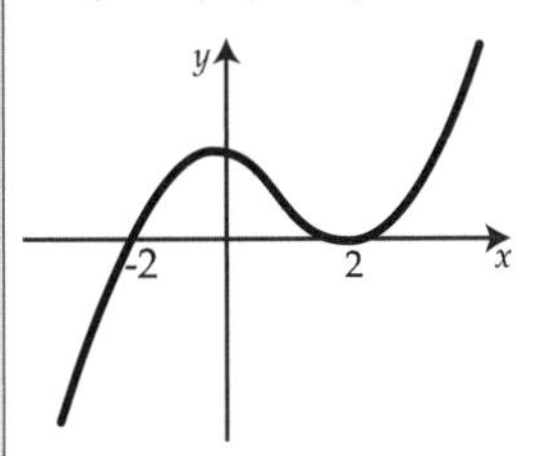

⑤ $-x^2(2x+3)(x-1)$

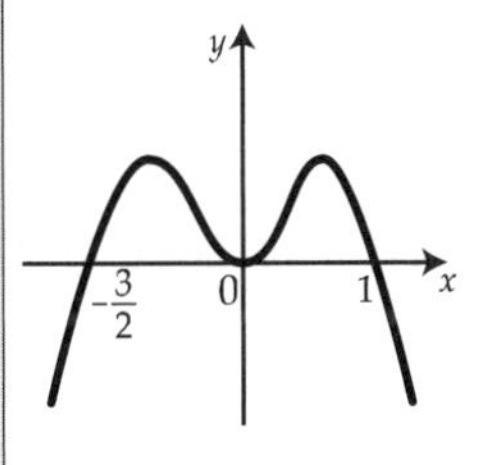

⑥ $x^2(x+2)(x^2-2x+4)$

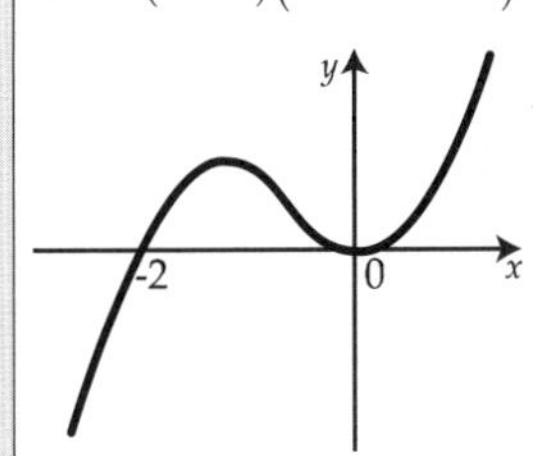

⑦ $(x-1)(x^2+x+1)(x+1)$

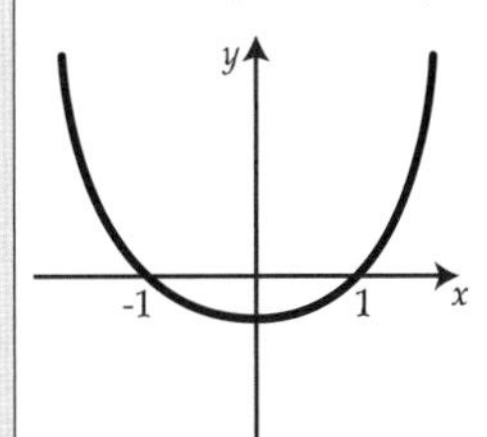

Ex5.

① $2(x+1)(x-3)^2$

② $-4x^2(x-2)$

③ $2x^2(x+3)(x+1)$

④ $-(x+1)(x-2)^3$

⑤ $-\frac{1}{2}x(x-1)(x-3)^3$

AP 1. B 2. D 3. B

2.2 Diving Polynomials

Ex1.

① $x-9+\frac{30x-54}{x^2-3}$

② $x+2+\frac{8x-2}{x^2-2x+2}$

③ $4x^2-20x+100-\frac{503}{x+5}$

④ $3x^2-8x+2+\frac{4x-2}{x^2+x+2}$

Ex2.

① x^2-6x+9

② $3x^2-7x+24-\frac{95}{x+4}$

③ $x^2-2x-4+\frac{10}{x+2}$

④ $5x^2-14x+42-\frac{129}{x+3}$

⑤ $x^3+4x^2+11x+44+\frac{166}{x-4}$

⑥ $2x^3-6x^2-12x-24-\frac{40}{x-2}$

	⑦ $x^2+\frac{5}{2}x+\frac{1}{4}+\frac{\frac{13}{4}}{2x-1}$ ⑧ $3x^2-\frac{1}{2}x+\frac{7}{4}+\frac{\frac{29}{4}}{2x-3}$
Ex3.	① 23 ② 26 ③ 3 ④ 6
Ex4.	① 21 ② 860 ③ $\frac{1}{3}$ ④ $\frac{49}{64}$
Ex5.	$\frac{3}{2}$
Ex6.	① Not Factor ② Factor ③ Factor ④ Not Factor ⑤ Factor
Ex7.	$\frac{1}{3}$
Ex8.	2
AP	1. D 2. A

2.3 Real Zeros of Poly

Ex1.	① $\pm1,\pm2,\pm\frac{1}{2},\pm\frac{1}{3},\pm\frac{2}{3},\pm\frac{1}{6}$ ② $\pm1,\pm2,\pm4,\pm8,\pm\frac{1}{2}$ ③ $\pm1,\pm2,\pm4,\pm8,\pm\frac{1}{5},\pm\frac{2}{5},\pm\frac{4}{5},\pm\frac{8}{5}$ ④ $\pm1,\pm7,\pm\frac{1}{2},\pm\frac{7}{2},\pm\frac{1}{4},\pm\frac{7}{4}$
Ex2.	① pos : 1 , neg : 2, 0 ② pos : 3, 1 , neg : 1 ③ pos : 3, 1 , neg : 0 ④ pos : 4, 2, 0 , neg : 3, 1
Ex3.	$(x-2)(2x-1)(x+3)$
Ex4.	① $(x-2)(x+2)(x+3)$ ② $(x+1)(x+2)(2x-1)$ ③ $(x+1)(2x^2+2x+3)$ ④ $(x-2)(2x^2-3x+3)$ ⑤ $(x+1)^2(x-4)(x+2)$ ⑥ $(x-1)(x+1)(x+3)(x+5)$

2.4 Fundamental Theorem of Algebra

Ex1.	① real :0 / img : 2 ② real :2 / img : 0 ③ real :3 / img : 0 ④ real :1 / img : 2 ⑤ real :2 (1 repeated) / img : 0 ⑥ real :3 (1 repeated) / img : 0 ⑦ real :2 / img : 2 ⑧ real :0 / img : 6 ⑨ real :1 / img : 4 ⑩ real :4 (1 repeated) / img : 0

Ex2.	① $x=0,\ \pm i$ $x(x+i)(x-i)$ ② $x=0,\ \pm 2i$ $x^2(x+2i)(x-2i)$ ③ $x=-2,\ \dfrac{-3\pm\sqrt{7}i}{2}$ $(x+2)\left(x-\dfrac{-3+\sqrt{7}i}{2}\right)\left(x-\dfrac{-3-\sqrt{7}i}{2}\right)$ ④ $x=1,\ 1\pm\sqrt{2}i$ $(x-1)(x-1+\sqrt{2}i)(x-1-\sqrt{2}i)$ ⑤ $x=3,\ \pm\sqrt{2}i$ $(x-3)(x-\sqrt{2}i)(x+\sqrt{2}i)$ ⑥ $x=\pm i,\ \pm 2\sqrt{2}i$ $(x+i)(x-i)(x+2\sqrt{2}i)(x-2\sqrt{2}i)$ ⑦ $x=\pm 2,\ \pm 2\sqrt{2}i$ $(x+2)(x-2)(x+2\sqrt{2}i)(x-2\sqrt{2}i)$
Ex3.	① x^2-2x+5 ② $x^2-6x+10$ ③ $x^3-2x^2-3x+10$ ④ x^3+x^2+9x+9 ⑤ $x^5+2x^4+5x^3+10x^2+4x+8$
Ex4.	$3, 2+3i$
AP	1. C 2. B 3. A

2.5 Rational Function

Ex1.	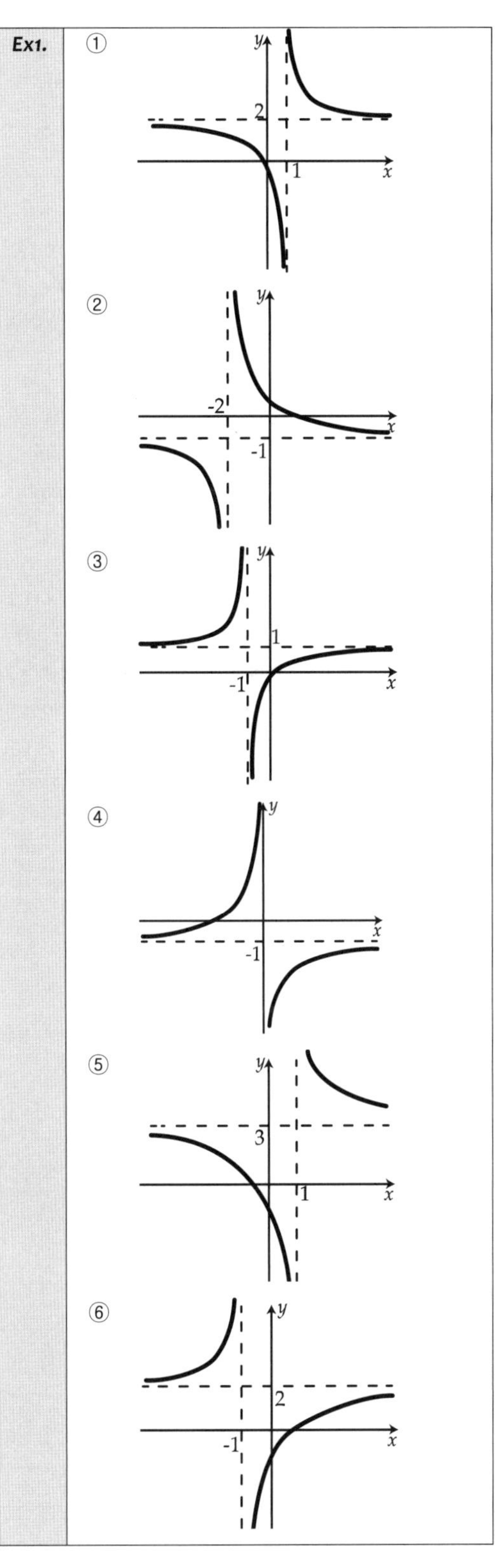

Ex2.

①

$D: \quad x \neq -4$
$V \cdot A: \quad x = -4$
$hole: \quad none$
$H \cdot A: \quad y = 3$

②

$D: \quad x \neq 3$
$V \cdot A: \quad x = 3$
$hole: \quad none$
$H \cdot A: \quad y = \frac{3}{2}$

③

$D: \quad x \neq 0, 3, -1$
$V \cdot A: \quad x = 0, -1$
$hole: \quad at\ x = 3$
$H \cdot A: \quad y = 0$

④

$D: \quad x \neq 0, 4$
$V \cdot A: \quad x = 0$
$hole: \quad at\ x = 4$
$H \cdot A: \quad y = -\frac{2}{3}$

⑤

$D: \quad x \neq 0, 2$
$V \cdot A: \quad x = 0$
$hole: \quad x = 2$
$H \cdot A: \quad none$

⑥

$D: \quad \mathbb{R}$
$V \cdot A: \quad none$
$hole: \quad none$
$H \cdot A: \quad y = 0$

⑦

$D: \quad \mathbb{R}$
$V \cdot A: \quad none$
$hole: \quad none$
$H \cdot A: \quad y = 0$

⑧

$D: \quad x \neq 0, 1$
$V \cdot A: \quad x = 0$
$hole: \quad at\ x = 1$
$H \cdot A: \quad none$

⑨

$D: \quad x \neq 1$
$V \cdot A: \quad x = 1$
$hole: \quad none$
$oblique: \quad y = x - 2$

⑩

$D: \quad x \neq 7, -2$
$V \cdot A: \quad x = 7, -2$
$hole: \quad none$
$oblique: \quad y = x + 5$

⑪

$D: \quad x \neq 2, 3$
$V \cdot A: \quad x = 3$
$hole: \quad x = 2$
$oblique: \quad y = x + 5$

Ex3.

① $x\text{int}: -\frac{5}{3},\ y\text{int}: -\frac{5}{6}$

② $x\text{int}: 2,\ y\text{int}: 4$

③ $x\text{int}: 0,\ y\text{int}: 0$

④ $x\text{int}: none,\ y\text{int}: -\frac{4}{3}$

⑤ $x\text{int}: -2,\ y\text{int}: none$

⑥ $x\text{int}: -\frac{1}{2},\ y\text{int}: none$

Ex4.

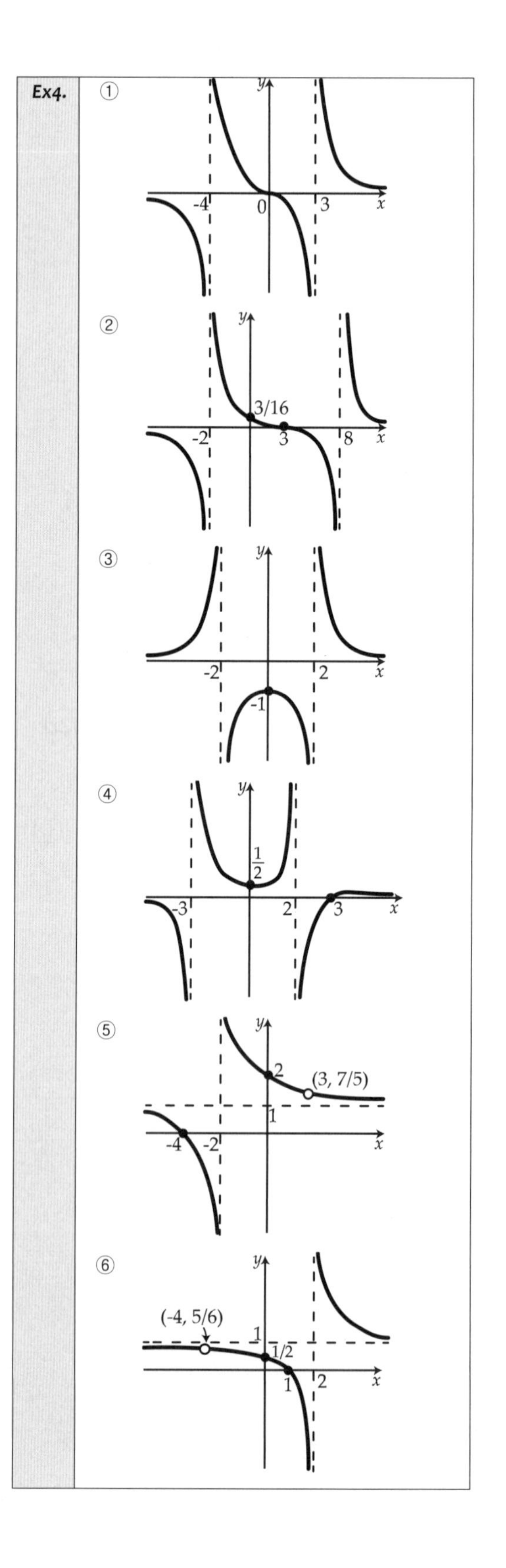

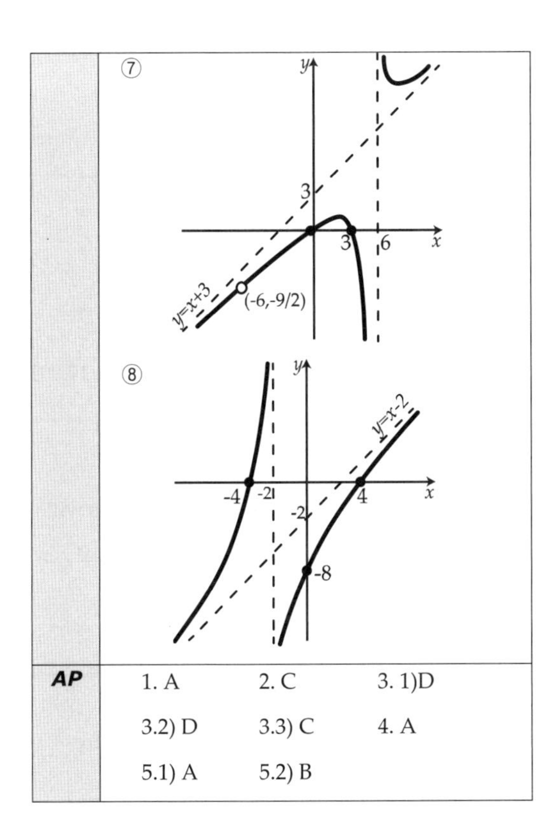

AP			
	1. A	2. C	3. 1)D
	3.2) D	3.3) C	4. A
	5.1) A	5.2) B	

2.6 Polynomial and Rational Function Inequalities

Ex1.	① $(-3,3)\cup(5,7)$
	② $(-\infty,-2]\cup[2,\infty)$
	③ 1
	④ $[-7,-3], x=0$
	⑤ (i) $(-\infty,-4)\cup(-4,10)$
	(ii) $(-\infty,-8)\cup(2,10)$
	⑥ (i) $(0,\infty)$
	(ii) $(-\infty,0]$
	⑦ (i) $(-\infty,-10], x=0$
	(ii) $[-8,-4]\cup[6,\infty)$
	⑧ (i) $[-\infty,-8]\cup[-2,3]$
	(ii) $(-9,0)\cup(0,5)$
Ex2.	① $(-\infty,-1)\cup(3,5)$
	② $(-2,0)\cup\left(\frac{1}{2},\infty\right)$
	③ $(-\infty,-2)\cup(-2,0)\cup(1,\infty)$
	④ $(-\infty,0)\cup(0,2)\cup(3,\infty)$
	⑤ $x=-1$, $[0,1]$, $x=3$
	⑥ $(-\infty,-2]\cup[0,\infty)$
	⑦ $x=-1$, $[0,\infty)$
	⑧ $(-\infty,1]$
	⑨ $(-\infty,-3)\cup(-2,2)\cup(3,\infty)$
	⑩ $(-2,0)\cup(9,\infty)$
	⑪ $(-6,0)\cup(0,6)$
	⑫ $[5,\infty)$, $x=0$
	⑬ $\left(-8,-\frac{5}{2}\right)\cup(2,\infty)$
	⑭ $[-5,-2]\cup[2,\infty)$
Ex3.	① $(-4,-2)\cup(4,\infty)$
	② $(-\infty,4)\cup(8,\infty)$
	③ $(-\infty,-2]\cup(4,8)\cup(8,\infty)$
	④ $(-2,0]$
Ex4.	① $(-\infty,-1)\cup[7,\infty)$
	② $(-\infty,-5)\cup(4,\infty)$
	③ $(-\infty,-6)$
	④ $(-\infty,-9)\cup(10,\infty)$
	⑤ $(-3,-2)$, $x=-1$, $[3,\infty)$

	⑥ $(-\infty, -1) \cup (-1, 1)$
	⑦ $(-\infty, 1) \cup (3, \infty)$
	⑧ $(7, 12]$, $x = -2$, $x = 0$
	⑨ $(-\infty, -2) \cup (0, 2)$
	⑩ $(-\infty, -2] \cup [2, \infty)$
	⑪ $[0, 2] \cup (5, \infty)$
	⑫ $(-1, 0) \cup (7, \infty)$
	⑬ $(-\infty, -19] \cup (1, 5)$
AP	1. B 2. C

2.7 Binomial Expansion

Ex1.	① $x^4 + 8x^3 + 24x^2 + 32x + 16$
	② $x^4 - 8x^3 + 24 - \frac{32}{x^2} + \frac{16}{x^4}$
	③ $32x^5 - 80x^4 + 80x^3 - 40x^2 + 10x - 1$
	④ $243 + 405x + 270x^2 + 90x^3 + 15x^4 + x^5$
	⑤ $x^{10} + 5x^7 + 10x^4 + 10x + \frac{5}{x^2} + \frac{1}{x^5}$
	⑥ $64x^3 - 144x^2y + 108xy^2 - 27y^3$
AP	1. C

Free Response Questions (from ch1-ch2)

1.	(a) $x = 3.384$
	(b) $f(1.5) = 16.3$
	$f(2.022) = 17.903$
	$f(2.7) = 14.092$
	Therefore, absolute maximum is 17.903 and absolute minimum is 14.092.
	(c) 2 imaginary zeros.
	$f(x)$ is 3rd degree polynomial, $f(x)$ will have 3 zeros. Since we have 1 real zero, there will be 2 imaginary zeros.
	(d) a = -5.571, -17.903
2.	(a) $f(x) = (x+1)(x+3)(x-2)$
	(b) Domain: All real numbers where $x \neq 3, -2, -1$
	$[or (-\infty, -2) \cup (-2, -1) \cup (-1, 3) \cup (3, \infty)]$
	Vertical Asymptotes: $x = 3$, $x = -2$
	(c) $y = \frac{1}{3}$; since the degree of the numerator and the denominator are the same.
	(d) $a = -1, b = \frac{1}{2}$

Answers

3. Exponential and Log Functions

3.1 Exponential Function

Ex1.	① 1 ② 2^{4n-3} ③ $x+y$ ④ $\frac{1}{x+y}$ ⑤ $x^{\frac{3}{4}}$ ⑥ $a^{\frac{7}{8}}$ ⑦ $a^{\frac{5}{3}}$ ⑧ $a^{\frac{1}{2}}$ ⑨ $\frac{1}{x^{\frac{1}{24}}}$ ⑩ $\frac{1}{x^{\frac{1}{24}}}$ ⑪ $a-b$ ⑫ $a-b$
Ex2.	① $2^n \cdot 9$ ② $2^3\left(2^n+1\right)$ ③ $2^n \cdot 12$ ④ 2^m ⑤ $\left(6^x-2\right)\left(6^x-9\right)$ ⑥ $\left(2^x-3\right)\left(2^x+2\right)$ ⑦ $\left(3^x+4\right)\left(3^x-1\right)$ ⑧ $\left(5^x-6\right)\left(5^x+1\right)$ ⑨ $\left(3^x+2^x\right)\left(3^x-2^x\right)$ ⑩ $\left(5-4^x\right)\left(5+4^x\right)$ ⑪ 2^n ⑫ 4^n
Ex3.	①  $\mathbb{R}$ $y>0$ $y=0$ 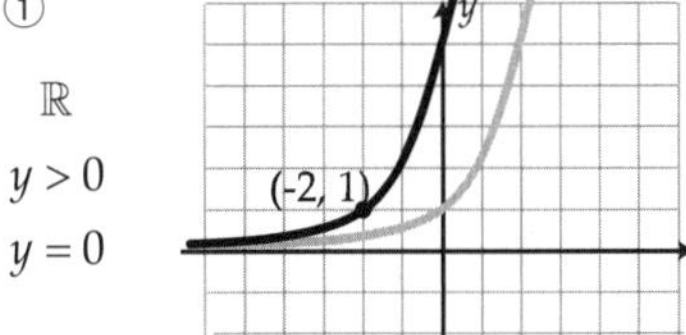 ② $\mathbb{R}$ $y>0$ $y=0$ 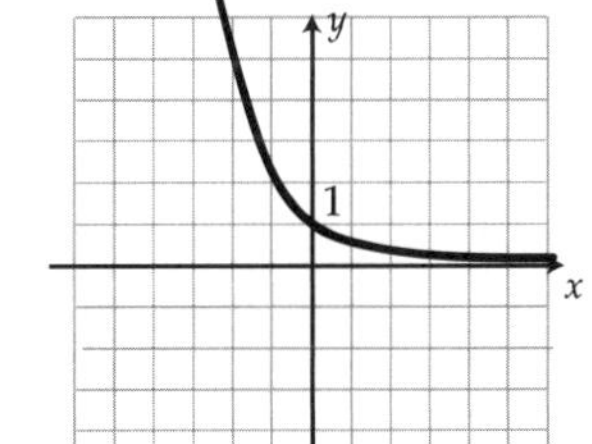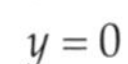 ③ $\mathbb{R}$ $y>-2$ $y=-2$ 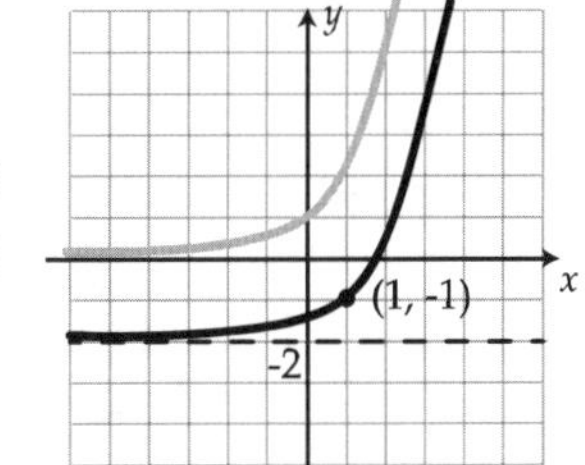 ④ $\mathbb{R}$ $y<1$ $y=1$ 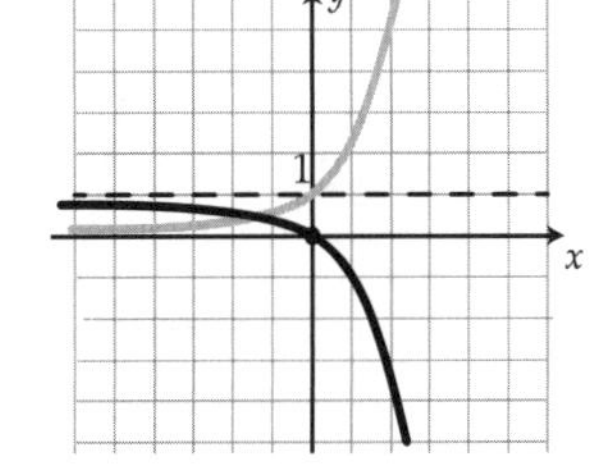 ⑤  $\mathbb{R}$ $y<1$ $y=1$ 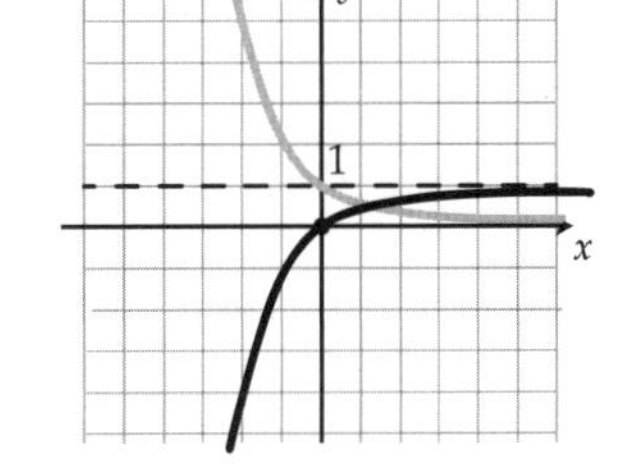

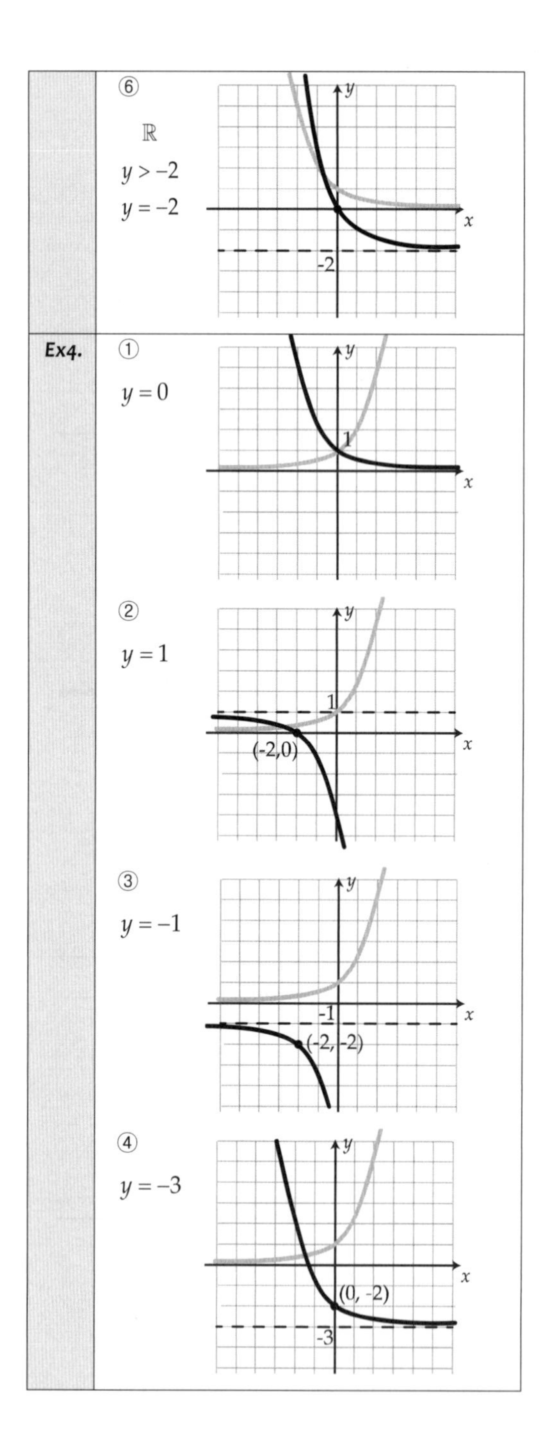

⑥

$\mathbb{R}$

$y > -2$

$y = -2$

Ex4.

① $y = 0$

② $y = 1$

③ $y = -1$

④ $y = -3$

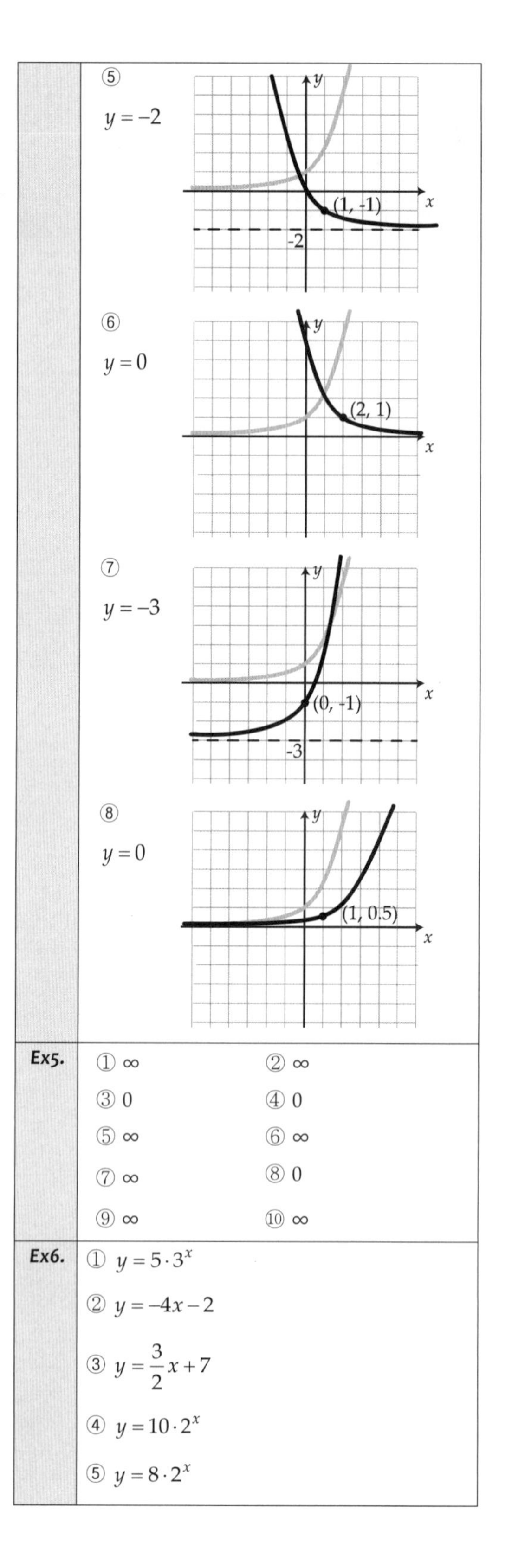

⑤ $y = -2$

⑥ $y = 0$

⑦ $y = -3$

⑧ $y = 0$

Ex5.

① ∞	② ∞
③ 0	④ 0
⑤ ∞	⑥ ∞
⑦ ∞	⑧ 0
⑨ ∞	⑩ ∞

Ex6.

① $y = 5 \cdot 3^x$

② $y = -4x - 2$

③ $y = \frac{3}{2}x + 7$

④ $y = 10 \cdot 2^x$

⑤ $y = 8 \cdot 2^x$

	⑥ $y = 3\cdot(\sqrt{3})^x$ ⑦ $y = 64\left(\frac{1}{2}\right)^x$ ⑧ $y = 45\left(\sqrt[3]{\frac{1}{3}}\right)^x$
AP	1. C 2. B 3. B 4. C 5. B 6. A 7. B

3.2 Compound Interest

Ex1.	① $1000(1+0.09)^8 = \$1992.56$ ② $1000\left(1+\frac{0.12}{2}\right)^{3(2)} = \$1418.52$ ③ $14000\left(1+\frac{0.14}{12}\right)^{13(12)} = \$85500.53$ ④ $480\left(1+\frac{0.09}{4}\right)^{5(4)} = \$749.04$ ⑤ $12000e^{0.06(4)} = \$15254.99$ ⑥ $4000e^{0.005(7)} = \$4142.48$
Ex2.	$2400(1+0.002)^{35} = 2574$
Ex3.	$32000(1-0.0001)^3 = 31990$
Ex4.	$\frac{1000}{(1+0.03/2)^3} = \$956.317$
Ex5.	$\frac{20000}{e^{0.09}} = \$18278.62$
AP	1. D 2. A 3. C

3.3 Logarithmic Function

Ex1.	① 5 ② 5 ③ –2 ④ –2 ⑤ –2 ⑥ –1 ⑦ $\frac{1}{2}$ ⑧ $\frac{1}{2}$ ⑨ 5 ⑩ $\frac{1}{4}$ ⑪ 125 ⑫ 4 ⑬ 6
Ex2.	① $x > -8$ ② $x < 2$ ③ $x < -5$ or $x > \frac{1}{2}$ ④ $x < -2$ or $x > -1$ ⑤ $x < -2$ ⑥ $x > -2$ ⑦ $-1 < x < 1$ ⑧ $-2 < x < 1$ or $x > 2$
Ex3.	① 3 ② $\frac{1}{2}$ ③ 11 ④ 0 ⑤ 2 ⑥ $\frac{1}{5}$ ⑦ –3 ⑧ 6

⑨ 25

⑩ $-\frac{1}{2}$

⑪ 4

⑫ $\frac{1}{2}$

⑬ 1

⑭ 5

⑮ kt

⑯ 2

Ex4.

① $1000 = 10^3$

② $x = \log_5 4000$

③ $\frac{1}{8} = e^x$

④ $27 = 3^3$

⑤ $\frac{1}{2} = \ln x$

⑥ $x = \log 2$

⑦ $x^y = e^2$

⑧ $2x = \log 5$

⑨ $x^2 - x = \log_2 3$

⑩ $4x^2 = \ln 7$

⑪ $10^y = x^2 - 4x$

Ex5.

① $f^{-1} = \log_2 x + 1$

② $f^{-1} = \log_5 (x-3)$

③ $f^{-1} = \ln\left(\frac{x+2}{3}\right)$

④ $f^{-1} = \log\frac{x}{3} + 2$

⑤ $f^{-1} = 5^x$

⑥ $f^{-1} = 7^{x-5}$

⑦ $f^{-1} = e^x - 2$

⑧ $f^{-1} = 10^{\frac{x+3}{2}}$

⑨ $f^{-1} = 10^{\frac{x-2}{3}}$

⑩ $f^{-1} = e^{x-1} + 1$

⑪ $f^{-1} = 3^{2(x-2)} + 1$

⑫ $f^{-1} = e^{2x+1} - 2$

Ex6.

① $x = 0$

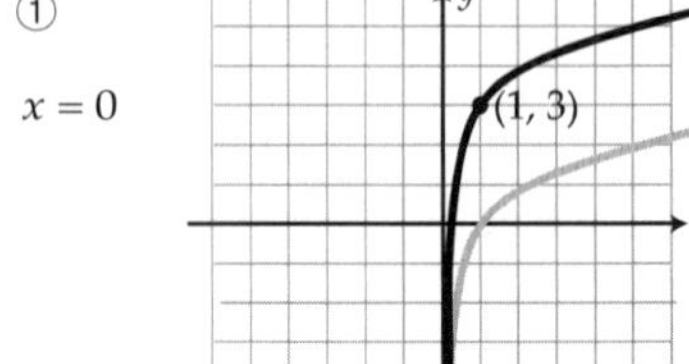

② $x = -3$

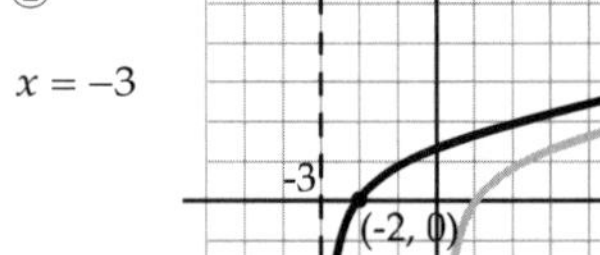

③ $x = 3$

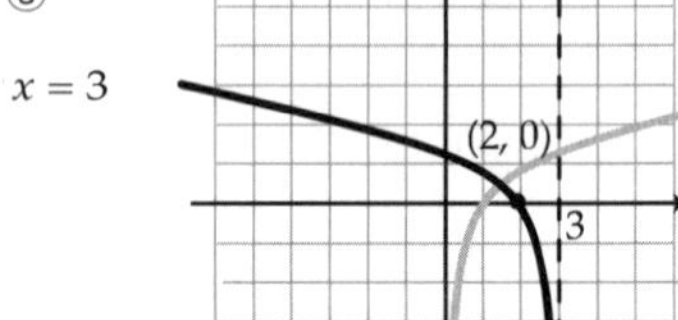

④ $x = 2$

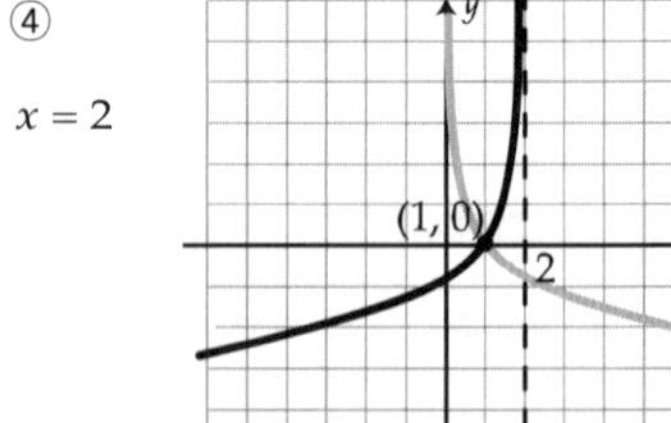

⑤ $x = 1$

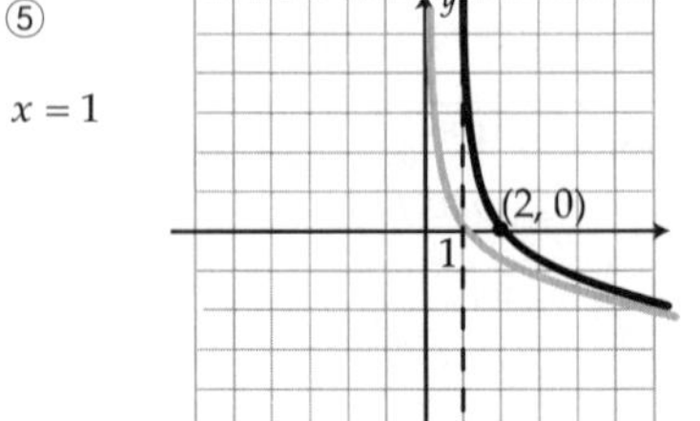

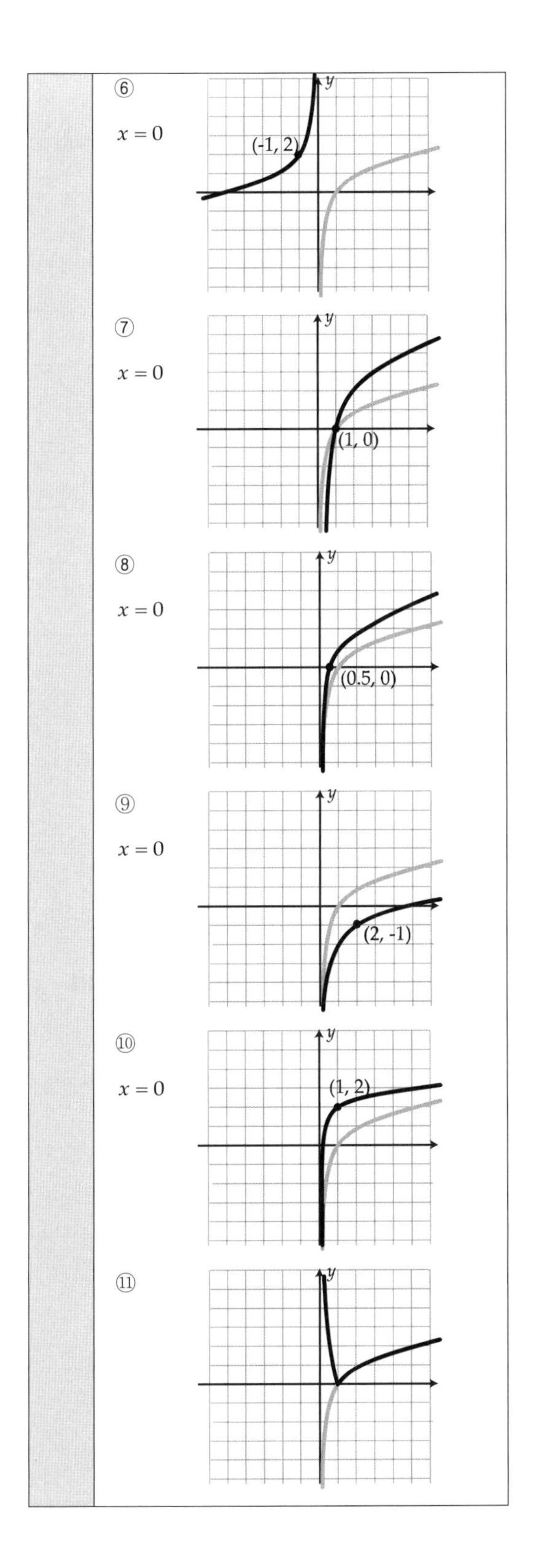

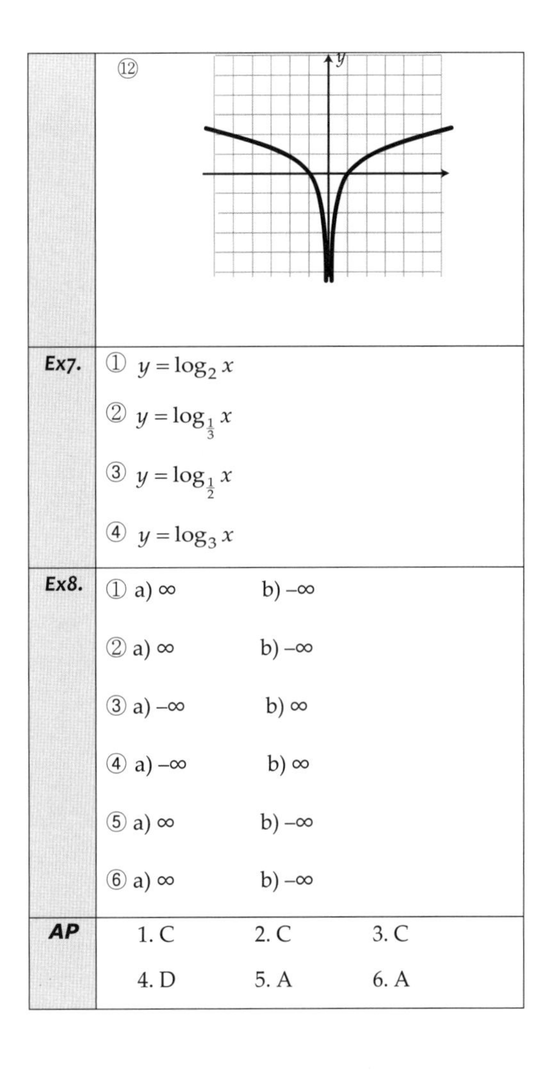

Ex7.	① $y = \log_2 x$		
	② $y = \log_{\frac{1}{3}} x$		
	③ $y = \log_{\frac{1}{2}} x$		
	④ $y = \log_3 x$		
Ex8.	① a) ∞	b) $-\infty$	
	② a) ∞	b) $-\infty$	
	③ a) $-\infty$	b) ∞	
	④ a) $-\infty$	b) ∞	
	⑤ a) ∞	b) $-\infty$	
	⑥ a) ∞	b) $-\infty$	
AP	1. C	2. C	3. C
	4. D	5. A	6. A

3.4 Properties of Logarithm

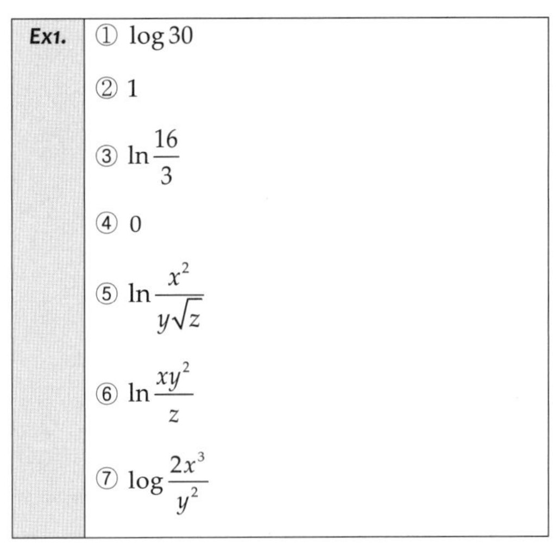

Ex1.	① $\log 30$
	② 1
	③ $\ln \frac{16}{3}$
	④ 0
	⑤ $\ln \frac{x^2}{y\sqrt{z}}$
	⑥ $\ln \frac{xy^2}{z}$
	⑦ $\log \frac{2x^3}{y^2}$

	⑧ $\log_5 \frac{x^2y^4}{z^6}$ ⑨ $\log_6 \frac{x-1}{x+1}$ ⑩ $\log(x-3)$ ⑪ 2 ⑫ $\frac{3}{2}$ ⑬ 2 ⑭ $\frac{\log 5}{\log 2}$ ⑮ $\frac{1}{\log 5}$ ⑯ $\frac{\log 8}{\log 5}$ ⑰ 25 ⑱ x^3 ⑲ 36 ⑳ 2 ㉑ $\log x^{\frac{3}{2}}$ ㉒ $\log_2 x^{\frac{5}{6}}$
Ex2.	① $a+3b$ ② $3(a+b)$ ③ $2a+b+\frac{1}{3}c$ ④ $2+\frac{1}{2}(a+b)$ ⑤ $1+2b-\frac{1}{2}c$ ⑥ $\frac{1}{2}a-3b-c$ ⑦ $\frac{3a}{2b}$ ⑧ $\frac{a}{2}+\frac{b}{4}$

	⑨ $\frac{2c+a}{6(a+b)}$ ⑩ $\frac{b+c}{2a}$
Ex3.	$\log_{13} 10-\frac{1}{5}\left[2\log_{13} x+3\log_{13}(4x+1)\right]$
Ex4.	$\log(x+y)+\log(x-y)$
Ex5.	6
Ex6.	III, V, VIII
Ex7.	$\log_3 2\cdot\log_3 7$
Ex8.	① 10! ② 110 ③ 10!
AP	1. B 2. A 3. D 4. B 5. B 6. C 7. C

3.5 Exp and Log Equations and Inequalities

Ex1.	① 24 ② $2+e^{\frac{3}{2}}$ ③ $\frac{2}{3^e}+1$ ④ 81 ⑤ 27 ⑥ $\frac{8}{7}$ ⑦ 1 ⑧ 4 ⑨ $3^{\frac{4}{3}}$

	⑩ 2^{10} ⑪ $\frac{7}{2}$ ⑫ $\frac{5}{4}$ ⑬ 6 ⑭ 1 ⑮ no solution ⑯ ± 1 ⑰ $\pm\sqrt{3}$ ⑱ 6 ⑲ 10,1000 ⑳ $10^{\frac{5}{2}}, 10^{-1}$ ㉑ $8, \frac{1}{8}$ ㉒ $10, 10^4$
Ex2.	$x = 3,\ y = 1$
Ex3.	① 3 ② −2 ③ 3 ④ −9 ⑤ $\frac{8}{7}$ ⑥ 12 ⑦ $\frac{\log 3}{\log 2}$ ⑧ $\frac{\log 8}{\log 3} - 1$ ⑨ $\frac{\ln 5}{2}$ ⑩ $\frac{1 + \ln 7}{3}$ ⑪ $\frac{2\log 5 + \log 3}{\log 5 - 3\log 3}$ ⑫ $\frac{\log 7 + \log 11}{\log 11 - 2\log 7}$

	⑬ $\frac{\log 7}{2\log 7 + \log 2} = \frac{\log 7}{\log 98}$ ⑭ $\frac{3\log 4}{\log 4 - 2\log 6} = -\frac{\log 8}{\log 3}$ ⑮ $\ln 2$ ⑯ $\ln 8, \ln 7$ ⑰ $\frac{\log 4}{\log 3} \left(= \log_3 4\right)$ ⑱ $0, \frac{\log 4}{\log 5} \left(= \log_5 4\right)$ ⑲ $0,\ \ln 5$ ⑳ $\ln 3$ ㉑ $\frac{\ln 4}{2}$ ㉒ $\frac{1}{3}\ln\frac{1}{2}$ ㉓ $0,\ -3$
Ex4.	$\frac{\log(8/5)}{\log 1.06} = 8.066$ years
Ex5.	$\frac{\log(124/80)}{4\log\left(1 + \frac{0.07}{4}\right)} = 6.315$ years
Ex6.	$\frac{\ln 2}{0.03} = 23.105$ years
Ex7.	① $x > \frac{\log 3}{\log 2} + 1$ ② $x < \frac{\ln 2 - 1}{2}$ ③ $\log 2 < x \le \log 4$ ④ $\log_3 8 \le x < 3$ ⑤ $x < 3$ ⑥ $-1 < x < 3$ ⑦ $0 < x < 2$ ⑧ $\frac{1}{2} < x < 1$ ⑨ $-\frac{1}{3} < x < 5$

	⑩ $x > e^5$
	⑪ $x \geq e^3$
	⑫ $\sqrt{10} < x < 100$
	⑬ $9 \leq x \leq 27$
AP	1. B 2. C 3. D 4. A 5. A 6. C 7. A

3.6 Exponential Growth and Modeling

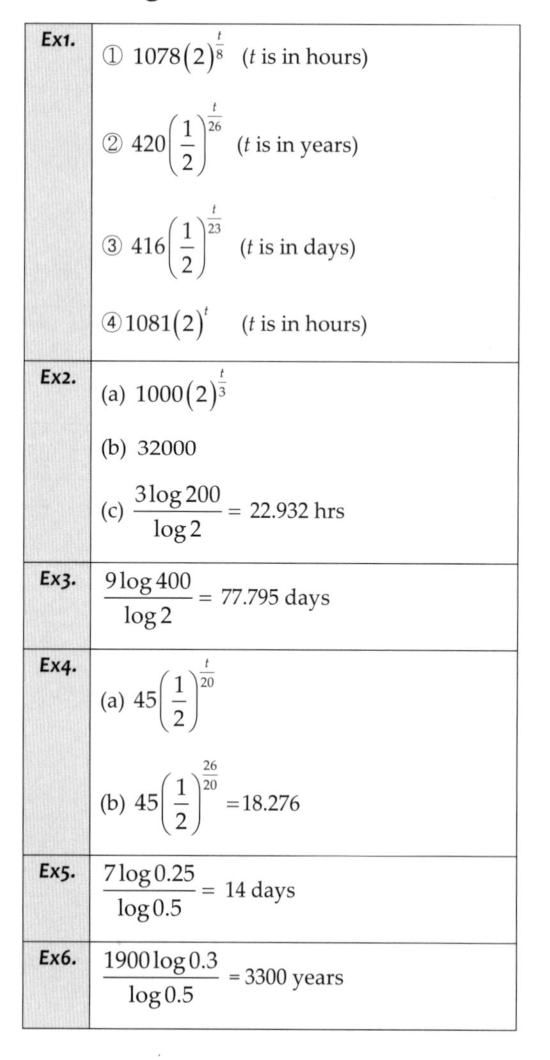

Ex1.	① $1078(2)^{\frac{t}{8}}$ (t is in hours) ② $420\left(\frac{1}{2}\right)^{\frac{t}{26}}$ (t is in years) ③ $416\left(\frac{1}{2}\right)^{\frac{t}{23}}$ (t is in days) ④ $1081(2)^{t}$ (t is in hours)
Ex2.	(a) $1000(2)^{\frac{t}{3}}$ (b) 32000 (c) $\frac{3\log 200}{\log 2} = 22.932$ hrs
Ex3.	$\frac{9\log 400}{\log 2} = 77.795$ days
Ex4.	(a) $45\left(\frac{1}{2}\right)^{\frac{t}{20}}$ (b) $45\left(\frac{1}{2}\right)^{\frac{26}{20}} = 18.276$
Ex5.	$\frac{7\log 0.25}{\log 0.5} = 14$ days
Ex6.	$\frac{1900\log 0.3}{\log 0.5} = 3300$ years

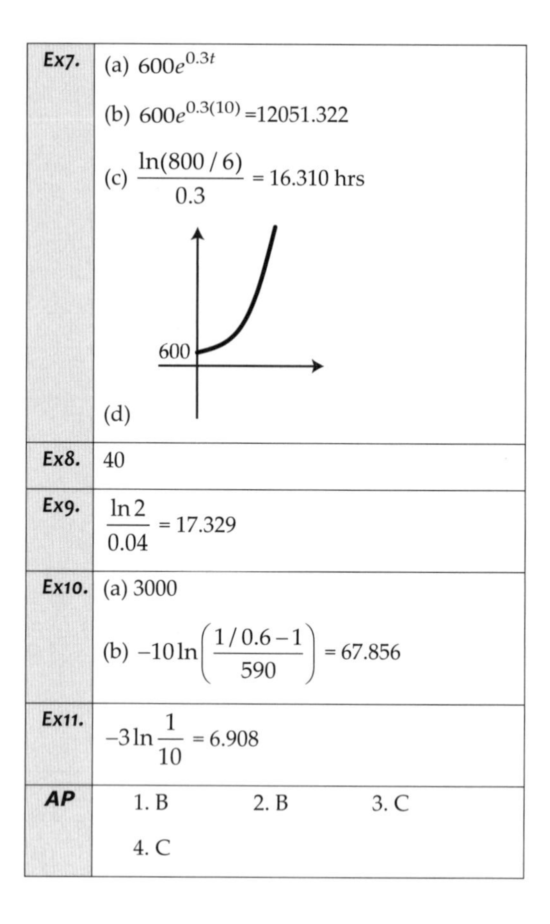

Ex7.	(a) $600e^{0.3t}$ (b) $600e^{0.3(10)} = 12051.322$ (c) $\frac{\ln(800/6)}{0.3} = 16.310$ hrs (d)
Ex8.	40
Ex9.	$\frac{\ln 2}{0.04} = 17.329$
Ex10.	(a) 3000 (b) $-10\ln\left(\frac{1/0.6-1}{590}\right) = 67.856$
Ex11.	$-3\ln\frac{1}{10} = 6.908$
AP	1. B 2. B 3. C 4. C

3.7 Sequences

Ex1.	① 5, 11, 29, 83 …… ② $-1, 0, 1, -2$ …… ③ $-1, \frac{1}{2}, -\frac{1}{4}, \frac{1}{8}$ …… ④ $2, \frac{3}{2}, \frac{4}{3}, \frac{5}{4}$ …… ⑤ 1, 9, 33, 105 …… ⑥ 2, 3, 5, 9 …… ⑦ $1, -1, -2, -1$ ……

	⑧ 1, 3, 4, 7 …… ⑨ $3, \frac{1}{3}, 3, \frac{1}{3}$ ……
Ex2.	① $a_n = \frac{1}{n}$ ② $a_n = \frac{n+1}{n}$ ③ $a_n = (-1)^n$ ④ $a_n = (-1)^{n+1} 2n$ ⑤ $a_n = (-1)^{n+1} \frac{1}{2n-1}$ ⑥ $a_n = (-1)^n 7n$ ⑦ $a_n = 4^{\frac{1}{2n+1}}$ ⑧ $a_n = x^{\frac{1}{n+1}}$ ⑨ $a_n = \frac{a^{n+1}}{n+3}$ ⑩ $a_n = \frac{5^n}{6^{n+1}}$ ⑪ $a_n = \frac{n}{x}$ ⑫ $a_n = x - n$ ⑬ $a_n = n(2n-1)$
Ex3.	① Arithmetic, $d = 3$ ② Arithmetic, $d = -2$ ③ Not arithmetic ④ Not arithmetic ⑤ Arithmetic, $d = \frac{1}{3}$ ⑥ Not arithmetic
Ex4.	① $d = 3,\ a_n = 2 + (n-1)3 = 3n - 1$ ② $d = -6,\ a_n = 7 + (n-1)(-6) = -6n + 13$ ③ $d = \frac{1}{2},\ a_n = 2 + (n-1)\frac{1}{2} = \frac{1}{2}n + \frac{3}{2}$ ④ $d = \frac{1}{4},\ a_n = 8 + (n-1)\frac{1}{4}$ ⑤ $d = -2,\ a_n = (x-1) + (n-1)(-2)$ $= x - 2n + 1$ ⑥ $d = 6,\ a_n = (2t+1) + (n-1)6$ ⑦ $d = 5\sqrt{3},\ a_n = 14\sqrt{3} + (n-1)5\sqrt{3}$ ⑧ $d = \ln 2,\ a_n = \ln 3 + (n-1)\ln 2$
Ex5.	① $7 + (n-1)2$ ② $3 + (n-1)5$
Ex6.	① $n = 18$ ② $n = 23$
Ex7.	① $r = -\frac{1}{2},\ a_n = 2\left(-\frac{1}{2}\right)^{n-1}$ ② $r = -2,\ a_n = -\frac{1}{4}(-2)^{n-1}$ ③ $r = \sqrt{2},\ a_n = \sqrt{2}\left(\sqrt{2}\right)^{n-1} = \left(\sqrt{2}\right)^n$ ④ $r = -1,\ a_n = 5(-1)^{n-1}$ ⑤ $r = -3,\ a_n = (x-3)(-3)^{n-1}$ ⑥ $r = 5,\ a_n = 2x(5)^{n-1}$ ⑦ $r = x^3,\ a_n = x^{a+2}\left(x^3\right)^{n-1} = x^{a+3n-1}$ ⑧ $r = x^{\frac{1}{3}},\ a_n = x^{\frac{1}{3}}(x^{\frac{1}{3}})^{n-1} = (x^{\frac{1}{3}})^n$ ⑨ $r = 3,\ a_n = 3^{n-1}\ln 2$ ⑩ $r = 3^{-\frac{1}{6}},\ a_n = 3^{\frac{1}{2}}\left(3^{-\frac{1}{6}}\right)^{n-1} = 3^{\frac{2}{3}-\frac{1}{6}n}$
Ex8.	① $a_n = \frac{1}{3^9}(3)^{n-1} = 3^{n-10}$ ② $a_n = \frac{1}{16}(2)^{n-1} = 2^{n-5}$
Ex9.	① $n = 9$ ② $n = 7$
Ex10.	$x = 4,\ y = 6$

AP			
	1. B	2. C	3. C
	4. A	5. C	6. D
	7. B		

3.8 Regression

Ex1.	b) $y = 0.342x + 10.871$ c) When the weight is increased by 1lb, the predicted height is increased by 0.342in. d) 66.310 in
Ex2.	b) $y = 0.095x^2 - 1.599x + 12.366$ c) 5.651 ft
Ex3.	b) $y = 1.538(2.165)^x$ c) 15613
Ex4.	b) $y = 6.953 + 7.749 \ln x$
Ex5.	① Residual plot does not show a pattern. So linear is appropriate. ② Residual plot does not show a pattern. So linear is appropriate. ③ 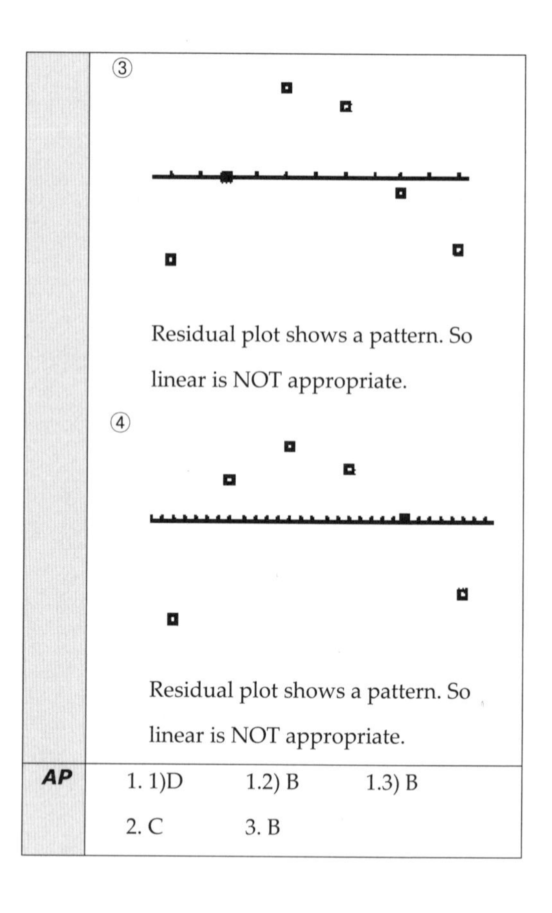 Residual plot shows a pattern. So linear is NOT appropriate. ④ Residual plot shows a pattern. So linear is NOT appropriate.

AP			
	1. 1)D	1.2) B	1.3) B
	2. C	3. B	

3.9 Semilog Plot

Ex1.	① f(x) is exponential since the semilog plot is linear. 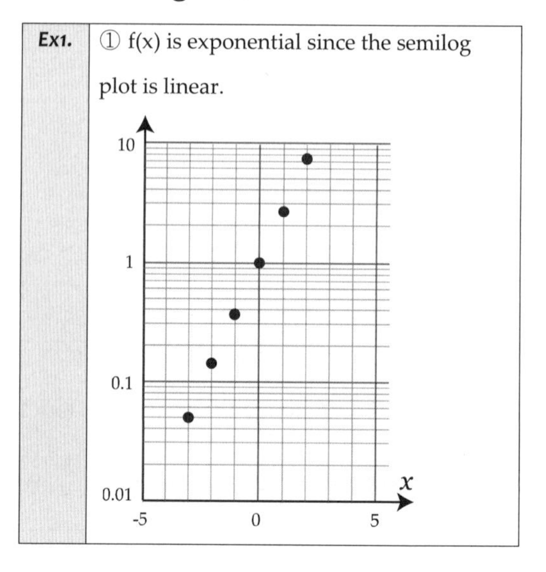

	② f(x) is exponential since the semilog plot is linear. 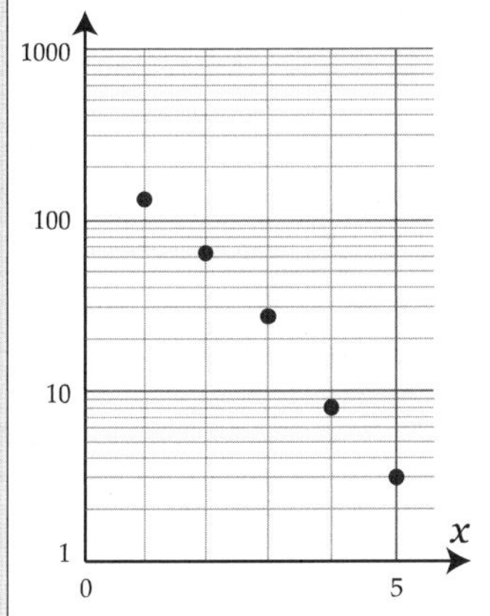 ③ f(x) is exponential since the semilog plot is linear. 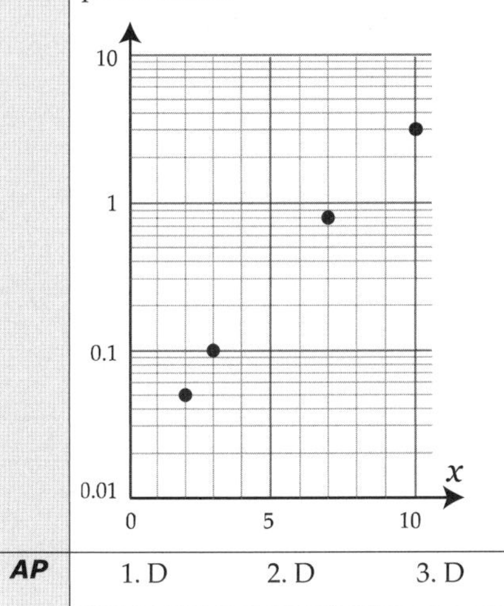
AP	1. D 2. D 3. D

Free Response Questions (from ch3)

1.	(a) 1, ∞ (b) Reflect over y axis followed by shifted to the right 1 unit. (*or* Reflect over y axis followed by horizontal translation by 1) (c) ∞, $-\infty$ (d) $a=-1, b=2, c=-1, d=1$ (e) $x=\ln\left(\frac{1}{2}\right)=-\ln 2$
2.	(a) $3e^{4k}=3(0.4)$ $k=\frac{\ln 0.4}{4}=-0.229$ (b) $3e^{-0.229072(6)}=0.759cc$ (c) $3e^{-0.229072t}=3(0.5)$ $\frac{\ln 0.5}{-0.229072}=3.026hr$ (d) $\frac{Q(6)-Q(4)}{6-4}=\frac{0.759-1.2}{2}$ $-0.221cc/hr$ When the time increased by 1 hour, dose of drug in the body decreased by 0221cc.

Answers

4. Trigonometry Definition and Graphs

4.1 Angles in Radian

Ex1.

① $\frac{\pi}{2}$

② $\frac{5\pi}{9}$

③ $-\frac{3\pi}{4}$

④ $\frac{5\pi}{6}$

⑤ $60°$

⑥ $300°$

⑦ $-240°$

⑧ $-15°$

⑨ $\frac{120°}{\pi}$

⑩ $\frac{1260°}{\pi}$

Ex2.

①

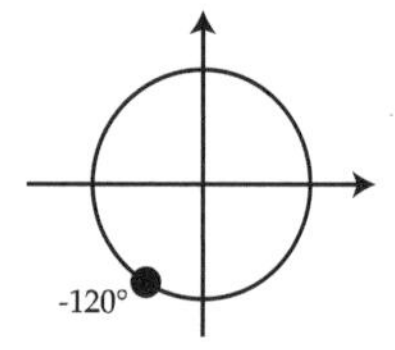

②

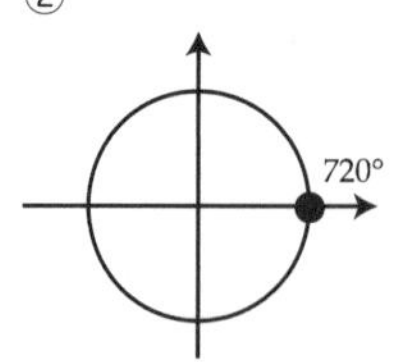

③ ④

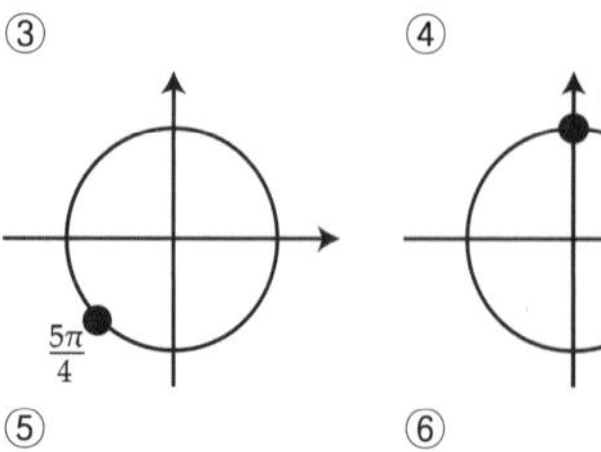

⑤ ⑥

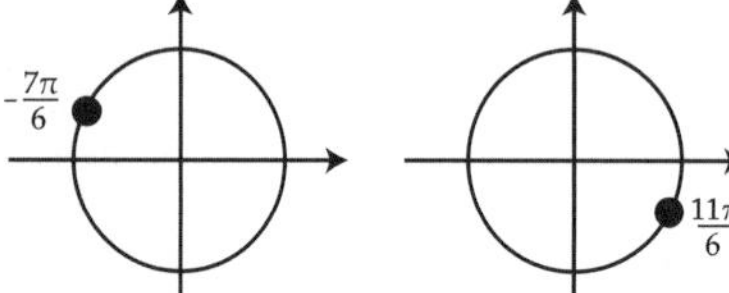

⑦

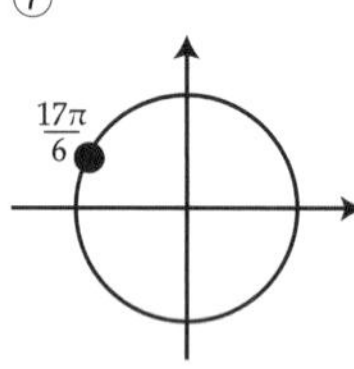

⑧

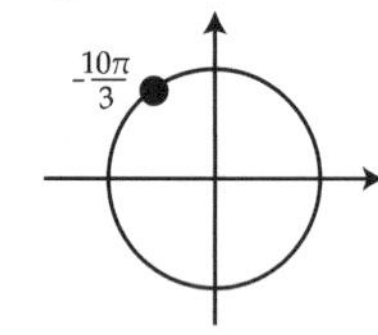

⑨

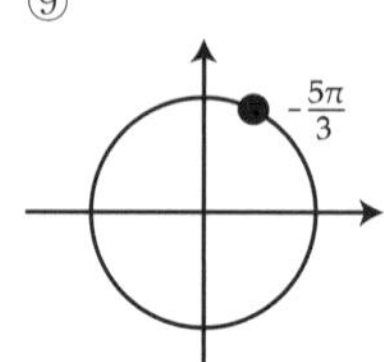

⑩

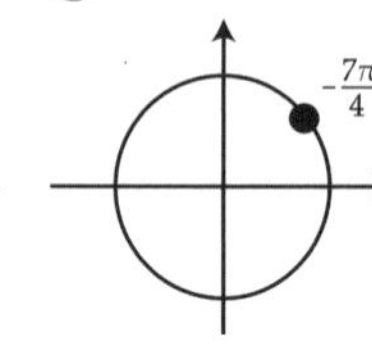

Ex3. $\pi-\theta,\ \pi+\theta,\ 2\pi-\theta$

Ex4.

① $480°$, $840°$, $-240°$, $-600°$

② $215°$, $575°$, $-505°$, $-865°$

③ $\frac{10\pi}{3}$, $\frac{16\pi}{3}$, $-\frac{2\pi}{3}$, $-\frac{8\pi}{3}$

④ $\frac{5\pi}{4}$, $\frac{13\pi}{4}$, $-\frac{11\pi}{4}$, $-\frac{19\pi}{4}$

Ex5.

① $S = 16\pi\, cm$, $A = 96\pi\, cm^2$

② $S = 4\pi\, in$, $A = 20\pi\, in^2$

③ $S = 10\pi\, in$, $A = 60\pi\, in^2$

④ $S = \frac{40\pi}{9}\, yd$, $A = \frac{160\pi}{9}\, yd^2$

⑤ $S = 3\pi\, m$, $A = 54\pi\, m^2$

⑥ $S = 26\pi\, m$, $A = 312\pi\, m^2$

Ex6.

① $5\ (rad)$

② $\frac{9}{5}\ (rad)$

③ $2\ (rad)$

④ $\frac{3}{2}\ (rad)$

Ex7.	(a) 5 cm (b) $10\pi - 10$ cm (c) 20 cm
Ex8.	$\dfrac{12}{2+\dfrac{5\pi}{18}} = 4.1773$
AP	1. B 2. B 3. C 4. D

4.2 Trigonometry of Right Triangles

Ex1.	① $\sin\theta = \frac{5}{13}$ $\csc\theta = \frac{13}{5}$ $\cos\theta = \frac{12}{13}$ $\sec\theta = \frac{13}{12}$ $\tan\theta = \frac{5}{12}$ $\cot\theta = \frac{12}{5}$ ② $\sin\theta = \frac{2}{\sqrt{13}}$ $\csc\theta = \frac{\sqrt{13}}{2}$ $\cos\theta = \frac{3}{\sqrt{13}}$ $\sec\theta = \frac{\sqrt{13}}{3}$ $\tan\theta = \frac{2}{3}$ $\cot\theta = \frac{3}{2}$ ③ $\sin\theta = \frac{1}{\sqrt{2}}$ $\sec\theta = \sqrt{2}$ $\tan\theta = 1$ ④ $\csc\theta = \sqrt{6}$ $\cos\theta = \frac{\sqrt{30}}{6}$ $\cot\theta = \sqrt{5}$ ⑤ $\csc\theta = \frac{7}{5}$ $\cos\theta = \frac{2\sqrt{6}}{7}$ $\cot\theta = \frac{2\sqrt{6}}{5}$ ⑥ $\sin\theta = \frac{24}{25}$ $\sec\theta = \frac{25}{7}$ $\tan\theta = \frac{24}{7}$
Ex2.	① $\frac{\sqrt{7}}{4}$ ② $2\sqrt{2}$ ③ $\frac{\sqrt{a^2-b^2}}{a}$ ④ $\frac{1}{\sqrt{a^2-1}}$ ⑤ $\sqrt{1+a^2}$
Ex3.	① $x=10\sin\theta$, $y=10\cos\theta$ ② $x=9\cos\theta$, $y=9\sin\theta$ ③ $x=\frac{4.5}{\tan\theta}$, $y=\frac{4.5}{\sin\theta}$ ④ $x=\frac{22}{\sin\theta}$, $y=\frac{22}{\tan\theta}$
Ex4.	① $\frac{\sqrt{3}}{2}$ ② $\frac{\sqrt{2}}{2}$ ③ $\frac{\sqrt{2}}{2}$ ④ $\frac{\sqrt{3}}{2}$ ⑤ $\sqrt{3}$ ⑥ 1 ⑦ 0 ⑧ $\frac{1}{2}$ ⑨ 1 ⑩ $\frac{1}{2}$ ⑪ $\frac{\sqrt{3}}{3}$ ⑫ 1 ⑬ $\frac{2\sqrt{3}}{3}$ ⑭ $\sqrt{2}$ ⑮ $\sqrt{2}$ ⑯ $\sqrt{3}$ ⑰ $\frac{\sqrt{3}}{3}$ ⑱ 2
Ex5.	$200\tan 57.6° =$ 315.150 ft

Ex6.	$\dfrac{2}{\tan 15°}=7.464\ mi$
Ex7.	$25\sin 35°=14.399\ ft$
Ex8.	height of the building = $500\tan 32°=312.435\,ft$ length of the flagpole = $500\tan 46°-500\tan 32°=205.33\ ft$
Ex9.	$6+100\tan 37°=81.355\ ft$
Ex10.	$1.295\ mi$
Ex11.	$821.534\ ft$

4.3 Trigonometry of Any Angles

Ex1.	① $\sin\theta=\dfrac{5}{13}$, $\cos\theta=\dfrac{12}{13}$, $\tan\theta=\dfrac{5}{12}$ ② $\sin\theta=-\dfrac{24}{25}$, $\cos\theta=\dfrac{7}{25}$, $\tan\theta=-\dfrac{24}{7}$ ③ $\csc\theta=-\dfrac{17}{15}$, $\cos\theta=\dfrac{8}{17}$, $\cot\theta=-\dfrac{8}{15}$ ④ $\csc\theta=-\dfrac{\sqrt{137}}{4}$, $\cos\theta=-\dfrac{11}{\sqrt{137}}=-\dfrac{11\sqrt{137}}{137}$, $\cot\theta=\dfrac{11}{4}$ ⑤ $\sin\theta=-\dfrac{4}{\sqrt{41}}=-\dfrac{4\sqrt{41}}{41}$, $\sec\theta=-\dfrac{\sqrt{41}}{5}$, $\tan\theta=\dfrac{4}{5}$ ⑥ $\sin\theta=\dfrac{1}{\sqrt{2}}=\dfrac{\sqrt{2}}{2}$, $\sec\theta=-\sqrt{2}$, $\tan\theta=-1$
Ex2.	① II ② III ③ IV ④ II ⑤ I ⑥ I ⑦ III ⑧ IV
Ex3.	① 60° ② 45° ③ $\dfrac{\pi}{3}$ ④ $\dfrac{\pi}{3}$ ⑤ $\dfrac{\pi}{6}$ ⑥ $\dfrac{\pi}{4}$
Ex4.	① $-\dfrac{\sqrt{2}}{2}$ ② $-\dfrac{\sqrt{3}}{2}$ ③ $-\dfrac{1}{\sqrt{3}}$ ④ 2 ⑤ $\dfrac{\sqrt{2}}{2}$ ⑥ $\dfrac{1}{2}$

	⑦ $-\frac{\sqrt{2}}{2}$ ⑧ $-\frac{\sqrt{3}}{3}$ ⑨ $-\frac{1}{2}$ ⑩ $-\frac{\sqrt{3}}{2}$ ⑪ $\sqrt{2}$ ⑫ $-\sqrt{2}$ ⑬ $\sqrt{3}$ ⑭ $-\frac{2\sqrt{3}}{3}$
Ex5.	① $\cos\theta=-\frac{3}{5}$ $\tan\theta=\frac{4}{3}$ ② $\sin\theta=\frac{2}{\sqrt{53}}=\frac{2\sqrt{53}}{53}$ $\cos\theta=-\frac{7}{\sqrt{53}}=-\frac{7\sqrt{53}}{53}$ ③ $\sec\theta=-\frac{4}{\sqrt{7}}=-\frac{4\sqrt{7}}{7}$ $\tan\theta=-\frac{3}{\sqrt{7}}=-\frac{3\sqrt{7}}{7}$ ④ $\sin\theta=-\frac{1}{\sqrt{5}}=-\frac{\sqrt{5}}{5}$ $\sec\theta=-\frac{\sqrt{5}}{2}$ ⑤ $\sin\theta=-\frac{12}{13}$ $\cot\theta=-\frac{5}{12}$ ⑥ $\sin\theta=-\frac{2\sqrt{6}}{7}$ $\cot\theta=\frac{5}{2\sqrt{6}}=\frac{5\sqrt{6}}{12}$
AP	1. B 2. D 3. D 4. C

4.4 Trigonometry in Unit Circle

Ex1.	① $-\frac{1}{2}$ ② $\frac{1}{2}$ ③ $-\frac{\sqrt{3}}{2}$ ④ $\frac{\sqrt{2}}{2}$ ⑤ -1 ⑥ $-\frac{\sqrt{3}}{3}$ ⑦ $-\frac{1}{2}$ ⑧ $-\frac{\sqrt{3}}{2}$ ⑨ $\frac{\sqrt{3}}{3}$ ⑩ $-\sqrt{3}$ ⑪ $\frac{1}{2}$ ⑫ $-\frac{\sqrt{3}}{2}$ ⑬ -2 ⑭ $-\sqrt{2}$ ⑮ 0 ⑯ 1 ⑰ 0 ⑱ -1 ⑲ -1 ⑳ 0 ㉑ 0 ㉒ und ㉓ und ㉔ 0
Ex2.	① 0.6 ② –0.6 ③ –0.6 ④ –0.6
Ex3.	① –0.4 ② –0.4 ③ 0.4 ④ 0.4
AP	1. D 2. B 3. C 4. A 5. B 6. A

4.5 Trigonometric Graphs for Sin,Cos

Ex1.

① –1

② 1

③ 2

④ 1

Ex2.

① $amp : 4 \quad per : \dfrac{2\pi}{5}$

$trans : left\dfrac{\pi}{40}, up\,3$

$range : -1 \le y \le 7$

② $amp : 1 \quad per : 6\pi$

$trans : right\,3\pi,\ down\,1,\ reflect\,x$

$range : -2 \le y \le 0$

③ $amp : 1 \quad per : 10\pi$

$trans : left\,15,\ up\,2,\ reflect\ \ x$

$range : 1 \le y \le 3$

④ $amp : 4 \quad per : \pi$

$trans : left\dfrac{3\pi}{2}, down\,2$

$range : -6 \le y \le 2$

⑤ $amp : \dfrac{1}{2} \quad per : 6$

$trans : right\dfrac{3}{5}, reflect\,x$

$range : -\dfrac{1}{2} \le y \le \dfrac{1}{2}$

⑥ $amp : 3 \quad per : \dfrac{2\pi^2}{3}$

$trans : left\dfrac{\pi^2}{3}, down\,3, reflect\,x$

$range : -6 \le y \le 0$

Ex3.

①

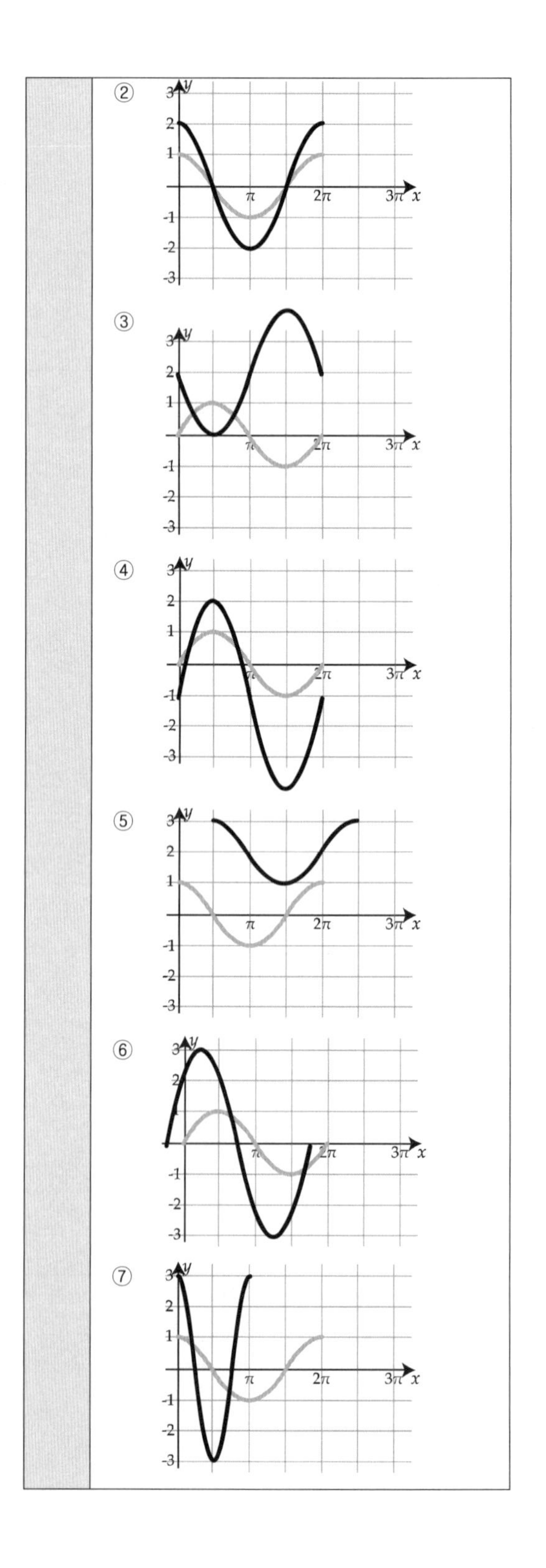

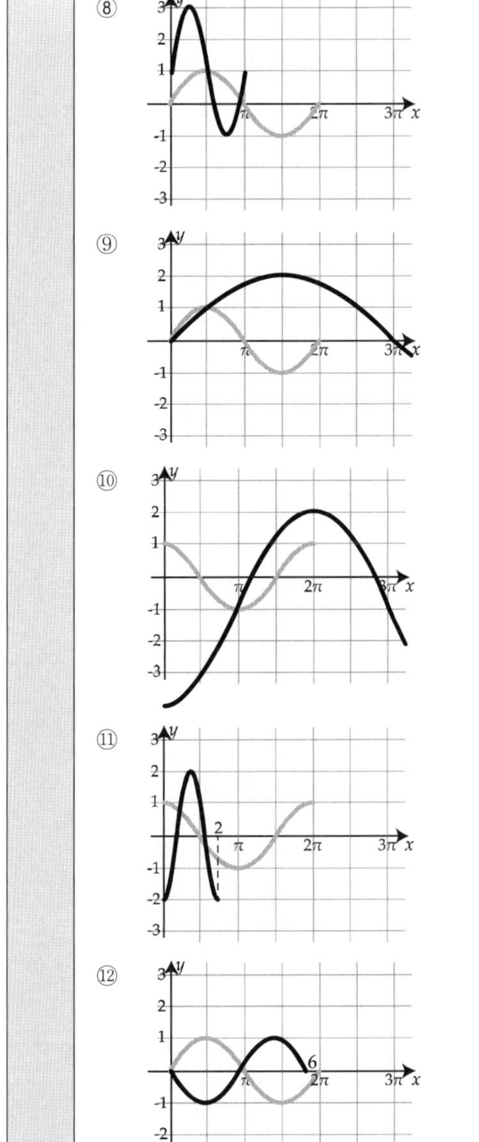

Ex4.

① $3\sin\frac{1}{2}x$

② $3\cos 2x$

③ $-3\cos 3x$

④ $2\sin\frac{1}{3}x$

⑤ $-\frac{1}{2}\sin 2\left(x+\frac{\pi}{3}\right)$

⑥ $-\frac{1}{5}\cos 3\left(x+\frac{\pi}{2}\right)$

⑦ $5\sin\frac{2\pi}{3}(x+1)$

⑧ $-10\sin\frac{5\pi}{2}\left(x+\frac{1}{5}\right)$

⑨ $5\sin\frac{\pi}{5}x+3$

⑩ $3\cos\frac{\pi}{6}x+1$

Ex5.

① 9

② 10

③ 6

④ 5

⑤ Not periodic

⑥ Not periodic

⑦ 5

⑧ 10

⑨ 6

⑩ 9

AP

1. D	2. A	3. B
4. A	5. A	6. C
7. D		

4.6 Modeling using Sin,Cos Functions

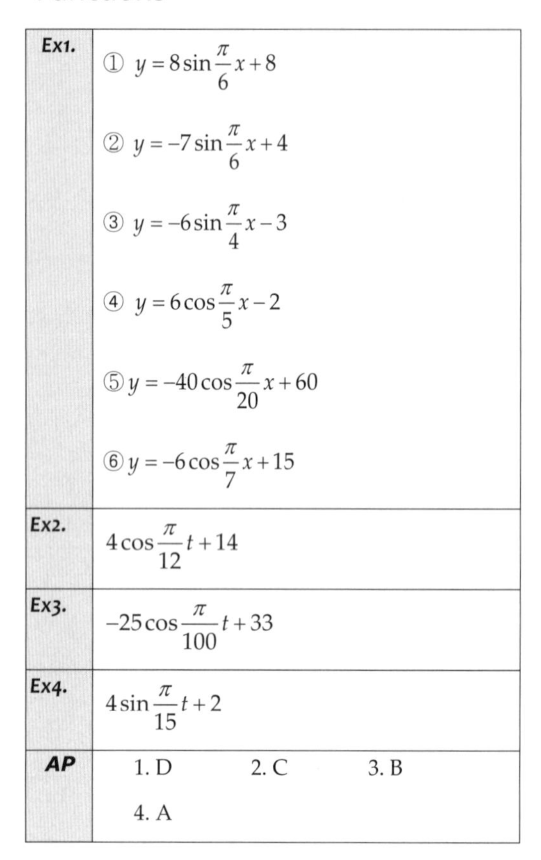

Ex1.	① $y = 8\sin\frac{\pi}{6}x + 8$ ② $y = -7\sin\frac{\pi}{6}x + 4$ ③ $y = -6\sin\frac{\pi}{4}x - 3$ ④ $y = 6\cos\frac{\pi}{5}x - 2$ ⑤ $y = -40\cos\frac{\pi}{20}x + 60$ ⑥ $y = -6\cos\frac{\pi}{7}x + 15$
Ex2.	$4\cos\frac{\pi}{12}t + 14$
Ex3.	$-25\cos\frac{\pi}{100}t + 33$
Ex4.	$4\sin\frac{\pi}{15}t + 2$
AP	1. D 2. C 3. B 4. A

4.7 Trigonometric Graphs for Others

Ex1.	① $per : \frac{\pi}{2}$ $trans : right\frac{\pi}{8}, up1$ ② $per : 4\pi$ $trans : left4\pi, up3$ ③ $per : 2\pi$ $trans : right4, down5$ ④ $per : \frac{1}{3}$ $trans : right\frac{1}{3}$

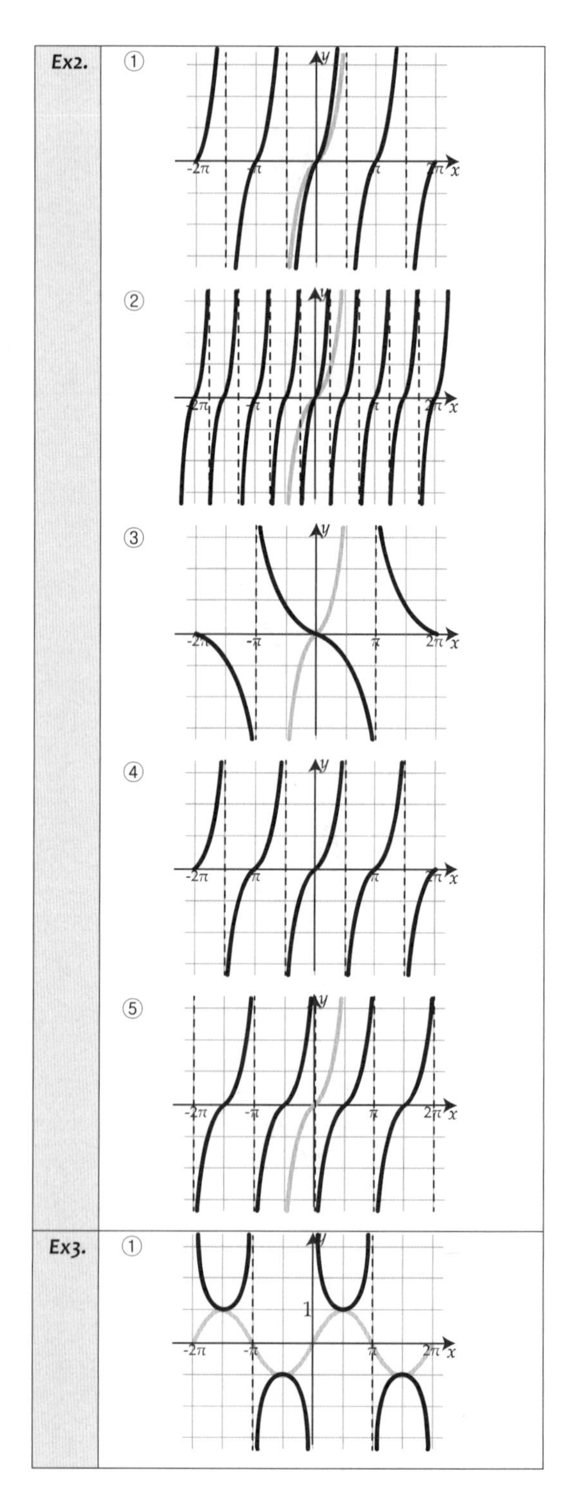

②

③

Ex4. ① ② ③ ④

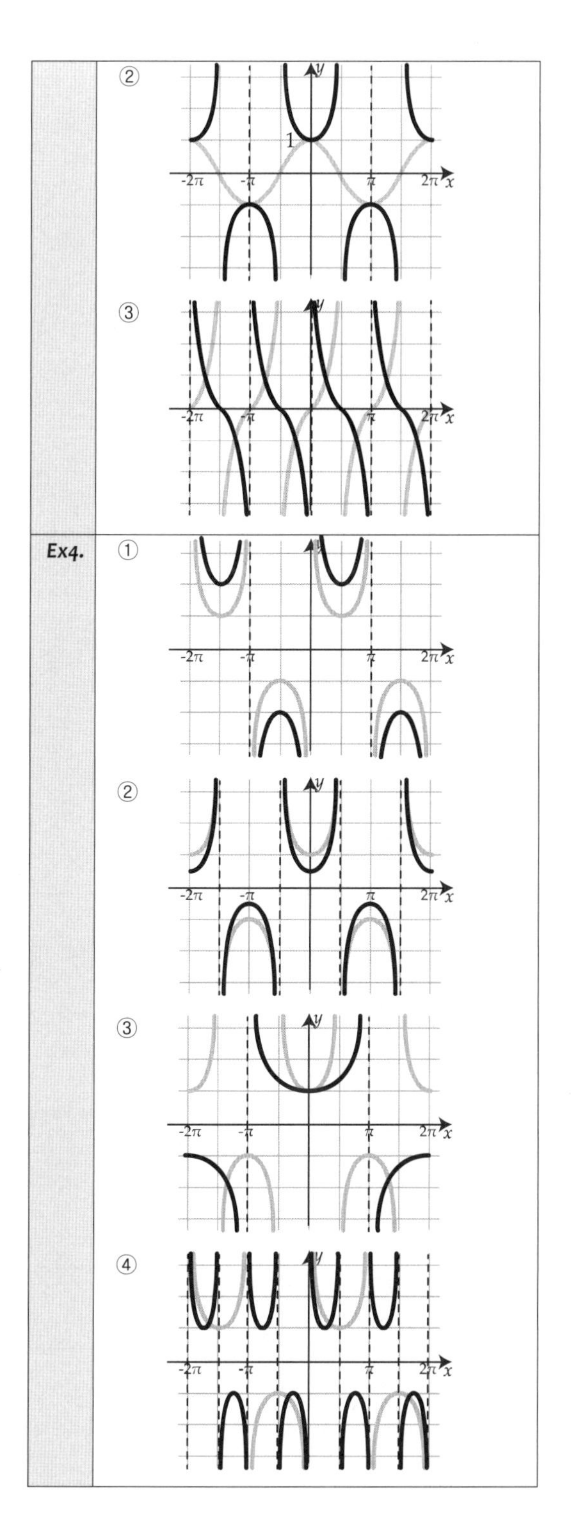

⑤

⑥

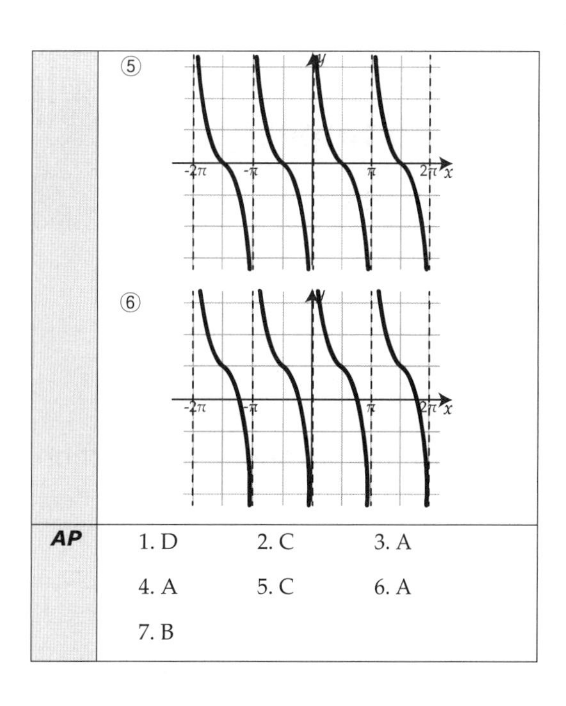

AP			
	1. D	2. C	3. A
	4. A	5. C	6. A
	7. B		

Answers

5. Trigonometry Identities

5.1 Inverse Trigonometry Function

Ex1.	① 42° ② 77° ③ 25° ④ 65°
Ex2.	$\sin^{-1}\dfrac{12}{40}=17.458°$
Ex3.	$\tan^{-1}\dfrac{51}{60}=40.365°$
Ex4.	① $\dfrac{\pi}{6}$ ② $\dfrac{\pi}{6}$ ③ $\dfrac{\pi}{4}$ ④ 0 ⑤ $\dfrac{\pi}{4}$ ⑥ $\dfrac{\pi}{2}$ ⑦ $\dfrac{\pi}{6}$ ⑧ 0 ⑨ $\dfrac{\pi}{6}$ ⑩ $\dfrac{\pi}{3}$ ⑪ $\dfrac{\pi}{4}$ ⑫ $\dfrac{\pi}{4}$ ⑬ $\dfrac{5\pi}{6}$ ⑭ $-\dfrac{\pi}{6}$ ⑮ $-\dfrac{\pi}{3}$ ⑯ $\dfrac{-\pi}{4}$ ⑰ $\dfrac{-\pi}{4}$ ⑱ $\dfrac{2\pi}{3}$
Ex5.	① 0.5 ② 0.2 ③ −0.4 ④ undefined ⑤ undefined ⑥ 0.8 ⑦ 2 ⑧ 3
Ex6.	① $\dfrac{\pi}{6}$ ② $\dfrac{\pi}{3}$ ③ $\dfrac{3\pi}{4}$ ④ $\dfrac{\pi}{3}$ ⑤ $\dfrac{\pi}{3}$ ⑥ $\dfrac{2\pi}{3}$ ⑦ $\dfrac{\pi}{4}$ ⑧ $-\dfrac{\pi}{6}$ ⑨ $\dfrac{\pi}{3}$ ⑩ $-\dfrac{\pi}{3}$ ⑪ $\dfrac{3\pi}{5}$ ⑫ $\dfrac{3\pi}{8}$
Ex7.	① $\dfrac{1}{\sqrt{5}}$ ② $-2\sqrt{2}$ ③ $\dfrac{4}{5}$ ④ $\dfrac{4}{5}$ ⑤ $-\dfrac{12}{5}$ ⑥ $-\dfrac{17}{8}$
Ex8.	① $\dfrac{x}{\sqrt{1-x^2}}$ ② $\sqrt{1-x^2}$ ③ x ④ x

	⑤ $\frac{x}{\sqrt{1-x^2}}$
	⑥ $\frac{\sqrt{1+x^2}}{x}$
	⑦ $\frac{x}{\sqrt{x^2+5}}$
	⑧ $\frac{3}{x}$
	⑨ $\frac{2}{\sqrt{x^2+4}}$
AP	1. C 2. B 3. D
	4. B 5. D 6. A

5.2 Basic Trigonometric Identities

Ex1.	① $\sin x$
	② 1
	③ $\cos u$
	④ $\tan\theta$
	⑤ $\sec\theta$
Ex2.	① $\tan^2 x$
	② $\sin\theta$
	③ $\cot x$
	④ $\sec x$
	⑤ $\tan\theta$
	⑥ –1
Ex3.	① –1
	② 1
	③ 1
	④ 1
	⑤ 1
	⑥ 1

Ex4.	① $-\frac{\sqrt{3}}{2}$
	② 0
	③ 0
	④ $-\frac{2\sqrt{3}}{3}$
	⑤ $\frac{\sqrt{2}}{2}$
	⑥ –1
	⑦ $-\sqrt{3}$
Ex5.	① $\cos x$
	② $\tan x$
	③ 2
	④ 1
	⑤ $\sec x$
	⑥ $\sin\theta$
	⑦ $-\sin x$
	⑧ $\cos x$
	⑨ $\tan x$
	⑩ $\tan x$
	⑪ $\sin x$
AP	1. C 2. A 3. D
	4. C

5.3 Verifying Trigonometric Identities

Ex1.

① $\cos^2 x - \sin^2 x$
$= 1 - \sin^2 x - \sin^2 x$
$= 1 - 2\sin^2 x$

② $\dfrac{\csc(-x)}{\sec(-x)} = \dfrac{-\csc x}{\sec x} = \dfrac{-\dfrac{1}{\sin x}}{\dfrac{1}{\cos x}}$
$= -\dfrac{\cos x}{\sin x} = -\cot x$

③ $\left(\dfrac{\sin x}{\cos x} + \dfrac{\cos x}{\sin x}\right)^4 = \left(\dfrac{1}{\sin x \cdot \cos x}\right)^4$
$= \csc^4 x \cdot \sec^4 x$

④ $\dfrac{\cos^2 x}{\sin^2 x} - \cos^2 x = \dfrac{\cos^2 x - \cos^2 x \cdot \sin^2 x}{\sin^2 x}$
$= \dfrac{\cos^2 x(1 - \sin^2 x)}{\sin^2 x} = \dfrac{\cos^2 x \cdot \cos^2 x}{\sin^2 x}$
$= \cot^2 x \cdot \cos^2 x$

⑤ $\dfrac{\cos\theta \cdot \dfrac{\cos\theta}{\sin\theta}}{1 - \sin\theta} - 1 = \dfrac{\dfrac{\cos^2\theta}{\sin\theta}}{1 - \sin\theta} - 1$
$= \dfrac{\dfrac{1 - \sin^2\theta}{\sin\theta}}{\dfrac{1 - \sin\theta}{1}} - 1 = \dfrac{1 + \sin\theta}{\sin\theta} - 1$
$= \csc\theta + 1 - 1 = \csc\theta$

⑥ $\dfrac{\dfrac{\sin x}{\cos x} - \sin x}{\dfrac{\sin x}{\cos x} \cdot \sin x} = \dfrac{\dfrac{\sin x - \sin x \cdot \cos x}{\cos x}}{\dfrac{\sin^2 x}{\cos x}}$
$= \dfrac{\sin x(1 - \cos x)}{\sin^2 x} = \dfrac{\sin x(1 - \cos x)}{1 - \cos^2 x}$
$= \dfrac{\sin x(1 - \cos x)}{(1 - \cos x)(1 + \cos x)}$
$= \dfrac{\sin x}{1 + \cos x} \cdot \dfrac{\tan x}{\tan x}$
$= \dfrac{\sin x \cdot \tan x}{\tan x + \sin x}$

⑦ $\dfrac{\dfrac{1}{\sin x} + \dfrac{1}{\cos x}}{\sin x + \cos x} = \dfrac{\dfrac{\cos x + \sin x}{\sin x \cdot \cos x}}{\dfrac{\sin x + \cos x}{1}}$
$= \dfrac{1}{\sin x \cdot \cos x} = \csc x \cdot \sec x$

⑧ $\dfrac{(\sin x + \cos x)(\sin^2 x - \sin x \cos x + \cos^2 x)}{(\sin x + \cos x)}$
$= 1 - \sin x \cdot \cos x$

⑨ $\dfrac{\cot u - \tan u}{(\cot u + \tan u)(\cot u - \tan u)}$
$= \dfrac{1}{\dfrac{\cos u}{\sin u} + \dfrac{\sin u}{\cos u}} = \dfrac{1}{\dfrac{1}{\sin u \cdot \cos u}}$
$= \sin u \cdot \cos u$

⑩ $\dfrac{(\sin x + \cos x)(\sin x - \cos x)}{(\sin x + \cos x)^2}$
$= \dfrac{(\sin x - \cos x)}{(\sin x + \cos x)}$
$= \dfrac{(\sin x - \cos x)^2}{(\sin x + \cos x)(\sin x - \cos x)}$
$= \dfrac{(\sin x - \cos x)^2}{\sin^2 x - \cos^2 x}$

⑪ $\dfrac{(1 + \sin\theta)^2 + \cos^2\theta}{\cos\theta(1 + \sin\theta)}$
$= \dfrac{1 + 2\sin\theta + \sin^2\theta + \cos^2\theta}{\cos\theta(1 + \sin\theta)}$
$= \dfrac{2(1 + \sin\theta)}{\cos\theta(1 + \sin\theta)} = 2\sec\theta$

⑫ $\dfrac{\csc\theta - \cot\theta + \csc\theta + \cot\theta}{\csc^2\theta - \cot^2\theta}$
$= \dfrac{2\csc\theta}{\csc^2\theta - \cot^2\theta} = 2\csc\theta$

⑬ $\dfrac{(\sec x - 1)(\sec x + 1)}{\tan x(\sec x + 1)}$
$= \dfrac{\sec^2 x - 1}{\tan x(\sec x + 1)}$
$= \dfrac{\tan^2 x}{\tan x(\sec x + 1)} = \dfrac{\tan x}{\sec x + 1}$

⑭ $\dfrac{(1+\sin x)^2}{(1-\sin x)(1+\sin x)} = \dfrac{(1+\sin x)^2}{1-\sin^2 x}$

$= \dfrac{(1+\sin x)^2}{\cos^2 x} = \left(\dfrac{1+\sin x}{\cos x}\right)^2$

$= (\sec x + \tan x)^2$

⑮ $\dfrac{\dfrac{\tan x}{\tan x \cdot \tan y} + \dfrac{\tan y}{\tan x \cdot \tan y}}{\dfrac{1}{\tan x \cdot \tan y} - \dfrac{\tan x \cdot \tan y}{\tan x \cdot \tan y}}$

$= \dfrac{\cot y + \cot x}{\cot x \cdot \cot y - 1}$

⑯ $\dfrac{2 + \dfrac{1}{\sin x \cdot \cos x}}{\dfrac{1}{\sin x \cdot \cos x}} = \dfrac{\dfrac{2\sin x \cdot \cos x + 1}{\sin x \cdot \cos x}}{\dfrac{1}{\sin x \cdot \cos x}}$

$= 1 + 2\sin x \cdot \cos x$

$= \sin^2 x + \cos^2 x + 2\sin x \cdot \cos x$

$= (\sin x + \cos x)^2$

⑰ $\dfrac{\sec^2 x - 1}{1 + \sec x} = \dfrac{(\sec x - 1)(\sec x + 1)}{1 + \sec x}$

$= \sec x - 1 = \dfrac{1}{\cos x} - \dfrac{\cos x}{\cos x}$

$= \dfrac{1 - \cos x}{\cos x}$

5.4 Sum and difference Identities

Ex1.

① $\dfrac{\sqrt{2}+\sqrt{6}}{4}$

② $\dfrac{\sqrt{6}-\sqrt{2}}{4}$

③ $\dfrac{\sqrt{6}+\sqrt{2}}{4}$

④ $\dfrac{(3+\sqrt{3})^2}{6} = \sqrt{3}+2$

⑤ $-\dfrac{(1+\sqrt{3})^2}{2} = -(\sqrt{3}+2)$

Ex2.

① $\dfrac{\sqrt{3}}{2}$

② $\dfrac{\sqrt{3}}{2}$

③ $-\dfrac{1}{2}$

④ 0

⑤ $-\dfrac{\sqrt{3}}{3}$

⑥ $-\sqrt{3}$

Ex3.

① $\cos x \cdot \cos\dfrac{\pi}{2} - \sin x \cdot \sin\dfrac{\pi}{2} = -\sin x$

② $\sin x \cdot \cos y + \cos x \cdot \sin y$
$+ \sin x \cdot \cos y - \cos x \cdot \sin y$
$= 2\sin x \cdot \cos y$

③ $\dfrac{\dfrac{\cos x \cdot \cos y - \sin x \cdot \sin y}{\cos x \cdot \cos y}}{\dfrac{\cos x \cdot \cos y + \sin x \cdot \sin y}{\cos x \cdot \cos y}}$

$= \dfrac{1 - \dfrac{\sin x \cdot \sin y}{\cos x \cdot \cos y}}{1 + \dfrac{\sin x \cdot \sin y}{\cos x \cdot \cos y}}$

$= \dfrac{1 - \tan x \cdot \tan y}{1 + \tan x \cdot \tan y}$

Ex4. -1

Ex5. $\tan 2A$

Ex6.

① $\dfrac{3+4\sqrt{3}}{10}$

② $\dfrac{3\sqrt{10}}{10}$

③ $-\dfrac{33}{65}$

④ $\dfrac{10+2\sqrt{5}}{15}$

⑤ 1

⑥ $\dfrac{x + x\sqrt{1-x^2}}{\sqrt{1+x^2}}$

Ex7.	$\frac{\pi}{4}$		
AP	1. B	2. B	3. C
	4. A	5.A	

5.5 Double-Angle Identity

Ex1.	① $\frac{\sqrt{3}}{2}$ ② $-\frac{1}{2}$ ③ $4\sin 3\alpha\cdot\cos 3\alpha$ ④ $2\sin 5B\cdot\cos 5B$ ⑤ $\cos^2 2A-\sin^2 2A$ $=2\cos^2 2A-1=1-2\sin^2 2A$ ⑥ $2\cos^2 4\beta-2\sin^2 4\beta$ $=4\cos^2 4\beta-2=2-4\sin^2 4\beta$ ⑦ $\frac{2\tan 2C}{1-\tan^2 2C}$
Ex2.	① $2\sin 2A$ ② $3\sin 4\beta$ ③ $\cos 6\alpha$ ④ $-\cos 4\beta$ ⑤ $-\cos 4\theta$ ⑥ $\cos A$ ⑦ $-2\cos\beta$ ⑧ $\cos 6\theta$ ⑨ $\tan 4\beta$ ⑩ $\tan 2C$
Ex3.	① $\sin 2(2x)=2\sin 2x\cdot\cos 2x$ $=2(2\sin x\cdot\cos x)(\cos^2 x-\sin^2 x)$ $=(4\sin x\cdot\cos x)(2\cos^2 x-1)$ ② $\cos 2(2x)=2\cos^2 2x-1$ $=2(1-2\sin^2 x)^2-1$ $=2(1-4\sin^2 x+4\sin^4 x)-1$ $=2-8\sin^2 x+8\sin^4 x-1$ $=1-8\sin^2 x+8\sin^4 x$ ③ $\frac{2\sin x\cdot\cos x+\sin x}{1+(2\cos^2 x-1)+\cos x}$ $=\frac{\sin x(2\cos x+1)}{\cos x(2\cos x+1)}=\tan x$ ④ $\tan 2x=\frac{2\tan x}{1-\tan^2 x}$ $=\frac{\frac{2\tan x}{\tan x}}{\frac{1}{\tan x}-\frac{\tan^2 x}{\tan x}}=\frac{2}{\cot x-\tan x}$
Ex4.	$8\cos^4 x-8\cos^2 x+1$
Ex5.	① $-\frac{240}{289}$ ② $-\frac{4\sqrt{6}}{25}$ ③ $-\frac{20}{101}$ ④ $-\frac{8}{17}$
Ex6.	① $\frac{120}{169}$ ② $-\frac{527}{625}$ ③ $\frac{25}{7}$ ④ $\frac{18\sqrt{11}}{55}$ ⑤ $2x\sqrt{1-x^2}$ ⑥ $2x^2-1$

Ex7.	① $\frac{1}{2}\cos 2x + 2\cos x + \frac{3}{2}$ ② $\frac{9}{2} - 4\sin x - \frac{1}{2}\cos 2x$ ③ $\frac{1}{4}\left[\frac{3}{2} - 2\cos 2x + \frac{1}{2}\cos 4x\right]$ ④ $\frac{1}{4}\left[\frac{3}{2} + 2\cos 2x + \frac{1}{2}\cos 4x\right]$ ⑤ $\frac{1}{8} - \frac{1}{8}\cos 4x$
AP	1. C 2. C 3. D 4. A 5.C 6. D

Answers

6. Trig Equations and Inequalities

6.1 Basic Trigonometric Equations

Ex1.	① *i)* $\frac{5\pi}{4}, \frac{7\pi}{4}$ *ii)* $\frac{5\pi}{4} + 2k\pi, \frac{7\pi}{4} + 2k\pi$ ② *i)* $\frac{\pi}{6}, \frac{11\pi}{6}$ *ii)* $\frac{\pi}{6} + 2k\pi, \frac{11\pi}{6} + 2k\pi$ ③ *i)* $\frac{5\pi}{6}, \frac{11\pi}{6}$ *ii)* $\frac{5\pi}{6} + 2k\pi, \frac{11\pi}{6} + 2k\pi$ ④ *i)* $\frac{\pi}{4}, \frac{5\pi}{4}$ *ii)* $\frac{\pi}{4} + 2k\pi, \frac{5\pi}{4} + 2k\pi$ ⑤ *i)* $\frac{\pi}{3}, \frac{5\pi}{3}$ *ii)* $\frac{\pi}{3} + 2k\pi, \frac{5\pi}{3} + 2k\pi$ ⑥ *i)* $\frac{5\pi}{4}, \frac{7\pi}{4}$ *ii)* $\frac{5\pi}{4} + 2k\pi, \frac{7\pi}{4} + 2k\pi$
Ex2.	① *i)* $\frac{\pi}{2}$ *ii)* $\frac{\pi}{2} + 2k\pi$ ② *i)* $0, \pi, 2\pi$ *ii)* $k\pi$ ③ *i)* $\frac{\pi}{2}, \frac{3\pi}{2}$ *ii)* $\frac{\pi}{2} + 2k\pi, \frac{3\pi}{2} + 2k\pi$

	④ $i)\ \pi$ $ii)\ \pi+2k\pi$
Ex3.	① $x=\sin^{-1}(0.8),\ \pi-\sin^{-1}(0.8)$ $(0.92729,\ \pi-0.92729)$ ② $x=\tan^{-1}(2.5),\ \pi+\tan^{-1}(2.5)$ $(1.19028,\ \pi+1.19028)$ ③ $x=\pi-\cos^{-1}(0.3),\ \pi+\cos^{-1}(0.3)$ $(\pi-1.2661,\ \pi+1.2661)$ ④ $x=\pi+\sin^{-1}\left(\frac{2}{3}\right),\ 2\pi-\sin^{-1}\left(\frac{2}{3}\right)$ $(\pi+0.72972,\ 2\pi-0.72972)$ ⑤ $x=\pi-\tan^{-1}(2),\ 2\pi-\tan^{-1}(2)$ $(\pi-1.10714, 2\pi-1.10714)$
Ex4.	① $\frac{2\pi}{3}, \frac{4\pi}{3}$ ② $\frac{\pi}{3}, \frac{2\pi}{3}, 0, \pi, 2\pi$ ③ $0, \pi, 2\pi, \cos^{-1}\frac{2}{3}, 2\pi-\cos^{-1}\frac{2}{3}$ $(0.84106,\ 2\pi-0.84106)$ ④ $0, \pi, 2\pi, \frac{\pi}{3}, \frac{5\pi}{3}$ ⑤ $\frac{\pi}{4}, \frac{3\pi}{4}, \frac{5\pi}{4}, \frac{7\pi}{4}$ ⑥ $\frac{\pi}{3}, \frac{4\pi}{3}, \frac{2\pi}{3}, \frac{5\pi}{3}$
Ex5.	$\frac{5\pi}{6}, \frac{11\pi}{6}, \frac{17\pi}{6}, \frac{23\pi}{6}$
Ex6.	$\frac{\pi}{4}, \frac{3\pi}{4}, -\frac{\pi}{4}, -\frac{3\pi}{4}$
AP	1. C 2. A 3. B

6.2 More Trigonometric Equations

Ex1.	① $\frac{\pi}{6}, \frac{5\pi}{6}$ ② $\frac{\pi}{6}, \frac{5\pi}{6}, \frac{7\pi}{6}, \frac{11\pi}{6}$ ③ $\frac{\pi}{4}, \frac{3\pi}{4}, \frac{5\pi}{4}, \frac{7\pi}{4}$ ④ $\frac{\pi}{3}, \frac{2\pi}{3}, \frac{4\pi}{3}, \frac{5\pi}{3}$ ⑤ $\pi+\sin^{-1}\frac{1}{5}, 2\pi-\sin^{-1}\frac{1}{5}$ ⑥ $\frac{2\pi}{3}, \frac{4\pi}{3}, 0, 2\pi$ ⑦ $0, \pi, 2\pi$ ⑧ $\frac{\pi}{2}, \frac{7\pi}{6}, \frac{11\pi}{6}$ ⑨ $\frac{\pi}{4}, \frac{5\pi}{4}$ ⑩ $\frac{\pi}{6}, \frac{7\pi}{6}$ ⑪ $0, 2\pi, \frac{2\pi}{3}, \frac{4\pi}{3}$ ⑫ $0, \pi, \frac{\pi}{3}, \frac{5\pi}{3}$ ⑬ $0, 2\pi$ ⑭ $\frac{\pi}{2}$ ⑮ $\frac{\pi}{2}, \pi$
Ex2.	① $\frac{\pi}{6}, \frac{5\pi}{6}, \frac{7\pi}{6}, \frac{11\pi}{6}$ ② $\frac{5\pi}{18}, \frac{7\pi}{18}, \frac{17\pi}{18}, \frac{19\pi}{18}, \frac{29\pi}{18}, \frac{31\pi}{18}$ ③ $\frac{\pi}{2}$ ④ $\frac{\pi}{8}, \frac{5\pi}{8}$

	⑤ $\frac{\pi}{8}, \frac{5\pi}{8}$ ⑥ $\frac{\pi}{12}, \frac{5\pi}{12}, \frac{13\pi}{12}, \frac{17\pi}{12}, \frac{\pi}{4}, \frac{5\pi}{4}$ ⑦ $0, \frac{\pi}{2}, 2\pi$
AP	1. B 2. B 3. D

6.3 Trigonometry Inequalities

Ex1.	① $\frac{\pi}{3} < \theta < \frac{2\pi}{3}$ ② $\frac{3\pi}{4} \le \theta \le \frac{9\pi}{4}$ ③ $-\frac{\pi}{4} \le \theta \le \frac{5\pi}{4}$ ④ $\frac{\pi}{3} < \theta < \frac{5\pi}{3}$ ⑤ $0 \le \theta < \frac{3\pi}{4}$ *or* $\frac{5\pi}{4} < \theta \le 2\pi$ ⑥ $\frac{4\pi}{3} < \theta < \frac{5\pi}{3}$ ⑦ $-\frac{2\pi}{3} < \theta < \frac{2\pi}{3}$ ⑧ $\frac{5\pi}{6} < \theta < \frac{7\pi}{6}$
Ex2.	① $-\frac{\pi}{2} < \theta < \frac{\pi}{4}$ *or* $\frac{\pi}{2} < \theta < \frac{5\pi}{4}$ ② $\frac{\pi}{4} < \theta < \frac{\pi}{2}$ *or* $\frac{5\pi}{4} < \theta < \frac{3\pi}{2}$ ③ $\frac{\pi}{2} < \theta < \frac{5\pi}{6}$ *or* $\frac{3\pi}{2} < \theta < \frac{11\pi}{6}$ ④ $-\frac{\pi}{6} \le \theta < \frac{\pi}{2}$ *or* $\frac{5\pi}{6} \le \theta < \frac{3\pi}{2}$
AP	1. A

Answers

7. Polar Curve

7.1 Polar Coordinates

Ex1.	① ② 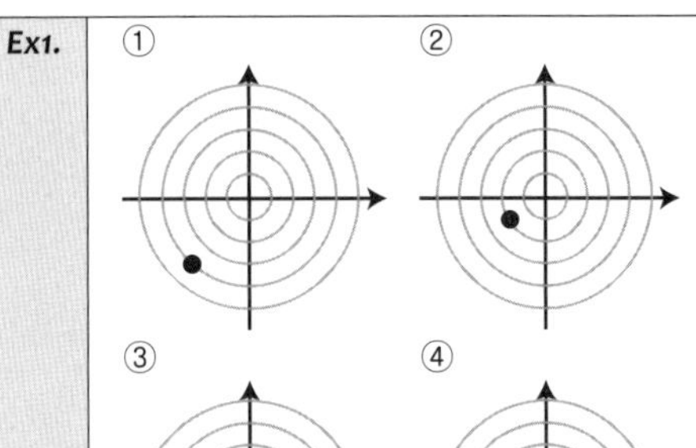 ③ ④ 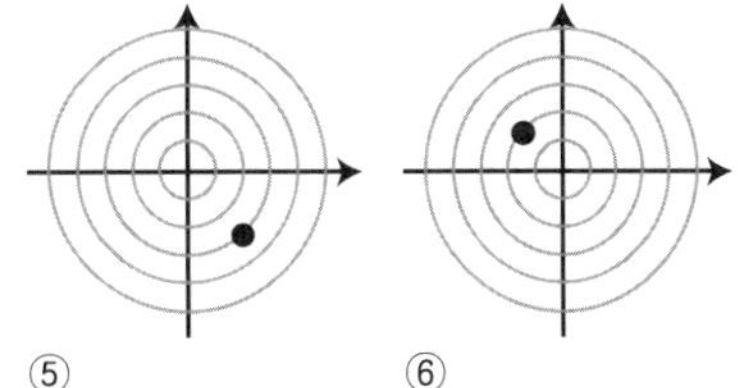 ⑤ ⑥ 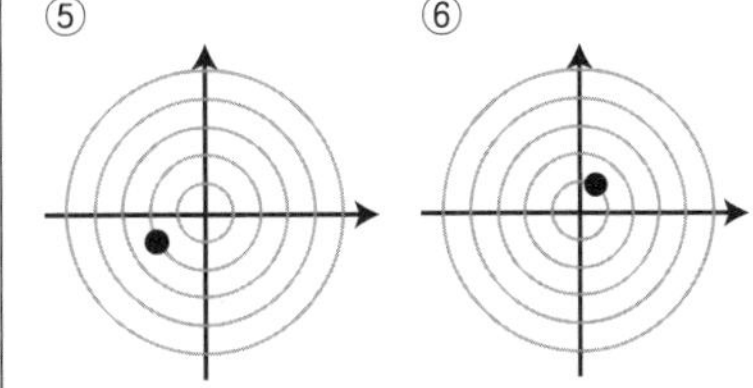 ⑦ ⑧ 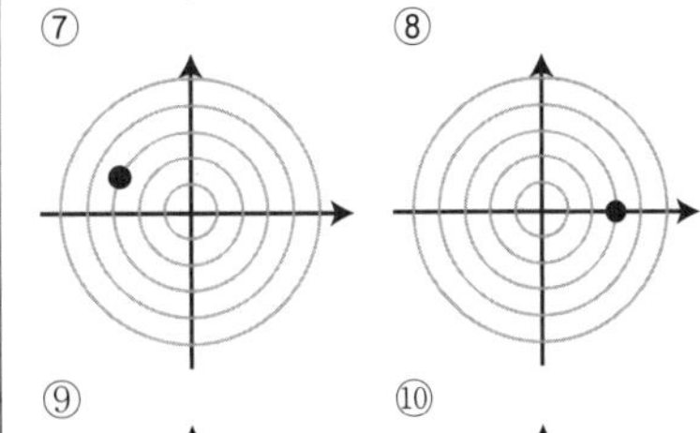 ⑨ ⑩ 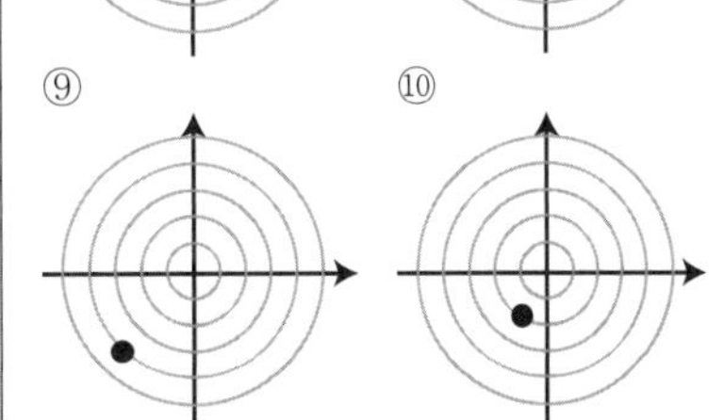
Ex2.	① $(3,-300°)$ $(-3,-120°)$ $(-3,240°)$ ② $(4,-270°)$ $(-4,-90°)$ $(-4,270°)$

	③ $\left(2,-\frac{7\pi}{6}\right)$ $\left(-2,-\frac{\pi}{6}\right)$ $\left(-2,\frac{11\pi}{6}\right)$ ④ $\left(3,-\frac{7\pi}{4}\right)$ $\left(-3,-\frac{3\pi}{4}\right)$ $\left(-3,\frac{5\pi}{4}\right)$
Ex3.	① $\left(-3\sqrt{2},-3\sqrt{2}\right)$ ② $\left(-\frac{7\sqrt{3}}{2},-\frac{7}{2}\right)$ ③ $\left(-\frac{7}{2},\frac{7\sqrt{3}}{2}\right)$ ④ $\left(-3\sqrt{2},-3\sqrt{2}\right)$ ⑤ $\left(\frac{3\sqrt{3}}{2},-\frac{3}{2}\right)$
Ex4.	① $\left(\sqrt{2},\frac{3\pi}{4}\right)$ ② $\left(12,\frac{5\pi}{3}\right)$ ③ $\left(10,\frac{\pi}{6}\right)$ ④ $(4,0)$ ⑤ $\left(3,\frac{3\pi}{2}\right)$ ⑥ $\left(2\sqrt{6},\frac{3\pi}{4}\right)$ ⑦ $\left(2\sqrt{5},\pi-\tan^{-1}2\right)=\left(2\sqrt{5},2.0344\right)$ ⑧ $\left(\sqrt{34},\pi-\tan^{-1}\frac{3}{5}\right)=\left(\sqrt{34},2.601\right)$

Ex5.	① $x^2+y^2=1$ ② $y=10$ ③ $\sqrt{x^2+y^2}+x=5$ ④ $x^2+y^2=y$ ⑤ $x^2+y^2-10y=0$ ⑥ $x-y=3$ ⑦ $4y+5x=2$ ⑧ $y=-2$ ⑨ $x=3$
Ex6.	① $r=\frac{7}{\sin\theta}=7\csc\theta$ ② $r=\frac{3}{\cos\theta+\sin\theta}$ ③ $r=\frac{4}{2\cos\theta-\sin\theta}$ ④ $r^2=4$ ⑤ $r^2=2\sec 2\theta$ ⑥ $r^2=-\sec 2\theta$ ⑦ $r^2=1$
AP	1. C 2. D 3. B 4. C

7.2 Graphs of Polar Equations

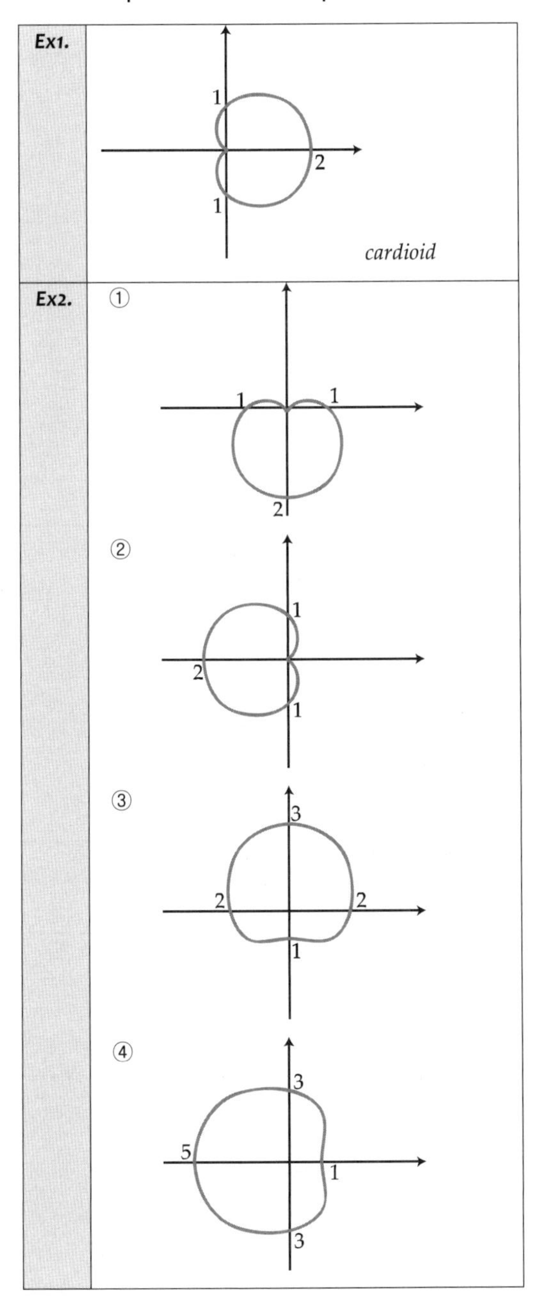

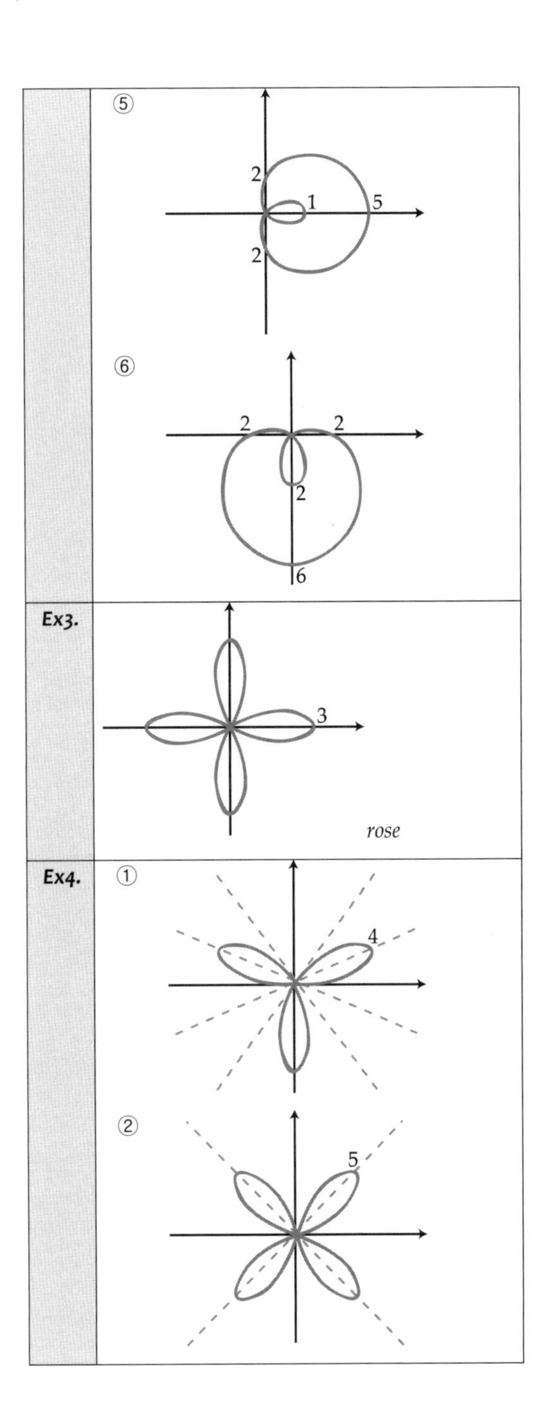

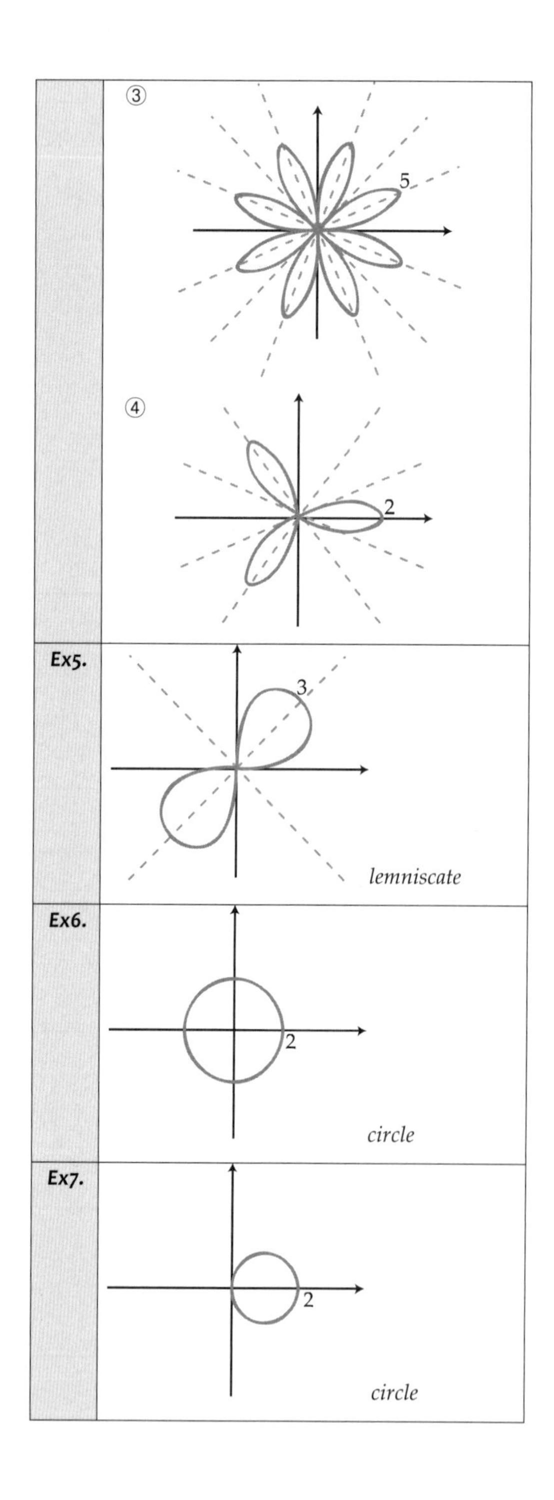

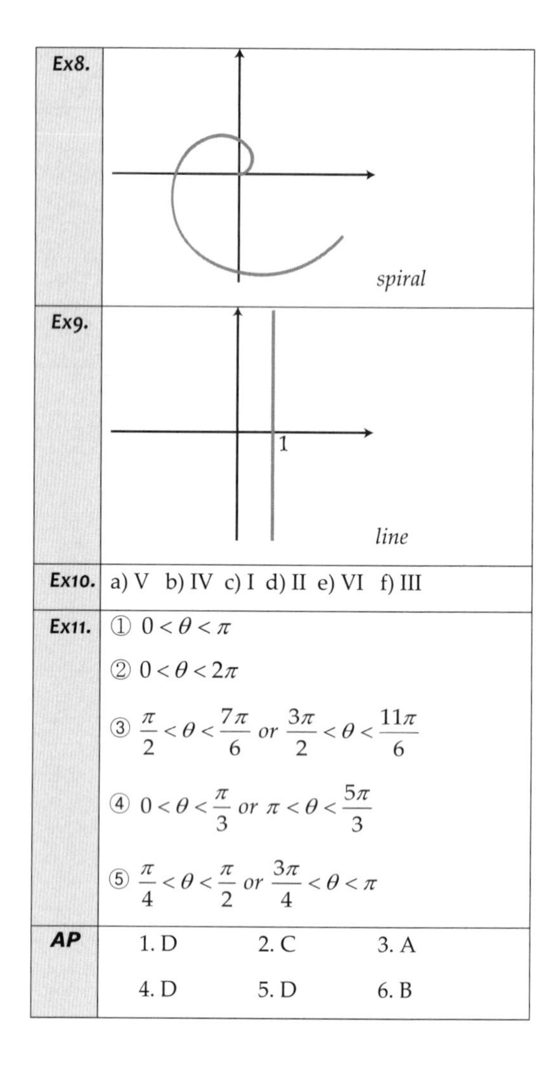

Ex10.	a) V b) IV c) I d) II e) VI f) III
Ex11.	① $0<\theta<\pi$ ② $0<\theta<2\pi$ ③ $\frac{\pi}{2}<\theta<\frac{7\pi}{6}$ *or* $\frac{3\pi}{2}<\theta<\frac{11\pi}{6}$ ④ $0<\theta<\frac{\pi}{3}$ *or* $\pi<\theta<\frac{5\pi}{3}$ ⑤ $\frac{\pi}{4}<\theta<\frac{\pi}{2}$ *or* $\frac{3\pi}{4}<\theta<\pi$
AP	1. D 2. C 3. A 4. D 5. D 6. B

7.3 Complex Numbers

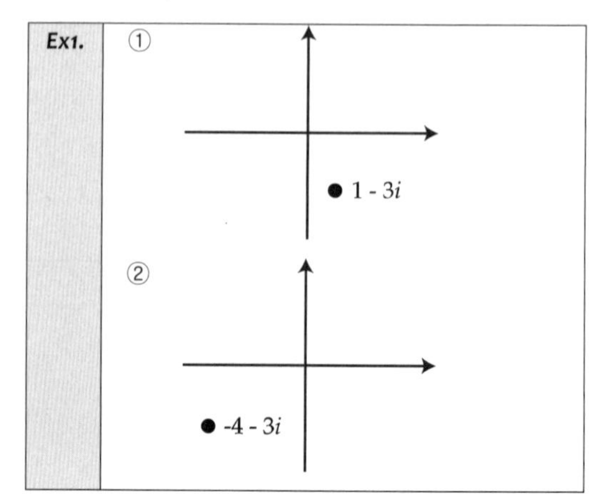

	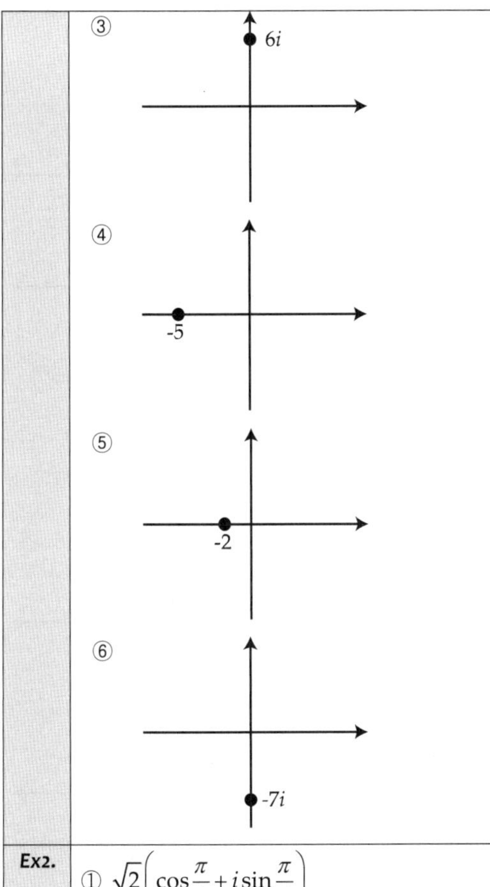
Ex2.	① $\sqrt{2}\left(\cos\frac{\pi}{4}+i\sin\frac{\pi}{4}\right)$ ② $2\left(\cos\frac{11\pi}{6}+i\sin\frac{11\pi}{6}\right)$ ③ $2\left(\cos\frac{5\pi}{3}+i\sin\frac{5\pi}{3}\right)$ ④ $2\sqrt{2}\left(\cos\frac{3\pi}{4}+i\sin\frac{3\pi}{4}\right)$ ⑤ $\pi-\tan^{-1}\frac{3}{2}=2.159$ $\sqrt{13}\left(\cos 2.159+i\sin 2.159\right)$ ⑥ $2\pi-\tan^{-1}\frac{1}{\sqrt{5}}=5.863$ $\sqrt{6}\left(\cos 5.863+i\sin 5.863\right)$

Ex3.	① $-1+\sqrt{3}i$ ② $-\frac{3\sqrt{3}}{2}-\frac{3}{2}i$ ③ $-\sqrt{3}-i$ ④ $-3i$ ⑤ -4 ⑥ 3
AP	1. A

Free Response Questions (from ch3)

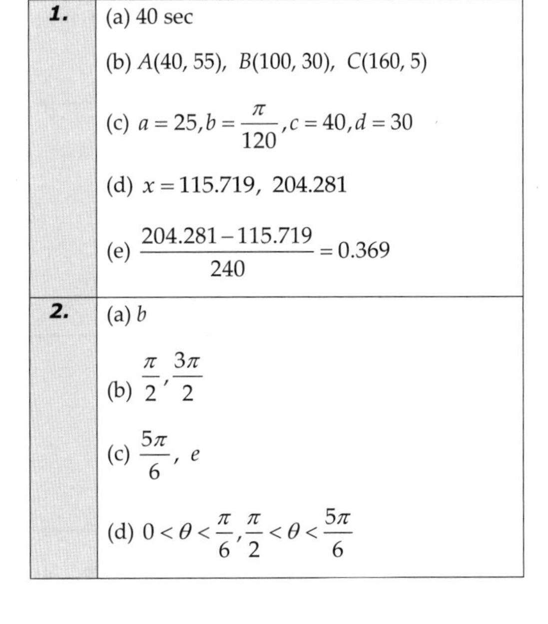

1.	(a) 40 sec (b) $A(40, 55),\ B(100, 30),\ C(160, 5)$ (c) $a=25, b=\frac{\pi}{120}, c=40, d=30$ (d) $x=115.719,\ 204.281$ (e) $\frac{204.281-115.719}{240}=0.369$
2.	(a) b (b) $\frac{\pi}{2}, \frac{3\pi}{2}$ (c) $\frac{5\pi}{6},\ e$ (d) $0<\theta<\frac{\pi}{6}, \frac{\pi}{2}<\theta<\frac{5\pi}{6}$

Answers

8. Parametric Equations

8.1 Parametric Equations

Ex1.	① $x=-3+2y \to line$ ② $y=\sqrt{3x} \to radical$ ③ $x=(-y+4)^2-1 \to parabola$ ④ $y=\left(\frac{x}{3}\right)^2-2 \to parabola$ ⑤ $\frac{y^2}{9}+\frac{x^2}{4}=1 \to ellipse$ ⑥ $\frac{y^2}{49}+\frac{x^2}{25}=1 \to ellipse$ ⑦ $\frac{x^2}{4}-\frac{y^2}{9}=1 \to hyperbola$ ⑧ $1+y^2=x^2 \to hyperbola$
Ex2.	① $D:[2,14]$ — (2,1), (14,5) ② $D:[-5,\infty)$ — (-5,12), (-1,12), (-3,4) ③ $D:[-1,3]$ — (-1,12), (1,0), (3,-4)

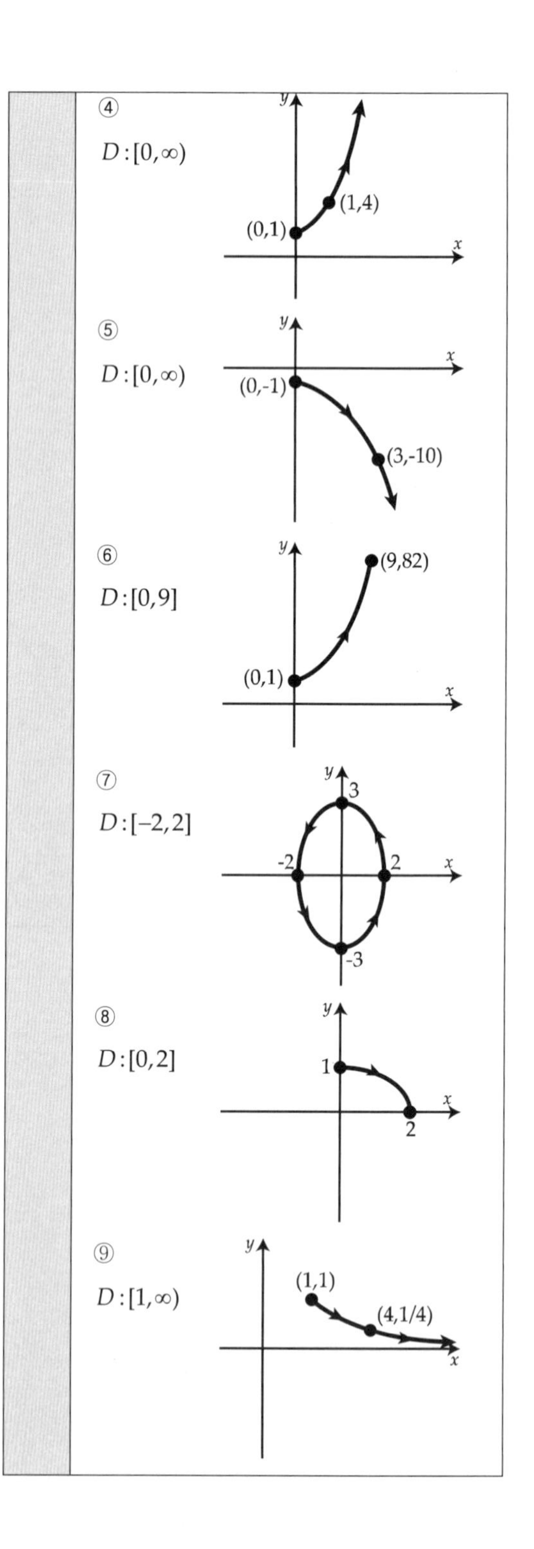

	⑩ $D:[0,1]$ 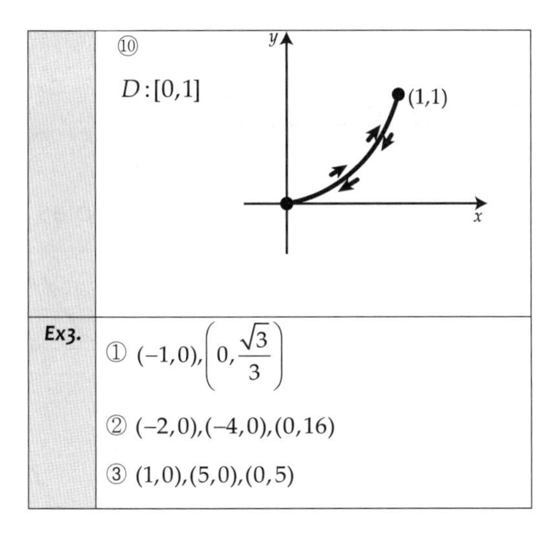
Ex3.	① $(-1,0),\left(0,\frac{\sqrt{3}}{3}\right)$ ② $(-2,0),(-4,0),(0,16)$ ③ $(1,0),(5,0),(0,5)$

8.2 Motions and Parametric Eq

Ex1.	① Farthest right: x = 1 Farthest down: y = -2 ② Farthest right: x = $\frac{1}{3}$ Farthest down: y = 2 ③ Farthest left: x = -1 Farthest up: y = 5 ④ Farthest left: x = -9 Farthest up: y = 7 ⑤ Farthest left: x = -2 Farthest down: y = -3 ⑥ Farthest left: x = -5 Farthest down: y = -7
Ex2.	① $\frac{1}{3}$ ② -3 ③ -4 ④ 9
Ex3.	II

8.3 Lines and Circles in Parametric Form

Ex1.	① $x=3t-4$ $y=-t+3$ ② $x=t-4$ $y=2t+3$ ③ $x=2t+2$ $y=-13t+5$ ④ $x=7t+5$ $y=3$
Ex2.	① $x=3\cos t-4$ $y=3\sin t+3$ ② $x=7\cos t+11$ $y=7\sin t-5$ ③ $x=7\cos t$ $y=7\sin t+3$ ④ $x=4\cos t-5$ $y=4\sin t$ ⑤ $x=5\cos t-1$ $y=5\sin t+2$ ⑥ $x=3\cos t-3$ $y=3\sin t$ ⑦ $x=\cos t$ $y=\sin t$
Ex3.	① (-1,3) ② (1,-1)

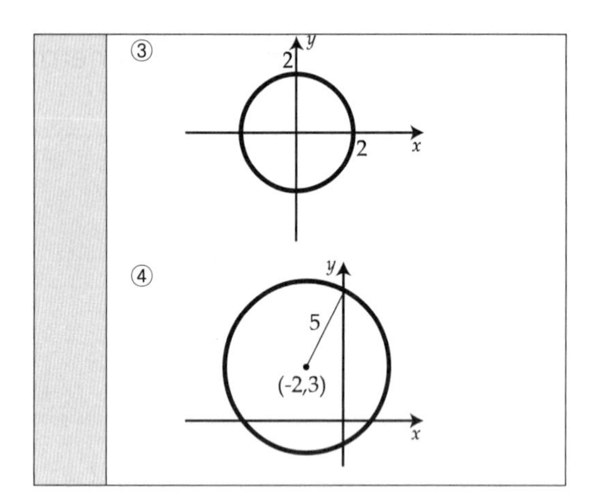

Answers

9. Conic Section

9.1 Conic Sections and Parabolas

Ex1.

①

$Focus:(0,1)$

$directrix: y=-1$

$f\ width: 4$

②

$Focus:(0,4)$

$directrix: y=-4$

$f\ width: 16$

③

$Focus:\left(\frac{1}{32},0\right)$

$directrix: x=-\frac{1}{32}$

$f\ width: \frac{1}{8}$

④

$Focus:(2,0)$

$directrix: x=-2$

$f\ width: 8$

⑤

$Focus:\left(-\frac{3}{2},0\right)$

$directrix: x=\frac{3}{2}$

$f\ width: 6$

⑥

$Focus:(0,-4)$

$directrix: y=4$

$f\ width: 16$

Ex2.

① $-8y = x^2$

② $12y = x^2$

③ $28x = y^2$

④ $-32x = y^2$

⑤ $12x = y^2$

⑥ $-16y = x^2$

⑦ $-8y = x^2$

9.2 Ellipses

Ex1.

①

Foci : $(0, \pm 3)$

Vertices : $(0, \pm 5)$

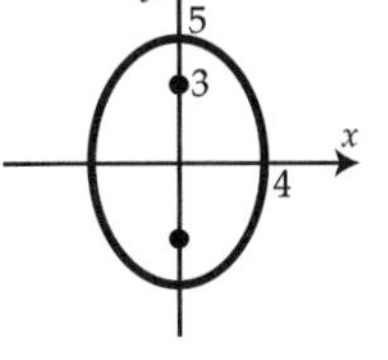

②

Foci : $(\pm\sqrt{46}, 0)$

Vertices : $(\pm 7, 0)$

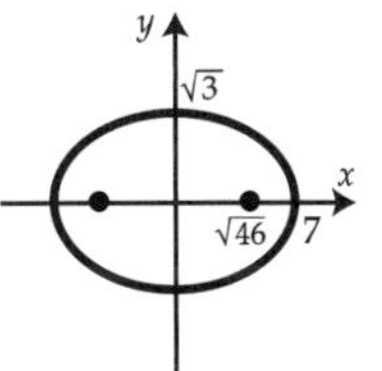

③

Foci : $(\pm 2\sqrt{3}, 0)$

Vertices : $(\pm 4, 0)$

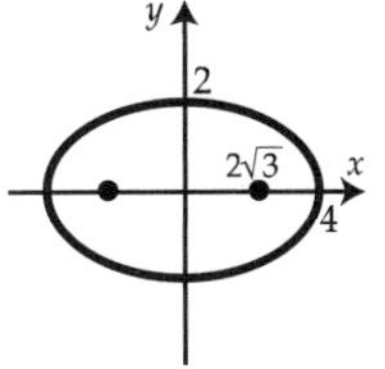

④

Foci : $(0, \pm\sqrt{7})$

Vertices : $(0, \pm 4)$

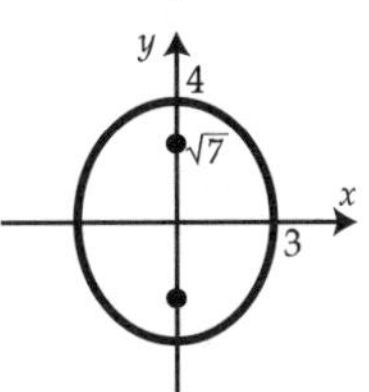

⑤

Foci : $\left(0, \pm\frac{\sqrt{3}}{4}\right)$

Vertices : $\left(0, \pm\frac{1}{2}\right)$

⑥

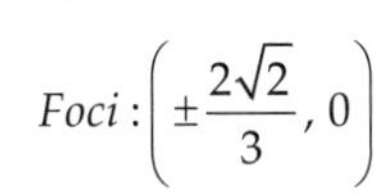

Foci : $\left(\pm\frac{2\sqrt{2}}{3}, 0\right)$

Vertices : $(\pm 1, 0)$

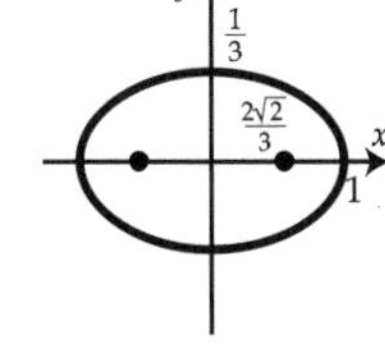

⑦

Foci : $(\pm 1, 0)$

Vertices : $\left(\pm\frac{\sqrt{5}}{2}, 0\right)$

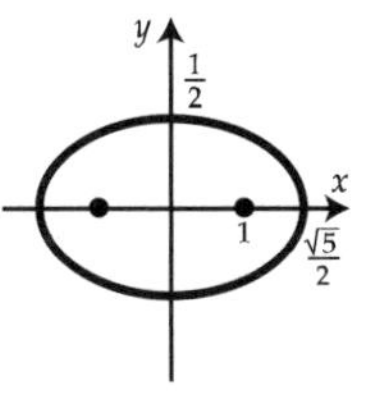

Ex2.

① $\frac{x^2}{100} + \frac{y^2}{75} = 1$

② $\frac{x^2}{16} + \frac{y^2}{7} = 1$

③ $\frac{x^2}{60} + \frac{y^2}{64} = 1$

④ $\frac{x^2}{5} + \frac{y^2}{9} = 1$

⑤ $\frac{x^2}{25} + \frac{y^2}{16} = 1$

⑥ $\frac{x^2}{25} + \frac{y^2}{74} = 1$

⑦ $\frac{x^2}{4} + \frac{y^2}{121} = 1$

⑧ $\frac{x^2}{9} + \frac{y^2}{64} = 1$

9.3 Hyperbolas

Ex1.

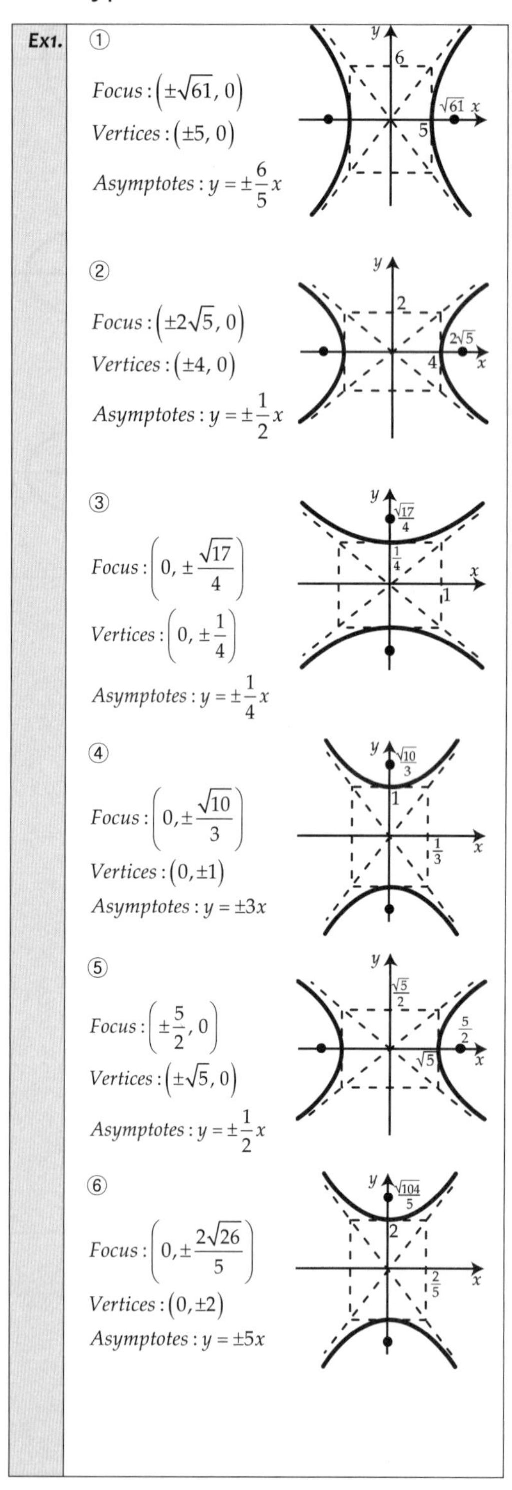

① Focus : $(\pm\sqrt{61}, 0)$

Vertices : $(\pm 5, 0)$

Asymptotes : $y = \pm\frac{6}{5}x$

② Focus : $(\pm 2\sqrt{5}, 0)$

Vertices : $(\pm 4, 0)$

Asymptotes : $y = \pm\frac{1}{2}x$

③ Focus : $\left(0, \pm\frac{\sqrt{17}}{4}\right)$

Vertices : $\left(0, \pm\frac{1}{4}\right)$

Asymptotes : $y = \pm\frac{1}{4}x$

④ Focus : $\left(0, \pm\frac{\sqrt{10}}{3}\right)$

Vertices : $(0, \pm 1)$

Asymptotes : $y = \pm 3x$

⑤ Focus : $\left(\pm\frac{5}{2}, 0\right)$

Vertices : $(\pm\sqrt{5}, 0)$

Asymptotes : $y = \pm\frac{1}{2}x$

⑥ Focus : $\left(0, \pm\frac{2\sqrt{26}}{5}\right)$

Vertices : $(0, \pm 2)$

Asymptotes : $y = \pm 5x$

Ex2.

① $\frac{x^2}{4} - \frac{y^2}{12} = 1$

② $\frac{y^2}{16} - \frac{x^2}{84} = 1$

③ $\frac{y^2}{25} - \frac{x^2}{56} = 1$

④ $\frac{x^2}{4} - \frac{y^2}{32} = 1$

⑤ $\frac{y^2}{16} - \frac{x^2}{100} = 1$

⑥ $\frac{x^2}{36} - \frac{y^2}{16} = 1$

9.4 Transformation of Conics

Ex1.

① Vertex : $(2, 3)$

Focus : $(2, 5)$

directrix : $y = 1$

② Vertex : $(-2, 5)$

Focus : $(2, 5)$

directrix : $x = -6$

③ Vertex : $(-2, -7)$

Focus : $\left(-\frac{5}{2}, -7\right)$

directrix : $x = -\frac{3}{2}$

④ Vertex : $(3, -1)$

Focus : $\left(3, -\frac{5}{4}\right)$

directrix : $y = -\frac{3}{4}$

Ex2.

① $8(x-1) = (y+9)^2$

② $12(y+12) = (x-4)^2$

③ $10(y-3) = (x-8)^2$

④ $-16(x-1) = (y+4)^2$

Ex3.	① Center : $(-4, 2)$ Foci : $(-4, 5), (-4, -1)$ Vertices : $(-4, 7), (-4, -3)$ ② Center : $(2, -1)$ Foci : $\left(2 \pm 3\sqrt{3}, -1\right)$ Vertices : $(8, -1), (-4, -1)$ ③ Center : $(2, -1)$ Foci : $\left(\frac{5}{2}, -1\right), \left(\frac{3}{2}, -1\right)$ Vertices : $\left(2 \pm \frac{\sqrt{2}}{2}, -1\right)$ ④ Center : $(-3, 2)$ Foci : $\left(-3 \pm \frac{\sqrt{5}}{6}, 2\right)$ Vertices : $\left(-\frac{7}{2}, 2\right), \left(-\frac{5}{2}, 2\right)$
Ex4.	① $\frac{(x-2)^2}{16} + \frac{(y-2)^2}{25} = 1$ ② $\frac{(x-2)^2}{25} + \frac{(y-1)^2}{16} = 1$ ③ $\frac{(x-3)^2}{16} + \frac{(y+4)^2}{9} = 1$
Ex5.	① Center : $(-5, 1)$ Foci : $(0, 1), (-10, 1)$ Vertices : $(-1, 1), (-9, 1)$ Asymptotes : $y - 1 = \pm\frac{3}{4}(x+5)$ ② Center : $(-4, -4)$ Foci : $\left(-4 \pm \sqrt{29}, -4\right)$ Vertices : $(-2, -4), (-6, -4)$ Asymptotes : $y + 4 = \pm\frac{5}{2}(x+4)$ ③ Center : $(1, 5)$ Foci : $\left(1, 5 \pm 2\sqrt{5}\right)$ Vertices : $(1, 7), (1, 3)$ Asymptotes : $y - 5 = \pm\frac{1}{2}(x-1)$ ④ Center : $(4, 3)$ Foci : $\left(4, 3 \pm \frac{\sqrt{5}}{2}\right)$ Vertices : $\left(4, \frac{7}{2}\right), \left(4, \frac{5}{2}\right)$ Asymptotes : $y - 3 = \pm\frac{1}{2}(x-4)$
Ex6.	① $\frac{(x-4)^2}{4} - \frac{(y+3)^2}{45} = 1$ ② $\frac{(y-5)^2}{9} - \frac{(x+2)^2}{40} = 1$ ③ $(x-6)^2 - \frac{(y+2)^2}{16} = 1$ ④ $\frac{(y-6)^2}{9} - \frac{(x+1)^2}{4} = 1$
Ex7.	① parabola Vertices : $(-3, -2)$ Focus : $(-3, -1)$ ② parabola Vertices : $(-1, -1)$ Focus : $\left(-\frac{3}{4}, -1\right)$ ③ Ellipse Vertices : $\left(2 \pm \sqrt{3}, -1\right)$ Foci : $(3, -1), (1, -1)$ ④ Ellipse Vertices : $(-1, 3), (-1, -1)$ Foci : $(-1, 2), (-1, 0)$ ⑤ hyperbola Vertices : $(2, -1), (0, -1)$ Foci : $\left(1 \pm \sqrt{2}, -1\right)$

	⑥ *hyperbola* *Vertices*: $(-1, 4), (-1, 0)$ *Foci*: $\left(-1, 2 \pm \sqrt{5}\right)$

9.5 Conic Sections in Parametric Form

Ex1.	① $x = -(t+2)^2$ $y = t$ ② $x = t$ $y = (t-1)^2$ ③ $x = t$ $y = 2(t+2)^2 - 1$ ④ $x = 2(t-2)^2 + 3$ $y = t$
Ex2.	① $x = 3\cos t - 4$ $y = 5\sin t + 3$ ② $x = 2\cos t$ $y = 7\sin t - 5$ ③ $x = 3\cos t - 2$ $y = \sin t$ ④ $x = 2\cos t + 5$ $y = 5\sin t + 2$ ⑤ $x = 2\cos t + 2$ $y = 5\sin t + 5$ ⑥ $x = 7\cos t - 7$ $y = 3\sin t - 3$
Ex3.	① (-2,3) (-2,0) (-2,-3) x y

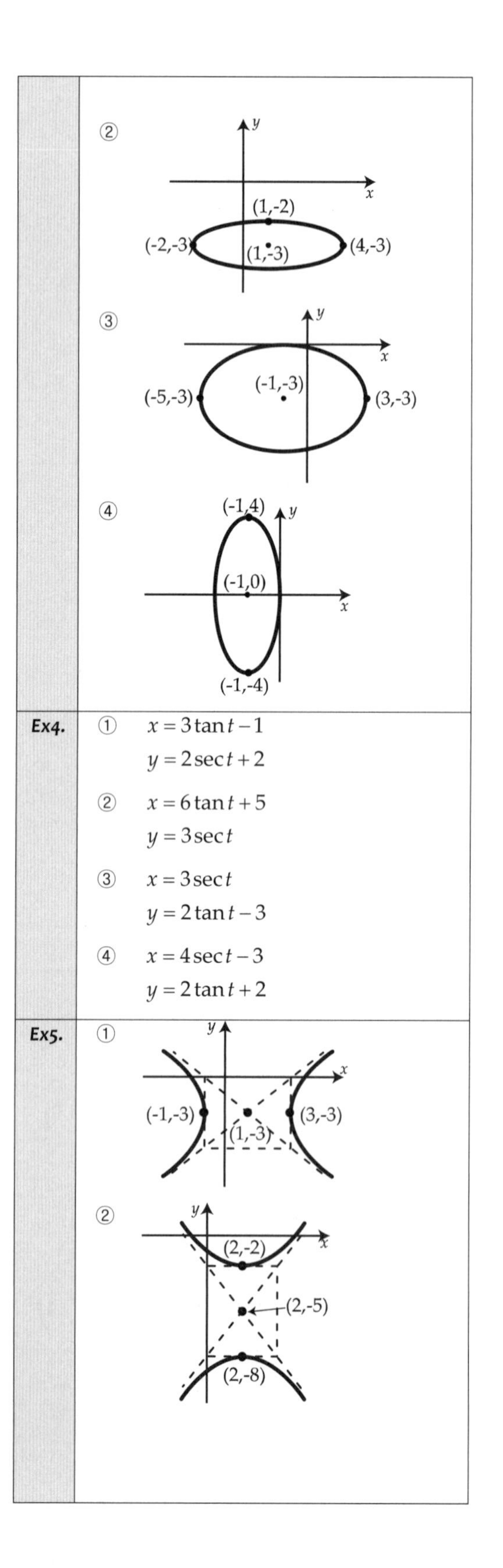

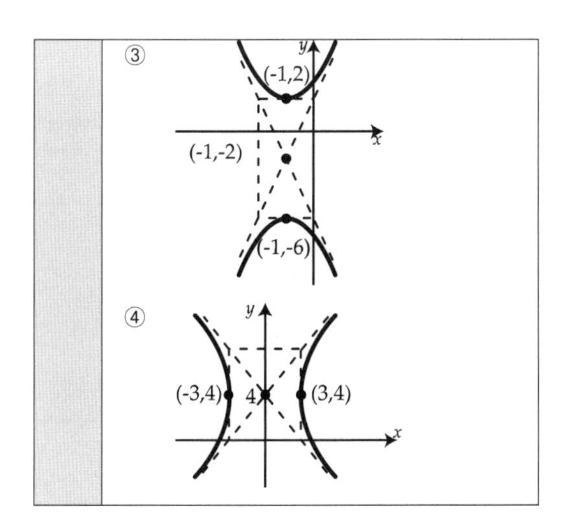

Answers

10. Vector

10.1 Vectors in Two Dimensions

Ex1.

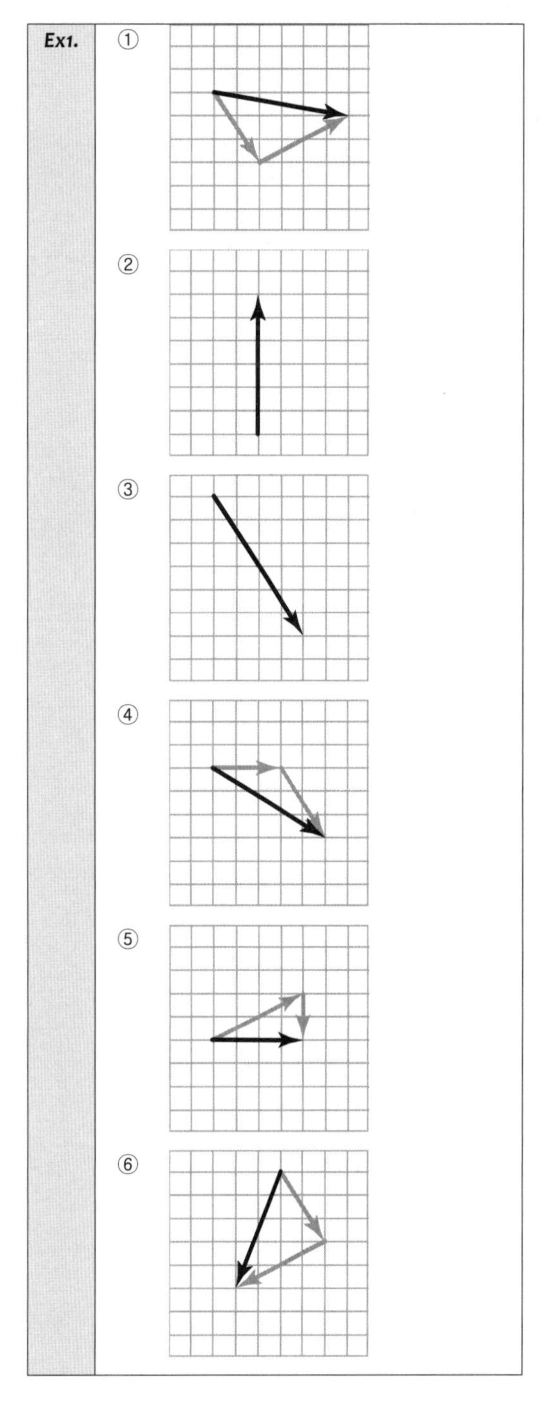

	⑦ ⑧ 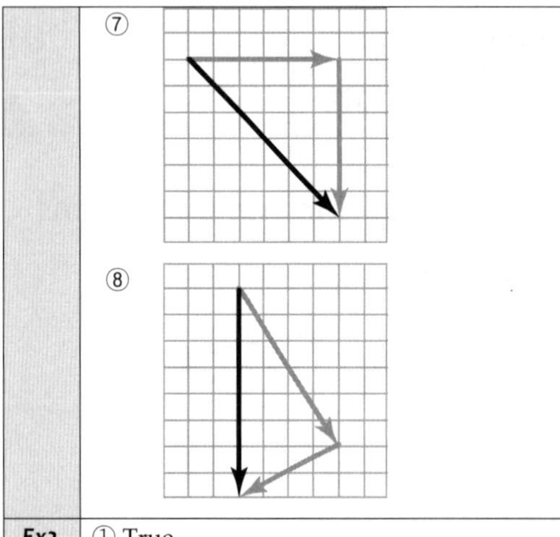
Ex2.	① True ② True ③ False ④ False ⑤ True ⑥ True ⑦ True
Ex3.	① $\langle 7,3\rangle = 7i+3j$ ② $\langle -6,0\rangle = -6i$ ③ $\langle 0,6\rangle = 6j$ ④ $\langle -5,-5\rangle = -5i-5j$
Ex4.	① $\langle 3,4\rangle$ ② $\langle 4,-5\rangle$ ③ $\langle 4,-5\rangle$ ④ $\langle 3,4\rangle$ ⑤ $\langle 1,-2\rangle$ ⑥ $\langle -4,5\rangle$ ⑦ $\langle -4,-2\rangle$ ⑧ $\langle 5,-7\rangle$ ⑨ $\langle 1,-9\rangle$ ⑩ $\langle 4,2\rangle$

	⑪ $\langle -5,7\rangle$
Ex5.	$(1,-1)$
Ex6.	$(8,-1)$
Ex7.	① $\langle 5,-1\rangle$ ② $\langle -1,-1\rangle$ ③ $\langle 13,-2\rangle$ ④ $\langle 10,1\rangle$
Ex8.	① $2i+2j$ ② $4i-4j$ ③ $i-11j$ ④ $6i-10j$
Ex9.	① $2, \langle 1,0\rangle$ ② $\sqrt{10}, \left\langle -\frac{3\sqrt{10}}{10}, \frac{\sqrt{10}}{10}\right\rangle$ ③ $13, \left\langle \frac{5}{13}, -\frac{12}{13}\right\rangle$ ④ $\sqrt{26}, \left\langle -\frac{\sqrt{26}}{26}, \frac{5\sqrt{26}}{26}\right\rangle$ ⑤ $\sqrt{2}, \left\langle \frac{\sqrt{2}}{2}, \frac{\sqrt{2}}{2}\right\rangle$ ⑥ $5, \left\langle \frac{3}{5}, \frac{4}{5}\right\rangle$
Ex10.	① $\frac{5}{2}i+\frac{5\sqrt{3}}{2}j$ ② $-\frac{3\sqrt{3}}{2}i+\frac{3}{2}j$ ③ $-\frac{3}{2}i-\frac{3\sqrt{3}}{2}j$ ④ $-12\sqrt{2}i-12\sqrt{2}j$
Ex11.	① $\langle 3\sqrt{2}\cos 45°, 3\sqrt{2}\sin 45°\rangle$ ② $\langle 2\cos 60°, 2\sin 60°\rangle$ ③ $\langle 6\cos 150°, 6\sin 150°\rangle$

	④ $\left\langle 5\sqrt{2}\cos 225°, 5\sqrt{2}\sin 225°\right\rangle$ ⑤ $180° + \tan^{-1} 5 = 258.69°$ $\left\langle \sqrt{26}\cos 258.69°, \sqrt{26}\sin 258.69°\right\rangle$ ⑥ $360° - \tan^{-1} 3 = 288.435°$ $\left\langle \sqrt{10}\cos 288.435°, \sqrt{10}\sin 288.435°\right\rangle$

10.2 Finding Resultant Vector using Trigonometry

Ex1.	① 15.024 ft ② 13.987 ③ 53.823° ④ 59.004°
Ex2.	① 53.121mi ② 20.999in ③ 26.998° ④ 137°
Ex3.	$9.561m/\sec$, 34.796° to the horizontal
Ex4.	$586.478km/hr$, $0.748 - 38 = -37.252°$ to the horizontal
Ex5.	$27.707mph$, 24.182° to the horizontal

10.3 The Dot Product

Ex1.	① 24 ② −20 ③ 20 ④ 16 ⑤ 24 ⑥ −2 ⑦ −3 ⑧ $-3\sqrt{3}$
Ex2.	① $\cos^{-1}\frac{4}{5} \simeq 37°$ ② $\cos^{-1}\left(-\frac{1}{\sqrt{26}}\right) \simeq 101°$ ③ $\cos^{-1} 0 \simeq 90°$ ④ $\cos^{-1}\frac{1}{5\sqrt{2}} \simeq 82°$ ⑤ $\cos^{-1}\frac{4}{\sqrt{17}} \simeq 14°$ ⑥ $\cos^{-1}\frac{1}{\sqrt{5}} \simeq 63°$
Ex3.	① orthogonal ② parallel ③ neither ④ orthogonal ⑤ parallel ⑥ neither ⑦ parallel

10.4 Motions in Vectors

Ex1.	① $\langle -2,1\rangle$ ② $\langle 2,-3\rangle$ ③ left ④ right ⑤ up ⑥ down ⑦ $\sqrt{5}$ ⑧ $\sqrt{13}$

Answers

11. Matrices

11.1 Algebra Of Matrices

Ex1.	① $\begin{bmatrix} 3 & 21 \\ -6 & 14 \end{bmatrix}$ ② $\begin{bmatrix} -13 & 7 & -1 \\ 0 & -8 & -3 \end{bmatrix}$ ③ $\begin{bmatrix} -2 & -21 \\ -16 & 7 \\ -17 & -19 \end{bmatrix}$ ④ Impossible ⑤ Impossible
Ex2.	① $\begin{bmatrix} 2 & 1 \\ -3 & 8 \end{bmatrix}$ ② $\begin{bmatrix} -2 & 14 & 3 \\ -4 & 1 & -11 \end{bmatrix}$
Ex3.	① $\begin{bmatrix} -2 & -8 \\ -12 & 20 \end{bmatrix}$ ② $\begin{bmatrix} -1 & 9 \\ -8 & 6 \end{bmatrix}$ ③ $\begin{bmatrix} 7 & 19 \end{bmatrix}$ ④ $\begin{bmatrix} 1 & -4 \\ 5 & 9 \end{bmatrix}$ ⑤ $\begin{bmatrix} 28 & 14 & -14 \\ -4 & 8 & -2 \end{bmatrix}$ ⑥ $\begin{bmatrix} -51 \end{bmatrix}$ ⑦ Impossible ⑧ Impossible
Ex4.	① Impossible ② 2×2 ③ 2×2 ④ 2×3 ⑤ 2×2 ⑥ 2×2 ⑦ 2×3
Ex5.	B, D
Ex6.	$A^2+AB+BA+B^2$
Ex7.	$A^2-AB+BA-B^2$
Ex8.	① $\begin{bmatrix} -1 & 0 \\ 2 & 4 \end{bmatrix}$ ② $\begin{bmatrix} 1 & 0 & 2 \\ 0 & 2 & -4 \\ 3 & -5 & 4 \end{bmatrix}$

11.2 Determinant and Inverse Matrix

Ex1.	① −1 ② 7 ③ 9 ④ 8 ⑤ 0 ⑥ 0
Ex2.	① $\begin{bmatrix} 2 & -3 \\ -3 & 5 \end{bmatrix}$ ② $\begin{bmatrix} 0 & -1 \\ 1 & 10 \end{bmatrix}$ ③ $\begin{bmatrix} 0 & \frac{1}{6} \\ -1 & -\frac{5}{6} \end{bmatrix}$ ④ $\begin{bmatrix} \frac{1}{2} & \frac{1}{2} \\ \frac{1}{2} & \frac{1}{4} \end{bmatrix}$ ⑤ Inverse does not exist (determinant is 0) ⑥ Inverse does not exist (determinant is 0)

Ex3.	① 27 ② 1 ③ 10 ④ 10

11.3 Linear Transformation

Ex1.	① $\langle 4,2\rangle$ ② $\langle -4,-2\rangle$ ③ $\langle -4,2\rangle$ ④ $\langle -2,4\rangle$ ⑤ $\langle 2+\sqrt{3},2\sqrt{3}-1\rangle$ ⑥ $\langle 2\sqrt{3}+1,2-\sqrt{3}\rangle$ ⑦ $\langle -2,18\rangle$ ⑧ $\langle -8,-6\rangle$
Ex2.	① $\langle 10,14\rangle$ ② $\langle -10,-8\rangle$ ③ $\langle -8,10\rangle$ ④ $\langle 2,-6\rangle$ ⑤ $\langle -2\sqrt{3}-1,2-\sqrt{3}\rangle$ ⑥ $\langle -2\sqrt{3}-1,2-\sqrt{3}\rangle$
Ex3.	① $\langle 4,2\rangle$ ② $\langle -4,-2\rangle$ ③ $\left\langle -\frac{1}{6},\frac{4}{3}\right\rangle$ ④ $\left\langle -2,-\frac{2}{3}\right\rangle$

11.4 Matrices Modeling Context

Ex1.	(a) 38% (b) 36.6%
Ex2.	(a) 4180 (b) about 4498

Calculator Skills

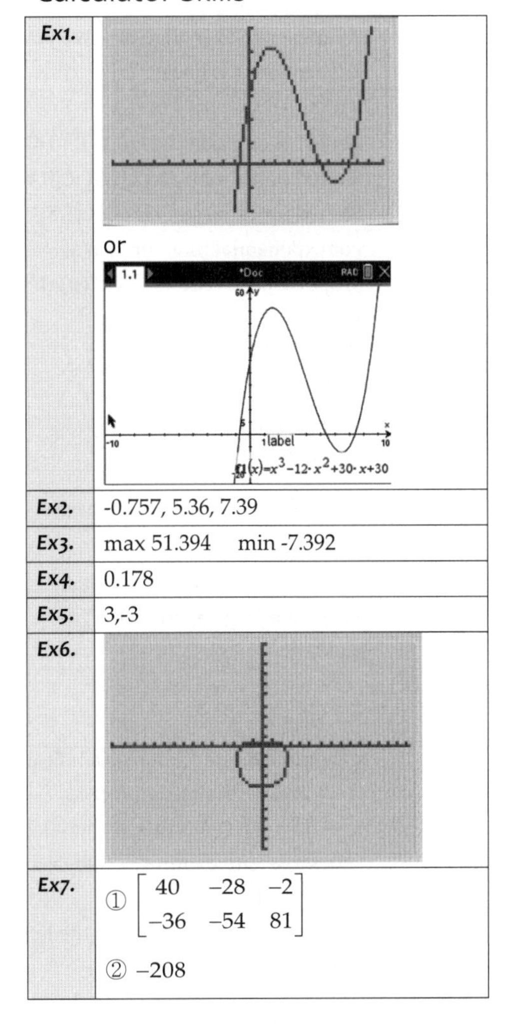

Ex1.	or 1.1 *Doc RAD $f1(x)=x^3-12\cdot x^2+30\cdot x+30$
Ex2.	-0.757, 5.36, 7.39
Ex3.	max 51.394 min -7.392
Ex4.	0.178
Ex5.	3,-3
Ex6.	
Ex7.	① $\begin{bmatrix} 40 & -28 & -2 \\ -36 & -54 & 81 \end{bmatrix}$ ② −208

Answers

Multiple Choice Practice Test

1.	A	Domain : $x^2-2x-8>0 \Rightarrow x<-2 or x>4$
2.	B	There are 5 zeros : $1-i, 1+i, -3i, 3i, 0$
3.	C	Average rate of change is a constant 2. Therefore g is linear.
4.	D	$f(-(x-1))+1$ Reflect over y axis ⇒ D:[0,2] R:[0,4] Shifted right 1 unit ⇒ D:[1,3] R:[0,4] Shifted up 1 unit ⇒ D:[1,3] R:[1,5]
5.	B	For exponential function, y is proportional over the equal length of x. $\frac{2000}{f(2)}=\frac{f(2)}{1000} \Rightarrow f(2)=1000\sqrt{2}$
6.	C	Sin and Tan has different sign. Cos and Sin has same sign. sin: +, −; cos: +, −; tan: −, + Tangent is only positive in 3rd quadrant.
7.	D	$y=\frac{x^2}{x(x-1)}=\frac{x}{x-1}$ x is canceled out ⇒ there is a hole at $x=0$ and vertical asymptote at $x=1$. Using test number 1.1 and 0.9, f goes to positive infinity on the right side of $x=1$, f goes to negative infinity on the left side of $x=1$.
8.	A	$2^{\log_2 27-\log_2 9}=2^{\log_2 \frac{27}{9}}$ $=2^{\log_2 3}=3$
9.	B	For exponential function, y is proportional over the equal length of x. A: logarithmic B: exponential C: logarithmic D: If semilog plot is not linear, then function is not exponential.
10.	A	$\frac{1}{\cos\left(\frac{\pi}{3}-\frac{\pi}{4}\right)}$ $=\frac{1}{\cos\frac{\pi}{3}\cos\frac{\pi}{4}+\sin\frac{\pi}{3}\sin\frac{\pi}{4}}$ $=\frac{1}{\frac{1}{2}\frac{\sqrt{2}}{2}+\frac{\sqrt{3}}{2}\frac{\sqrt{2}}{2}}=\frac{4}{\sqrt{2}+\sqrt{6}}$ $=\sqrt{6}-\sqrt{2}$
11.	D	$(x-3)^2(x+1)(x-5)^3\leq 0, x\neq -1, 5$
12.	A	$y=-2(1.2)^x$
13.	D	$y=-\ln x$
14.	A	$\cos\theta=\frac{1}{2}\Rightarrow\theta=\frac{\pi}{3},\frac{5\pi}{3}$
15.	C	tan A is negative, tan B is positive. tan A < tan B
16.	B	$y=x^2+5$ vertical dilation by 3 ⇒ $y=3(x^2+5)$ shifted down 1 ⇒ $y=3x^2+14$

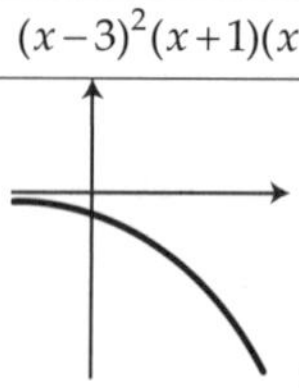

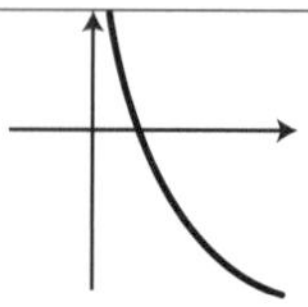

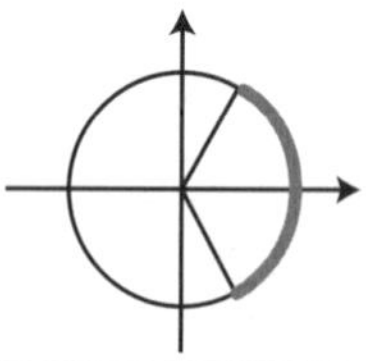

No.	Answer	Explanation
17.	B	$a = \dfrac{4-(-2)}{2} = 3$ $period = \pi = \dfrac{2\pi}{b} \Rightarrow b = 2$ $a = \dfrac{4+(-2)}{2} = 1$
18.	C	Rate is negative $\Rightarrow$ graph is decreasing Rate is decreasing $\Rightarrow$ graph is concave down
19.	B	$P(-k) = 0$ $\Rightarrow (-k)^2 + (k+1)(-k) - 3 = 0$ $\Rightarrow k = -3$
20.	B	$x = \dfrac{3\pi}{2}$ is vertical asymptote of tan.
21.	B	Range of $\cos^{-1} 2x \Rightarrow 0 \le y \le \pi$ Range of $2\cos^{-1} 2x \Rightarrow 0 \le y \le 2\pi$
22.	A	$\log(2x-1) + \log x = 1$ $\Rightarrow \log x(2x-1) = 1$ $\Rightarrow x(2x-1) = 10$ $\Rightarrow x = \dfrac{5}{2}, -2$ -2 is not possible, therefore $x = \dfrac{5}{2}$
23.	D	Amplitude = 17 $\dfrac{2\pi}{b} = 3 \Rightarrow b = \dfrac{2\pi}{3}$ shifted up 5 units
24.	C	When the degree of denominator is bigger that the degree of numerator, then $\lim\limits_{x\to\infty} f(x) = 0$
25.	C	Graph is a rose that passes through $(3,0) \Rightarrow r = 3\cos b\theta$ There are 5 petals $\Rightarrow r = 3\cos 5\theta$
26.	D	$\tan^2 2x = 3 \Rightarrow \tan 2x = \pm\sqrt{3}$ $\Rightarrow \tan A = \pm\sqrt{3}, -\pi \le A \le \pi$ $\Rightarrow A = -\dfrac{2\pi}{3}, -\dfrac{\pi}{3}, \dfrac{\pi}{3}, \dfrac{2\pi}{3}$ $\Rightarrow x = -\dfrac{\pi}{3}, -\dfrac{\pi}{6}, \dfrac{\pi}{6}, \dfrac{\pi}{3}$
27.	C	$\dfrac{b_8}{b_2} = \dfrac{ar^7}{ar} = r^6 = 64$ $\dfrac{b_{33}}{b_{30}} = \dfrac{ar^{32}}{ar^{29}} = r^3 = 8$
28.	A	$x^4 + 4x^3\left(\frac{2}{x}\right) + 6x^2\left(\frac{2}{x}\right)^2 + 4x\left(\frac{2}{x}\right)^3 + \left(\frac{2}{x}\right)^4$
29.	B	
30.	A	$P(0.5)^{\frac{t}{72}} = (0.2)P$ $\Rightarrow t = \dfrac{72\ln 0.2}{\ln 0.5}$
31.	C	$R = \dfrac{(x-4)(x+2)}{x(x-7)(x+2)} \Rightarrow$ hole at x = -2 $\lim\limits_{x\to -2} \dfrac{(x-4)(x+2)}{x(x-7)(x+2)} = \lim\limits_{x\to -2} \dfrac{(x-4)}{x(x-7)} = -\dfrac{1}{3}$
32.	D	$(\sin^2 x - 1)(1 + \tan^2 x) = -\cos^2 x \sec^2 x$ $= -1$
33.	B	Quadratic Residual has no pattern. $\Rightarrow$ Quadratic model is appropriate.
34.	A	$y(2) = -7.4$ $y(2.65) = -9.23788$ $y(3) - 8.6$
35.	B	When $t = 24$
36.	B	Since leading term is $-x^5$, the end behavior will be ↑ ↓.
37.	C	Using calculator; 2.669 –(–0.473) =3.142

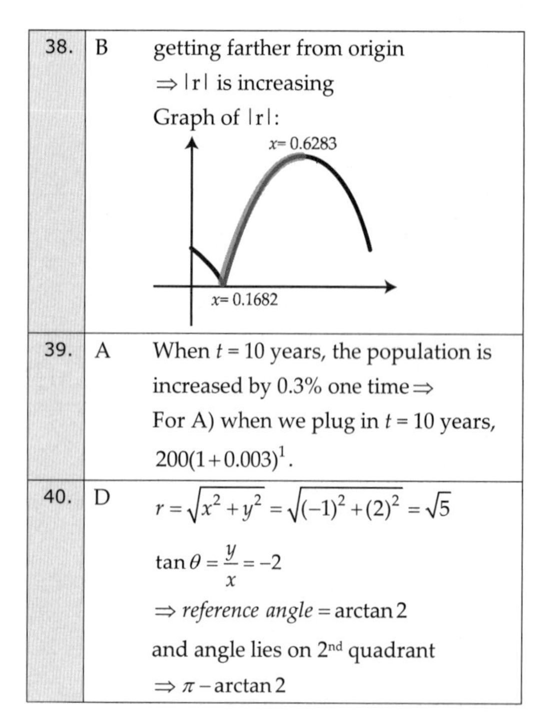

38.	B	getting farther from origin $\Rightarrow$ |r| is increasing Graph of |r|: x= 0.6283 x= 0.1682
39.	A	When $t = 10$ years, the population is increased by 0.3% one time$\Rightarrow$ For A) when we plug in $t = 10$ years, $200(1+0.003)^1$.
40.	D	$r = \sqrt{x^2+y^2} = \sqrt{(-1)^2+(2)^2} = \sqrt{5}$ $\tan\theta = \dfrac{y}{x} = -2$ $\Rightarrow$ *reference angle* $= \arctan 2$ and angle lies on 2nd quadrant $\Rightarrow \pi - \arctan 2$